PHYSICS OF SUSTAINABLE ENERGY II: USING ENERGY EFFICIENTLY AND PRODUCING IT RENEWABLY

PHYSICS OF SUSTAINABLE ENERGY II: USING ENERGY EFFICIENTLY AND PRODUCING IT RENEWABLY

Berkeley, California, USA   5 – 6 March 2011

EDITORS

David Hafemeister
California Polytechnic State University, San Luis Obispo, CA

Daniel Kammen
World Bank, Washington, DC

Barbara G. Levi
Physics Today, Santa Barbara, CA

Peter Schwartz
California Polytechnic State University, San Luis Obispo, CA

SPONSORING ORGANIZATIONS
American Physical Society Forum on Physics and Society
American Association of Physics Teachers
American Physical Society Topical Group on Energy Research and Applications

Melville, New York, 2011
AIP | CONFERENCE PROCEEDINGS ■ 1401

Editors

David Hafemeister
553 Serrano Dr.
San Luis Obispo, CA 93405

E-mail: dhafemei@calpoly.edu

Daniel Kammen
1818 H. St. NW
Washington, DC 20433

E-mail: dkammen@worldbank.org

Barbara G. Levi
1616 LaVista del Oceano
Santa Barbara, CA 93109

E-mail: bglevi@msn.com

Peter Schwartz
Physics Department
California Polytechnic State University
San Luis Obispo, CA 93407

E-mail: pschwart@calpoly.edu

L.C. Catalog Card No. 2011938628
ISBN 978-0-7354-0972-9
ISSN 0094-243X
Printed in the United States of America

AIP Conference Proceedings, Volume 1401
Physics of Sustainable Energy II: Using Energy Efficiently and Producing it Renewably

Table of Contents

SESSION D: ENHANCED EFFICIENCY OF BUILDINGS

SESSION E: RENEWABLE ENERGY

APPENDICES

PREFACE:
Physics of Sustainable Energy II: Using Energy Efficiently and Producing it Renewably

There are many reasons *to use energy efficiently and produce it renewably.* Following Elizabeth Barrett Browning, let us count the ways:

- reduced anthropogenic carbon dioxide
- less dependence on imported oil from the unstable Middle East and other foreign sources
- reduced need for agricultural competition between food, fuel, and fiber
- reduced need for fission and fusion
- reduced trace gas and particulate pollution of urban and rural air
- reduced environmental damage associated with coal mining, mountain top removal, oil drilling, transport of coal, oil and gas
- in many cases, substantial monetary savings, and
- increased job creation relative to fossil-fuel based energy systems

As members of the American Physical Society's Forum on Physics and Society, we are concerned with the need to produce and use energy more wisely. Our hope is to assist in the education of fellow physicists, especially those who teach in our colleges and universities, about the technical details of some of the more promising techniques for efficient and renewable energy. A PhD in physics leads to both depth and breath in a wide range of subfields and specializations built on the fundamentals of classical and modern physics. Up until recently, there has not been much time in this curriculum to address applied energy topics. While this traditional physical education is immensely valuable, and defines a unique contribution to the global body of knowledge, we would like to help enhance the background of the young scientists who want to work on the applied energy topics.

To that end, we organized two short courses on *Physics of Sustainable Energy: Using Energy Efficiently and Producing It Renewably* in 2008 and 2011 at UC-Berkeley, each attracting 200 conferees consisting of physics professors, post-docs and graduate students. The chapters in the two 400-page AIP proceedings are written versions of the talks at the short courses with additional chapters to supplement. Those attending the two recent Berkeley conferences benefited from the speakers' pro-bono donations of time and travel. The meetings were sponsored by the APS Forum on Physics/Society, APS Topical Group on Energy Research/Applications and the American Association of Physics Teachers.

This book is the fourth in a series of books resulting from APS-sponsored conferences on energy, all of which have been published by the American Institute of Physics. The first was the 1975 *American Institute of Physics Conference Proceedings 25*, titled *Efficient Use of Energy*. In the wake of the 1973-74 oil embargo, the APS sponsored a summer study to examine enhanced end-use energy, realizing that it is easier to save a kilowatt-hour than it is to produce a kilowatt-hour. AIP25 launched the

Physics of Sustainable Energy II: Using Energy Efficiently and Producing it Renewably
AIP Conf. Proc. 1401, 1-4 (2011); doi: 10.1063/1.3653841

energy careers of such physicists as David Claridge, Rob Socolow, Art Rosenfeld and Marc Ross, and it spurred the establishment of energy programs at Lawrence Berkeley National Laboratory and at Princeton. The second energy book appeared twenty years ago. It resulted from a short course organized by the *APS Forum on Physics/Society, called Energy Sources: Conservation and Renewables*. The 700–page book, *AIP135,* resulting from that course, became a valuable reference in physics libraries.

The chapters for the present book are organized in five sections: (A) energy policy, (B) environmental effects of fossil fuels, (C) decarbonizing transportation, (D) enhanced efficiency buildings and (E) renewable energy. PowerPoint presentations of the talks are available at http://rael.berkeley.edu/apsenergy2011, along with the 2008 talks at http://rael.berkeley.edu/files/apsenergy. In addition, the chapters of AIP25, AIP135, AIP1044 and this one are available in electronic form at http://proceedings.aip.org/proceedings/cpreissue.jsp.

David Hafemeister	Daniel Kammen	Barbara Levi	Peter Schwartz
Cal Poly U.	World Bank/UCB	*Physics Today*	Cal Poly U.

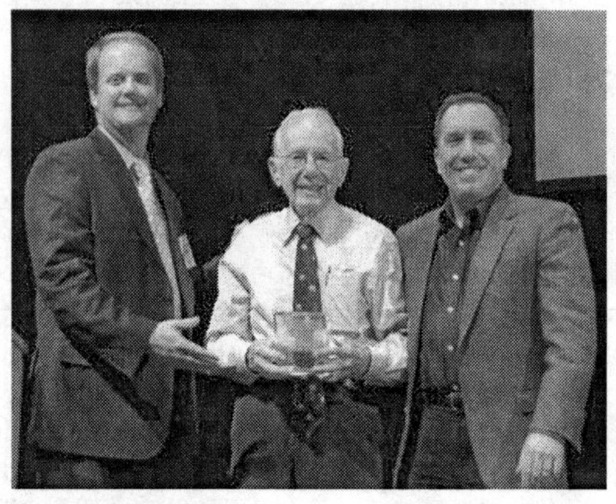

Art Rosenfeld receives an award commemorating the introduction of the *Rosenfeld* energy efficiency power unit. [Chris Calwell, Art, Jon Koomey at UC-Davis, March 9, 2010. Photo by A. Gottlieb]

We dedicate this book to Art Rosenfeld, who has led the way towards energy solutions that reduced the waste of energy. Following the oil embargo of 1973-74, Art shifted his research from particle physics to enhancing end-use energy efficiency, over the years participating in and promoting such developments as energy simulations, electronic ballasts for lighting, white roofs, reduced stand-by power, energy efficient appliances and standards to regulate them, daylighting in buildings, indoor air quality and the blower door, clean water in the developing world, and economic analysis of enhanced energy efficiency. Art has been recognized by the prestigious Fermi, Szilard and Carnot awards, but more importantly he has been a personal role model and a very fun friend. Chapter 2 defines the Rosenfeld unit of electricity savings in honor of Art, and is coauthored by sixty of Art's friends and colleagues.

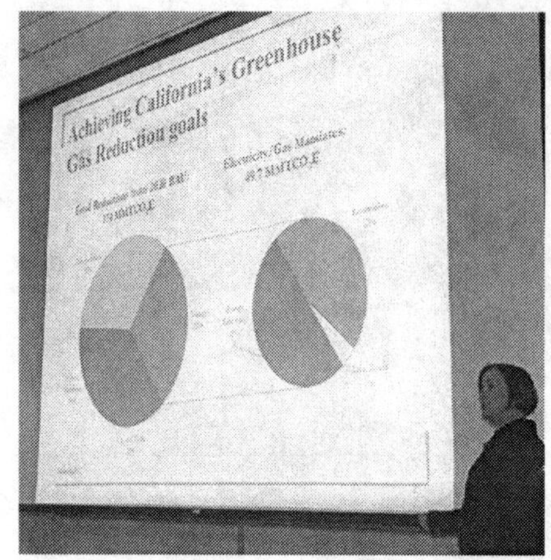
Achieving California's Greenhouse
Gas Reduction goals

IN OFFICE BUILDINGS

Clockwise: Dian Grueneich (former CA Public Utility Commissioner), Dave Claridge (Texas AM), Art Rosenfeld, and Dan Kammen (World Bank and UC Berkeley). [Photos courtesy of Richard Cohen]

SESSION A: POLICIES FOR SUSTAINABLE ENERGY

California Enhances Energy Efficiency

Arthur H. Rosenfeld

Former Commissioner, California Energy Commission..
Distinguished Scientist Emeritus, Lawrence Berkeley National Laboratory
Professor of Physics (emeritus), University of California at Berkeley

Abstract. This article will discuss how my colleagues and I have promoted energy efficiency over the last 40 years. Our efforts have involved thousands of people from many different areas of expertise. The work has proceeded in several areas:

- Investigating the science and engineering of energy end-use,
- Assessing the potential and theoretical opportunities for energy efficiency,
- Developing analytic and economic models to quantify opportunities,
- Researching and developing new equipment and processes to bring these opportunities to fruition,
- Participating in the development of California and later federal standards for energy performance in buildings and appliances,
- Ensuring that market incentives were aligned with policies, and
- Designing clear and convincing graphics to convey opportunities and results to all stakeholders.

ENERGY EFFICIENCY IN CALIFORNIA

Here I tell the story of how we developed and combined these efforts by developing a conceptual framework, calculating costs and benefits, and deploying our findings in a way that would achieve maximum persuasive impact. This article is built around a collection of favorite graphs that my colleagues and I have used over the years to support the campaign for efficiency.[1, 2] I begin with two graphs that illustrate general concepts of energy efficiency and energy intensity in order to illustrate the amazing savings available from improvements in energy use. Next, a series of figures chronicle how we used and continue to use technical and economic data to substantiate our arguments for an effective energy efficiency policy. I have chosen several examples of innovation that have contributed substantially to efficiency improvements over the long term: refrigerators, electronic lighting ballasts, computer applications that simulate building energy performance, and valuation methods for conserved energy. These cases are not necessarily the most recent—some are based on research performed many decades ago—but each one illustrates the complex web of challenges in engineering, economics, and policy that is typical of the efficiency field, and each one continues to bear fruit.

The cases discussed in this article all originated in my home state of California before they went on to influence energy efficiency strategy at the national or global level. California has been, and remains, the main arena for my efforts, and after four

Physics of Sustainable Energy II: Using Energy Efficiently and Producing it Renewably
AIP Conf. Proc. 1401, 7-25 (2011); doi: 10.1063/1.3653842

decades of innovation we are a leader in energy efficiency. The gap between our lower per capita electricity use and national consumption has been dubbed the "California Effect." How much of this effect can be credited to our efficiency efforts, as opposed to advantages in climate and industrial mix, is a point of debate. I give my own analysis here.

In conclusion, I will describe an exciting new policy development that represents the culmination of many years of multi-pronged, interdisciplinary ground- work. In September 2008, the California Public Utilities Commission (CPUC) released California's Long-Term Energy Efficiency Strategic Plan, which was followed in September 2009 by the announcement of a $3.1 billion budget for the first three-year stage of implementation. [3] The Strategic Plan is a crucial component of the state's effort to roll back GHG emissions to 1990 levels by the year 2020. Achieving this goal, as set forth in the landmark Global Warming Solutions Act of 2006 (Assembly Bill 32), will bring California into near-compliance with the Kyoto Protocols. More importantly, the plan's detailed and entirely feasible program of increased energy savings, paired with job creation, provides a much-needed road map for a nationwide "green economy" stimulus.

Figure 1 displays the U.S. Energy Intensity, the ratio of energy consumed divided by the Gross Domestic Product (E/GDP) versus time between 1949 and 2007. In the high-growth decades following World War II, primary energy use, gross domestic product, and CO2 emissions from combustion increased nearly in lockstep. Between 1949 and 1973, energy intensity barely changed. In the years preceding the first OPEC oil embargo, the American consumer had not just scarce but diminishing motivation to reduce energy usage. The average retail price of electricity hovered below 2 cents/kWh through the late 1960s and early 1970s; in fact, the real price (in fixed 2000 dollars) actually declined.

Beginning in 1973, however, the rising price of oil changed the U.S. perspective on energy, spurring California and then other states to adopt energy efficiency standards for buildings and appliances. After 1973, as Figure 1 shows, energy use grew much more slowly than the GDP, and energy intensity improved rapidly. Many factors contributed to these changes, including the increasing cost of energy and the implementation of federal Corporate Average Fuel Economy (CAFE) standards in the transportation sector. We also note that even as energy intensity was improving, U.S. energy use and emissions were increasing though at nowhere near the rate of the 1950s and 1960s.

A central concept of energy efficiency is that it can be measured as a source of energy. Every unit of energy we avoid using thanks to a more efficient device has its equivalent in a unit of fossil fuel that need not be prospected and combusted, or on a macro scale, a power plant that need not be built. After the oil embargo of 1973-74, the lower curve representing the E/GDP data shows a drop of 30% between 1973 and 1985. The drop over the extended period of 1973 to 2007 is 2%/year, about 5 times the historic rate.

Figure 2 displays the U.S. and California per capita electrical energy consumption (kWh/person-year) versus time between the years 1960 and 2008. The sizable gap between the U.S. and California curves amounts to 5,300 kWh/person-year.

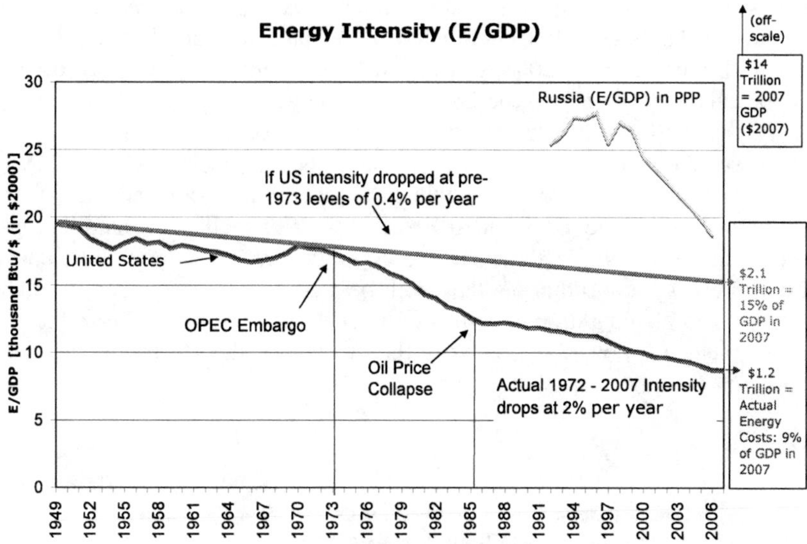

FIGURE 1. U.S. Energy Intensity (E/GDP) vs time (1949 to 2007). Energy is given in Thousands of BTU's with GDP in $2000. The top line shows the drop in E/GDP at the pre-1973 rate of 0.4%/year. After the oil embargo of 1973-74, the lower curve representing the E/GDP data shows a drop of 30% between 1973 and 1985. The drop over the extended period of 1973 to 2007 is 2%/year, about 5 times the historic rate. Russian Energy Intensity is in the upper right corner.

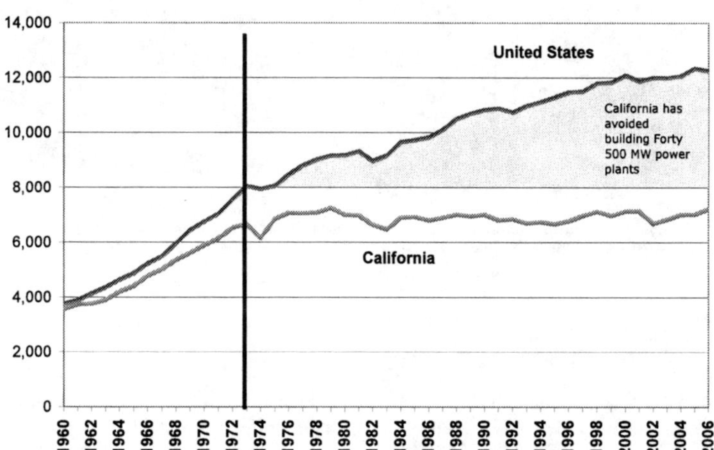

FIGURE 2. U.S. and California percapita electrical energy consumption (kWh/person-year) versus time (1960-2008). Note that California consumption uses 40% less electricity than the U.S.

Figures 3 and 4 show an especially dramatic example that reaches far beyond California. The Three Gorges Dam in China is the largest hydroelectric power station in the world, completed in 2008 at a cost of $30 billion. The left side of Figure 3 shows the amount of energy the dam can generate, compared to the amount of electricity that will be avoided in 2020 as a result of China's mandatory energy efficiency standards for refrigerators and air conditioners, launched in 1999 and revised every four to five years. The right side of the graph compares the dollar value of generation and saved or avoided electricity. We spotlight China because the Chinese example points to the amazing opportunities for energy efficiency. The quantity of energy consumption that will be avoided from greater efficiency in refrigerators and air conditioners will total over double the output from the nation's largest hydroelectric power station. And the "value" of the electricity saved will be over four times that of the power station.

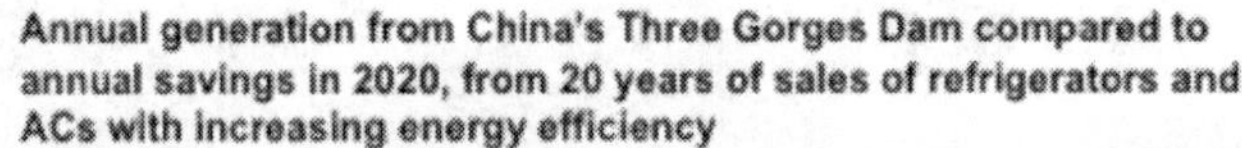

Annual generation from China's Three Gorges Dam compared to annual savings in 2020, from 20 years of sales of refrigerators and ACs with increasing energy efficiency

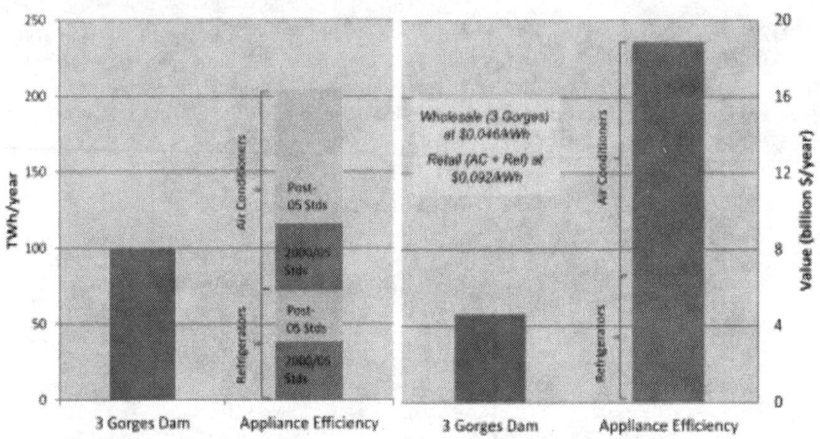

Appliance efficiency savings are calculated on the basis of annual savings in 2020. "Post-05" standards accounts for China's periodic standards revision schedule of 4 to 5 years.

FIGURE 3. Comparison of Three Gorges Dam in China to Refrigerator and AC Efficiency Improvements. The left side compares energy of the dam to savings in AC/refrigerators, the right side compares them economically, since delivered electricity is more expensive than the buss-bar cost.

When considering all 36 types of equipment in China subject to mandatory energy efficiency standards, the numbers are even more striking. By 2020, avoided electricity consumption will be more than five times the output of the Three Gorges Dam, while the value of the electricity will be ten times greater.

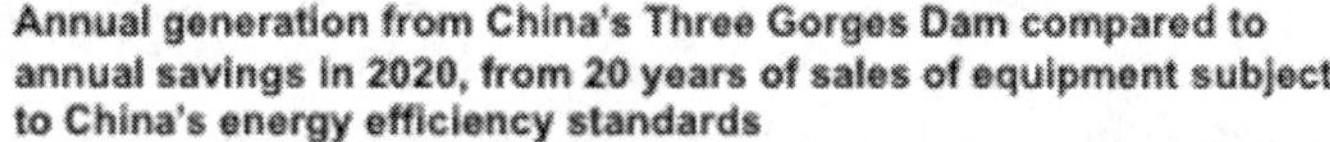

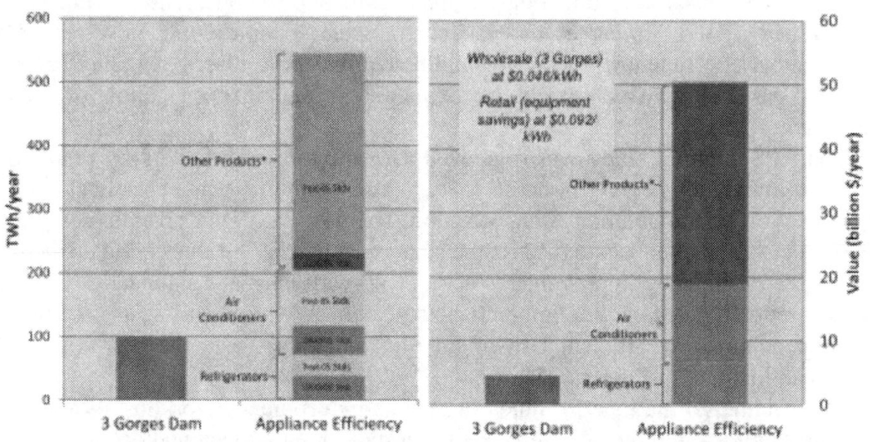

FIGURE 4. Comparison of the annual production from 3 Gorges Dam to the annual savings twenty years after adoption of energy standards on 36 equipment types, summarized above. The left side compares energy (TWh/year) and the right side compares monetary value (billions of dollars/year).

"INVENTING" ENERGY EFFICIENCY

The price spikes of the 1973 OPEC embargo drew nationwide attention to energy end-use, but in rapidly-growing California, already sensitized to environmental issues such as smog and water shortages, the problem caused particular concern. I can't claim any great personal prescience: at the time of the crisis, my data set on energy consumption consisted of exactly two points, both gleaned from my European colleagues. First, European cars got an average of 27 miles per gallon, compared to our average of 14 mpg. Furthermore, Western Europeans used on average half as much energy per capita as their American counterparts, but I knew that they weren't "freezing in the dark" (the typical phrase used at the time by anti- conservation naysayers). I had stumbled upon the idea that per-capita energy use could be reduced without deprivation.

My learning curve spiked in 1974 when I served as a co-leader of a month-long workshop on energy efficiency, convened by the American Physical Society (APS) at Princeton University. Our first realization, which soon became a slogan for the field, was "what's cheap as dirt gets treated like dirt." In the world's other advanced economies, a higher dependence on expensive imported fuels made energy costs a critical factor in long-range economic strategy (on tax policy, balance of trade, and national security). Consumer psychology was also affected by higher energy prices: whereas Americans made their purchasing decisions largely on first cost (sticker price), the Europeans and Japanese were more likely to incorporate life-cycle cost (sticker price plus future operating costs) into their decisions. The soaring price of energy had a silver lining as a teachable moment: people could now realize that adopting better efficiency practices would be equivalent to discovering huge domestic oil and gas fields, which could be extracted at pennies per gallon of gasoline equivalent.

The APS summer study was organized as a mixture of briefings by practitioners from commercial sectors where energy consumption was a salient concern (construction, manufacturing, transportation, utilities, etc.), and analytic sessions led by physicists and chemists to discuss the state of research. Our overriding concern was to focus on efficiency improvements achievable with current technology, rather than on theoretically elegant but impractical research. We published our findings and recommendations in Efficient Use of Energy, for many years the best-seller of the American Institute of Physics. [4]

The volume set the tone for much of the energy efficiency work to follow, with its mixture of pure and directed research, its incorporation of social and economic factors into the engineering analysis, and its emphasis on feasibility. We were also aware that we had to illustrate our findings with concrete examples that would convey the importance of efficiency to a non-expert public (and government). For example, one-third of Efficient Use of Energy was devoted to discussion of recent advances in window technology, such as thin films of low-emissivity (low-E) semi-conductor material; when applied to the inside surface of double-glazed windows, they doubled the thermal resistance.

Like much of the volume, this section was highly technical and inaccessible to the lay reader, and yet it contained highly practical implications that we wanted to convey to the public. It was written just as the last environmental objections to the Trans-Alaska Pipeline were overruled in favor of construction. The section's authors calculated that low-E windows, installed nationwide, would save the equivalent of half the oil produced in the Prudhoe Bay oilfields. In combination with other modest efficiency measures, these windows could have eliminated the need for the pipeline; it was this simple memorable fact, rather than the painstaking calculations, that became the public angle for the book.

STATE AND SCIENCE IN CALIFORNIA

Returning to California after the APS Efficiency Study, I took what was intended as a temporary leave from particle physics in order to teach, conduct research, and proselytize about energy efficiency. It seemed logical to focus on buildings and appliances rather than the transportation sector, since the latter was already under the oversight of the Department of Transportation, whereas work on the former was virtually tabula rasa. After a few years, it was clear that my sabbatical from physics had turned into a permanent defection. Worse yet, I coaxed a number of other scientists away from traditional career paths in physics or chemistry in favor of the risks of an upstart field. Colleagues including Sam Berman, Will Siri, Mark Levine, and Steve Selkowitz joined me in the process of redirecting our skills from basic research to the mixture of science, economics, and policy that efficiency work entailed. My most promising physics graduate students, David Goldstein and Ashok Gadgil, also joined us.

I do not wish to suggest that California was the only locus of innovation in energy efficiency. Colleagues in other parts of the country made the same career shift and did important early work, including Marc Ross at the University of Michigan and Rob Socolow at Princeton. The critical difference was that we were graced with optimal conditions for our ventures. Lawrence Berkeley National Laboratory (LBNL) had recently come under the fresh leadership of Andrew Sessler, who signaled the lab's intention to engage with society's most pressing problems by creating a new Energy and Environment Division as his first act as director in 1973. The division was a natural host for my Energy Efficient Buildings Program (later known as the Center for Building Science), and sheltered it from much of the instability and administrative strife faced by similar programs at other institutions. At the same time, the University of California at Berkeley launched a doctoral program in Energy and Resources under the visionary leadership of John Holdren. Because this unique program created a talent pool with the necessary interdisciplinary skill set in policy, economics, and science, we were able to take on more ambitious projects than other institutions.

Finally, and most importantly, our California community of efficiency scientists formed just as the state's first efficiency legislation came into effect. A proposal to establish state oversight of energy supply and demand had been languishing on Governor Reagan's desk since 1973, opposed by utility companies, appliance manufacturers, and the building industry. However, in the atmosphere of crisis following the OPEC embargo, the governor was compelled to act, and the Warren-Alquist Act was signed into law in 1974. The Act established the California Energy Commission (CEC), which had the authority to approve or deny site applications for new power plants, to write energy performance standards for new buildings, to fund research and development, and to support investment in efficiency programs. Soon thereafter, the commission's mandate was expanded to include major appliances. The first generation of state appliance performance standards (Title 20) was published in 1976, followed in 1978 by a building standard (Title 24).

The establishment of the CEC created a market for our research, which in turn made the commission effective. This fortunate convergence of policy requirements and scientific knowledge was a key factor behind California's leadership in energy efficiency. In the years before the commission's in-house research capability was developed, it relied upon local scientists for data, forecasts, testing protocols, and analytic tools. One example was the creation of a computer application to simulate the thermal performance of buildings. In early drafts of Title 24 (residential building standards), the commission proposed limiting window area to 15 percent of wall area, based on the (erroneous) belief that larger window areas would waste heat in winter or "coolth" in summer. No allowance was made for the compass orientation of the windows; indeed, I don't think the sun was even mentioned.

The staff had used a computer simulation that ran on a "fixed-thermostat" assumption, maintaining indoor temperature at 72° F (22° C) year round. Keeping to this exact mark required heating or cooling—or both—every day of the year! We saw the need for a simulation that allowed a "floating temperature" mode, permitting indoor temperature to rise slightly during the day, as solar heat entered and was stored in the building's mass, and then float down at night, as the house coasted on stored heat. Such a model could demonstrate that in many situations, expanded window area would actually lower energy demand, supporting the inclusion of passive solar methods in the state building code. Unfortunately, the existing public-domain programs were too awkward and bug-ridden to handle more complex and realistic thermal simulations. I immediately sat down with my colleague Ed Dean, a professor of architecture, to write a residential thermal simulator, which we dubbed Two-Zone because it distinguished between the north and south halves of the house. The CEC was soon convinced to drop the proposed limit on non-north-facing windows, and the concept of passive solar heating was included in Title 24, years before the term itself was in common use.

Two-Zone became the progenitor of a generation of public-domain building performance simulators. When the federal Department of Energy (DOE) was formed in 1976, it funded further development of the software through a collaboration of the national labs at Berkeley, Argonne, and Los Alamos. Since that time, the program, known as DOE-2, has been an essential tool for evaluating energy use in complex systems. Although similar proprietary programs were also developed, the public availability of DOE-2 allowed extensive feedback, which fed the increasing sophistication of the model. While enabling tools such as DOE-2 do not in themselves save energy, without them it would not be possible to write appropriate state and federal buildings standards, or to establish high-profile certification programs such the Green Building Council's Leadership in Energy and Environmental Design (LEED). [5]

Improved HVAC (heating, ventilation, and air-conditioning) performance in buildings has been one of the most profitable and uncontroversial ways for society to save energy and money. It would be tedious to calculate exactly how much of these savings can be attributed to the DOE-2 program, since standards were implemented gradually across the states, and some technical improvements occurred independently of implementation. My own guesstimate is that annual U.S. savings in buildings energy use (compared to pre-standards performance) are roughly $10 billion per year, and that the modest allocation of public funds to support the creation of a viable public-domain modeling tool advanced the adoption of standards by 1-3 years.

THE POLITICS OF DEMAND FORECASTING

Another early task of the California Energy Commission was to determine an appropriate balance between increasing generation capacity through granting permits for new plants and extracting more "service" from the existing supply. Often as not, these decisions took place against a politically charged backdrop. Proposition 15, scheduled to go to California voters in March 1976, proposed to halt the construction of all nuclear power plants. My graduate student David Goldstein and I were determined to cut through the noise surrounding this hot-button issue with the first rigorous study of peak demand forecasts. We hoped that if the rising demand for electricity could be slowed through more efficient performance standards then the contentious issue of new power plants might be avoided.

The left side of Figure 4 shows the actual supply curve during the high-growth decade leading up to 1974, when peak production capacity reached about 30 gigawatts (GW). The right side of the figure compares two future (post-1974) scenarios. Under the "business as usual" (BAU) scenario assumed by the utilities, demand would continue to grow at 5 percent per annum, requiring the construction of an average of two large (one-GW) power plants every year, mainly nuclear or fossil fueled. More than half of that new electricity (i.e., more than one plant per year) would be used to supply electricity to new construction. In the days before Title 24, two of the most egregious sources of waste were widespread electric resistance heating in residences and 24/7 lighting in commercial buildings. When we calculated the potential savings from eliminating these practices, we came to the remarkable conclusion that the state's annual growth rate could drop to 1.2 percent. This scenario would eliminate the need not only for the contentious nuclear plants but also for planned fossil fueled plants. When we demonstrated our findings at a State Assembly hearing in December 1975, the utility companies were so skeptical that Pacific Gas & Electric (PG&E) called Director Sessler to suggest that I be fired on the grounds that physicists were unqualified to forecast electricity demand.

Over the course of the later 1970s and early 1980s, our vision was slowly vindicated and the hostility of PG&E was gradually replaced with a productive collaboration. After 1975, the actual growth of peak demand dropped to 2.2 percent per annum, much closer to our forecast than to that of the utilities. (For purposes of comparison, we later added this actual growth curve to the original version of Figure 5.) The fall from favor of nuclear power plants due to a combination of public opposition and unexpectedly high costs is well known, but in fact no application to build any kind of large power plant (nuclear, coal, or gas) was filed in California between 1974 and 1998. Demand continued to grow during that time, of course, but new supply came from small independent producers and co-generators, from renewables (hydroelectric, geothermal, and wind resources), and from sources outside the state. Improved efficiency was the largest single source of new electric services during that period.

After the deregulation of California's energy supply system in the late 1990s, and the ensuing electricity "crisis" of 2001, policies were put in place to encourage the procurement of a "reserve margin" large enough to guarantee reliability. In response to state incentives, investments in both power plants and efficiency accelerated. Fortunately, the benefits of efficiency were not forgotten in the rush to increase capacity. In 2003, the CPUC and the CEC issued the first Energy Action Plan (EAP I) to guide energy policy decisions. A major function of EAP I and of subsequent updates has been to prescribe a "loading order" of energy supplies whenever increased demand needs to be satisfied. For immediate demand crises, demand response (e.g., shutting off unnecessary load) should take precedence over costly purchase of peaking generation from the market. For longer-term supply planning, investments in efficiency should "load" into the supply system before investments in generation; when new generation is necessary, renewable generation should load before fossil generation. From 2001 to 2009, over 15,000 MW of generation resources, including renewables, have been built in California, yet efficiency investments are increasing.

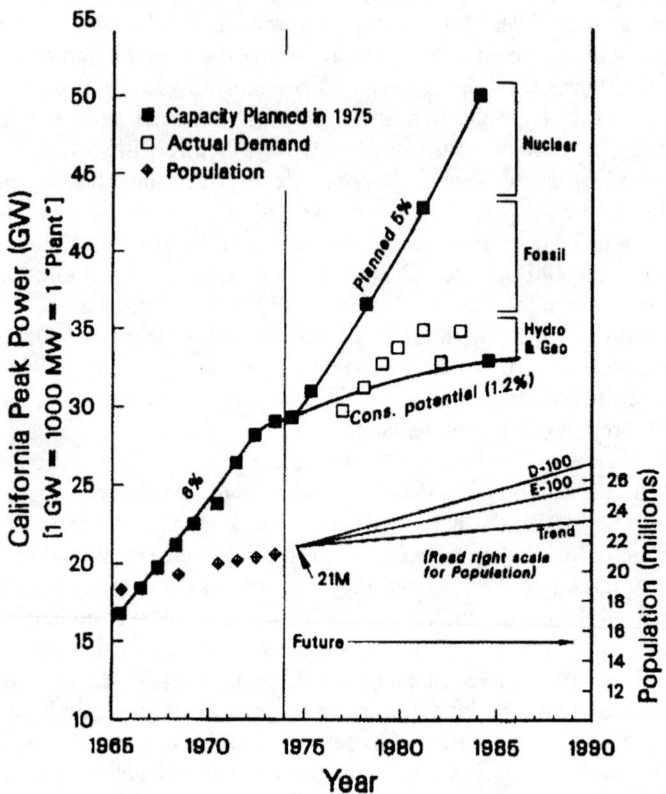

FIGURE 5. 1975 Projection of California Power Demand: Business as Usual versus Goldstein and Rosenfeld Efficiency Scenario.

INITIATING APPLIANCE STANDARDS IN CALIFORNIA

Whereas gaining acceptance for state oversight of energy standards in buildings was relatively straightforward, creating a state appliance standard proved more controversial. [6, 7] Since manufacturers usually sold to the national market, federal responsibility seemed more appropriate and effective to most people. In addition, the appliance industry was more concentrated and organized than the construction sector, and thus better able to mount opposition to changes. This did not deter David

Goldstein and me from satisfying our curiosity about the correlation between refrigerator price and efficiency. Our interest in the refrigerator was motivated by its place as the most energy-thirsty appliance in the family home: in the 1970s it accounted for more than a quarter of the typical residential electricity bill. We tested 22 units from model year 1975, expecting to see some correlation between higher sticker price and higher performance, defined as the cooling service delivered per energy input. In other words, if we could establish a correlation between sticker price and efficiency, we could support informed consumer choices based on payback time (how long it takes to offset a high purchase price with lower energy bills) and life-cycle cost (purchase price plus lifetime operating costs).

Results of our refrigerator tests are shown as a "scatter chart," the only feasible choice of format given that the data were truly scattered! Despite our efforts to control for every factor imaginable (volume, door configuration, options, etc), there was very poor correlation between purchase price and performance. Some of the lowest priced models showed the same or even cheaper life-cycle costs than models costing $100 to $200 more. We quickly realized that if the less efficient half of the model group were deemed unfit for the market, the consumer would not perceive any change in the market range of prices or options while being "forced" to save on average $350 over the 16-year service life of a refrigerator. Presumably, as performance standards spurred further technical improvements, these savings would grow. The macroeconomic conclusion was even more exciting: since statewide energy use by refrigerators alone already accounted for about five GW, implementing even mild state standards could avoid the need to construct numerous power plants. In 1976 California Governor Jerry Brown was looking for a way to avoid approving Sundesert, the only application still pending for a one-GW nuclear power plant. I took advantage of a chance meeting at the Berkeley Faculty Club to sketch out Figure 4 for him on a napkin. Thinking our findings too good to be true, the governor called Energy Commissioner Gene Varanini for corroboration. I believe his exact words were, "Is this guy Rosenfeld for real?" Commissioner Varanini vouched for us, Sundesert was cancelled, and California's Appliance Efficiency Regulations (Title 20) were implemented later that year.

REFRIGERATORS: AN EFFICIENCY SUCCESS STORY

The dramatic improvement in refrigerator energy efficiency over the last half-century is illustrated in Figure 6, which shows electricity use by new U.S. refrigerators for the model years 1947–2001. The heavy line with dark squares represents the annual kWh use of the sales- weighted average new refrigerator. Note that the energy consumption of new models has declined steeply in absolute terms, even though this line is not adjusted for increasing volume. In fact, the volume of the average model grew from 8 cubic feet to 20 during this period, as shown by the line marked with open circles; if the consumption line were adjusted for volume, the efficiency gains would look even more impressive. The right-hand scale shows the number of large (one-GW) base-load (5,000 hours/year) power plants required to power 150 million average refrigerator-freezer units. The difference between the annual energy consumption of an average

1974 model (1,800 kWh) and an average 2001 model (450 kWh) is 1350 kWH. The energy savings from this 1,350 kWh/year difference, multiplied by 150 million units, is 200 TWh/year, equivalent to the output of 50 avoided one-GW plants. The monetary savings of course depends on the price of electricity, which varies considerably. To give a rough sense of the magnitude of savings, at 8 cents/kWh, the avoided annual expense to consumers is $16 billion.

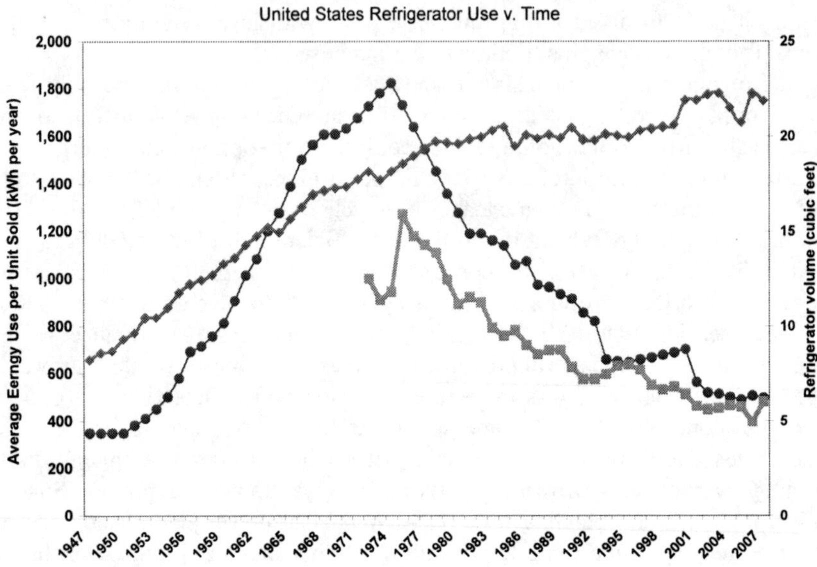

FIGURE 6. U.S. Refrigerators 1947-2007: Energy Use (kWh/y), Volume (cubic feet), Price (1993$).

The other factor contributing to the sudden drop in refrigerator energy use in the mid-1970s was the advent of a new manufacturing technology, blown-in foam insulation. The coincidence of California's first performance standards with the market entry of better-performing models began a positive reinforcing cycle that continues to this day. Targeted, government-assisted R&D helps make possible the introduction of increasingly efficient new models, which themselves become the basis for tightening the efficiency standards, because they demonstrate that meeting a tighter standard is technologically feasible. When California standards were tightened in 1980 and 1987, followed by federal standards in 1990, 1993, and 2001, manufacturers were able and willing to meet the challenge, an example of government-industry partnership that has served society very well.

BRINGING THE UTILITY COMPANIES ON BOARD

Turning the utility companies from opponents of energy conservation into stakeholders was a key part of California's innovation in energy efficiency. As mentioned earlier, the encouraging results of initial efficiency policies gradually changed a contentious relationship into a collaborative one. High oil prices lasting through the late 1970s until 1985 helped PG&E and other companies perceive that their interests might lie in supporting affordable conservation rather than in pursuing expensive new energy supplies. However, telling utilities to promote efficiency was essentially asking them to sell less of their primary product and thus to lose revenue, at least according to a traditional business model.

A new business model aligning market incentives with policy objectives was needed. The CEC, the CPUC, and the Natural Resources Defense Council created a new utility business model disconnecting profits from the amount of energy generated. A compensatory revenue stream from public goods charges was awarded to companies that agreed to promote efficiency through consumer education programs or fluorescent lightbulb subsidies. The technique of disconnecting utility company revenue from sales became known as "decoupling." Working out the details of decoupling was, and remains, a complex process.

One serious obstacle to the innovation of decoupling was the inability to easily compare conventional energy supplies with the potential of conservation. The value of the utilities' efficiency programs could not be established without a standardized way to set equivalencies in cost and scale. Conventional energy supplies tend to be large and concentrated, thus easy to measure, whereas conservation practices tend to be small and diffuse, thus difficult to measure in aggregate. Our task as scientists was to provide data to counter the skeptics who argued that the granular nature of efficiency—a lightbulb here, a new refrigerator there—could not possibly add up to a significant "supply." Alan Meier and Jan Wright of the Lawrence Berkeley National Labs unraveled this tangled methodological problem in the late 1970s by standardizing "bookkeeping" methods for avoided use, and creating a new investment metric, "the cost of conserved energy." [8, 9] This allowed us to aggregate the energy and cost impacts of scattered conservation steps into a unified supply curve.

The basic assumption when calculating the cost of conserved energy (CCE) is that any conservation measure begins with an initial investment, which then creates a stream of energy savings for the lifetime of the measure. Thus:

$$\text{CCE} = [\text{annualized investment cost}]/[\text{annual energy savings}]$$

The equitable yearly repayment to an investor (e.g., the utility) should be the annualized cost of energy conserved. In the case of avoided electricity use, the energy savings can be expressed in units of $/kWh, or in other cases in units for gas ($/MBtu), or wind, or geothermal. Since the CCE does not depend on a particular local price or type of displaced energy, the comparisons have the virtue of "portability" across regional price variations and types of supply.

Supply Curves of Conserved Energy give rigorous efficiency bookkeeping methods to determine if a mitigation method is cost-effective. Furthermore, the supply curves of conserved energy provided a simple way to compare proposed new energy technologies with energy-saving actions. The challenge of creating reliable supply curves is that deriving sound "macro" estimates from the "micro" contributions of individual changes rests on the painstaking collection of data on population, household size, and consumer purchasing practices, along with lightbulb cost, performance, and life span, and much more. Working out the proper energy accounting methods is the core of this work.

ELECTRONIC BALLASTS

The development of electronic ballasts for fluorescent lamps is the key technical innovation behind the recently burgeoning use of compact fluorescent lights (CFLs), which has resulted in tremendous energy savings. The story of electronic ballasts (also known as "high-frequency" or "solid state" ballasts) is a typical example of how innovations in engineering, policy, and commerce need to be aligned to achieve efficiency improvements.

At the APS Efficiency Summer Study in 1974, we considered the feasibility of creating an electronic ballast that would boost current to 1,000 times that delivered by the power line. We knew that such a device would increase the efficiency of fluorescent lights by 10 percent to 15 percent, and also eliminate the annoying buzz that was a major obstacle to replacing quiet but wasteful incandescent bulbs in residential settings. Moreover, electronic ballasts would enable miniaturization, dimming, remote control, and other user friendly, energy-saving features not possible with magnetic ballasts.

Around that time, the major ballast manufacturing firms did, in fact, consider developing an electronic ballast, but rejected the idea due to the substantial capital investment required and the losses from early retirement of existing infrastructure. As is often the case in overly concentrated sectors—two large firms accounted for 90 percent of the ballast industry—the market provided more disincentives than incentives for innovation. It was clear to us that the impetus for R&D would have to come from elsewhere. In the wake of the APS study, Sam Berman resigned a tenured post at Stanford University to lead LBNL's research on solid-state ballasts (as well as the low-E windows discussed earlier).

Fortunately, the newly-formed DOE included a small Office of Conservation and Solar Energy, which was willing to fund both these projects. From 1977 to 1981, the DOE supported the development, evaluation, and introduction of electronic ballasts into the U.S. marketplace. Basic research took place at LBNL and two subcontracting laboratories. Three small, innovative firms new to the ballast field were awarded cost-sharing contracts to carry out development. Berman shepherded the prototypes through UL certification and persuaded PG&E to host a critical field test in its San Francisco skyscraper, which demonstrated electricity savings of greater than 30 percent over magnetic ballasts.

When the first electronic ballasts came to market in the late 1980s, they were so clearly superior that the major lighting manufacturers felt compelled to adopt and continue to develop the technology. Philips, in particular, reasoned that if large electronic ballasts were effective for traditional tubular fluorescent lamps, they could miniaturize ballasts to produce very efficient CFLs. The appearance of products such as Philips' 16-W CFLs, radiating as much light as a 70-W incandescent light and lasting 10,000 hours instead of 750, was a turning point in the penetration of fluorescent lamps into the residential market.

The risk and expense of converting lighting plants to manufacture a new generation of ballasts was an important difference from the earlier case of improving refrigerators. Converting to blown-in foam insulation was comparatively simple, and invisible to the end-user, so it required no consumer re-education. It is unlikely that the large manufacturers would have taken this step without the assurance of market success afforded by DOE-funded research. In the case of electronic ballasts, it was much harder to launch a positive reinforcing cycle of tightening standards and improving technologies. States did promulgate efficiency standards for fluorescent ballasts (California in 1983, New York in 1986, Massachusetts and Connecticut in 1988, and Florida in 1989). By themselves, however, state standards could not drive market transformation, since they could be satisfied by conventional magnetic ballasts (which, not coincidentally, improved once the electronic ballasts were developed). The experience suggests that in some cases, the seeding effect of publicly funded research is essential for market transformation.

IS THERE A "CALIFORNIA EFFECT"?

There is little doubt that California's energy efficiency policies have been successful. [10] How successful, exactly, remains an open question. There is an ongoing debate about how much of California's lower per capita electricity consumption is due to policy differences, and how much to climate or the comparatively low level of heavy industry. As the need to reduce energy consumption and CO_2 emissions becomes more urgent, the so-called "California Effect" is coming under increasing scrutiny. Whether or not to emulate California's efforts hangs on the question of their efficacy. In 1960, California's per-capita consumption was within 5 percent of the national average. The curves gradually diverged between 1960 and the mid 1970s, but the difference was still only about 15 percent at the time of the OPEC embargo. By 2006, however, Californians were using over 40 percent less electricity per capita than the national average—and only about 10 percent more than they had in 1975.

Calculating the proportion of electricity savings directly traceable to our efficiency efforts is a complicated task. Our best conservative estimate, shown by the middle line in is that at least 25 percent of the observed difference can be directly attributed to policy—an estimate that does not include any secondary effects due to changes in building practices, and appliance markets. Differences in climate and industrial mix, electricity price, demographic trends, and other factors help explain some of the difference, but other trends have been at work as well. In California, for example, building standards and electricity prices have discouraged the use of electric water heating in favor of natural gas, which reduces electricity consumption relative to the national average.

At the same time, most new housing has been built in the hotter inland valley and desert areas, dramatically increasing energy consumption for air conditioning. Also, most appliance standards initiated in California were eventually adopted nationally, so the policy impacts of appliance standards also affect the national per-capita consumption average, an effect that is not captured by the difference in per-capita consumption. Thus, for a variety of reasons, electricity use in California has been essentially flat and should either continue or even decrease as California extends standards to new devices, accelerates building performance requirements, and expands programs aimed at improving efficiency.

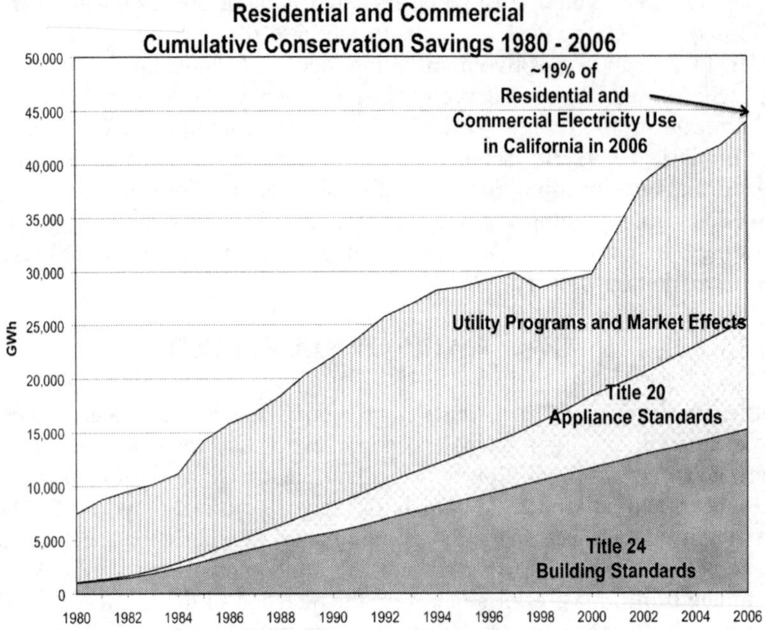

FIGURE 7. California's Annual Energy Savings Attributed to Efficiency Measures.

Figure 7 shows California's savings in greater detail, breaking down the part of the consumption gap that can be attributed to efficiency efforts. Performance standards for buildings and appliances, which as noted have been progressively strengthened every few years, account for roughly half the savings. The other half has resulted from utility company programs that promote adoption of energy efficient technologies, such as commercial lighting retrofit incentives and residential appliance rebates. Through 2003, these measures have resulted in about 40,000 GWh of annual energy savings and have avoided 12,000 megawatts (MW) of demand—the same as 24 500-MW power plants (the MW data is not shown in the graph). These savings have reduced CO_2 emissions from the electricity generation sector by nearly 20 percent compared to what otherwise might have happened without these programs and standards. This equates to an avoidance of CO_2 emissions in the state as a whole of about four percent, due to historical energy efficiency programs and standards. These savings will only continue to grow.

The effect of efficiency policies is even more pronounced at peak load. Peak loads are a serious concern in California, as in other Sunbelt states and many fast- growing economies around the world. Air conditioning loads on hot afternoons can greatly increase system demand—as much 30 percent in California. Reducing the magnitude of these warm-season spikes is one of the most pressing items on the efficiency agenda. Building standards that focus on minimizing heat gain and thermal transfer and appliance standards that set minimum efficiency levels for air conditioning equipment can reduce peak demand. This in turn lowers the customer's immediate cooling costs as well as the system-wide costs of maintaining underutilized peaking capacity year round; both measures contribute to lower bills. The 12,000 MWs of capacity provided by efficiency measures have effectively avoided the need to build additional power plants to meet that demand.

FROM "INNOVATION" TO "BUSINESS AS USUAL": THE LONG-TERM ENERGY EFFICIENCY STRATEGIC PLAN

When the campaign for energy efficiency in California began four decades ago, the goal was simply to reduce the expense, pollution, and political turmoil resulting from over-dependence on generating energy from fossil fuels. However, as awareness of the climate-changing effects of GHGs grew, so too did recognition of efficiency as a low-cost, low-impact, reliable source of energy. Now that our environmental concerns must share the stage with the current economic crisis, efficiency has suddenly become something of a mantra. Since efficiency investments have some of the fastest payback times in the "green economy," and since efficiency improvements are based on currently available technology, implementation offers a uniquely practical opportunity to stimulate economic growth and reduce GHG emissions at the same time.

A year ago, the California Public Utilities Commission (CPUC) issued California's Long-Term Energy Efficiency Strategic Plan, mapping out the steps toward meeting the state's GHG reduction goals by 2020. The commission estimates that the Strategic Plan will create annual energy savings of close to 7,000 gigawatt hours, 1,500 megawatts, and 150 million metric therms of natural gas. This is roughly equal to the avoided construction of three 500-megawatt power plants. Avoided emission of GHGs is expected to reach three million tons per year by 2012, equivalent to the emissions of nearly 600,000 cars. It is hoped that new efficiency programs will create between 15,000 and 18,000 jobs, in areas ranging from construction to education. The plan has four "Big and Bold" goals:

- All new residential construction in California will be zero net energy by 2020.
- All new commercial construction in California will be zero net energy by 2030.
- The Heating, Ventilation, and Air Conditioning (HVAC) industry will be reshaped to ensure optimal equipment performance.
- All eligible low-income homes will be energy efficient by 2020.

The budget for just the first three years of the Strategic Plan was recently set at $3.1 billion, making it the largest-ever state commitment to efficiency. Funding will support a wide variety of programs in pursuit of the overarching goals, including the four examples below:

- CalSPREE, the largest residential retrofit effort in the United States, will cut energy use by 20 percent for up to 130,000 existing homes by 2012.
- $175 million will go to programs to deliver "zero net energy" homes and commercial buildings.
- $260 million will go to 64 local agencies (city, county, and regional) that would otherwise lack the expertise to create more energy-efficient public buildings.
- More than $100 million will go to for education and training programs at all levels of the education system.

From my perspective as a veteran of the efficiency campaign, the Strategic Plan presents a fascinating combination of old lessons and new ambitions. Although the overall scope of the plan is far more comprehensive and coordinated than anything yet seen, clearly the content of the programs is based on many years of experience in buildings and appliance standards. Furthermore, the plan was developed in collaboration with more than 500 stakeholder groups, including the state's major investor-owned utilities (IOUs): Pacific Gas and Electric Co., San Diego Gas and Electric Co., Southern California Edison, and Southern California Gas Co. The IOUs will be responsible for actually implementing the programs in their respective regions. The budget for the programs comes from the increased public goods charges authorized by the CPUC, on the condition that the funds be invested in efficiency. The slightly increased costs to ratepayers will be quickly offset by their reduced consumption. Of course, this process of coordinating best engineering practices with policy goals and utility market mechanisms has its origins in our forays into collaboration in the early 1980s.

The most ambitious and innovative aspect of the Long-Term Energy Efficiency Strategic Plan is its insistence on re-branding the practice of energy efficiency as normative behavior rather than crisis response. Commissioners Michael Peevey and Dian Grueneich have frequently spoken of "making efficiency a way of life." If successful, this would mean a reversal of the prevailing mindset. For many years, my graphs of energy supply/demand forecasts displayed competing scenarios labeled respectively "with efficiency measures" and "business as usual." Business as usual was understood to mean "without efficiency measures." If California's Strategic Plan succeeds, the comprehensive approach to energy efficiency that we have been pursuing for over 30 years will have finally become "business as usual."

ACKNOWLEDGEMENTS

Thanks to Dave Hafemeister, who turned a talk into a chapter, to Deborah Poskanzer for editing, and to Pat McAuliffe, who both edited and produced and/or updated most of the figures.

REFERENCES

1. Rosenfeld and D. Poskanzer, "A Graph is Worth a Thousand Gigawatt-Hours," *Innovations* (fall) 2009. This contains a more complete list of references.
2. A.H Rosenfeld, "The Art of Energy Efficiency: Protecting the Environment with Better Technology," *Annual Review of Energy and the Environment*, 24, 33–82 (1999).
3. California Public Utilities Commission and California Energy Commission, September 2008, *California Long-Term Energy Efficiency Strategic Plan*: www.cpuc.ca.gov/PUC/energy/electric/Energy+Efficiency.
4. American Institute of Physics, *Efficient Use of Energy: The APS Studies on the Technical Aspects of the More Efficient Use of Energy*, *AIP Conference Proceedings*, 25 (1975).
5. P. Haves, "Energy Simulation Tools for Buildings," this volume.
6. H.S. Geller and D.B. Goldstein, 1999. "Equipment Efficiency Standards: Mitigating Global Climate Change at a Profit," *Physics and Society* 28(2).
7. L. Desrouches, "Appliance Standards and Advanced Technologies," this volume.
8. Meier, J. Wright and A.H. Rosenfeld, 1983. Supplying Energy Through Greater Efficiency: The Potential for Conservation in California's Residential Sector, Berkeley, CA: University of California Press.
9. L. Desrouches, see data on cost of conserved energy.
10. "Energy Research at DOE: Was It Worth It?," in *Energy Efficiency and Fossil Energy Research 1978 to 2000*, Appendix E, pp 100-104. Washington, DC: National Academies Press, 2001.

Defining a Standard Metric for Electricity Savings

Jonathan Koomey[a], Hashem Akbari, Carl Blumstein, Marilyn Brown, Richard Brown, Robert Budnitz, Chris Calwell, Sheryl Carter, Ralph Cavanagh, Audrey Chang, David Claridge, Paul Craig, Rick Diamond, Joseph H. Eto, William J. Fisk, William Fulkerson, Ashok Gadgil, Howard Geller, José Goldemberg, Chuck Goldman, David B. Goldstein, Steve Greenberg, David Hafemeister, Jeff Harris, Hal Harvey, Eric Heitz, Eric Hirst, Holmes Hummel, Dan Kammen, Henry Kelly, Skip Laitner, Mark Levine, Amory Lovins, Gil Masters, Pat McAuliffe, James E. McMahon, Alan Meier, Michael Messenger, John Millhone, Evan Mills, Steve Nadel, Bruce Nordman, Lynn Price, Joe Romm, Marc Ross, Michael Rufo, Jayant Sathaye, Lee Schipper, Stephen H. Schneider, Robert H. Socolow, James L. Sweeney, Malcolm Verdict, Alexandra von Meier, Diana Vorsatz, Devra Wang, Carl Weinberg, Richard Wilk, John Wilson, Jane Woodward, and Ernst Worrell

[a]Consulting Professor, Stanford University. JGKoomey@stanford.edu, <http://www.koomey.com>, PO Box 1545, Burlingame, CA 94011-1545. Other addresses not given because coauthors are too numerous

Abstract. The growing investment by governments and electric utilities in energy efficiency programs highlights the need for simple tools to help assess and explain the size of the potential resource. One technique that is commonly used in that effort is to characterize electricity savings in terms of avoided power plants, because it is easier for people to visualize a power plant than it is to understand an abstraction like billions of kilowatt-hours. Unfortunately, there is no standardization around the characteristics of such power plants.

In this article we define parameters for a standard avoided power plant that have physical meaning and intuitive plausibility, for use in back-of-the-envelope calculations. For the prototypical plant this article settles on a 500-megawatt existing coal plant operating at a 70% capacity factor with 7% T&D losses. Displacing such a plant for one year would save 3 billion kWh/year at the meter and reduce emissions by 3 million metric tons of CO_2 per year. The proposed name for this metric is the *Rosenfeld*, in keeping with the tradition among scientists of naming units in honor of the person most responsible for the discovery and widespread adoption of the underlying scientific principle in question – Dr. Arthur H. Rosenfeld.

Physics of Sustainable Energy II: Using Energy Efficiently and Producing it Renewably
AIP Conf. Proc. 1401, 26-43 (2011); doi: 10.1063/1.3653843
2011 American Institute of Physics 978-0-7354-0972-9/$30.00

INTRODUCTION

In the three decades since the energy crises of the 1970s we have learned a great deal about the potential for energy efficiency and the means to deliver it cost effectively and reliably [1]. Back then, many analysts still held to the now discredited "ironclad link" between energy use and economic activity, which implied that any reduction in energy use would make our society less wealthy [2, 3, 4, 5]. Now we know (from cross-country comparisons and technical analysis) that there are many ways to produce and consume goods and services, some energy efficient and others not [6, 7, 8, 9, 10]. And we know that the available efficiency resource is enormous, inexpensive, and largely untapped (particularly if whole-system clean-slate redesign is employed), making it an important option for reducing climate risks and improving energy security [11, 12, 13, 14, 15, 16]. Finally, we know that tapping these resources requires more than getting energy prices right—we'll also need to further develop and implement cost-benefit-tested non-price policies like minimum efficiency standards, Energy Star labeling programs, utility rebates, "Golden Carrot" incentives, research and development, tax credits, feebates, and other programs whose goal is to align private financial incentives with the economic and environmental interests of society as a whole [12, 13, 15, 17, 18, 19, 20, 21, 22, 23].

The increased focus on energy efficiency for shaping our energy future highlights the need for simple tools to help understand and explain the size of the potential resource. One technique that is commonly used in that effort is to characterize electricity savings in terms of avoided power plants, because it is easier for people to visualize a power plant than it is to understand an abstract concept like billions of kilowatt-hours. Unfortunately, there is no standardization around the size and operational characteristics of such plants.

In this article we propose standard characteristics for an avoided power plant that have physical meaning and intuitive plausibility, for use in back-of-the-envelope calculations and characterizing energy savings results. We also propose naming the annual energy savings of such a plant as a new unit in Art Rosenfeld's honor (the *Rosenfeld*) because Dr. Rosenfeld continues to be the most prominent advocate of characterizing efficiency savings in terms of avoided power plants.

ARTHUR H. ROSENFELD'S CONTRIBUTIONS

Dr. Rosenfeld (**Figure 1**) made a transition from particle physics to studying energy efficiency at the time of the first oil embargo [1]. Over the past 35 years he has been at the forefront of efforts to improve the efficiency of energy use around the world and has devoted special care to making the results of complex energy analysis understandable to a lay audience. For years, Dr. Rosenfeld has characterized oil savings in terms of "Arctic Refuges saved" and electricity savings in terms of "avoided power plants" to emphasize that supply and demand side policy options are fungible and that replacing power plants with more efficient energy technologies would be beneficial for consumers' energy bills and for the environment.

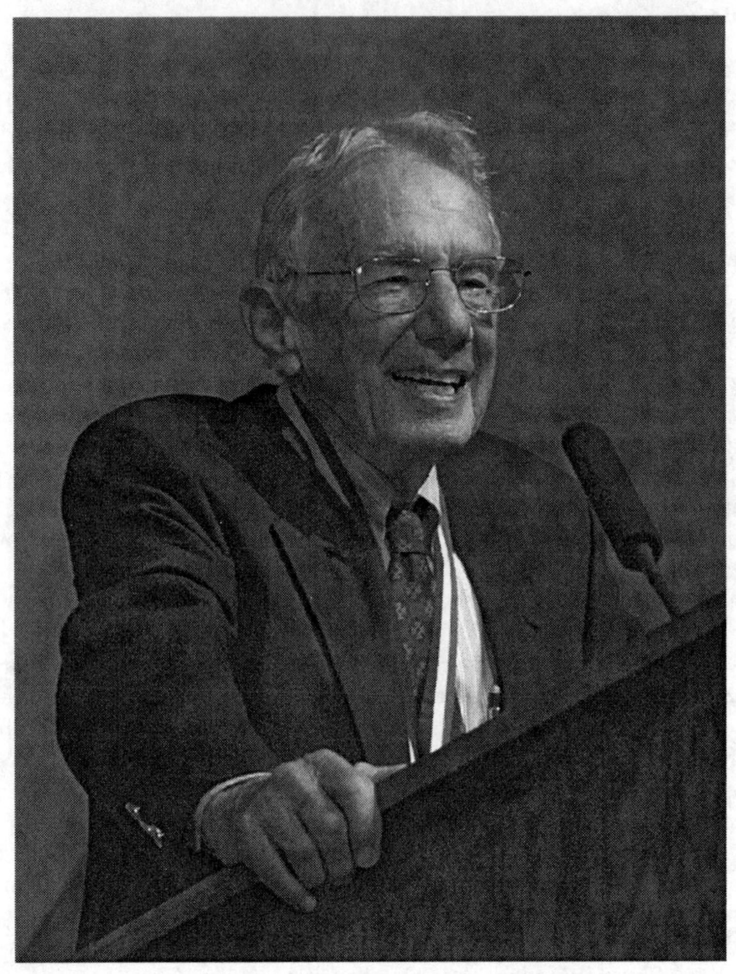

FIGURE 1. Arthur H. Rosenfeld.

Dr. Rosenfeld has in the past most commonly used a 1000 MW power plant operating at a 60 or 65% capacity factor as the standard power plant avoided by energy efficiency. These assumptions mirrored the capacity and operational characteristics of typical U.S. nuclear power plants circa 1990, but since that time the capacity factors of such plants have increased to about 90%. No new nuclear plants have been completed in the U.S. since 1996, so the appropriateness of this choice of assumptions has decreased over time. More recently, Dr. Rosenfeld has used a 500 MW plant operating 5,000 hours per year as his standard avoided plant [24].

CRITERIA

Choosing characteristics of a typical avoided power plant is inevitably somewhat arbitrary—there is no single correct answer. In our view, those choices should meet the following criteria:

1) *Simplicity of presentation and ease of recall*: Round numbers of one significant figure should be preferred to more accurate numbers with several decimal places of precision because they are easier to remember and use. Moreover, "average" power plant sizes and capacity factors change each year, so a value with several decimal places of precision would have no longevity in any case.

2) *Intuitive plausibility*: The parameters chosen should reflect people's general understanding of power plants and their operation in the utility system.

3) *Physical meaning*: The chosen characteristics should reflect real-world attributes of the physical systems in which power plants are embedded and should be expressed as savings *at the meter* (to account for transmission and distribution losses).

4) *Policy relevance*: The main result for avoided power plants would be electricity savings (which is an important metric for energy policy). Carbon savings associated with those energy savings (reflecting climate change, the most important environmental challenge facing humanity) should also be estimated, but electricity savings are the key focus. Costs and non-CO_2 emissions for avoided power plants vary greatly by technology, by country, and over time, so including them would make this task needlessly complicated.

The next step is to assess the key parameters for characterizing power plants to see which choices might meet those criteria. To make that assessment easier, we add two additional constraints:

1) *We focus on power plants avoidable in the long run.* Utility emissions savings can be the result of either short run operational changes or long run retirement and construction decisions. Emissions savings from operational changes are much more difficult to characterize in a general way than are long-term changes (analyzing the former is very situation dependent and typically requires complicated production-cost/dispatch simulation modeling).

2) *We assume that the standard avoided power plant should be coal-fired.* Between 2000 and 2007, 151 new coal-fired power plants were proposed in the United States; 10 have been completed, 25 more are under construction, and 59 have been canceled or indefinitely deferred [25]. In 2007, existing coal plants totaled more than 300 GW (out of almost 1,000 GW total installed capacity in the U.S.).

Coal plants generate about half of all U.S. electricity and were responsible for about one third of total U.S. carbon emissions in 2007. They are also ubiquitous in other countries responsible for substantial percentages of world carbon emissions (e.g., China and India). Truly facing the climate challenge will require the retirement or displacement of hundreds or thousands of such plants [26, 27, 28, 29]. Finally, the capacity factors of coal plants are relatively insensitive to fuel price changes (compared to natural gas plants) so their operational characteristics are more predictable than for some other plants.

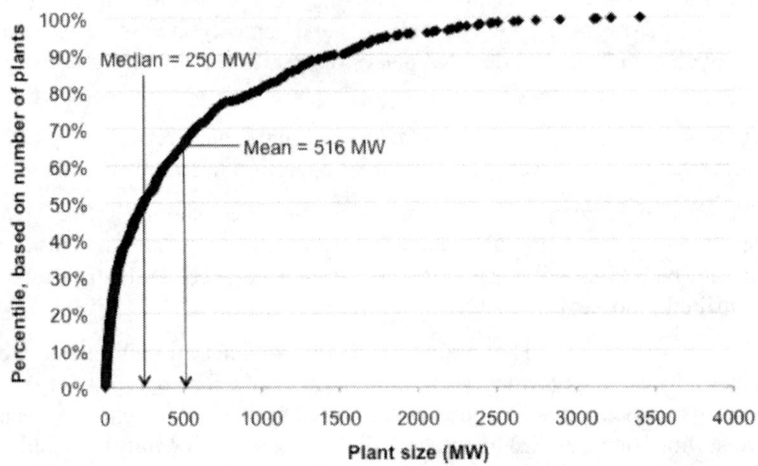

FIGURE 2. Cumulative distribution of capacity for existing U.S. coal-fired power plants in 2007. Source: U.S. Energy Information Administration, Form EIA-860 Annual Generator Report Database <http://www.eia.gov/cneaf/electricity/page/eia860.html>.

CHARACTERISTICS OF COAL-FIRED POWER PLANTS

This section describes our review of the literature for each key characteristic of coal-fired power plants in advance of choosing parameters for a typical plant.

Capacity

Power plants vary greatly in their capacity (measured in Megawatts, MW, or million watts), which can be expressed as a nameplate (nominal) rating or as net capacity after subtracting out power needed to run the plant.

The Energy Information Administration or EIA [30] gives characteristics of new conventional power plants for the US. Pulverized coal plants with scrubbers fall at 600 MW.

As shown in **Figure 2**[1] the median nameplate capacity for existing non-cogenerating U.S. coal plants in 2007 was 250 MW, with a mean of about 500 MW

[1] See Koomey and Hultman [31] for similar cumulative distribution graphs describing historical data on nuclear power plants in the U.S.

and a range of less than a MW to about 3500 MW, <http://www.eia.gov/cneaf/electricity/page/eia860.html>. Most of the smaller plants (less than 200 MW) tend to be older (1960s or earlier), while the larger plants tend to be newer (1970s or later).

EIA's *Electric Power Annual 2007 + 2009* [32, 33] show that total capacity for U.S. coal fired power generation was remarkably stable over the period 1996 to 2009, starting and ending at just over 300 GW (**Table 1**). There have been a few retirements and new plants constructed, but the U.S. has seen no significant change in total coal capacity over this period.

Capacity Factors

The capacity factor is defined as

$$\text{Capacity factor} = \frac{\text{Actual generation/year (BkWh)}}{\text{Maximum generation/year (BkWh)}} \tag{1}$$

Dividing numerator and denominator by the number of hours per year (8766 hours when averaged across leap and non-leap years) we get

$$\text{Capacity factor} = \frac{\text{Average output capacity (MW)}}{\text{Rated (maximum) capacity (MW)}} \tag{2}$$

Coal plants can have a wide range of capacity factors: they are usually operated for baseload electricity but are flexible enough to serve all but the lowest of intermediate loads as well. Their capacity factors are relatively insensitive to coal prices though they can be influenced when the price for the main competing fuel in the power sector (natural gas) fluctuates greatly.

New coal plants typically have high capacity factors (up to 90%). Capacity factors for existing plants in the U.S. averaged about 70% over the 1996 to 2009 period (as shown in **Table 1**). The stock of existing plants includes many older plants that are smaller, less efficient, and more polluting than new plants. They have long since been depreciated, so utilities have an incentive to keep them running as long as the marginal costs are not too high (and as long as environmental regulations do not impose additional costs or constraints that make them uneconomic).

TABLE 1. Characteristics of existing U.S. coal-fired power plants.

	Coal fired capacity GW	Net generation TWh	Capacity factor %	Coal consumed in million short tons	Heat content of utility coal MBtu/short ton	Average HHV efficiency %
1996	313	1,795	65.2%	907	20.55	32.9%
1997	314	1,845	67.2%	932	20.52	32.9%
1998	316	1,874	67.7%	946	20.52	32.9%
1999	315	1,881	68.1%	950	20.49	33.0%
2000	315	1,966	71.0%	995	20.51	32.9%
2001	314	1,904	69.2%	973	20.34	32.8%
2002	315	1,933	70.0%	988	20.24	33.0%
2003	313	1,974	72.0%	1014	20.08	33.1%
2004	313	1,978	71.9%	1021	19.98	33.1%
2005	313	2,013	73.3%	1041	19.99	33.0%
2006	313	1,991	72.6%	1031	19.93	33.1%
2007	313	2,016	73.6%	1047	19.91	33.0%
2008	313	1,986	72.2%	1042	19.71	33.0%
2009	314	1,756	63.8%	935	19.54	32.8%
Average			69.8%			33.0%

(1) Coal consumed, capacity, and net generation include all coal-fired power plants in the U.S., including utility and non-utility central station plants as well as industrial cogeneration plants.
(2) Coal fired capacity, net generation and coal consumed taken from US DOE [32] through 2007 and US DOE [33] for 2008 and 2009. Heat content of coal taken from Table A-5 in US DOE [34] through 2007 and from US DOE [35] for 2008 and 2009. MBtu = million Btus.
(3) Capacity factor calculated from capacity and net generation assuming 8760 hours for non-leap years and 8784 hours for leap years.
(4) Power plant efficiency (higher heating value) calculated by converting net generation to Btus assuming 3412 Btus/kWh and then dividing by the product of coal consumed and heat content of utility coal.

Transmission and Distribution Losses

Table 2 shows data from EIA's *Electric Power Annual 2007* [32] on the supply and disposition of electricity in the U.S. from 1995 to 2007 ([33] has the same data for 2008 and 2009). Losses are expressed as a percentage of the sum of electricity sales, direct use by power plants, and exports. These losses range from 5.7% to 7.4% with a simple average of 6.8% over that period.

TABLE 2. U.S. average transmission and distribution (T&D losses) over time.

	Total electric industry sales TWh	Direct use TWh	Total exports TWh	Losses and unaccounted for TWh	T&D losses %
1996	3101	153	3	231	7.1%
1997	3146	156	9	224	6.8%
1998	3264	161	14	221	6.4%
1999	3312	172	14	240	6.9%
2000	3421	171	15	244	6.8%
2001	3394	163	16	202	5.7%
2002	3465	166	16	248	6.8%
2003	3494	168	24	228	6.2%
2004	3547	168	23	266	7.1%
2005	3661	150	20	269	7.0%
2006	3670	147	24	266	6.9%
2007	3765	159	20	264	6.7%
2008	3733	132	24	287	7.4%
2009	3597	127	18	261	7.0%
Average					6.8%

1) Data on electric industry sales, direct use, exports, and losses are taken from US DOE [32] through 2007 and from US DOE [33] for 2008 and 2009.
2) T&D losses calculated as a percentage of sales plus direct use plus exports.

TABLE 3. Direct carbon emissions factors for fuels used by utilities to generate electricity.

	M tons C/quad	kg C/GJ	gC/kWh.f	*Index NG = 1.0*
Natural gas	14.47	13.7	49.4	*1.00*
Distillate oil	19.95	18.9	68.1	*1.38*
Residual oil	21.29	20.2	72.6	*1.47*
Coal	25.83	24.5	88.1	*1.78*

(1) Carbon emissions factors (Mt-C/quadrillion Btus) taken from EIA data for 2006 (downloaded from <http://www.eia.gov/environment>). It is unclear if these data have already built in a combustion fraction but we assume so. Combustion fractions are typically very close to 1.0 for fossil fuels in utility plants in any case.
(2) All energy values based on higher heating value (HHV) of the fuels.
(3) kWh.f = energy content of fuel converted to kWh using 3412 Btu/kWh.

Carbon Emissions Factors for Fossil Fuels

The EIA <http://www.eia.gov/environment> gives historical data on the carbon content of fuels for U.S. electric utilities. The data for 2006, expressed in higher heating value (HHV) terms, are shown in **Table 3**. Coal emits almost 80% more carbon than natural gas per unit of heat released.

Power Plant Efficiencies

Large coal steam plants have HHV efficiencies of 30-40%, depending on their age, level of pollution control, and technology type. For typical new 600 MW coal plants in 2008, EIA gives an estimate of 37% HHV efficiency [30]. The average efficiency of existing coal steam plants in the U.S. for the period 1996 to 2009, derived using the heat content of utility coal from US DOE [34, 35] and the other parameters in **Table 1**, is 33%, which doesn't vary much over this period.

DEFINING THE ROSENFELD

We experimented with different combinations of plant capacities and capacity factors to meet the criteria listed above, focusing mainly on the characteristics of existing U.S. coal plants. We choose this approach because of the rich data characterizing these plants and because most existing coal plants will need to be retired if we're to substantially reduce carbon emissions by the middle of this century, as climate stabilization requires.

As summarized in Table 4 and Figure 3, we've defined the Rosenfeld unit assuming the average coal plant capacity of 500 MW, a capacity factor of 70% (the average capacity factor of existing U.S. coal plants from 1996 to 2009), and system-wide T&D losses of 7% (rounded up from 6.8% for ease of recall). This combination of parameters yields annual electricity delivered at the meter of about 3 BkWh/year. Using the carbon burden for U.S. utility coal and the efficiency of average existing coal steam plants, the emissions saved are almost exactly 3 million metric tons of CO_2 (Mt CO_2) per year.

If measured in terms of site energy, there are 100 Rosenfeld-years per exajoule, and in primary energy terms there are about 30 Rosenfeld-years per exajoule. Another nice equivalence factor that emerges from these numbers is that each kilowatt-hour of coal-fired electricity delivered to the meter emits about 1 kg of CO_2.

USING THE ROSENFELD

This simplification aids in the creation of quick calculations and cogent interpretation of analysis results from studies of energy efficiency. To use the Rosenfeld, analysts have to remember the numbers associated with the power plant characteristics (500 MW, 70% capacity factor, 7% T&D losses, 33% HHV efficiency), and the number 3 (which evokes 3 billion kWh per year saved at the meter, 3 million metric tons of carbon dioxide avoided per year, and 30 Rosenfeld-years per exajoule of primary energy).

Consider the recent authoritative study on energy efficiency by the American Physical Society [15]. Figure 25 in that study shows potential U.S. residential sector efficiency savings of almost 600 billion kWh/year in 2030. What does that number mean in terms of power plants avoided?

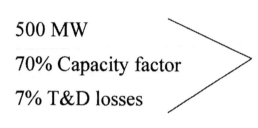

500 MW

70% Capacity factor

7% T&D losses

3 billion kWh/yr of coal-fired electricity saved at the meter

+

3 million metric tons of carbon dioxide avoided per year

1 Exajoule (primary energy) ~ 30 Rosenfeld-years

1 Pacala/Socolow Wedge ~ 30,000 Rosenfeld-years = 600 Rosenfelds each year for 50 years

FIGURE 3. Summary of the Rosenfeld unit.

TABLE 4. Estimating electricity delivered and carbon emitted from a typical coal plant in the U.S.

	Units	Value	Notes
Electricity generated			
Capacity	MW	500	1
Capacity factor	%	70%	2
Hours per year	hours	8766	3
Assumed T&D losses	%	7%	4
Total electricity generated at the busbar	Billion kWh/year	3.07	5
Total electricity delivered to the meter	Billion kWh/year	2.87	6
Site energy (HHV)	Quadrillion Btus/year	0.010	7
	Exajoules/year	0.010	8
Primary energy (HHV)	Quadrillion Btus/year	0.032	9
	Exajoules/year	0.034	8
Carbon emitted			
Coal carbon burden (based on HHV)	gC/kWh.fuel	88.1	10
Efficiency (based on HHV)	%	33%	11
Carbon burden at the busbar	gC/kWh.elect generated	267	12
Carbon burden at the meter	gC/kWh.elect delivered	286	13
Carbon emissions	Million metric tons C/yr	0.82	14
	Million metric tons CO_2/yr	3.01	15

(1) Capacity is based on average existing U.S. coal plants from EIA-860 survey results as summarized in Figure 2 <http://www.eia.gov/cneaf/electricity/page/eia860.html>.
(2) Capacity factor is the average for existing US coal plants from 1996 to 2009 from Table 1.
(3) Hours per year is an average over leap years and non-leap years.
(4) T&D (Transmission and distribution) losses are typical for the U.S. utility system (from Table 2), rounded up to 7% for ease of recall.
(5) Total electricity generated at the busbar is the product of capacity, capacity factor, and hours per year, expressed using the American notation of billion equaling 10^9.
(6) Total electricity delivered to the meter is total electricity generated divided by (1 + percentage T&D losses).
(7) Site energy in quadrillion Btu/year calculated by multiplying kWh per Rosenfeld at the meter by 3412 Btus/kWh.
(8) Quadrillion btus converted to exajoules using the factor 1055.1 joules/Btu.
(9) Primary energy in quadrillion Btu/year calculated by converting the efficiency described in footnote 11 to a heat rate (primary energy per kWh), then multiplying that heat rate times (1 + percentage T&D losses) and multiplying again by the number of kWh per Rosenfeld.
(10) The carbon burden of coal is expressed in grams of carbon (C) per kWh of fuel (fuel converted to kWh assuming 3412 Btus/kWh). This carbon burden is taken from EIA for 2006, as described in Table 3.
(11) Power plant efficiency, in Higher Heating Value (HHV) terms, is the average for existing US coal plants from 1996 to 2009 from Table 1.
(12) Carbon (C) burden at the busbar (calculated in grams of carbon per kWh generated) is calculated as the ratio of the coal C burden from Table 3 and the power plant efficiency (both in HHV terms).
(13) C burden at the meter is the carbon burden at the busbar times (1 + percentage T&D losses).
(14) C emissions in million metric tons are the product of electricity consumed at the meter and the C burden at the meter.
(15) Carbon dioxide emissions in million metric tonnes are equal to C emissions times the ratio of molecular weights of carbon dioxide (44) and carbon (12).

Six hundred billion kWh/year is the equivalent of about 200 Rosenfelds (600/3), or 200 typical coal fired power plants, which together emit 600 million metric tons of CO_2 per year. This simple calculation adds real physical meaning to the electricity savings (but it's no substitute for more sophisticated approaches). Other important studies that would have benefitted from using this approximation include Brown et al. [16], EPRI [36], Koomey et al. [37], Meier et al. [38] Rosenfeld and Hafemeister [39], Rosenfeld et al. [40], and any other efficiency potentials studies that don't include a full integrated analysis of supply and demand-side options.

Another widely used approximation for understanding carbon reductions is that of the "Stabilization wedge", popularized by Pacala and Socolow [41]. Each wedge represents cumulative carbon reductions over a 50-year period of 25 billion metric tons of carbon, or 91.7 billion metric tons of CO_2. Each Rosenfeld saves 3 million tons of CO_2 per year, so a full wedge is equivalent to 91,700/3 or about 30,000 Rosenfeld-years (equivalent to fully eliminating 600 coal-fired power plants [or 300 GW] for their entire 50-year lifetimes).

For those situations where the avoided carbon emissions would be quite different from those of a coal plant, we show **Table 5**, which gives the relationship between the carbon emissions factors for a coal plant and average emissions factors for different power plant technologies (from US DOE [30]), for the power sectors of different countries (from US DOE [32] for the U.S. and Wheeler and Ummel [42] for other countries), and for California (from Mahone et al. [43]). Natural gas plants are significantly less carbon intensive than coal. In places where the avoided power plant is an advanced natural gas combined cycle (typical for recently constructed gas plants) the emissions per kWh are 63% lower than that of an existing coal plant, resulting in annual emissions displaced of about 1 million metric tons of CO2 per year for one Rosenfeld-year of electricity savings. In addition, Table 5 shows that China and India, two of the largest and most rapidly growing economies, have average power sector carbon emissions factors that are close to that of the existing coal plant used in this study, indicating that most of their electricity generation comes from coal.

LIMITATIONS

All simplifications are imperfect, and this one is no exception. The specific characteristics of electricity systems (like power plant capacity factors, efficiencies, coal carbon content, and line losses) all vary greatly around the world. Thus, no single number will apply everywhere, and trying to create an approximation that perfectly characterizes all situations is futile and antithetical to the spirit of this entire exercise. So we accept that this simplification is useful, but limited.

The Rosenfeld is most useful when applied to studies of energy efficiency in isolation from the electricity supply side, because it lends context to such studies that otherwise would require a detailed analysis of avoided power plants. Even given the limitations of an approximation like this, the contextual depth and conceptual understanding that it can bring to energy efficiency studies make it well worth applying.

TABLE 5. Carbon emission factors for electricity delivered to the meter.

	Fuel	Efficiency HHV	Emissions factor gC/kWh fuel	Emissions factor gC/kWh elect delivered	Index Existing coal = 1.0	Notes
Existing plants						
Steam turbine	Coal	33.0%	88.1	286	1.00	1, 2
Steam turbine	Residual oil	32.8%	72.6	237	0.83	1, 3
Steam turbine	Distillate oil	32.8%	68.1	222	0.78	1, 3
Steam turbine	Natural gas	32.6%	49.4	162	0.57	1, 3
Gas turbine	Distillate oil	25.8%	68.1	282	0.99	1, 3
Gas turbine	Natural gas	29.8%	49.4	177	0.62	1, 3
Combined cycle	Distillate oil	31.0%	68.1	235	0.82	1, 3
Combined cycle	Natural gas	45.8%	49.4	115	0.40	1, 3
New plants						
Steam turbine, scrubbed	Coal	37.1%	88.1	254	0.89	1, 4
Advanced combined cycle	Natural gas	50.5%	49.4	105	0.37	1, 4
Average power sector carbon emissions factor by country 2007 [Rank in 2007 power sector emissions in square brackets]						
China [1]				279	0.98	5
United States [2]				174	0.61	6
India [3]				259	0.91	5
Russia [4]				156	0.54	5
Germany [5]				197	0.69	5
Japan [6]				117	0.41	5
United Kingdom [7]				179	0.63	5
Australia [8]				287	1.00	5
South Africa [9]				296	1.03	5
South Korea [10]				143	0.50	5
Indonesia [18]				213	0.74	5
France [27]				28	0.10	5
Brazil [44]				16	0.06	5
World average 2007				175	0.61	5
California 2008						
Average				119	0.42	7
Marginal				156	0.55	7

(1) Emissions factors for fossil fuels taken from Table 3.
(2) Steam turbine efficiency for average existing US coal plants from 1996-2009 taken from Table 1.
(3) Steam turbine, gas turbine, and combined cycle efficiencies for existing oil and gas plants calculated from higher heating value (HHV) heat rates in the Electric Power Annual 2007 [32], Table A7, which represent an average for existing plants in 2007. The Electric Power Annual table does not differentiate between residual oil and distillate oil steam turbine efficiencies so we assume these are the same.
(4) Efficiencies for 2008 new plants derived from heat rates in Assumptions to the AEO 2009 [30], Table 8.2.
(5) Carbon emissions factors for the power sectors in different countries and the world in 2007 taken from the CARMA database <http:// www.carma.org>, documented in Wheeler and Ummel [42]. We apply 7% T&D losses to the CARMA emissions factors to bring them back to the meter, fully cognizant of the substantial differences in line losses between these countries but lacking any consistent data source for those losses. The total power sector emissions for the top 10 countries in 2007 represents about 77% of the world power sector total.
(6) Average carbon emissions factors for the U.S. in 2007 derived from CO_2 emissions for central station and combined heat and power plants reported by the Electric Power Annual 2007 [32] and the sum of utility sales, electricity exports, and internal electricity use for industrial customers from Table 2 (also taken from Electric Power Annual 2007).
(7) California average and marginal power sector emissions for 2008 derived as a simple average from the typical hourly average and marginal emissions factors in the model documented in Mahone [43] and corrected for 7% transmission and distribution losses to estimate the emissions factor at the meter.

One of the most important caveats to the use of this simplification relates to the load shape impacts of efficiency options, which are typically summarized in terms of conservation load factor or CLF [44, 45]. The Rosenfeld approximation is most accurately applied to electricity savings from a broad efficiency portfolio with CLFs between 50% and 100%.[2] It should not be used for efficiency options with low CLFs

[2] Studies that estimate peak demand impacts for a broad range of efficiency options typically calculate aggregate CLFs close to the average utility load factor of about 60%. For example, the comprehensive study by Rufo and Coito [46], which estimated

that save electricity mostly at times of peak load (like those for air conditioners), because the avoided power plants are more likely to be gas-fired peaking plants with characteristics quite different from those of coal plants.

It is most appropriate to apply the Rosenfeld to annual electricity savings. To fully displace a power plant, which typically lasts for fifty years, efficiency savings will need to continue for the life of that plant. Analysts should use caution when treating cumulative electricity savings over time with this approximation.

Policy studies assessing the emissions reductions from efficiency and supply side options will generally distinguish between the average and marginal emissions factors for the power system. The marginal emissions factor is the reduction in emissions from decreased power generation divided by the amount of electricity savings driving those reductions (it can be calculated for either the short or long run). The estimated long-run marginal emissions savings may or may not equal the emissions savings for coal plants calculated above (and they vary greatly by utility, state, or country, as shown in Table 5). Care must therefore be used when applying the Rosenfeld to the results from emissions reduction studies.

To retire a power plant, the most important condition is that there be a resource to displace the generation of that plant, be it energy efficiency or another power plant. Of course, the choice of *which* power plant to retire is a function of economics—more specifically, it is a function of the economic incentives facing the electric utility, and the utility's incentives may or may not be aligned with the optimal outcome for society. Many existing coal plants are fully depreciated and their marginal costs are low. In the absence of a change in policy, the utility won't retire these plants—instead, new resources will be deferred or other, higher marginal cost resources will be displaced.

The amount of carbon savings calculated in this article for one Rosenfeld (based on an avoided existing coal plant) assumes that one or more additional things happen to affect this economic calculus:

1) A price on carbon emissions will be put in place that significantly raises the marginal cost of coal plants;

2) Increased regulation of criteria pollutant emissions will create large retrofit costs or increased marginal costs (many existing coal plants have up until now been "grandfathered" so that they are allowed to emit many more criteria pollutants than new coal plants); and/or

3) Retiring coal plants will become an explicit policy goal and incentives or standards will be put in place to encourage this outcome.

Because of the urgency of the climate problem and because of coal's significant contribution to it, we believe these changes are likely for many countries in the coming decade. Each of these actions represents a significant shift from the status quo, but more importantly, they represent an internalization of societal costs that heretofore have not been included in the operational and investment decisions of electric utilities.

CLFs for electricity efficiency options throughout the California economy, found the aggregate CLFs in the various scenarios to range between 57% and 66%.

They are not by themselves sufficient to guarantee significant coal plant retirements, but in combination with investments in energy efficiency or new low carbon power generation resources (which would be the driving force for such retirements) they would allow that outcome.

CONCLUSIONS

The Rosenfeld can best be used in rough back-of-the-envelope calculations and high-level summaries of analysis results for less technical audiences. If an efficiency technology or policy would save 3 BkWh per year at the meter, it saves one Rosenfeld, or one 500 MW coal plant operating at 70% capacity factor in that year (assuming 7% T&D losses). It also saves 3 million metric tons of CO_2/year (assuming all the savings come from conventional coal plants). In addition, avoiding 600 coal-fired power plants of this size over their 50 year lifetimes (i.e. 50 x 600 or 30,000 Rosenfeld-years) saves the same amount of carbon dioxide (about 90,000 $MtCO_2$) as one Pacala/Socolow wedge, which is a nice link to another widely used analytical simplification of this type.

These parameters satisfy the initial criteria of simplicity of presentation, ease of recall, intuitive plausibility, physical meaning, and policy relevance. We encourage other analysts to use this new unit as a way to increase conceptual understanding of the scope of the climate challenge and to honor Art Rosenfeld, whose efforts to create a more hopeful and sustainable future continue to inspire us all.

ACKNOWLEDGEMENTS

The original idea for creating the Rosenfeld unit came from Chris Calwell of Ecos Consulting. Jonathan Koomey, who was a Ph.D. student of Dr. Rosenfeld's from 1985 to 1990, conducted the analysis and the writing, with comments and other contributions from the coauthors (who are all friends and colleagues of Art's). This article has been updated from an earlier version that appeared as Jonathan Koomey et al. 2010. "Defining a standard metric for electricity savings." *Environmental Research Letters*. vol. 5 014017, no. 1 January-March. <http://iopscience.iop.org/1748-9326/5/1/014017>. The parameters for the Rosenfeld unit are the same in both articles, although some of the underlying data have been extended to later years. Some additional coauthors were also added in this version. Many colleagues reviewed the drafts, including two anonymous reviewers, and we owe them our thanks. Glenn McGrath and Channele Wirman at EIA deserve our special thanks for supplying data from and timely explanations about the EIA-860 and EIA-906/920/923 data files. All errors and omissions are the authors' responsibility alone.

REFERENCES

1. Rosenfeld, Arthur H. 1999. "The Art of Energy Efficiency: Protecting the Environment with Better Technology." In *Annual Review of Energy and the Environment 1999*. Edited by J. M. Hollander. Palo Alto, CA: Annual Reviews, Inc. pp. 33-82.
2. Lovins, Amory B. 1979. *Soft Energy Paths: Toward a Durable Peace*. New York, NY: Harper Colophon Books.
3. Koomey, Jonathan G. 1984. *Energy Policy in Transition: The Rise of Conservation*. A.B. Honors Thesis, History of Science Department, Harvard University.
4. Levine, Mark D., and Paul P. Craig. 1985. "A Decade of United States Energy Policy." In *Annual Review of Energy 1985*. Edited by J. M. Hollander. Palo Alto, CA: Annual Reviews, Inc. pp. 557-587.
5. Craig, Paul, Ashok Gadgil, and Jonathan Koomey. 2002. "What Can History Teach Us?: A Retrospective Analysis of Long-term Energy Forecasts for the U.S." In *Annual Review of Energy and the Environment 2002*. Edited by R. H. Socolow, D. Anderson and J. Harte. Palo Alto, CA: Annual Reviews, Inc. (also LBNL-50498). pp. 83-118.
6. AIP. 1975. *Efficient Use of Energy: The American Physical Society Studies on the Technical Aspects of the More Efficient Use of Energy (AIP Conference Proceedings No. 25)*. New York, NY: American Institute of Physics.
7. Schipper, Lee, and Allan J. Lichtenberg. 1976. "Efficient Energy Use and Well-Being: The Swedish Example." *Science*. vol. 194, no. 4269. December 3. pp. 1001 - 1013. <http://www.sciencemag.org/cgi/content/citation/194/4269/1001>
8. Darmstadter, Joel, Joy Dunkerley, and Jack Alterman. 1977. *How industrial societies use energy*. Baltimore, MD and London, UK: Johns Hopkins University Press.
9. Schipper, Lee, Stephen Meyers, Richard Howarth, and Ruth Steiner. 1992. *Energy Efficiency and Human Activity: Past Trends, Future Prospects*. New York, NY: Cambridge University Press.
10. IEA. 1997. *Indicators of Energy Use and Efficiency: Understanding the Link Between Energy and Human Activity*. Paris: Organization of Economic Cooperation and Development (OECD), International Energy Agency.
11. Brohard, Grant J., Merwin L. Brown, Ralph Cavanagh, Lance E. Elberling, George R. Hernandez, Amory Lovins, and Art Rosenfeld. 1998. *Advanced Customer Technology Test for Maximum Energy Efficiency (ACT2) Project: The Final Report*. San Francisco, CA: PG&E. <http://207.67.203.54/Qelibrary4_p40007_documents/ACT2/act2fnl.pdf>
12. Brown, Marilyn A., Mark D. Levine, Walter Short, and Jonathan G. Koomey. 2001. "Scenarios for a Clean Energy Future." *Energy Policy (Also LBNL-48031)*. vol. 29, no. 14. November. pp. 1179-1196.
13. Lovins, Amory B., E. Kyle Datta, Odd-Even Bustnes, Jonathan G. Koomey, and Nathan J. Glasgow. 2004. *Winning the Oil Endgame: Innovation for Profits, Jobs, and Security*. Old Snowmass, Colorado: Rocky Mountain Institute. September. <http://www.oilendgame.com>
14. Lovins, Amory B. 2005. *Energy End-Use Efficiency*. Rocky Mountain Institute for InterAcademy Council (Amsterdam). September 19. <http://www.rmi.org/images/other/Energy/E05-16_EnergyEndUseEff.pdf>
15. APS. 2008. *Energy Future: Think Efficiency*. College Park, MD: American Physical Society. September. <http://www.aps.org/energyefficiencyreport/>
16. Brown, Rich, Sam Borgeson, Jonathan Koomey, and Peter Biermayer. 2008. *U.S. Building-Sector Energy Efficiency Potential*. Berkeley, CA: Lawrence Berkeley National Laboratory. LBNL-1096E. September. <http://enduse.lbl.gov/info/1096E-abstract.html>
17. Krause, Florentin, and Joseph Eto. 1988. *Least-Cost Utility Planning: A Handbook for Public Utility Commissioners (v.2): The Demand Side: Conceptual and Methodological Issues*. National Association of Regulatory Utility Commissioners, Washington, DC. December.
18. Koomey, Jonathan G. 1990. *Energy Efficiency Choices in New Office Buildings: An Investigation of Market Failures and Corrective Policies*. PhD Thesis, Energy and Resources Group, University of California, Berkeley. <http://enduse.lbl.gov/Projects/EfficiencyGap.html>

19. Lovins, Amory B. 1992. *Energy-Efficient Buildings: Institutional Barriers and Opportunities*. E-Source. Strategic Issues Paper. December.
20. Krause, Florentin, Eric Haites, Richard Howarth, and Jonathan G. Koomey. 1993. *Cutting Carbon Emissions–Burden or Benefit?: The Economics of Energy-Tax and Non-Price Policies*. El Cerrito, CA: International Project for Sustainable Energy Paths. <http://files.me.com/jgkoomey/2sxhmd>
21. Krause, Florentin, David Olivier, and Jonathan Koomey. 1995. *Negawatt Power: The Cost and Potential of Low-Carbon Resource Options in Western Europe*. El Cerrito, CA: International Project for Sustainable Energy Paths. <http://files.me.com/jgkoomey/u6fltp>
22. Koomey, Jonathan G., Alan H. Sanstad, and Leslie J. Shown. 1996. "Energy-Efficient Lighting: Market Data, Market Imperfections, and Policy Success." *Contemporary Economic Policy*. vol. XIV, no. 3. July (Also LBL-37702.REV). pp. 98-111. <http://enduse.lbl.gov/Info/37702-abstract.html>
23. Koomey, Jonathan G., Carrie A. Webber, Celina S. Atkinson, and Andrew Nicholls. 2001. "Addressing Energy-Related Challenges for the U.S. Buildings Sector: Results from the Clean Energy Futures Study." *Energy Policy (also LBNL-47356)*. vol. 29, no. 14. November. pp. 1209-1222.
24. Rosenfeld, Arthur H., and Satish Kumar. 2001. *Tables to Convert Energy or CO2 (saved or used) to Familiar Equivalents - Cars, Homes, or Power Plants (US Average Data for 1999)*. Sacramento, CA: California Energy Commission. May. <http://www.energy.ca.gov/commissioners/rosenfeld_docs/Equivalence-Matrix_2001-05.pdf>
25. Calwell, Chris, and Laura Moorefield. 2008. *Efficiency in a Climate-Constrained World: Are We Aiming High Enough?* Asilomar, CA: American Council for an Energy Efficient Economy. August 2008.
26. Krause, Florentin, Wilfred Bach, and Jonathan G. Koomey. 1992. *Energy Policy in the Greenhouse*. NY, NY: John Wiley and Sons.
27. Caldeira, Ken, Atul K. Jain, and Martin I. Hoffert. 2003. "Climate Sensitivity Uncertainty and the Need for Energy Without CO_2 Emission " *Science*. vol. 299, no. 5615. pp. 2052-2054. <http://www.sciencemag.org/cgi/content/abstract/299/5615/2052>
28. Black, Richard. 2009. *'Safe' climate means 'no to coal'* BBC News, April 29 2009 [cited <http://news.bbc.co.uk/2/hi/science/nature/8023072.stm>
29. Meinshausen, Malte, Nicolai Meinshausen, William Hare, Sarah C. B. Raper, Katja Frieler, Reto Knutti, David J. Frame, and Myles R. Allen. 2009. "Greenhouse-gas emission targets for limiting global warming to 2 degrees C." *Nature*. vol. 458, April 30. pp. 1158-1162. <http://www.nature.com/nature/journal/v458/n7242/full/nature08017.html>
30. US DOE. 2009. *Assumptions to the Annual Energy Outlook 2009, with Projections to 2030*. Washington, DC: Energy Information Administration, U.S. Department of Energy. DOE/EIA-0554(2009). March. <http://www.eia.doe.gov/oiaf/aeo/assumption/_>
31. Koomey, Jonathan G., and Nathan E. Hultman. 2007. "A reactor-level analysis of busbar costs for U.S. nuclear plants, 1970-2005." *Energy Policy*. vol. 35, no. 11. November. pp. 5630-5642. <http://dx.doi.org/10.1016/j.enpol.2007.06.005>
32. US DOE. 2009. *Electric Power Annual 2007*. Washington, DC: Energy Information Administration, U.S. Department of Energy. DOE/EIA-0348(2007). January 21. <http://www.eia.doe.gov/cneaf/electricity/epa/epa_sum.html>
33. US DOE. 2011. *Electric Power Annual 2009*. Washington, DC: Energy Information Administration, U.S. Department of Energy. DOE/EIA-0348(2009). April. <http://www.eia.doe.gov/cneaf/electricity/epa/epa_sum.html>
34. US DOE. 2008. *Annual Energy Review 2007*. Washington, DC: Energy Information Administration, U.S. Department of Energy. DOE/EIA-0384(2007). June. <http://www.eia.doe.gov/aer/>
35. US DOE. 2010. *Annual Energy Review 2009*. Washington, DC: Energy Information Administration, U.S. Department of Energy. DOE/EIA-0384(2009). August. <http://www.eia.doe.gov/aer/>
36. EPRI. 2009. *Assessment of Achievable Potential from Energy Efficiency and Demand Response Programs in the U.S. (2010 - 2030)*. Palo Alto, CA: Electric Power Research Institute. 1016987. January 14. <http://mydocs.epri.com/docs/public/000000000001018363.pdf>

37. Koomey, Jonathan G., Celina Atkinson, Alan Meier, James E. McMahon, Stan Boghosian, Barbara Atkinson, Isaac Turiel, Mark D. Levine, Bruce Nordman, and Peter Chan. 1991. *The Potential for Electricity Efficiency Improvements in the U.S. Residential Sector*. Lawrence Berkeley Laboratory. LBL-30477. July. <http://enduse.lbl.gov/info/30477-abstract.html>
38. Meier, Alan, Jan Wright, and Arthur H. Rosenfeld. 1983. *Supplying Energy Through Greater Efficiency*. Berkeley, CA: University of California Press.
39. Rosenfeld, Arthur H., and David Hafemeister. 1988. "Energy-efficient Buildings." In *Scientific American*. April. pp. 78-85.
40. Rosenfeld, Arthur, Celina Atkinson, Jonathan G. Koomey, Alan Meier, Robert Mowris, and Lynn Price. 1993. "Conserved Energy Supply Curves." *Contemporary Policy Issues*. vol. XI, no. 1. January (also LBL-31700). pp. 45-68.
41. Pacala, S., and Rob Socolow. 2004. "Stabilization Wedges: Solving the Climate Problem for the Next 50 Years with Current Technologies " *Science*. vol. 305, no. 5686. August 13. pp. 968-972. <http://www.sciencemag.org/cgi/content/abstract/305/5686/968>
42. Wheeler, David, and Kevin Ummel. 2008. *Calculating CARMA: Global Estimation of CO2 Emissions From the Power Sector*. Center for Global Development. Working Paper 145. May. <http://www.cgdev.org/content/publications/detail/16101/>
43. Mahone, Amber, Snuller Price, and William Morrow. 2009. *Developing a Greenhouse Gas Tool for Buildings in California: Methodology and User's Manual*. San Francisco, CA: Energy and Environmental Economics, Inc. January. <http://www.ethree.com/E3_Public_Docs.html>
44. Koomey, Jonathan G., Arthur H. Rosenfeld, and Ashok K. Gadgil. 1990. *Conservation Screening Curves to Compare Efficiency Investments to Power Plants: Applications to Commercial Sector Conservation Programs*. Proceedings of the 1990 ACEEE Summer Study on Energy Efficiency in Buildings. Asilomar, CA: American Council for an Energy Efficient Economy.
45. Koomey, Jonathan G., Arthur H. Rosenfeld, and Ashok K. Gadgil. 1990. "Conservation Screening Curves to Compare Efficiency Investments to Power Plants." *Energy Policy*. vol. 18, no. 8. October. pp. 774-782.
46. Rufo, Michael, and Fred Coito. 2002. *California's Secret Energy Surplus: the Potential for Energy Efficiency*. San Francisco, CA: The Energy Foundation and The Hewlett Foundation. <http://www.ef.org/documents/Secret_Surplus.pdf>

Listening to the Planet and Building A Sustainable Energy Economy

Daniel M. Kammen

Chief Technical Specialist for Renewable Energy and Energy Efficiency
The World Bank
Washington, D.C.

PART 1: THE CHALLENGE OF DEALNG WITH A PROBLEM FOR WHICH WE LACK A SUITABLE LANGUAGE

Over the past two years, we have seen remarkable swings of pessimism and guarded optimism, and then pessimism again, on our individual, national, and collective global ability to make progress on implementing the sustainable energy agenda. COP15 in Copenhagen was a step backwards, while COP16 in Cancun was a guarded step forward. On the positive side, the United Kingdom has taken a significant step by establishing a floor price and an escalation schedule for carbon. Mexico and Brazil have launched ambitions energy efficiency and clean energy development plans. China has launched experimental regional carbon cap and trade schemes. California, has upheld a greenhouse gas plan and on January 1, 2012, will launch a carbon market.

On the negative side, several nations have examined significant advances in clean energy markets, and have retreated, or at least so far, failed to act. And, while several large-scale, clean energy projects being designed, costs and logistics are proving significant hurdles. This mixed story is unfortunate, but not surprising given that we still lack metrics, methods and a currency to describe, reflect and manage a world driven by sustainable energy.

The sad fact is that we as a global society have wasted many good years – decades actually -- during which we could have launched an economy based on job creation and investments in human capacity and creativity. Instead, we have ignored the changing signals that nature has been sending us about the status and stresses we are placing on our rivers, oceans, skies, mountains and on the health of every ecosystem. There are many metrics we might cite that consistently tell us that we are approaching, or that we are at, or that we are beyond the carrying capacity of the planet in terms of

Physics of Sustainable Energy II: Using Energy Efficiently and Producing it Renewably
AIP Conf. Proc. 1401, 44-53 (2011); doi: 10.1063/1.3653844

flows of pollutants, and our need for resources [1]. That is not to say we do not have options. We do, but we need to listen to the planet, and ourselves, and put our ideas into practice.

I am a physicist by training who wandered into the world of energy science and policy quite by accident. I did so because the most interesting people I met were consistently rebelling against disciplinary boundaries and I wanted to see what I could contribute in that exciting and challenging environment. It was and remains to me clear where to apply such tools; energy. The reason is simple: energy is the largest (legal) piece of the global economy, by a factor of two over the global food industry. Several of my advisors said that an easy way to pick problems to tackle is to start with issues that matter most, and this one does.

How big is the problem? Not only is energy the dominant part of our economy, but its impact on the planetary system can be seen in a few historical trends which are summarized in Figure 1.

The Intergovernmental Panel on Climate Change – which shared the 2007 Nobel Peace Prize – has been issuing reports on the state of climate science since 1990. Several thousand climate and energy experts participate in the IPCC assessment and reporting process during any one assessment report (AR). I have been involved in several since my first 'special report' on technology transfer in 1999 [2].

Figure 1 shows the consensus data on three broad climate indicators. It also shows the timing of each of the first four assessment reports.

In 1990, the first assessment report concluded that it would take another decade at least to see clear signs of climate change in the natural record. In Figure 1 the first assessment report, ***FAR*** is indicated by the green vertical line in 1990. The key language – hotly debated and worked over most carefully – is shown to the right of the figure. This is as far as the first report could go.

Successive reports –the *Second Assessment Report (**SAR**)* shown as a black vertical line in Figure 1, five years later, in 1995 – advanced the clarity with a consensus statement that, "the balance of evidence suggests discernible human influence". This was clearer than the ***FAR*,** but still far from sealed.

The **Third** (blue vertical) and **Fourth** (red vertical in Figure 1) assessment reports clarified things considerably, assigning an analytic confidence (66% in the ***TAR***) to the likelihood that the environmental change in the Figure is due to human activities (energy generation, agriculture, and forest disruption.

Finally, but the Fourth Assessment Report attributes *most* of the warming is very likely (90% confidence) is due to human activity. At this point, the climate discussion expanded to address social and economic implications, and arrived at the conclusion that the consequences of global warming *will most strongly and most quickly affect the world's poor.* Sadly, if climate change is framed as a poverty issue, it is more likely to be neglected, given past performance of global efforts to address poverty. This record of environmental change is already significant, however, and has only just begun. What is needed in response is an ambitious set of individuals and networks of people and institutions committed to changing this *status quo* of our energy economy. There are basic scientific breakthroughs needed, for example in energy storage and solar cells. There are business opportunities in smart-grid energy services sectors, and in electric vehicles, and there are tremendous political wins for the elected officials who truly champion the clean energy economy.

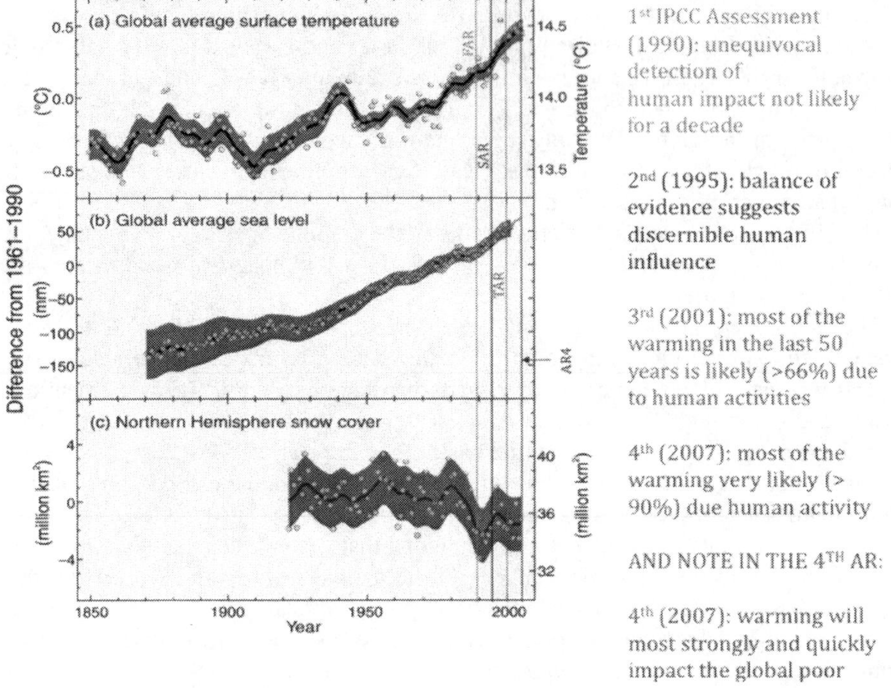

FIGURE 1. Changes in three key environmental indicators, 1850 – 2010, and vertical lines indicating the dates of the first through the fourth *Intergovernmental Panel on Climate Change* (IPCC) assessment reports. At right are key findings, in brief negotiated text, of each of these successive Assessment Reports.

My challenge to you, then, is to find ways to work to change these trends. Collectively, we need to put more of our good ideas into practice: as an academic community and as a entrepreneurial private sector, and as a civil society, and as a community of public servants, we need to do better – not next year or next month, but right now. First, we need a sustained and vibrant research base to understand our energy options and their resulting climate impacts [3].

We also need something that such common practice that it seems mundane [4] to people in fields outside energy, such as agriculture and education: it is 'extension services'. Farmers rely on information networks to plan their activities. Universities, community colleges and night schools all focus on making continuing education an available extension service. We need a mechanism to bring these innovations from the laboratory to the market. In agriculture, almost all countries have networks of agricultural extension services. Perhaps because much of the 'modern' energy system has been managed by large, centralized utilities, the energy sector has not had such an extension network. It is time to build one.

Costing Energy Choices

One of the most important tools that we need to develop is an economy-wide "appreciation" of the costs and benefits of our energy choices. That, in fact, is why I felt it was so critical for me to work at the World Bank at this time. First, the World Bank is significantly increasing its commitment and investment in energy efficiency and in clean and sustainable energy. Since 2003, the World Bank Group has invested about $17 billion in low carbon investments, of which $14.2 billion were in renewable energy and energy efficiency. Excluding large hydropower, new renewable energy investments alone contributed $4.9 billion. Financing of renewable energy and energy efficiency projects and programs in developing countries rose seven percent in 2010 to reach a record $3.6 billion.

Second, and of vital importance, banks are focused on the bottom line, pure and simple. This has the virtue of *clarity*. Our society, however, is so far only *casually* and *vaguely* interested in sustainability. Beyond specific technical and policy innovations there have been some important insights. These include the realization that distributed networks of energy suppliers and consumers (some of whom may be one and the same) could not only complement large, traditional energy systems, but that in some ways, they may be superior. I liken this to a transition from old to new thinking. Old thinking viewed the energy grid as a one-way flow of energy *to* consumers. New thinking sees it instead as more of an eBay: where anyone can buy and/or sell power, with the job of the utility – and the network regulator – being to provide fair and transparent rules for these transactions.

These are vital innovations. But innovations can languish if our interest in sustainability, that is, our interest in accessing the resources we need without degrading the opportunities of future generations, remains only casual and unclear.

This casualness is not the same as indifference. People are worried about the world their children will inhabit. Poll after poll shows that when asked this question in isolation, people respond that they are very concerned about the future. Nor are we afraid to make hard decisions for what we want – people and even governments (much to the surprise of some skeptics) do this all the time. People in government generally work hard for the public good. Civil society and non-governmental groups put exceptional effort into innovating and giving voice to the watchdog role that every community, large or small, critically needs.

It is true that immediate gratification versus long-term quality of life (and thus we must also address the complicated issue of discounting and of undervaluing the world in which we and then our children will live [5]) remains a problem for most people to keep squarely in mind. This issue, however, relates to the clarity that honest banks – and in particular the value of having a clear and well-understood *currency* – can bring to our planning for the future.

As a society we are only vaguely interested in sustainability in part because we collectively do not speak a language that permits us to value the world in which we live, except when we cut it down, spoil our waters with human and industrial effluent, or poison the skies with the waste of our energy generation. In contemplating this situation, my friend and colleague George Lakoff made a remarkable and chilling observation.

It is clear to the environmental science community that nature is being degraded [6], in fact destroyed by the current course of human action and neglect. It is also true that the so-called industrialized nations emitted the majority of the greenhouse gas emissions if we go back to the beginning of the industrial revolution. So, by this measure, they should 'pay'. And yet, the so-called developing nations will emit the bulk of the greenhouse gases over the coming decades, so they, arguably, should 'pay it forward', if you will. What is more important, however, is that we will *all* live in the world of the future. So while we can argue about that all day, the solution must be a collective one.

We have established that anthropogenic climate change is the act of degrading our collective home, the planet. Yet the environment is a *complex system* that responds in ways that are not always predictable. Professor Lakoff's observation is that we actually lack a means to express – and thus to fully understand this situation. In other words, no language has a simple verb form that captures the effect of a system acting on the individual [7]. Certainly, there are ways to express this idea – notably if we become anthropomorphic and refer to the planetary system as an entity: *Mother Nature*, for example. However, if we move beyond this view of the planet as a single, coordinated, entity to the complex system that it is, we are not equipped to understand the process of this collective causality in terms of how it relates to us as individuals.

So, we are in a difficult place. First, we lack a *language* and, second, we lack something else vital to understand the planet. We lack a *currency* in which to *value* the planet, in which to value a clean energy economy, and in which to value our future.

Placing a price on greenhouse gas emissions to the atmosphere will not solve global climate change and environmental destruction by itself, but it gives us a language in which to express our values. Given that humans are social creatures that communicate constantly about every aspect of our individual and collective existence, to be without a means to communicate about our future is not only shortsighted, it is simply unacceptable.

Now, those who study economics and are interested in valuing the planet will correct me here and say, "actually, things are worse than you say." They, in fact, would be right.

The true story is *not* that we don't value the environment in a positive way, but quite the contrary, that we reward damaging the environment, and hence ourselves. Rewarding waste is in fact, *placing a negative value on the planet and saying financially that sustainability is a bad thing.* This is not to say that we are *intentionally* damaging our 'nest', but that through our inaction, we are in fact sending the economic and political signal that individual profit is more important that our collective well-being and that of the natural ecosystem. We can, for example, choose to invest in local job creation by supporting people and companies that provide energy services without spoiling nature. We know the job creation benefits are real in terms of the *higher* numbers of jobs created in clean energy areas relative to polluting sectors [8]. This is not because 'clean energy' is inherently superior, it is simply because when ramping up a new field, greater investments in infrastructure and hence in jobs is needed. This means that when our energy dollars are put to productive use, and not simply used to increase our debt to the environment, we gain an added benefit. When adding up these environmental advantages of 'going green' it is often hard to see why this transition is so hard, and yet it clearly has been. In this regard, I am reminded of a perplexing cartoon (Figure 2), particularly because we know, as we can see in Figure 1, that climate change is not a 'big hoax'.

FIGURE 2. What value is there in saving the planet?

PART 2: PUTTING OUR NEW LANGUAGE AND ENERGY AND CLIMATE INTO ACTION

Let us hope that we are able to build and use this language of energy and environmental clarity. In this respect, the story begins to get brighter and brighter. Once we recognize that our language and financial metrics are (or were) lacking, action becomes not only clearer, but in my view, easier than many people think.

How can this be?

First is the observation that not only is sustaining the planet good for us in the long-term (that should go without saying!), but there are many other positive returns on wasting less and in polluting less.

One of the most important lessons of the rapidly-expanding mix of energy efficiency, solar, wind, biofuels, and other low-carbon technologies is that the costs of deployment are lower than many forecasts, and at the same time, the benefits are larger than expected.

This seeming 'win-win' claim deserves examination, and continued verification, of course.

Over the past decade, the solar and wind energy markets have been growing at rates over 30% per year, and in the last several years growth rates of over 50% per year have taken place in the solar energy sector. This explosive and sustained growth has meant that costs have fallen steadily, and that an increasingly diverse set of innovative technologies and companies have been formed. Government policies in an increasing number of cities, states, and nations are finding creative and cost-effective ways to build these markets still further.

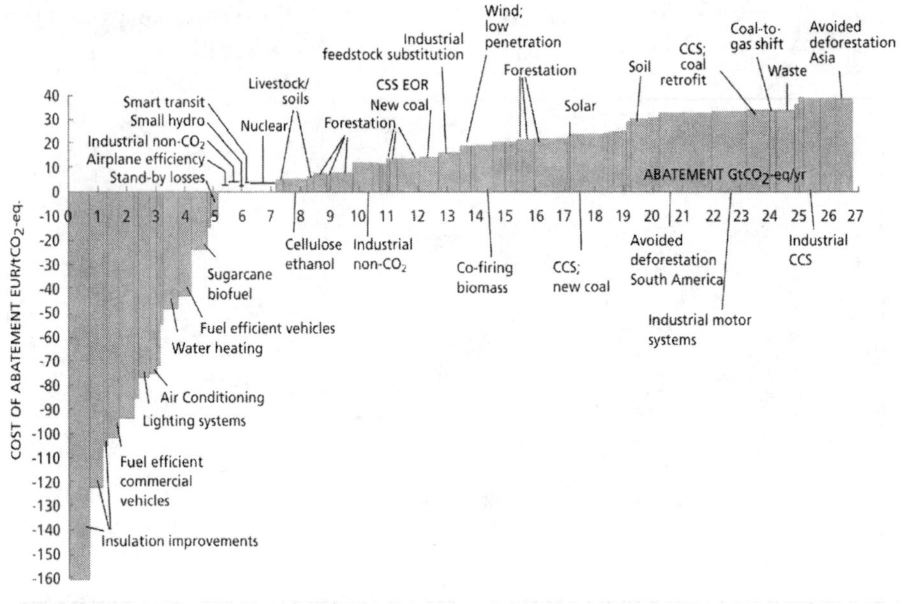

FIGURE 3. A 'carbon abatement curve' showing financial cost (+) or savings (-) for a range of efforts, projects, and programs with the unit financial impact per ton of carbon dioxide *not* emitted to the atmosphere. Examples exist for a range of countries, including Brazil, China, Mexico, the United States, and the United Kingdom, with the list expanding every day.

At the same time that a diverse set of low-carbon technologies are finding their way to the market, energy efficiency technologies (e.g. 'smart' windows, energy efficient lighting and heating/ventilation systems, weatherization products, and efficient appliances) and practices are all in increasingly widespread deployment. Many of these energy efficiency innovations demonstrate negative costs over time, meaning that when the full range of benefits (including improved quality of energy services, improved health, and worker productivity) are tabulated, some energy efficiency investments are vehicles for net creation of social benefits over time.

Marginal Carbon Abatement

Carbon abatement curves have become famous since the Swedish power company Vattenfall collaborated with the McKinsey Company to develop a set of estimates on the costs to deploy and operate a range of energy efficiency, land use, and energy generation technologies. (They are actually just knock-offs of marginal pricing curves used in the electricity industry for decades, but context and timing is everything!) These costs of conserved carbon curves depict the costs (or savings, in the case of a number of 'negative cost' options such as building efficiency) as well as the magnitude (in giga-tonnes) of abatement potential at a projected future time.

In Figure 3 such an abatement curve for the entire world for 2030 is presented. The basic message: saving money often saves carbon emissions, if you are strategic about where to invest. A World Bank-supported low-carbon development study shows that Mexico can reduce carbon emissions by 42% more than its target of 1,137 metric tons by 2030—477 million tons, to be precise—by decisive action on multiple fronts. It can achieve this by moving in key areas such as improving bus systems, road and rail freight logistics, fuel economy standards, and vehicle inspection at the border, among others.

This is exciting news. It shows that significant—even dramatic—carbon reductions can be achieved by adjusting use of existing technologies. Such adjustments can reduce costs too. These conclusions emerged from calculations based on a marginal abatement cost curve, or MAC, an analytical tool developed in 2008 by McKinsey & Company, and used by a team of experts studying Mexico's climate challenges headed by the World Bank.

The study in which this methodology was used, *Low-Carbon Development for Mexico*, by Johnson, Alatorre, Romo and Liu [9], is one of a series of such studies financed by the Energy Sector Management Assistance Program (ESMAP) that also includes Brazil and Nigeria (forthcoming).

This same MAC tool has now been applied, with promising results, to two tiny communities of 1,100 people on Nicaragua's Atlantic coast. Results of a study published in November 2010 in *Science* Magazine demonstrate that low-carbon rural energy services can be delivered at cost savings in cases where communities utilize diesel-powered, isolated, electricity grids.

The study, on which I worked with Christian Casillas, will, we hope, spur efforts elsewhere to build similar community-level carbon abatement and energy service tools. This could mean that communities often ignored or lumped together as "those billions without modern energy" can create their own locally appropriate development goals, and groups working with them can develop energy solutions at a price lower than the one they're paying now.

In 2009, the rural Nicaraguan communities of Orinoco and Marshall Point, which share a diesel micro-grid, partnered with the national government and an NGO to implement energy efficiency measures including metering, which prompted residents to reduce wasteful use of electricity. Compact fluorescent light bulbs were also introduced, as well as more efficient outdoor lighting, and replacement of part of the diesel power with biogas from dung.

After the government installed meters, energy use dropped by 28%, and people's electric bills dropped proportionately. The NGO, blueEnergy, based in San Francisco, which offered the compact fluorescent light bulbs (CFL), was able thereby to cut household energy use by another 17%.

The net result was reduced burning of diesel, even allowing for the fact that the community's reduced energy needs allowed the local energy supplier to run its generators two extra hours each day, providing longer service to customers. In the month after the conservation campaign, energy costs per household had dropped by 37 percent.

That the MAC curve can be used to analyze energy use in the community and pinpoint areas where investments would save the most energy and the most money for homeowners is something of a breakthrough. Until now, the model has been used mostly on a global or country-wide scale to target areas for carbon abatement. But now it has gone local. That means some of the world's poorest communities can reduce their energy costs by local action which, multiplied worldwide, could produce global change in reduced carbon emissions.

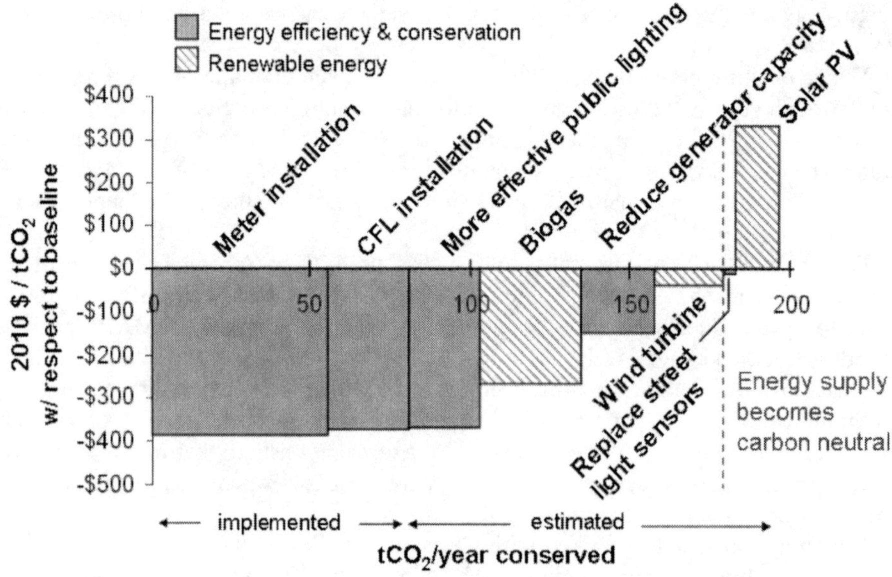

FIGURE 4. Marginal abatement curve for greenhouse gas emissions for a rural community on the Atlantic coast of Nicaragua. Source: Casillas and Kammen (2010) [10].

These curves illustrate the range of low-carbon options that exist, and that if we can continue to build a menu of options that have been tested, vetted, and implemented, a new paradigm of clean energy development has a solid economic footing in a wide range of national, city, and community environments.

Finally, let me conclude with a brief note on building the business model for clean energy. This is a piece of the story that gets left behind in many discussions: creating a new energy economy cannot be a battle between environmentalists saying we must 'go green' and the business community saying we 'cannot go green' today, or not that rapidly. In fact there is a great deal of emerging data – such as these marginal abatement curves -- that if one manages the process of innovation and implementation well, we can find ways to both grow the economy and make it dramatically greener.

The German experience in wind and solar, and sound urban and agricultural planning is a great example of doing both well. In fact, Germany is finding that an emerging green economy can deliver export earnings and job creation, as well as a stable economy in a time of oil and gas shocks.

REFERENCES

1. Rokström, J. *et al,* "A safe operating space for humanity," *Nature*, 461, 472 - 475 (2009).
2. Intergovernmental Panel on Climate Change Working Groups II and III (2000), *Methodological and Technological Issues in Technology Transfer* (Cambridge University Press: Cambridge, UK and New York, NY).
3. Nemet, G. F. and D. M. Kammen, "U.S. energy research and development: Declining investment, increasing need, and the feasibility of expansion," *Energy Policy*, 35(1), 746 - 755 (2007).
4. Kammen, D. M. and Dove, "The virtues of Mundane Science," *Environment.* 39 (6), 10 - 15, 38 - 41 (1997).
5. Schelling, T. C. , "The cost of combating global warming: Facing the tradeoffs," *Foreign Affairs*, **76**, 8 - 14 (1997).
6. Intergovernmental Panel on Climate Change (AR4), *Climate Change 2007: The Physical Science Basis.* Contribution of Working Group I to the Fourth Assessment Report of the Intergovernmental Panel on Climate Change. Cambridge University Press, 2007 (available online at http://www.ipcc.ch).
7. Lakoff, G., *Women, Fire, and Dangerous Things: What Categories Reveal about the Mind* (University of Chicago Press, 1987).
8. Wei, M., Patadia, S. and D.M. Kammen, "Putting renewables and energy efficiency to work: How many jobs can the clean energy industry generate in the U. S.?" *Energy Policy*, 38, 919 - 931 (2010).
9. Johnson, T., C. Alatorre, Z. Romo, and F. Liu, (2009), *México: Estudio Sobre las disminicución de emisiones de carbono* (The World Bank: Washington, DC).
10. Casillas, C. and D.M. Kammen, "The energy-poverty-climate nexus," *Science*, 330, 1182 - 1182 (2010).

Energy in the Developing World

Ashok Gadgil[a,b], David Fridley[a], Nina Zheng[a], Andree Sosler[b,c], Thomas Kirchstetter[a], and Amol Phadke[a]

[a]*Environmental Energy Technologies Division*
Lawrence Berkeley National Laboratory
1 Cyclotron Road
Berkeley, CA 94720

[b]*UC Berkeley*
100 Blum Hall
Berkeley, CA 94720

[c]*Darfur Stoves Project*
2150 Allston Way, Suite 300
Berkeley, CA 94704

Abstract. The five billion persons at the lower economic levels are not only poor, but commonly use technologies that are less efficient and more polluting, wasting their money, hurting their health, polluting their cites, and increasing carbon dioxide in the atmosphere. Many first-world researchers, including the authors, are seeking to help these persons achieve a better life by collaborating on need-driven solutions to energy problems. Here we examine three specific examples of solutions to energy problems, and mitigation strategies in the developing world:

(1) *Energy Efficiency Standards and Labeling in China.* Between 1990 and 2025, China will add 675 million new urban residents, all of whom expect housing, electricity, water, transportation, and other energy services. Policies and institutions must be rapidly set up to manage the anticipated rapid rise in household and commercial energy consumption. This process has progressed from legislating, and setting up oversight of minimum energy performance standards in 1989 (now on 30 products) to voluntary efficiency labels in 1999 (now on 40 products) and to mandatory energy labels in 2005 (now on 21 products). The savings from just the standards and labels in place by 2007 would result in cumulative savings of 1188 teraWatt–hours (TWh) between 2000-2020. By 2020, China would save 110 TWh/yr, or the equivalent of 12 gigaWatts (GW) of power operating continuously.

(2) *Fuel-efficient biomass cookstoves* to reduce energy consumption and reduce pollution. Compared to traditional cooking methods in Darfur, the BDS cooks faster, reduces fuel requirement, and emits less carbon monoxide air pollution. A 2010 survey of 100 households showed that users reduced spending on fuelwood in North Darfur camps from 1/2 of household non-fuelwood budget to less than 1/4 of that budget. The survey showed that each $20 stove puts $330/year in the pocket of the women using the stove, worth $1600 over the stove-life of 5 years. Per capita income of these households is about $300/year.

(3) *Super Efficient Appliance Deployment.* Global domestic electricity consumption is expected to double in 25 years, from 5,700 TWh/yr in 2005 to 11,500 TWh/yr in 2030. The four appliances using largest shares of domestic electricity (lighting, refrigeration, air-

Physics of Sustainable Energy II: Using Energy Efficiently and Producing it Renewably
AIP Conf. Proc. 1401, 54-74 (2011); doi: 10.1063/1.3653845

conditioning, television) would use some 5,000 TWh/yr in 2030, or 43% of the total, in the baseline scenario. More than 50% of this consumption will be in China, India, European Union and US. We outline efforts to save up to 1.5 gigatons of carbon dioxide emissions per year in 2030 by helping deploy the most efficient commercially available technologies in these four categories. Furthermore, if this effort is extended to twenty-four categories of appliances and equipment, the projected savings in CO2 emissions increase to 6.7 gigatons per year by 2030.

HUMAN DEVELOPMENT INDEX

The five billion persons at the lower economic levels are not only poor, but commonly use technologies that are less efficient and more polluting, wasting their money, hurting their health, polluting their cites, and increasing carbon dioxide in the atmosphere. From their perspective, managing carbon in the atmosphere is not a priority. This section shows how the aspirations of these persons connect to energy and CO_2. It also puts into context the three illustrative projects in this chapter, which benefit the people in the developing countries and help reduce carbon emissions to the atmosphere.

The people at the bottom of the economic pyramid are inevitably raising their real incomes. What do these people want? They want "development". To understand development and its connection to energy, it helps to use its quantitative metric, the Human Development Index (HDI). HDI is a quantitative measure of development – or human well-being constructed for the UN by experts from developing countries. The HDI is built up from metrics for three broad categories:

(1) Economic well-being or prosperity,

(2) Life expectancy, public health and health care, and

(3) Literacy and education.

The United Nations publishes HDI values for countries, with range in magnitude from a minimum of zero to a maximum of one. Fig. 1 shows that the industrialized countries of United States, Canada, Australia, New Zeeland, Japan and Western Europe have an HDI of over 0.95. This is followed with HDI's from 0.8 to 0.95 for countries like Argentina, Brazil, Lebanon, Mexico, Russian Federation, Turkey, Thailand, and Saudi Arabia. Below these are HDI values of 0.6 to 0.8, for China, Iraq, Iran, and South Africa, and 0.55 for India, and between 0.35 and 0.5 for most African countries south of the Sahara desert.

The HDI is very closely related to the consumption of modern energy. In Fig. 2 we have plotted 2007 data for HDI against the per-capita consumption of electrical energy in kWh/yr, which ranges from below 1000 kWh/yr for Zambia to 25,000 kWh/year for Norway.

The industrial countries are members of Organization for Economic Cooperation and Development (OECD). So, OECD membership is a convenient way to separate industrial from non-industrial countries. Fig. 3 displays the projections of the International Energy Agency for energy use in exajoules (10^{18} J) for the OECD and non-OECD countries. The latter is separated between the BRIC countries (Brazil,

Russia, India and China) and rest of the world (ROW). We see that OECD countries accounted for 50% of energy consumed in 1990, and their contribution drops to about 30% in 2030. Thus, we see that BRIC and ROW are going to dramatically increase their use of energy, as well as their share of the global energy use.

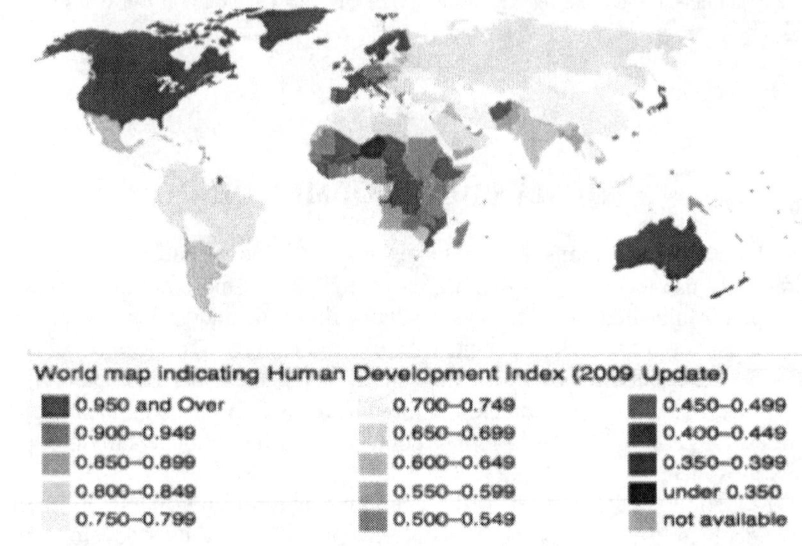

FIGURE 1. The Human Development Index (HDI) is a quantitative measure of human well-being, calculated by the United Nations. It incorporates metrics for three factors: (1) Economic well-being or prosperity, (2) Life expectancy, public health and health care, and (3) Literacy and education.

Fig. 4 plots the time-trajectories of several countries as GDP per capita (in units of PPP 2000 US$) versus CO_2 emissions per capita (in units of metric tonnes) for 1990-2008. PPP stands for "Purchasing Power Parity" i.e., corrected in terms of prices of local goods and services. For the US, per capita CO_2 emission is about 20 metric tonnes per year, while India is at about 2 t/yr and China is now about 5 t/year. The upper left corner of the figure gives most recent available number (for 2008) for carbon intensity (kg CO_2 emission per PPP dollar of GDP, in units of 2000 US$) for the US at 0.47, with Japan at 0.32 and UK at 0.28. In these same units, the carbon intensity for China is 0.62 and for India 0.33. However, in nominal dollars, the GDPs for the developing countries decrease sharply, and hence their carbon intensities (measured as kg of CO_2 per nominal dollar of GDP) increase sharply. For example the 2008 carbon intensities for China and India computed with GDP using nominal exchange rates become 1.14 and 1.04. The higher values for China and India are probably because these younger economies have much more heavy industrial activity relative to the mature economies of the US and Europe. GDP values of all countries in both nominal and in PPP dollar units are published by many international institutions and are available on Wikipedia. These graphs in Fiure 4 gloss over the highly diverse nature of energy use in India and China.

Wealthy Indians and Chinese live at standards comparable to those of average West Europeans or Americans. The poor segments of India and China live in poverty and want.

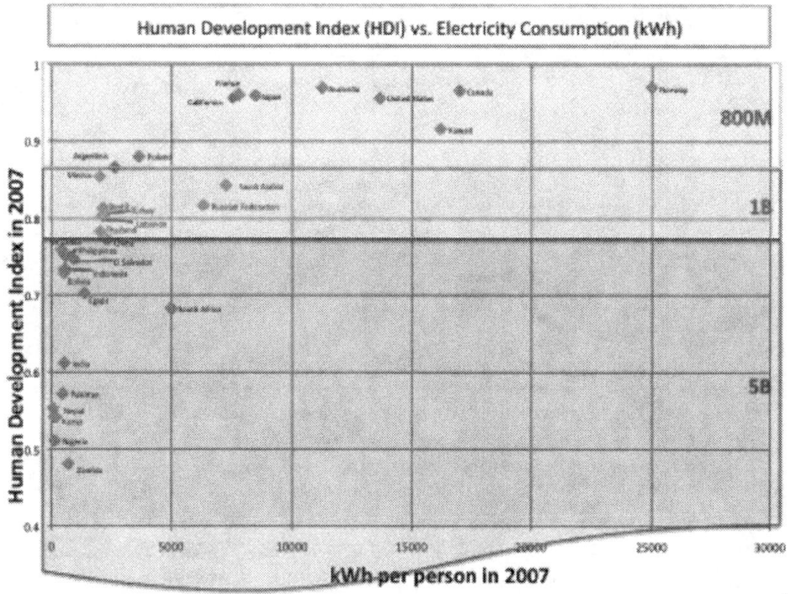

FIGURE 2. The HDI is very closely related to the consumption of modern energy. This graph plots the per-capita consumption of electrical energy in kWh/yr versus the HDI values.

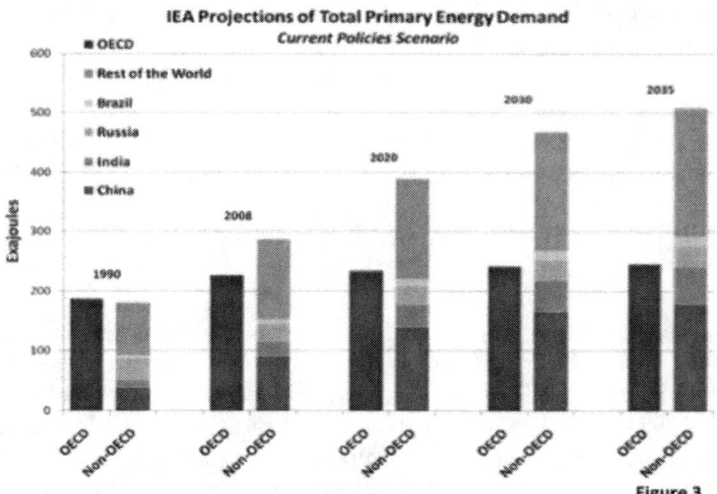

FIGURE 3. The International Energy Agency estimates are given in exajoules (10^{18} J) for the OECD nations and rest of the world and the BRIC nations (Brazil, Russia, India and China). The OECD accounted for 50% of energy consumed in 1990, which drops to 30% in 2030.

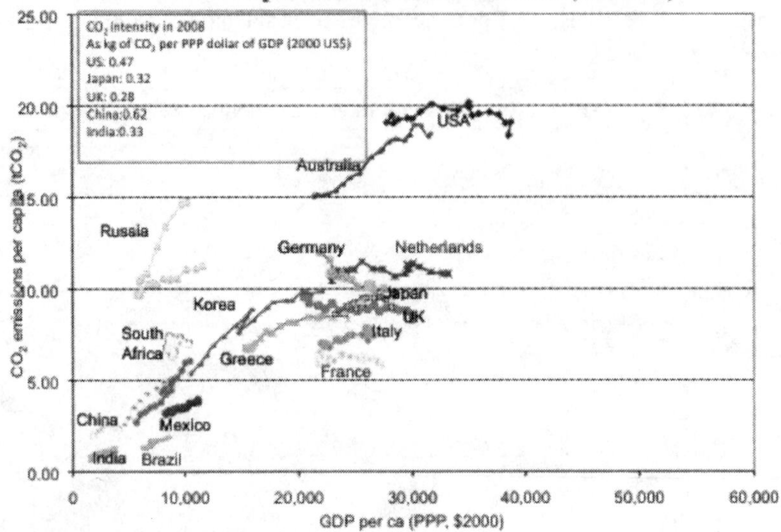

FIGURE 4. The GDP per capita in 2000 PPP US$ is plotted versus CO_2 emissions in metric tons per capita. This plot shows the trends over time for the various countries. The US per capita carbon emission is about 20 metric tons per year, while India is about 2 t/yr and China is now about 5 t/year. The upper left corner of the figure shows the carbon intensity for the US at 0.51 kg CO_2 per 2000 PPP US$, with Japan at 0.24 and UK at 0.32.

Since space is limited, we limit this chapter to three current projects to illustrate what can be done, and what amazing impact is possible.

(1) Energy Efficiency Standards and Labeling in China – focus on middle and upper income households in China

(2) Fuel-Efficient Biomass Cookstoves – focus on below-poverty-line households in the developing world

(3) Super Efficient Appliance Deployment – a strategy with potentially huge future global impact across the world

ENERGY-EFFICIENCY STANDARDS AND LABELING IN CHINA

Let us start with an illustration of what the U.S. has accomplished via standards for refrigerators. Fig. 5 shows that new models of US domestic refrigerators consumed 350 kWh/yr in 1947, which then rapidly rose by more than a factor of five to 1850 kWh/yr by 1973. The 1973-74 oil embargo put pressure on the U.S. to do better, which we did, steadily lowering consumption to 450 kWh/yr in 2002. This was accomplished while the size of the average new domestic refrigerator increased from 9 cubic feet in 1947 to 22 cubic feet in 2002, and concurrently the price dropped from $1,270 to $462 (in 1983 US$). This progress came about through improved engineering and improved manufacturing driven first by California then by Federal standards for minimal acceptable refrigerator efficiency. The logic behind this approach to standards is that individual decisions on refrigerator purchases (and therefore designs of models offered by manufacturers) are driven by need to have lower first-costs and more features, not by the overall societal benefits that accrue from long-term energy saved from more efficient designs. Thus the more efficient designs needed to be supported and the lowest efficiency designs needed be disallowed by an external regulatory agency acting in the overall societal interest. In California, this role has been played by the California Energy Commission (CEC), and for the U.S. overall, by the US Department of Energy. (In China, this role is held by the China Administration for Quality, Supervision, Inspection and Quarantine (AQSIQ), supported by the China National Institute of Standardization (CNIS)).

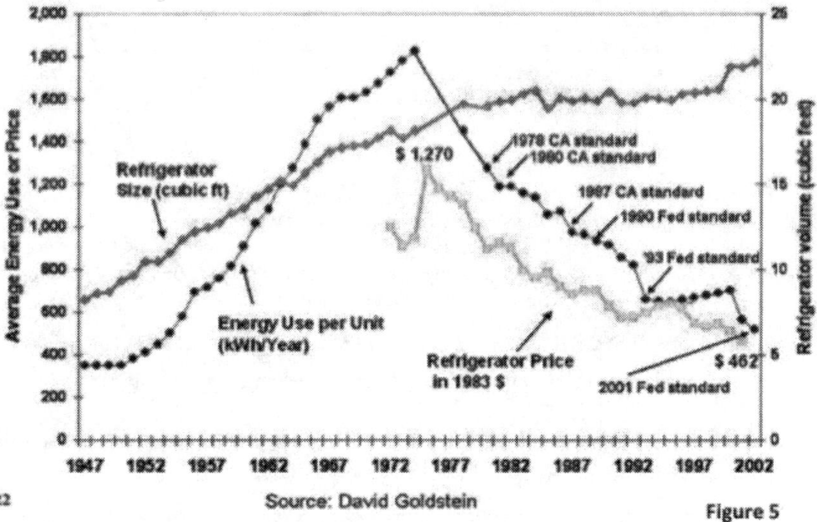

FIGURE 5. US refrigerators consumed 350 kWh/yr in 1947, which then rose by a factor of five to 1850 kWh/yr by 1973. The 1973-74 oil embargo put pressure on the manufacturers to do better, which they did, lowering consumption to 450 kWh/yr in 2002. This was accomplished while the size of the refrigerators increased from 9 cubic feet in 1947 to 22 cubic feet in 2002, while the price dropped from $1,270 (1983$) to $462 (1983$).

Figure 6 shows the remarkable rise of appliance ownership in urban China. From essentially no appliance ownership in 1981, urban ownership of color TVs has risen to more than 135 per 100 households in 2008, and that of clothes washers, refrigerators and room air-conditioners is close to 100%. Ownership of personal computers has risen to 60 per 100 urban households and cars to 10 per 100 urban households. This is a dramatic change in just 27 years.

Appliance ownerships in Chinese rural households lags that in urban households by 10 to 20 years, but the rural fraction of total households is declining, and will drop from 73% in 1990 to 36% in 2025, as shown in Fig. 7. Over this period of 35 years, urban population will have risen by 675 million people, all of whom expect housing, energy, water, transportation, and other energy services.

Why Standards in China?

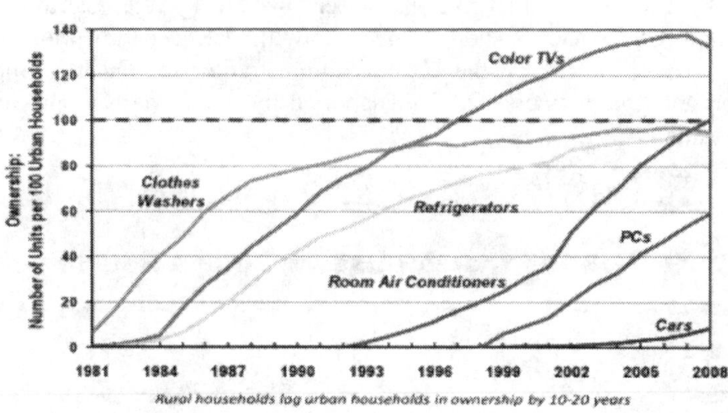

Figure 6

FIGURE 6. The rise of appliance ownership in urban China has been remarkable, from essentially no modern appliances in 1981 to color TV ownership of 130 per 100 urban household in 2008, while clothes washers, refrigerators and room air-conditioners are in almost all urban household. Personal computers have risen to 60 per 100 urban household and cars to 10 per 100 urban households.

Prospects for additional growth are enormous as urbanization rises

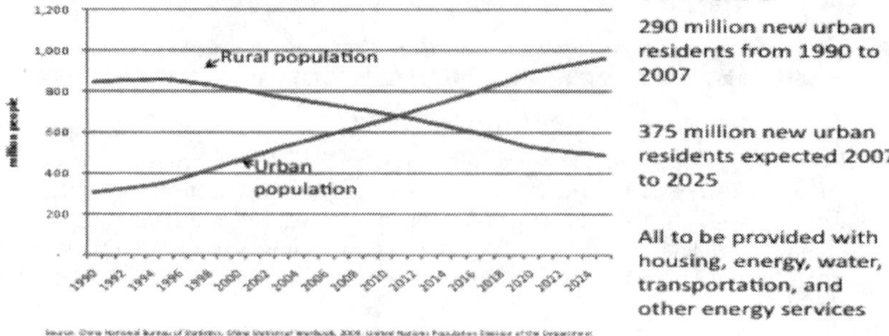

Figure 7

FIGURE 7. Rural households lag urban households in ownership by 10-20 years. The rural fraction of total households will drop from 73% in 1990 to 36% in 2026, as shown above. During this period of 36 years, there will be 675 million new urban residents.

All of the above goes to show why successful appliance standards programs in China are of great importance to their economy, as well as of benefit to the world. Lawrence Berkeley National Laboratory (LBNL) initiated working with the People's Republic of China around 1990 on energy-efficient technologies. LBL's work with China has been broad and inclusive, with the governments of US and China, with foundations, other national laboratories, international organizations, non-governmental organizations and LBNL's own counterparts in China. During the period of initial engagement from 1998 to 2003, focus was primarily on training and implementation, which included:

∞ Training in the DOE analysis toolkit for appliance standards setting (technical analysis, energy impact, economic impact, consumer impact),

∞ Development of initial standards for major domestic appliances such as refrigerators and air conditioners

∞ Transfer of the modeling techniques used by Energy Star to estimate program savings, and

∞ Development of the first set of efficiency-labeled products.

From 2003 to 2007, cooperation built on the successes of the initial phase expanded the scope and coverage of the standards and labeling program through:

∞ Introduction of mandatory energy information labels,

∞ Development of "reach" (tiered) standards that detailed future, more stringent, standard levels,

∞ Linkage of voluntary labels to government procurement, and

∞ International harmonization (this refers to using the same testing protocols and standards as used in large international markets, to avoid retesting and recertification of products for each market separately).

Since 2008, in addition to ongoing technical cooperation, there has been significant institutional strengthening of the standards and labeling work with:

∞ Check-testing of appliances in provincial markets to assess and assure compliance,

∞ Laboratory-based round-robin testing of appliances in multiple laboratories, and

∞ Developing standards and labeling regulations for new products.

China has the most thorough and comprehensive standards and labeling program in the developing world and is more aggressive than programs in many industrial countries. More products are covered under the Chinese programs than in the EU. The Chinese government has placed growing emphasis on the importance of strong labeling and standards regulations and has expanded monitoring and verification to ensure that the anticipated savings actually get realized.[1] Standards have progressed from first implementation of minimum energy performance standards in 1989 to encompassing 30 products today. Voluntary energy efficiency labels (similar to Energy Star in the US) were first introduced in 1999 and now extend to

40 products. Mandatory energy labels (similar to the program in EU) were first implemented in 2005; and by 2010 extended to 21 products.

The potential for energy savings is tremendous. As shown in Fig. 8, the cumulative savings between 2000 and 2020 will be 1188 TWh on just eight of the 30 products now subject to standards.[2] To put this into perspective, a 1 GWe plant operating continuously (24 hours a day, seven days a week) during the year would generate 8766 hours/year x 10^9 Watts = 8.8 TWh/year (in reality, the output can range from 2 TWh/year for peaking plants to 6 TWh/year for baseload plants). In 2020, China would save about 110 TWh/year, or the equivalent of 18 1-GW coal-fired baseload power plants each producing 6 TWh/year. As further shown in Fig. 9, additional avoided CO_2 emissions from post-2009 standards would amount to 340 million metric tons by 2020 and 683 million metric tons by 2030.[3] For more details about LBNL work to support energy efficiency in China visit:china.lbl.gov

Impacts

Savings from existing standards in China:

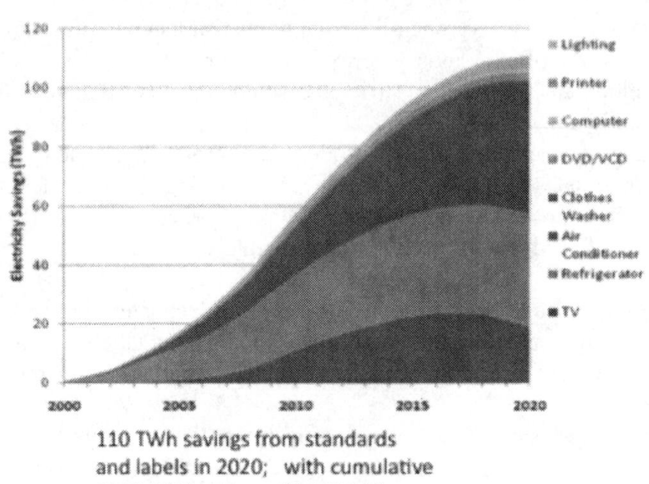

110 TWh savings from standards and labels in 2020; with cumulative 2000-2020 savings of 1188 TWh.

Figure 8

FIGURE 8. The potential cumulative savings between 2000-2020 from standards and labels are projected to be 1188 terra-Watt hours. In 2020, China will save 110 TWh/yr, or the equivalent of 12 GW of power operating continuously with the annual carbon dioxide savings of 340 million metric tons by 2020 and 683 million metric tons by 2030.

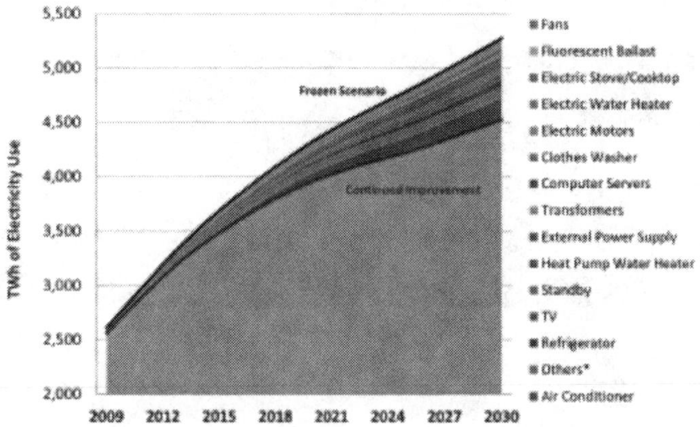

FIGURE 9. Additional potential savings from post-2009 standards in China.

FUEL-EFFICIENT BIOMASS COOKSTOVES: APPLICATION FOR DISPLACED WOMEN IN DARFUR

More than 2 billion persons around the world cook their daily meals with diverse biomass fuels, including wood, twigs, agricultural waste, animal dung, and charcoal. They cook in diverse ways with diverse foods in diverse pots across many cultures, often on three-stone fires (TSF). A TSF is a traditional cookstove and (refers to three stones supporting a pot under which a fire is lighted). Cooking with traditional cookstoves is mostly inefficient and grossly polluting. Women and newborn infants kept close to them are at the greatest risk, as the women spend countless hours tending to these polluting and hazardous stoves. The World Health Organization (WHO) estimates more than a million deaths annually from indoor exposure to biofuel smoke from cooking. Furthermore, residential biofuels' contribution to the total sunlight absorbing black carbon emissions is also substantial - comprising 63% of the total in India, 30% in China, 67% in Africa and 33% globally[4]. Thus airborne pollution from biomass cookstoves contributes significantly to indoor air pollution locally, and also to global warming. The maps in Fig. 10 from [5] show that a considerable reduction in black carbon pollution would be obtained if the emissions from present cookstoves could be reduced with new designs.

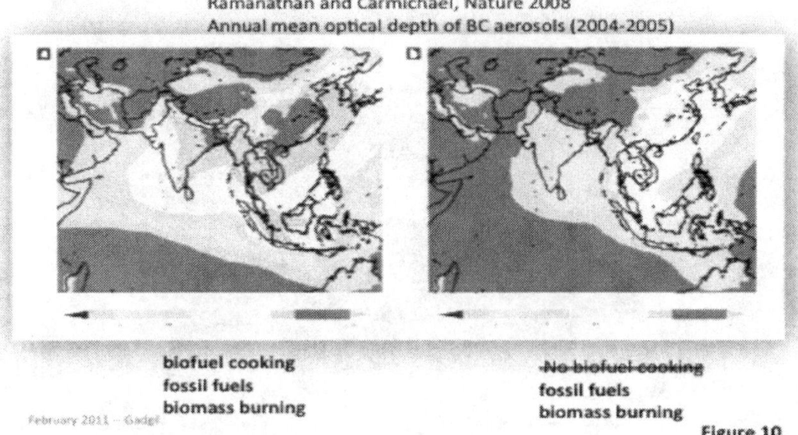

FIGURE 10. Black carbon (soot) emissions over Asia (Ramanathan and Carmichael, Nature, 2008) in terms of the annual mean optical depth of black carbon (2004-2005). The left map totals the contributions from biofuel cooking, fossil fuel burning and biomass burning. The right includes fossil fuel and biomass burning, but does include biofuel cooking, giving a much smaller impact on Asia.

In addition, the plight of Internally Displaced Persons (IDPs) in Darfur, Sudan is extreme, particularly for women and girls, who routinely risk rape and mutilation when they must leave the safety of their camps to gather fuelwood for their daily meals. A typical wood-gathering trip lasts 7 hours, as observed at the Kalma and Otash Camps, Darfur, in 2005.[6]

Subsequent to that first Darfur visit, scientific and technical staff at LBNL and UC Berkeley students designed and tested improved cookstoves suitable for the IDP women, as seen in Figure 11. Recognizing that our smartest scientific efforts are worthless unless the users accept the "improved stove," we have worked hard to keep the users at the center of testing and implementation. The current design iteration (completed in 2009), known as the Berkeley-Darfur Stove™ Version-14 (BDS v14), is stable and simpler to build, and is now undergoing limited mass production. Based on lab-tests, we had initially estimated that each stove would save $250 per year in fuelwood costs, and last 5 years, saving $1250 over its 5-year life. Each stove would double the disposable income of the refugee woman using it.

We qualified an Indian factory to make "flat-kits" of our design, which can be punched out of sheet metal and assembled into stoves in Darfur. The factory can produce 5,000 flat-kits per month on a single-shift production line, with a cost of US$14 per flat-kit. We signed a memorandum of understanding with Oxfam America to assemble stoves in Darfur from these flat-kits and provided all the tools and training for the assembly shop. In October 2009, the workers (actually trained

IDPs from the camps) in Darfur built the first 1,000 stoves in the assembly shop. Output capacity of the assembly shop is a stove every 5 minutes, or 2,000 stoves per month, single-shift, assuming availability of flat kits and a sufficient level of security to keep the shop in operation.

At LBNL, we tested the BDS and TSF for efficiency and emissions using a surrogate cooking task (based on field observations of cooking) of bringing to boil 2.5 L water starting from room temperature, and simmering it for an additional 15 minutes. The BDS cooks faster; the time to heat water to boiling is faster with the BDS design at 5.8 ^{o}C/minute, as compared to the TSF design at 3.6 ^{o}C/minute, as shown in Fig. 12.

The BDS uses less fuel; the TSF consumed on average 564 grams of fuelwood in this task compared to the BDS which consumed on average 368 grams of fuelwood, a savings of 35% compared to TSF, as shown in Fig. 13. The carbon-monoxide emissions from BDS for the same cooking task were only 59% of that from the TSF, as shown in Fig. 14. Thus the BDS emits less CO than the savings in fuelwood. The BDS also emits less black carbon than the TSF (results not shown for brevity), but the reduction in black carbon is smaller than the reduction in fuel use. Further details of testing are provided in Kirchstetter *et al.*[7]

Summer 2006

BERKELEY DARFUR STOVE
AKA "V5 STOVE"
Berkeley-Darfur stove redesigned for production in Darfur conditions "V5"

Members of LBNL, UC Berkeley and EWB-SFP team behind the Berkeley-Darfur stove redesign
Figure 11

FIGURE 11. LBNL, members of Engineers Without Borders San Francisco Professional Chapter (EWB-SFP) and UC Berkeley students designed and tested improved cooking-stove for Darfur.

It Cooks Quicker

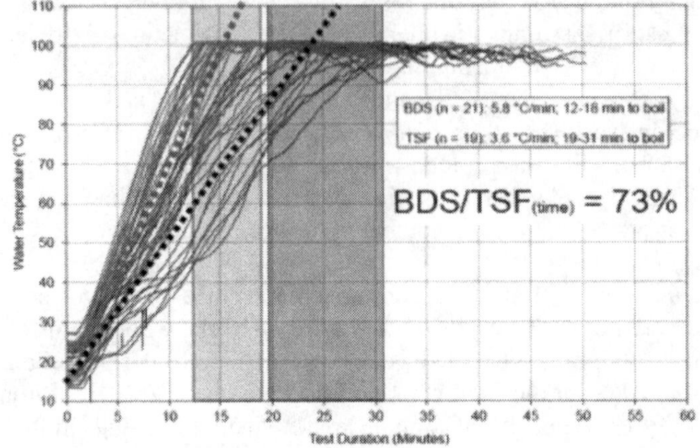

Test = Heat 2.5L water in Darfur pot to 100C, simmer for 15 min, end test

February 2011 -- Gadgil

Figure 12

FIGURE 12. The time to heat water to boiling is faster with the BDS design at 5.8 °C/minute, as compared to the TSF design at 3.6 °C/minute.

It Uses Less Fuel

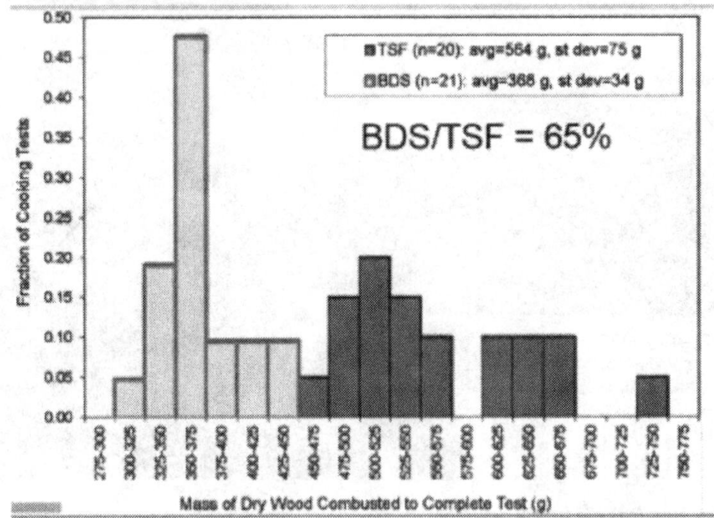

Test = Heat 2.5L water in Darfur pot to 100C, simmer for 15 min, end test

February 2011 -- Gadgil

Figure 13

FIGURE 13. The TSF stove consumed on average 564 grams to heat 2.5 liters to boiling, plus 15 minutes of simmering. The BDS stove consumed on average 368 grams, 65% of the TSF stove consumption, for a savings of 35%.

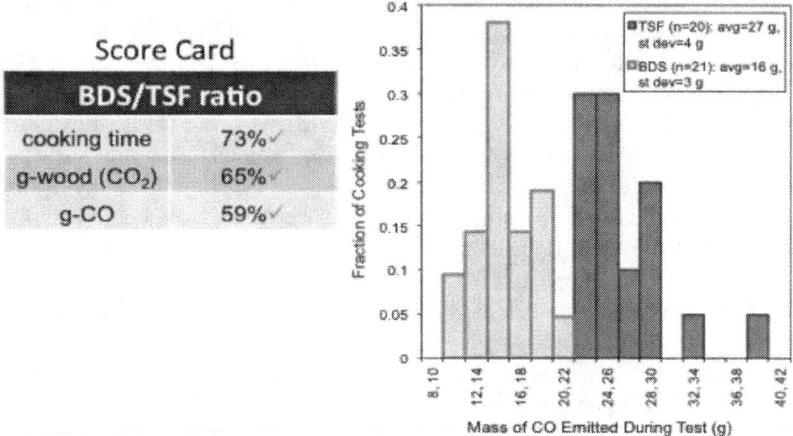

FIGURE 14. The BDS carbon emissions were only 57% of that of the TSF stove.

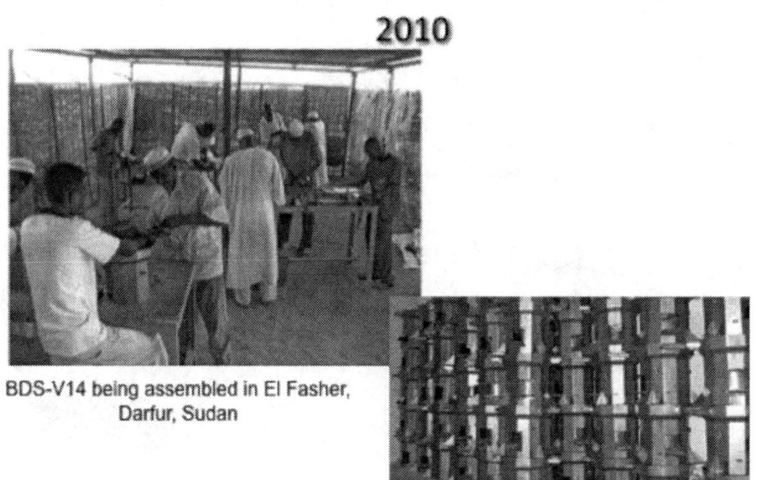

2010
BDS-V14 being assembled in El Fasher, Darfur, Sudan
Stack of Berkeley-Darfur Stoves in Darfur assembly shop. By Feb. 2011, 16,000 of these stoves had been built in Darfur and distributed locally.
February 2011 -- Gadgil
Figure 15

FIGURE 15. Trained refugees from camps assembling the Berkeley-Darfur Stoves in Darfur, Sudan.

In 2010, we conducted a baseline survey of 100 households using TSF stoves for cooking in North Darfur's Zamzam camp, and a follow-up survey 8 months after the women had switched to using BDS stoves for cooking. There was no change in fuelwood prices during this interval. Preliminary results showed that this group reduced its spending on fuelwood from 1/2 of non-fuel spending (using TSF), to less than 1/4 of their non-fuel spending (using the BDS). The survey showed that each $20 stove puts $330/year in the pocket of the women using the stove; over the estimated stove-life of 5 years this is worth $1600 (note that these savings are larger than our earlier lab-based estimates). As of February 2011, nearly 16,000 stoves have been distributed in Darfur, helping more than 100,000 women and their dependents, worth more than $25 million to their recipients. Ten thousand additional stoves are planned for 2011. The 2.7 million IDPs in Darfur need an estimated 400,000 stoves. So, we have a good start, but a lot more remains to be done. Figure 15 shows the stoves assembly shop in Darfur, and some of the stoves assembled at the shop. Further resources about Darfur Stoves Project as well as updated news of its progress are available online at: www.darfurstoves.org

SUPER-EFFICIENT APPLIANCE DEPLOYMENT (SEAD)

At the Copenhagen meeting of Conference of Parties (COP) in December 2009, DOE Secretary Chu announced U.S. initiative and leadership of the Super-Efficient Appliance Deployment (SEAD) program with the goal of global market transformation in favor of super-efficient appliances.

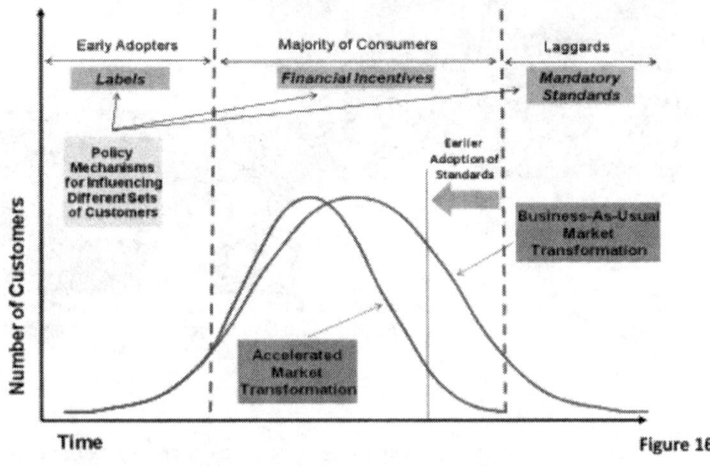

FIGURE 16. A symbolic model for the adoption of energy-efficient appliances. Early adopters of energy-efficient appliances are encouraged with labels, while the majority of consumers need some financial incentives, and laggards need mandatory standards to purchase.

A symbolic model for the adoption of energy-efficient appliances is shown in Figure 16. Social scientists who study spread of technology innovations in societies commonly divide the society into different categories according to the time-delay with which the innovation is taken up. The adoption curve is commonly approximated as a Gaussian, and Fig. 16 focuses on three population segments (shown) and highlighted in yellow above each segment are the social motivators that can induce each segment to adopt. The curve in blue represents the normal adoption process, and the one in red represents a desired "market acceleration" for earlier adoption. For the leading minority, "Early Adopters," labels distinguishing energy-efficient appliances are adequate. For the majority of consumers labels are not enough by themselves; they need some financial incentives. The trailing minority, the "laggards" need mandatory standards to purchase energy efficient appliances (standards mandate inefficient appliances to be removed from the shelves). With these various measures in place to accelerate market adoption, the blue curve shifts to red curve.

SEAD is an ambitious project to coordinate across countries efforts to accelerate market transformation in favor of energy efficient appliances. Actions can be broadly categorized into three types. The first group of actions "raises the ceiling" for highly efficient appliances – this refers to creating a large global market for such appliances to encourage many manufacturers to enter the field by coordinating incentives across countries for highly efficient devices. The second group of actions "raise the floor" so as to systematically exclude the very poor performers from markets across many countries – so it is no longer worthwhile to manufacture them for small market segments. The last group of actions "strengthens the foundations" for this market transformation with coordination across countries for testing, certification, data sharing, and technical support.

The motivation for SEAD is illustrated graphically in Figure 17. Global domestic electricity consumption is expected to double in 25 years, from 5,700 TWh/yr to 11,500 TWh/yr in 2030. Under the baseline scenario, the four largest appliances (lighting, refrigeration, air-conditioning, television) would use some 5,000 TWh/yr in 2030, or 43% of the total, as shown in Fig. 17. These consumption estimates for 2010 to 2030 are disaggregated for 13 countries and 3 appliances in Fig. 18. The key point of Figs. 17 and 18 is to note how only a few countries or regions constitute a large portion of the global consumption of these products and a few suppliers dominate the global market for these products (just fifteen suppliers produce 75% of global production of major appliances). So, coordination and collaboration on policies and programs even among few countries or regions on select products to improve their efficiency will provide a strong and consistent signal to major global manufacturers to produce efficient products and will reduce global electricity consumption significantly [8].

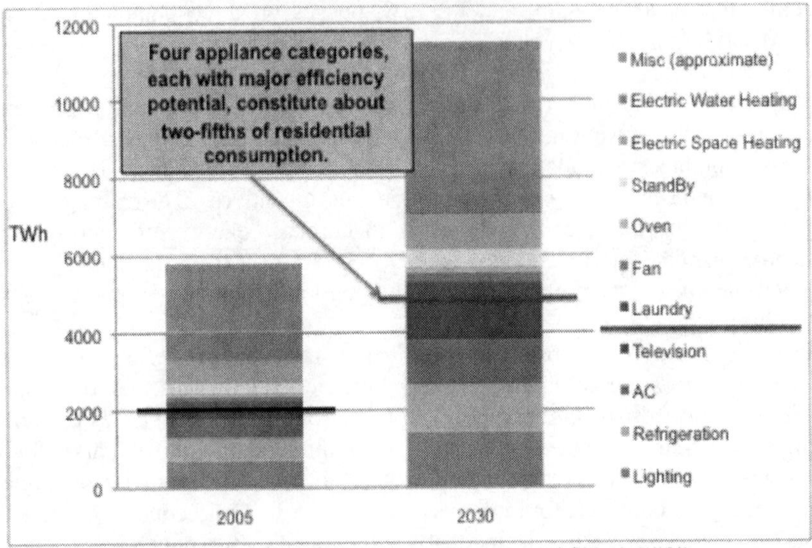

Why SEAD?
Global electricity consumption expected to double in 25 years
Four appliance categories, each with major efficiency potential, constitute about two-fifths of residential consumption.
12000
10000
8000
TWh
6000
4000
2000
0
2005
2030
Misc (approximate)
Electric Water Heating
Electric Space Heating
StandBy
Oven
Fan
Laundry
Television
AC
Refrigeration
Lighting
Source: Consumption data based on analysis using the BUENAS model. For the data and methodology used in the BUENAS model- McNeil, M. A., V. E. Letschert and S. de la Rue du Can (2008). "Global Potential of Energy Efficiency Standards and Labeling Programs". Lawrence Berkeley National Laboratory Report LBNL 760-E Available at
Figure 17

FIGURE 17. Global domestic electricity consumption is expected to double in 25 years, from 5,700 TWh/yr to 11,500 TWh/yr in 2030. The four largest appliances (lighting, refrigeration, air-conditioning, television) would use some 5,000 TWh/yr in 2030, or 43% of the total.

The maximum on-power drawn by television sets with screen sizes ranging from 19 inches to 60 inches (top X-axis shows screen size; bottom X-axis shows screen area) for 4 different energy standards is shown in Fig. 19. For the largest size of 60 inches, the power varies from 130 watts to 240 watts. Data points show that there are at least a few commercially available models that are substantially more efficient than the Energy Star 5.1 requirement, which is one of the most stringent energy efficiency label requirements in the world that will come into effect only in 2012. These efficiency levels are achieved by adopting technologies such as LED back lighting, more efficient optical films, and local dimming [9]. SEAD will lead to much more rapid adoption of such technologies compared to the business-as-usual (BAU) situation.

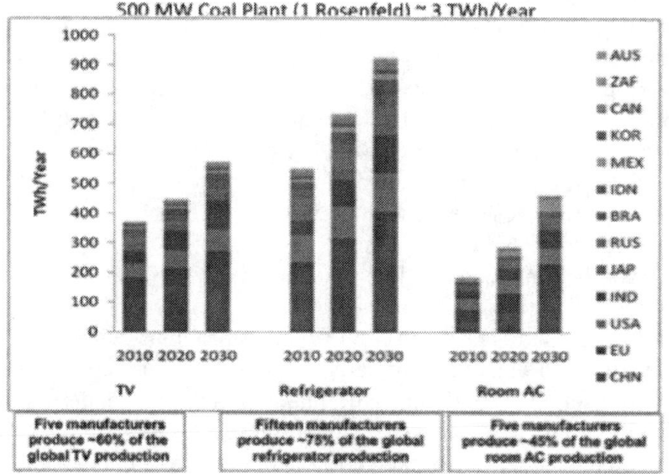

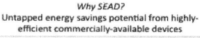

FIGURE 18. Consumption estimates for 2010 to 2030 are disaggregated for 13 countries and 3 appliances.

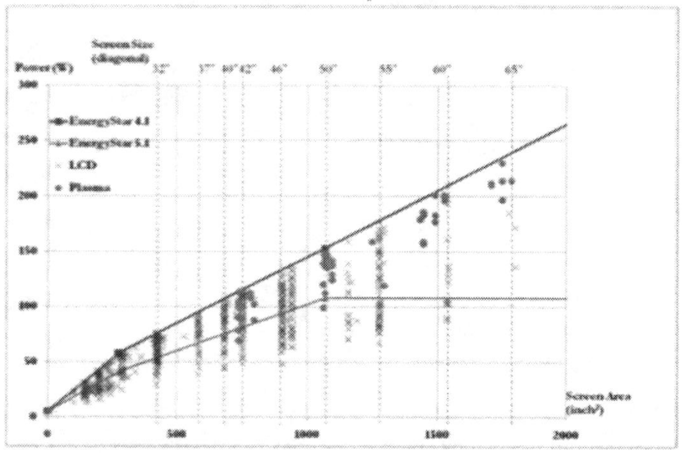

FIGURE 19. The power needed for television sets varies with screen size and standards, shown for screens from 19 inches to 75 inches in size, with 4 different energy standards. For the largest screen size of 75 inches, the power varies from 100 watts to 500 watts.

In India, ceiling fans are the second largest domestic use of electricity after lighting. The typical Indian ceiling fan at full power typically uses 75 watts, and ceiling fans that qualify for India's Energy Star rating use 65 watts. However, far larger improvements are technically possible, and can't reach the market without support such as that envisaged by SEAD. A fan with a high-efficiency induction motor would use 45 Watts at full power, and one with a super-efficient brushless DC motor would use only 35 Watts[10]. The collaboration envisioned in the SEAD program will facilitate the availability of such products in the market at affordable prices owing to lower costs from volume production that will be possible only with policy support for market transformation.

A study by the International Energy Agency shows that to keep global CO_2 concentrations at 550 ppm by 2030, we need emissions reduction by 12 gigatonnes / year relative to the BAU [11]. To keep CO_2 concentrations at 450 ppm by 2030, we need reductions of 15 gigatonnes of carbon dioxide annually compared to the BAU. Figure 20 summarizes the findings of this report which underlines that energy efficiency will be a hugely important part of these CO_2 reductions. A recent study shows that global adoption of the world's best commercially available end-use technology for the twenty four most energy consuming appliances and equipment, would lead to annual global savings potential in 2030 of 11,000 TWh of electricity demand and 6.7 gigatons of CO_2 emissions [12]. SEAD envisages enabling savings on this enormous scale by coordinating across countries in favor of superior energy efficiency in appliances and equipment with largest energy footprint.

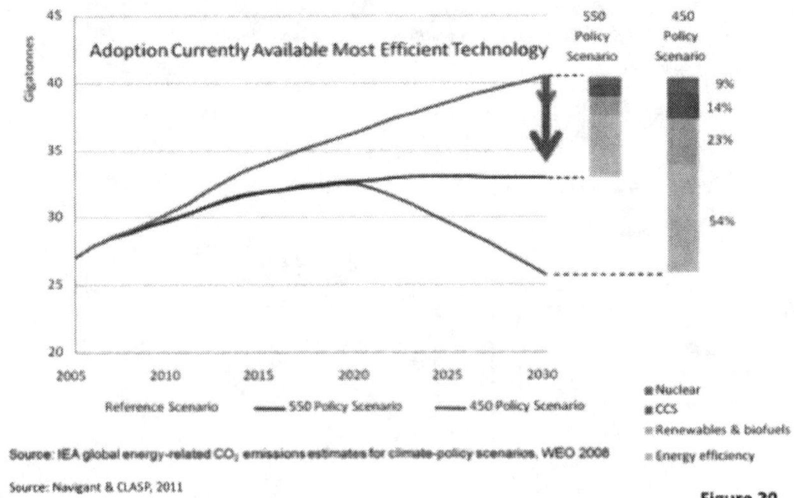

FIGURE 20. Two approaches to reducing carbon dioxide emissions by using the most efficient technologies.

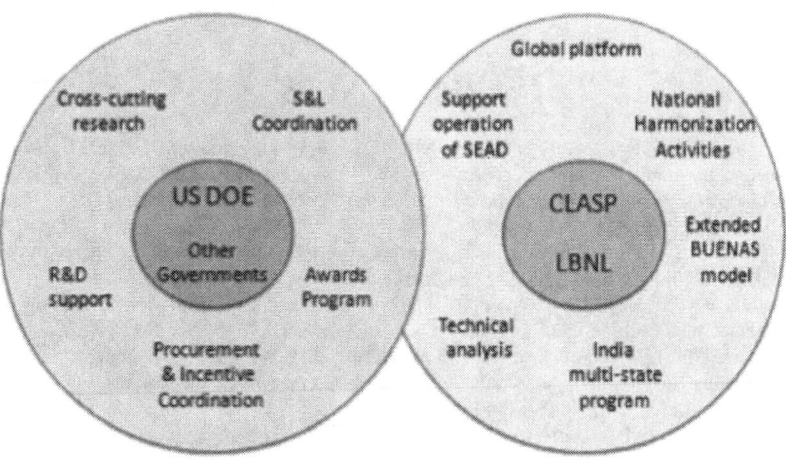

Figure 21

FIGURE 21. The organizational chart for SEAD is shown with its complexity.

SUMMARY

As the 5 billion people outside the first world industrialize, and inevitably increase their income in real terms, they too aspire for development, which comes with higher per capita use of modern energy and higher per capita CO_2 emissions. Helping the world achieve a higher standard of living while concurrently using energy far more efficiently is in their interest as well as ours – and in the interest of the planet. In response to the rising energy demand in developing countries, and given the high diversity of income levels and energy end-uses, we present three specific examples to illustrate the large energy and CO_2 savings possible with application of modern science and technology, in addition to new policy insights, to the rising energy demand in the developing countries.

ACKNOWLEDGEMENTS

The work was funded by the Department of Energy under Contract No. DE-AC02-05CH11231. The work described in this chapter is the combined effort of many individuals, several of them at LBNL, but many also at our collaborating institutions and funding agencies here and abroad. We gratefully acknowledge their support and contributions without which this work would not have been possible. In particular, our many LBNL colleagues – too numerous to name here individually –

have contributed substantially to the accomplishments described above. We are grateful to Susan Addy and David Hafemeister, whose helpful comments have improved the manuscript.

REFERENCES

1. N. Zhou, D. Fridley, A. Pierrot and Y. Saheb, "Compliance and Verification of Standards and Labeling Programs in China: Lessons Learned," *Proceedings of the 2010 American Council for an Energy Efficient Economy's Summer Study on Energy Efficiency in Buildings,* Washington, DC: ACEEE, 2010.
2. D. Fridley, et al, "Impacts of China's Current Appliance Standards and Labeling Program to 2020", *LBNL-62802, Berkeley, CA: Lawrence Berkeley National Lab,* 2007.
3. N. Zhou, et al, "Analysis of Potential Energy Saving and CO2 Emission Reduction of Home Appliances and Commercial Equipment in China", *Energy Policy,* May 2011 (forthcoming).
4. T.C. Bond, D.G. Streets, K.F. Yarber, S.M. Nelson, J.H. Woo and Z. Klimont, "A technology-based global inventory of black and organic carbon emissions from combustion", *J. Geophys. Res., 109,* D14203, doi: 10/1029/2003JD003697, 2004.
5. V. Ramanathan and G. Carmichael, "Global and regional climate changes due to black carbon", *Nat. Geosci., 1,* 221-227, 2008.
6. C. Galitsky, A.J. Gadgil, M. Jacobs and Y.M. Lee, "Report from a field trip to Darfur", *Lawrence Berkeley Laboratory Report LBNL-59540,* 2006.
7. T. W. Kirchstetter, O.L. Hadley, C.V. Preble and A.J. Gadgil, "Emission Rates of Pollutants Emitted from a Traditional and an Improved Wood-burning Cookstove", *Proceedings of the 12th International Conference on Indoor Air Quality and Climate, Austin, Texas,* 5-10 June 2011.
8. A. Phadke, G. Sant, R. Bharvirkar, R. Liberman and J. Sathaye, "Accelerating the Deployment of Super-Efficient Appliances and Equipment with Multi Country Collaboration", *American Council for an Energy Efficient Economy (ACEEE), Summer Study,* 2010. Available online at *eec.ucdavis.edu/ACEEE/2010/data/papers/2100.pdf*
9. W. Park, A. Phadke, N. Shah and V. Letschert, "Energy Consumption Trends and Efficiency Improvement Opportunities in Televisions" *Lawrence Berkeley National Laboratory Draft Report,* publication due June 2011.
10. A. Phadke, N. Sathaye and A. Pednekar, "Cost Effectiveness and Global Electricity Saving Potential of Efficiency Improvement Options for Ceiling Fans", *Lawrence Berkeley National Laboratory Draft Report,* publication due June 2011.
11. International Energy Agency (IEA), *World Energy Outlook 2008,* Paris, 2008, available online at http://www.worldenergyoutlook.org/2008.asp
12. P. Waide, "Opportunities for Success and CO2 Savings From Appliance Efficiency Harmonization", *Navigant Consulting and CLASP,* 2011., Available online at http://www.clasponline.org/clasp.online.resource.php?disdoc=781#opportunities

The Co-Created Guatemalan Field School: Carbon Reduction with Appropriate Technology

Peter V. Schwartz

Physics Department
California Polytechnic State University
San Luis Obispo, CA 93407

Abstract. We are exploring a collaborative development model where US students study in a developing country with local students, in this case in San Pablo, Guatemala – a village of 800 at elevation 3000 m near the Mexican Boarder. The Cal Poly summer study abroad program "Guateca", to commence July 1, 2011, was jointly conceived with San Pablo leadership on August, 2010, and has since grown through input from both Cal Poly and San Pablo communities. The program aims to build cross-cultural community and explore choices both societies have in the context of the rapidly changing energy landscape to develop in a way that preserves the environment and builds independence from increasingly expensive conventional energy sources.

DEVELOPMENT IN INDUSTRIALIZED AND IN DEVELOPING COUNTRIES

I will define "development" as the establishment of technological and societal infrastructure intended to meet the needs of the people. The present industrialized countries developed in an era of inexpensive fossil fuels with little accountability for the "external costs" of energy use including pollution and climate change. Most of the world is not industrialized and does not have the option to develop in the same way that the United States (for instance) did, as resources are now scarcer. Additionally, increased concern about the local and global impacts of unrestrained growth is propagating attention to sustainability in both industrialized and developing countries. While the economic context is different, the same strategies may successfully answer the same needs in both societies. Our intention is to work together with communities in developing countries to jointly educate ourselves, promoting novel technologies and community practices.

There is a near universal human drive to increase wealth, energy consumption, and the resulting emission of pollutants, including greenhouse gases, well past the point that this increased "development" improves any measure of human well-being. For instance, take child mortality: while 20% of the children in the poorest African nations die before the age of 5 years, there is no correlation between child mortality and income once per capita income exceeds $20,000 US per year – less than half the per capita income of the United States. Psychological studies also indicate that once human needs are met, well-being does not increase with increased wealth [1]. Yet, most societies, people, and governments strive to maximize income regardless of

Physics of Sustainable Energy II: Using Energy Efficiently and Producing it Renewably
AIP Conf. Proc. 1401, 75-83 (2011); doi: 10.1063/1.3653846

present economic prosperity. With this seemingly without-benefit excess prosperity comes a near proportional increase in CO_2 emissions, as seen in Fig. 4 of the paper submitted by Ashok Gadgil. If happiness is our societal goal, then we *should* be able to achieve it within the carrying capacity of the planet partially by decreasing our collective economic goals. While I recognize the societal challenge in this goal, I am compelled nonetheless to recognize it as the obvious conclusion from the data at hand.

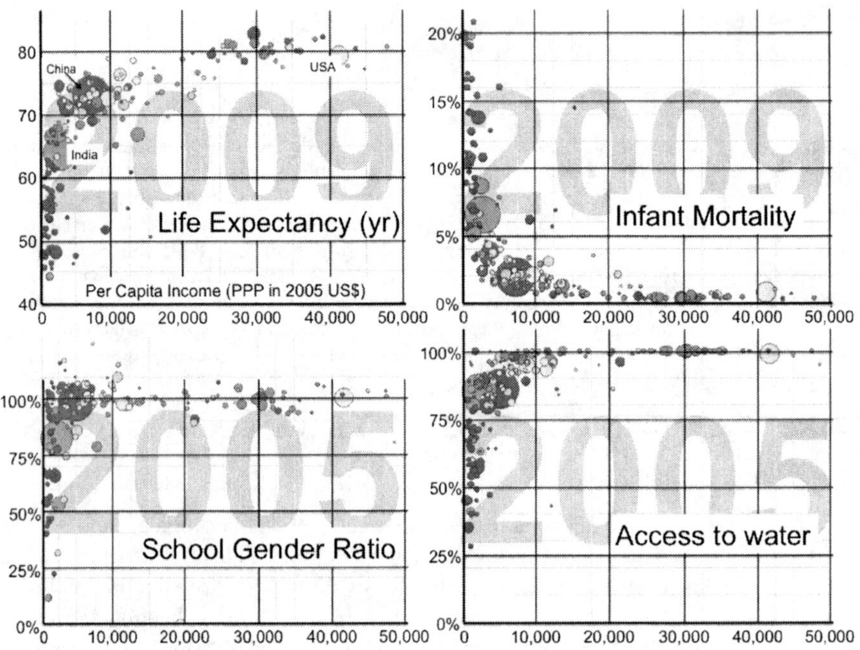

FIGURE 1. Four Happiness (and Unhappiness) Indicators as a function of per capita income in 2005 United States Dollars corrected for Purchasing Power Parity. Each dot represents a country, and the area is proportional to the population. These statistics and many more are free material available at www.gapminder.org. [2]

Many past development efforts have used industrialized development as a model for developing countries and imposed solutions with minimal input from those "to be developed". The results are often that technologies do not work as planned, or have adverse unintended consequences. Usually it is the very poor rural people who bear the brunt of the unintended consequences, as their viewpoints are left out of development planning processes most consistently. For example, building hydroelectric dams in developing countries usually means destroying the livelihood of up to millions of people who live on the riverbanks;[3] whereas in the United States these poor rural, indigenous communities have already been nearly destroyed or moved to cities, so the immediate threat to livelihoods is not a necessary consideration. We have as one of our goals to be working with the rural poor so that none of our actions will affect them negatively. Our group is only beginning to build collaborative knowledge through our joint program, with the intention of making us a more effective agent of empowerment, as well as provide insight to us on how to make better choices for our own lives.

The past decade has seen an increased respect for people in developing communities recognizing what is best for themselves and being the most appropriate consultant for their own empowerment.[4] This has lead to a more collaborative outreach and development process. We embrace this direction, recognizing that while we have access to significant mechanical, electrical, and agricultural technologies, this knowledge is applicable where we learned it, in American Universities. In the culture, land, and infrastructure of a developing country, our technological solutions may not work, as they are not "appropriate" for the setting. Thus, we begin a study together without well-established technological goals, but with a commitment to an educational process including inventory of the community's needs as well as the collective physical and academic resources that each side has.

Vision Statement

Guateca will become a model for cross-cultural collaborative education, emphasizing sustainable enterprises and fostering international community and local well-being.

Mission Statement

Guateca is a collaborative education program between Cal Poly and San Pablo students, faculty, and citizens in San Pablo, a village of 800 in the Guatemalan mountains. This co-developed education program has the following goals:

1. Builds cross-cultural community with the needs and interests of both communities in mind.
2. Fosters technological and social development by encouraging curiosity and empowering innovation.
3. Advances language and cultural fluency, while studying energy and innovation of sustainable enterprises.
4. Develops sustainable technologies to meet the needs of San Pablo as well as generate income locally.
5. Our intention is to work openly together, sharing resources and ideas, embracing new challenges, goals, and resources.

The Guateca Summer Study Abroad Program

Beginning what is a five-year program commitment, the first group of Cal Poly students will go to Guatemala for collaborative study this summer. After a week-long orientation, 15 Cal Poly students from a broad distribution of majors will travel to San Pablo to be joined by 15-20 San Pablo college students for two months of culture, language, sustainable technology, and service learning. We will stay with local families, study, and work on projects with the community. Supported by summer tuition from the Cal Poly students, three classes will be taught by instructors from Cal Poly, and supported by various US and Guatemalan institutions:

1) Energy, Society, and Environment.

2) Language. With a 1-to-1 student-teacher ratio, between the equal number of US and Guatemalan students, Cal Poly students will study Spanish while San Pablo students study English. Spanish curriculum and assessment will be conducted by Cal Poly for US students. We seek a Guatemalan university to recognize the English curriculum.

3) Development of Sustainable Enterprises – Instruction will be supported by local experts in appropriate engineering fields, business, and agriculture.

Program History

In Cal Poly's project-based classes, an interdisciplinary team of faculty and guest instructors mentors 30 – 40 students across all majors. For four years, students have explored the causes of poverty (UNIV-391) and innovated prototypes to address technical needs (UNIV-392). Additionally, this year, we are exploring appropriate business models in Agriculture Business (AGB-450). While the classes have disseminated important global information and fostered creativity, there has lacked a connection to a real community partner to give the projects meaning.

San Pablo, Tacana (population ~800) has a history of community and social well-being that make it particularly well suited for the Guateca endeavor. Jesuit influence since 1960 is seen in the terraced landscape, litter-free countryside, and strong community organization. The community successfully combats alcoholism, and prioritizes education. They have developed a K-12 school system offering hands-on experiences through weekend workshops and plan to enlarge their high school program with small business development and pre-university courses. San Pablo has organized Guateca program participants (15-20 university students and 5-7 community leaders), who have begun collaborating with their Cal Poly counterparts, additional instructors, and local universities, as we develop a plan for the coming program this summer.

Last summer (2010), we visited San Pablo for four days and Met with community leaders. Together we started designing a collaboration model and also identified technologies that would have the greatest probability of success and best fulfill local needs. During December 2010, 12 Cal Poly students from UNIV-391 spent 10 days in San Pablo refining ideas for their projects. In winter of 2011, a new group of Cal Poly students in UNIV-392 did more work on technology development for the program while we identified and hired two English teachers to both teach English to San Pablo students and to facilitate communication between San Pablo and Cal Poly. A group of 4 Cal Poly students forming the Guateca advisory board has worked with Schwartz,

Cal Poly's Continuing Education, and the interested parties in San Pablo and Cal Poly administration to shape this summer's program and recruit students.

CARBON REDUCTION MECHANISMS

Guatemala's Carbon Footprint

Guatemala's 15 million citizens presently average slightly less than one metric ton of greenhouse gases per year (see Fig. 2). While this figure is more than 4 times what it was 60 years ago, it is still only 5% of the near constant 20 tons per person that the U.S. has maintained during the last 60 years. The economic intention of developing countries is to emulate the US economy, which would greatly increase global emissions as developing countries represent the bulk of the world's population. Our intention is to foster an alternate means to human prosperity (as indicated by a happiness index of needs met, rather than per capita income) to emulating the high carbon practices in the U.S. Besides serving global interests, a low-carbon development alternative may also serve the adopting poor countries because today, a society that develops will do so without the benefit (or handicap as the case may have been) of inexpensive energy. In any case, we recognize the autonomy of each Guatemalan to make these choices for themselves. To empower these choices, we recognize education as the most effective way out of poverty [5] as we ourselves explore and model what we (US and Guatemalan participants) find to be compelling energy practices. The summer school, funded by US students, is a model that provides education that is presently out of reach of most Guatemalans.

Guateca's Carbon Footprint

We have begun an ongoing emissions assessment for the Guateca summer school itself and for the technologies that we are exploring. Given the distance to Guatemala and the mileage of a 747, we estimate about 500 kg of CO_2 emitted from the round trip flight to Guatemala, and another 100 kg from bus and automobile travel, and about 0.2 metric tons for living as a Guatemalan for two months. At the same time, 2 months of US emissions will be displaced: about 3 metric tons. With this rough estimate, the carbon footprint of the program should be about *negative* 2.2 metric tons per student. The calculation has several uncertainties. An average college student likely has less than average US American emissions, San Pablo emissions are certainly well below the Guatemalan average, and we may have considerable travel on the road in Guatemala. In the coming summer, we will monitor activities to improve this estimate.

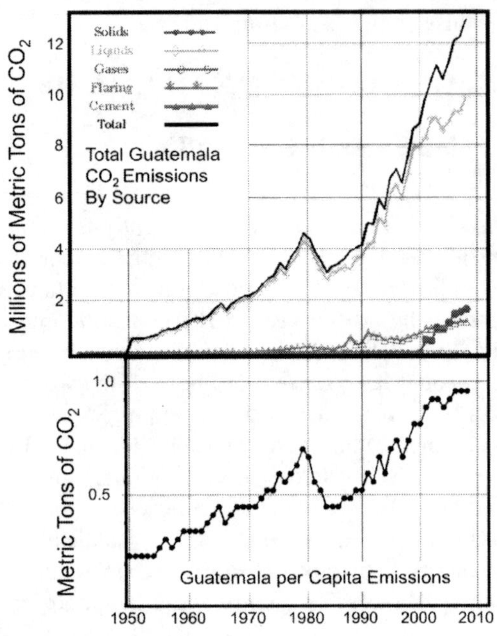

FIGURE 2. Guatemalan CO_2 emissions by source (above), and per capita (below).[6] In comparison, US per capita emissions during the same period have remained relatively stable at about 20 tons/person.

Introduction of Low Carbon Practices and Consumer Goods

The sustainable technologies and enterprises Guateca will explore will also have an impact on carbon emissions – both in the developing and industrialized world. During the summer school, students will estimate the lifecycle emissions of the introduced enterprise and compare it with that which the new enterprise will displace. For a rough example, a family uses 40 gallons (150 kg) of hot water a day. If the water needs to be heated by 40YC, about 7 kWh are consumed. Thirteen percent of Guatemala's electricity is produced from coal.[6] We presume that all the available hydro electricity is used, and that any additional load to the grid will be met with (the marginal generation of) coal-fired electricity with an efficiency of roughly 1/3. The Guatemalan grid has a loss of 14%, resulting in a carbon intensity of the consumed electricity of about 1.1 kg of CO_2 per kWh. Heating this water with grid-supplied, coal-generated electricity would result in about 8 kg of CO_2 emissions, or about three metric tons of CO_2 per year. Seven kWh of heat could be extracted on a sunny day with 2 m^2 of solar thermal panels, but we might play it safer by installing a 4 m^2 section of a roof with a clear plastic panel over the preexisting roof, connecting it to a 100 liter tank and requiring 20 m of plastic tubing for thermal transfer. Such a device might entail about 300 kg of embedded carbon dioxide emissions, indicating a carbon payback time of less than two months. At present, we do not know if the proposed technology will work on site, what the required materials will ultimately be, and how the technologies will be used. These calculations are part of the course work for the present classes taught at Cal Poly [7] and will be part of the Guateca projects during the summer.

Supporting and Showcasing a Community that Already Prioritizes Sustainability

San Pablo is already a model sustainable community, by having developed a very high standard of living through education and community responsibility. Our program is already receiving significant attention in Guatemala. Other communities are interested in learning how they can become involved. Our response is an observation of the building of a successful community that San Pablo achieved before we arrived. Thus the mere presence of our program elevates San Pablo to be a model for other poor rural communities in implementing practices that prioritize healthy, sustainable living.

Providing an Alternative to "Becoming US-Americans"

As stated above, the US presently has a per capita carbon footprint that is 20 times that of Guatemala and other developing countries. Ten percent of Guatemala's GDP is remittance payments from mostly young men working in the US to their families in Guatemala [8]. In the process, they often leave their children for many years, and often do not return at all. The pursuit of an American lifestyle is very apparent. The numbers beckon the question, "can we prevent them from becoming like us?" Certainly, we are collectively embarrassed by the arrogance and hypocrisy in such a statement. We fully recognize the need for the US to reduce our carbon footprint and the right of each person in a developing country to achieve their desired lives. We reframe the idea into the statement, "while we at home in the US endorse carbon reduction strategies, can we foster more compelling solutions for Guatemalans to choose from besides emigration to the US and/or becoming an extension of the US economy through remittance employment." We have already found a community (San Pablo) where most of the young people wish to remain. They have expressed an interest to develop a more effervescent professional community in San Pablo in connection with the interest to retain educated youth. We will explore if the presence of Guateca fosters the creation of such an atmosphere and if the technologies introduced and the associated technological challenges has an effect on retaining educated youth in the village where they benefit from both the economic and social support of their community.

Providing a Test Site and Model of Alternative and Sustainable Energy Practices for Industrialized Nations

Consumer acceptance is a major obstacle for any technology, especially for one that can be perceived as compromising, or requiring a change in lifestyle. While industrial countries have indicated a commitment to adopt sustainable technologies and practices, our everyday needs are met by our present carbon-intense practices. For example, while people universally express approval of my outdoor solar shower in the 7 years I've had it, none pursued building one. These people already receive the services of conventional fossil fuel heated showers. In contrast, we did a "willingness to pay" survey for various technologies in San Pablo. We asked the question, "How much do you presently pay for hot water?" A common answer was, "nothing, we don't have hot water." In developing countries the carbon-intense technologies are often not an option because they are also more expensive. Because an experimental technology here may provide the difference of meeting a need, the barrier to adoption may be much lower. We anticipate that large-scale adoption of some successful appropriate technologies will provide a means to test and learn about the technologies and how people implement them. Additionally, San Pablo can also provide a non-industrial model of development for communities in both developing countries as well as in industrialized countries.

CONCLUSION

We recognize that this contribution lacks substantial data. We have presented an idea and plan:

1) Given that development efforts have often failed due to lack of input from the people being targeted, Cal Poly and San Pablo will begin with a co-development model based on equal input and participation from both sides.
2) Given the strength of education as a means to walk out of poverty and promote social change, our program will be centered around education.
3) Given the importance of economic and environmental sustainability, we will focus on sustainable technologies and practices.

Our school begins this month. Data to follow.

ACKNOWLEDGMENTS

Many thanks to the community of San Pablo for the insight and hard work they have demonstrated in the creation of a progressive modern community providing an inspiration to exit poverty by leveraging education and community organization; and for including us in their program. I thank Luz Marina Delgado for bringing me to San Pablo. I thank the many students – both Guatemalan and American, especially Jamie Cignetti and Kristian Velásquez Pérez for their insight and honesty and Julio Velásquez Roblero, a San Pablo community leader. I thank the many instructors that have taught the appropriate technology classes with me especially Kevin Williams, Sema Alptekin, Patrice Engle, and Andy Kreamer, an activist for social progress living in Berkeley, California.

REFERENCES

1. Martin E. P. Seligman, *Authentic Happiness*, New York, NY: The Free Press, 2002
2. Free material from www.gapminder.org.
3. Helmut Kloos, "Development, Drought, and Famine in the Awash Valley of Ethiopia", *African Studies Review*, Vol. 25, No. 4 (Dec., 1982), pp. 21-48
4. P. Polak, *Out of Poverty: What Works When Traditional Approaches Fail*, San Francisco: Berrett-Koehler Publishers Inc., 2008
5. *Moving out of Poverty, Cross-Disciplinary Perspecives on Mobility*, edited by D. Narayan, P. Petesch, Ther International Bank for Reconstruction and Development / The World Bank, 2007
6. Carbon Dioxide Information Analysis Center, Oakridge National Laboratory. http://cdiac.ornl.gov/trends/emis/gut.html
7. http://appropriatetechnology.wikispaces.com/Univ+392+Winter+2011
8. World Bank Working Paper No. 86. 2006, *The US-Guatemala Remittance Corridor, Understanding better the Drivers of Remittances Intermediation* http://siteresources.worldbank.org/INTAML/Resources/US-Guatamala.pdf

THE NEXUS OF ENERGY AND WATER IN THE UNITED STATES

MICHAEL E. WEBBER

Associate Director, Center for International Energy & Environmental Policy and Mechanical Engineering Department
University of Texas at Austin
Austin, Texas

Abstract. This manuscript presents an overview and a relevant framework for thinking about the nexus of energy and water. Here are the key points of this article:

∞ Energy and water are interrelated; we use energy for water and water for energy

∞ The Energy-water relationship is under strain, and that strain introduces cross-sectoral vulnerabilities (that is, a water constraint can become an energy constraint, and an energy constraint can induce a water constraint)

∞ Trends imply that this strain will be exacerbated because of 1) growth in total demand for energy and water, primarily driven by population growth, 2) growth in per capita demand for energy and water, primarily driven by economic growth, 3) global climate change, which will distort the availability of water, and 4) policy choices, by which we are selecting more water-intensive energy and more energy-intensive water

INTRODUCTION

Energy and water are both fundamental ingredients of modern civilization and are precious resources. They are key inputs to our agricultural systems, factories, and buildings, and it can be argued that they are even more fundamental than food, shelter, healthcare and education.

Energy and water are also closely interconnected and under strain. Consequently, the nexus of the two has been the subject of recent attention by the scientific community [3-5], popular media [1-2, 6-9], and Congress [10]. This nexus manifests itself in society in many ways. For example, water provides electric power and plays a growing role for irrigation of energy crops to produce biofuels such as ethanol. And, the thermoelectric sector is the largest user of water in the U.S., for cooling. In parallel, the water industry uses power for moving, pumping, treating, and heating water. On top of this relationship, the parts of the world with high-expected rates for population growth and economic expansion are also often places where water sources are scarce. Combining these trends with projections for more irrigation implies rapid growth for water demands that localities might satisfy with desalination or wastewater treatment, both of which are very energy-intensive.

Physics of Sustainable Energy II: Using Energy Efficiently and Producing it Renewably
AIP Conf. Proc. 1401, 84-106 (2011); doi: 10.1063/1.3653847

Despite all these advances, approximately 2.4 billion people live in highly water-stressed areas. Furthermore, the largest public health problem globally remains the more 1.1 billion people without access to improved freshwater sources [11, 12] for drinking, cooking and washing (100 million people in China alone [13]), and 2.6 billion remain vulnerable to water-borne diseases because they lack access to wastewater treatment (e.g. sanitation). [11, 12] Consequently, it has been noted that improving water quality is a significant way to improve public health worldwide [14-16]; however, creating universal access to clean water will require a lot of energy for treatment and moving it to where it is needed. Thus, it is fair to say that solving the world's public health crises begins at the nexus of energy and water.

Despite the importance of each and the close relationship of energy and water, the funding, policymaking, and oversight of these resources are typically performed by different people in separate agencies. Also, energy planners often assume they will have the water they need and water planners assume they will have the energy they need—if one of these assumptions fails, the consequences will be dramatic. By bringing scientific and engineering expertise to bear on this vastly understudied problem, this scenario might be avoided.

This manuscript brings some of the relevant information together to present an overview and a relevant framework for thinking about the nexus of energy and water. Here are the key points of this article:

- Energy and water are interrelated; we use energy for water and water for energy
- The Energy-water relationship is under strain, and that strain introduces cross-sectoral vulnerabilities (that is, a water constraint can become an energy constraint, and an energy constraint can induce a water constraint)
- Trends imply that this strain will be exacerbated because of 1) growth in total demand for energy and water, primarily driven by population growth, 2) growth in per capita demand for energy and water, primarily driven by economic growth, 3) global climate change, which will distort the availability of water, and 4) policy choices, by which we are selecting more water-intensive energy and more energy-intensive water

This manuscript examines each of these points in detail and then closes with policy recommendations for how to mitigate the most vulnerable aspects of the energy-water nexus.

Energy & Water Are Interrelated

In addition to these two resources being the most essential ingredients of modern civilization, they are also highly interconnected. That is, we use water for energy and energy for water. For example, water is a direct source of energy through hydroelectric dams, which provide about 7% of total US electricity generation (268 million MWh in 2006), or approximately 3% of total energy consumption) in the US.[17]

In addition to direct power generation, water indirectly enables power generation through the cooling it provides for thermoelectric power plants, which provide more than 90% of the electricity in the US (approximately 3,500 million MWh).[17] As a result of the large cooling needs for power plants, the thermoelectric power sector is the single largest user of water in the nation, responsible for nearly half of all water withdrawals (about 200 billion gallons per day, when including seawater), ahead of even agriculture.[19] When considering only freshwater withdrawals, then the power plants and agriculture are approximately tied for first place among users.

An important feature of water use is the distinction between water *withdrawals* and *consumption.* Nearly all of the water used for power plants is returned to the source (typically a river or cooling pond), though at a different temperature and quality. As a result of these returns, power plants are responsible for a small portion of national water consumption, despite being the leading sector for withdrawals. On average across the thermoelectric power sector, 21 gallons of water are withdrawn and 0.5 gallons consumed for every kilowatt-hour of electricity that is generated (please see Table 1).[20] Note that the hydroelectric sector is listed in Table 1 for comparison. Hydroelectric dams are associated with a significant amount of water consumption for power generation primarily because the increased surface area of man-made reservoirs beyond the nominal run-of-river accelerates the evaporation rates from river basins.[21] Notably, the estimates for this increased evaporation depend significantly on regional location. Furthermore, whether all the evaporation should be attributed to power generation is not clear, as reservoirs serve multiple purposes, including water storage, flood control, and recreation.

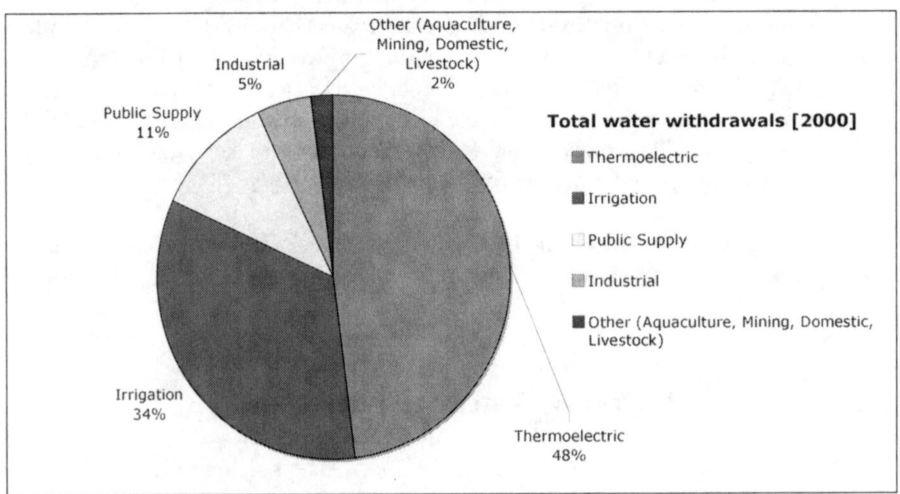

FIGURE 1. Total water withdrawals in the US by sector, including fresh and saline water.[19]

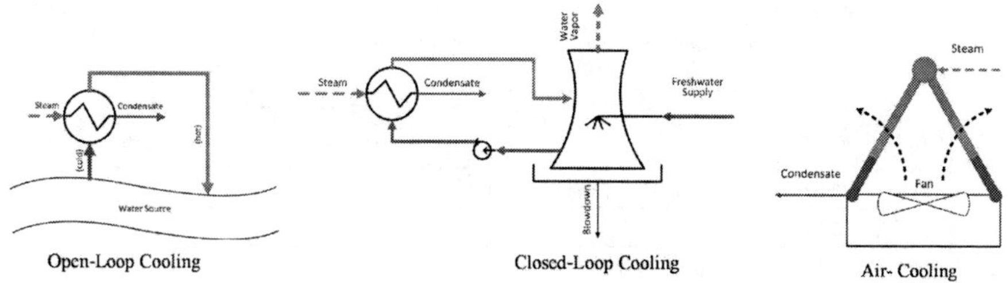

FIGURE 2. Power-plants typically use three types of cooling: open-loop, closed-loop and air cooling.[78]

The amount of water that is withdrawn and consumed by thermal power plants is driven primarily by 1) the type of fuel and power cycle that a power plant uses (for example fossil fuels or nuclear fuels with steam cycles, natural gas with combined cycle, etc.) and 2) the cooling method. There are three basic cooling methods: open-loop, closed-loop, and air-cooling (see Figure 2).[22] For figures on water withdrawal and consumption by power plants, please see Table 1 for national averages and Table 2 for a breakdown by fuel and cooling type.

Open-loop, or once-through, cooling withdraws large volumes of surface water, fresh and saline, for one-time use and returns nearly all the water to the source with little of the overall water being consumed due to evaporation. While open-loop cooling is energy efficient and low in infrastructure and operational costs, the discharged water is warmer than ambient water, causing thermal pollution, which can kill fish and harm aquatic ecosystems. Thus, environmental agencies regulate discharge temperatures, taking into account a water body's heat dissipation capacity. Closed-loop cooling requires less water withdrawal, since the water is recirculated through use of cooling towers or evaporation ponds. However, since the cooling is essentially achieved through evaporation, closed-loop cooling results in higher water consumption (See Table 2). The alternative, air-cooling, does not require water, but instead cools by use of fans that move air over a radiator similar to that in automobiles. However, power plant efficiency for air-cooling is lower, up-front capital costs are higher, and real estate requirements are sometimes larger, often making this option less attractive economically unless water resources are scarce.

Even though power plants return most of the water they withdraw, the need for such large amounts of water at the right temperature for cooling introduces vulnerabilities for the power plants. If a severe drought or heat wave reduces the availability of water or restricts its effectiveness for cooling due to heat transfer inhibitions or thermal pollution limits, the fact that the power plant consumes so little water becomes less important than the fact that it needs the water in the first place.

TABLE 1. The water used (withdrawn and/or consumed) for different types of power plants (thermal and hydro). The national average from thermo- and hydro-electric sources is that 2 gallons of water is consumed per kWh of electricity that is generated.[20]

Power Source	Withdrawals [gal/kWh]	Consumption [gal/kWh]
Thermoelectric (National Average)	21	0.5
Hydroelectric	18	18
National Average (all types)	---	2

Just as significant volumes of water are used for energy, a significant fraction of national energy is used for water. Specifically, to heat, treat and move water, sometimes across vast distances. For example, on the west coast, where snowmelt is moved across two mountain ranges to thirsty Southern Californians, a single aqueduct is the largest electricity customer in the state.[27, 28] The pithy maxim that "Water flows uphill towards wealth" reflects the amount of effort and money that is required to move water due to its density. In addition, hot water is needed for hygiene reasons such as showers, cooking and cleaning. Consequently, 9% of all residential electricity use is just for water heating alone,[29] which does not include the direct use of natural gas on-site in the residential or commercial sectors for water heating. The wastewater treatment and water sectors also consume vast amounts of energy, primarily in the form of electricity, for water supply and treatment.[17, 30] The old adage that "sewage flows downhill" partly reflects the reality of dense, solid-laden water streams flowing downward from mountaintop castles to the villages below is a nod towards the vast energy requirements that would be used to move wastewater the other direction.

TABLE 2. The water used (withdrawn and/or consumed) varies for different fuels (coal, nuclear, natural gas, solar and wind), power cycles (combined and open cycle), and cooling methods (open-loop, closed-loop and air-cooling).[78]

		Water Use and Cooling Technologies			
		Closed-Loop (cooling tower)		Open-Loop	
		Withdrawals [gal/kWh]	Consumption [gal/kWh]	Withdrawals [gal/kWh]	Consumption [gal/kWh]
Fuels & Technologies	Nuclear	1	0.7	42.5	0.4
	Solar CSP	0.8	0.8	---	---
	Coal	0.5	0.5	35	0.3
	Natural Gas *(Combined Cycle)*	0.23	0.18	13.8	0.1
	Natural Gas *(Combustion Turbine)*	Negligible	Negligible	Negligible	Negligible
	Solar PV	Negligible	Negligible	Negligible	Negligible
	Wind	Negligible	Negligible	Negligible	Negligible

The energy required to produce, treat and distribute water varies depending on the source (see Table 3). Surface water (e.g. from lakes and rivers), is the easiest and least energy-intensive to treat. However, note that for states like California, water conveyance can require as little as 0 kWh per million gallons (for gravity-fed systems) to as much as 14,000 kWh per million gallons for long-haul systems.[27, 28] Groundwater (e.g. from aquifers) requires more energy, primarily for pumping water to the surface for treatment and distribution. For example, water collection alone from a depth of 120 feet requires 540 kWh/million gallons, while a depth of 400 feet requires 2000 kWh/million gallons, in addition to treatment energy use.[31, 32, 78]

As fresh water supplies become strained, many have turned to water sources once considered unusable, including brackish groundwater and seawater.[79] While use of these water sources helps mitigate constraints on drinking water supplies, treatment of brackish groundwater and seawater requires use of advanced filtration (e.g. reverse osmosis membranes), specialty materials, and high pressure pumps for desalting. Overall, treatment of these water sources can require as much as 16,500 kWh per/Mgal,[31] or 10-12 times the energy use of standard water treatment. The theoretical minimum energy requirement for desalination using reverse osmosis systems is 2650 kWh/Mgal.[33]

TABLE 3. Water and wastewater treatment and conveyance requires vast amounts of energy. Average US figures for water production are listed below, and include the energy use for distribution. [27, 30,31,78]

Water Type	Source / Treatment Type	Energy Use (kWh/million gal)
Water	Surface Water	1,400
	Groundwater	1,800
	Brackish Groundwater	3,900-9,750
	Seawater	9,780-16,500
Wastewater	Trickling Filter	955
	Activated Sludge	1,300
	Advanced Treatment without Nitrification	1,500
	Advanced Treatment with Nitrification	1,900

Wastewater treatment also requires large amounts of energy. Wastewater treatment in the US is primarily conducted by the over 16,000 Publicly Owned Treatment Works (POTWs).[34] POTWs are often the largest local consumer of energy, generally requiring 1 to 3% of a community's total energy use, and a non-negligible fraction of national energy consumption[17, 35]. Stricter discharge regulations in the United States led to implementation of more energy-intensive treatment technologies. Trickling filter treatment, which uses a biologically active substrate for aerobic treatment, is a reasonably passive system, consuming over 950 kWh/Mgal on average.[31, 78] Diffused air aeration as part of activated sludge processing is a more energy intensive form of wastewater treatment around, requiring 1,300 kWh/Mgal due to blowers and gas transfer equipment.[31, 78] More advanced wastewater treatment, utilizing filtration and the option of nitrification, requires 1,500-1,900 kWh/Mgal.[31, 78] In fact, more advanced sludge treatment and processing can consume energy in the range of 30-80% of total wastewater plant energy use.[36] POTWs that treat wastewater sludge through anaerobic digestion can also produce energy through the creation of methane-rich biogas, a renewable fuel that can be used to generate up to 50% of the POTW's electricity needs[30,37].

Looking at just the public supply of water, which is primarily for the residential and commercial sectors, approximately 4.7% of the nation's annual primary energy and 6.1% of national electricity consumption, respectively, is required for water.[80] End-use energy requirements associated with water for municipal, industrial, and self-supplied sectors (i.e. agriculture, thermoelectric, mining, etc.) represents another 5% or more of national energy consumption.

Globally the Energy-Water Relationship Is Already Under Strain

The interrelationship of energy and water and the strains on both resources manifests itself in tough choices at the local level. For example, low water levels in hydroelectric reservoirs can force power plants to turn off. Fifty-eight percent of US hydroelectricity is generated in California, Oregon, and Washington alone, making the power supply vulnerable to regional changes in water availability. Though hydroelectric power is attractive for many reasons, it is least reliable during droughts when the need to use water for other purposes (e.g. for drinking, irrigation, etc.) may take precedence over hydroelectricity. For example, without a change in water usage patterns, Lakes Mead and Powell along the Colorado River, which are used for hydroelectric power and municipal supply, are projected to have a 50% chance of running dry by 2021.[18] Outside Las Vegas, Lake Mead has dropped 100 feet in six years, making people worry whether the massive hydroelectric turbines inside Hoover Dam would no longer turn and the city would have to ration water use.[52-54] Cities in Uruguay must choose whether they want the water in their reservoir to be used for drinking or electricity.[51]

The problem is not just limited to hydroelectric reservoirs. Since thermoelectric power plants also require vast amounts of water, they are vulnerable to droughts or heat waves restricting their output. Heat waves in France in 2003 also caused power plants to draw down their output because of limits on rejection temperatures that are imposed for environmental reasons.[6, 24, 25] That severe heat wave and drought killed approximately 15,000 people and created river water levels that were too hot for effective power plant cooling. As a consequence, many nuclear power plants had to operate at much reduced capacity and an environmental exemption was enacted to allow the rejection temperatures of cooling water from power plants to exceed prior limits.[24, 25] At the same time, 20% of hydropower capacity was not available because of low river levels.[6] That is, just as demand for electricity was spiking for air conditioning in response to the heat, supplies were being cut back. The dilemma has also shown up in the United States. For example power plants in Atlanta during the winter 2008 drought were within daof shutting off because the vast amounts of cooling water were at risk from diversion for other priorities such as municipal use for drinking water.[23, 55, 56]

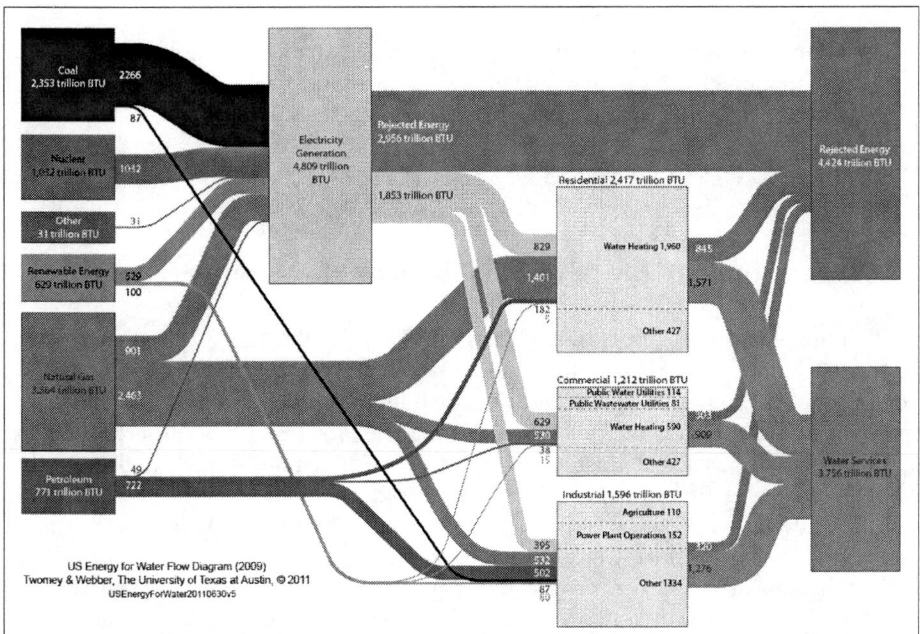

FIGURE 3. This diagram summarizes the energy flows for the public water supply system in the United States. Fuels (on the left) are used directly and indirectly via electricity generation for different purposes (on the right). The thickness of the flows is proportional to the amount of energy consumed. About 60% of the total energy consumption is lost as waste heat. Only energy consumption related to the conveyance, treatment, distribution and heating (in the commercial and residential sectors) and public water and wastewater treatment distributed in the US public water supply is included. Self-supplied sectors, including agriculture and industry are not included.[80]

While water limitations can restrict energy, energy limitations can also restrict water. For example, power outages (due to storms or intentional acts) at water and wastewater treatment plants puts the water system at risk of disruptions due to energy shortages. And, this tradeoff becomes a strategic question for some countries. Saudi Arabia uses a lot of its own best products—crude oil and natural gas—to get what it doesn't have—freshwater—facing the choice about whether it's better to sell the energy resources at record prices or have enough freshwater available to maintain municipal needs.[59, 60]

TRENDS IMPLY STRAIN IN THE ENERGY & WATER RELATIONSHIP WILL BE EXACERBATED

While there is already strain in the energy-water relationship, trends imply that this strain will be exacerbated because of:

1) Growth in total demand for energy and water, primarily driven by population growth,

2) Growth in per capita demand for energy and water, primarily driven by economic growth,
3) Global climate change, which will distort the availability of water, and
4) Policy choices, by which we are selecting more water-intensive energy and more energy-intensive water.

These different trends are each discussed below.

Trends Imply Growth In Total Demand for Energy & Water

Though the global fuel mix is quite diverse in total, fossil fuels (oil, coal and natural gas) satisfy more than 80% of the world's needs for primary energy resources.[38] Total energy consumption, including traditional biomass such as wood and dung, was approximately 500 quads in 2008.[38, 39] Estimates for fossil fuel resources suggest that they should last for at least a century more, with some estimates indicating that the total resource base can last for a few centuries. The total resource of water in the world, while quite vast, is primarily undrinkable, with 97.5% in the form of saltwater.[40] Of the freshwater resources, more than two-thirds is held in ice and permanent snow cover, with most of the rest in groundwater (e.g. underground aquifers), leaving very little that is easily accessible at the surface.[40] Surface water is also mostly renewable from rainfall, though groundwater resources are often finite and slowly replenished. Total water withdrawals and consumption in 2000 were approximately 1.05 and 0.52 quadrillion gallons, respectively.[41]

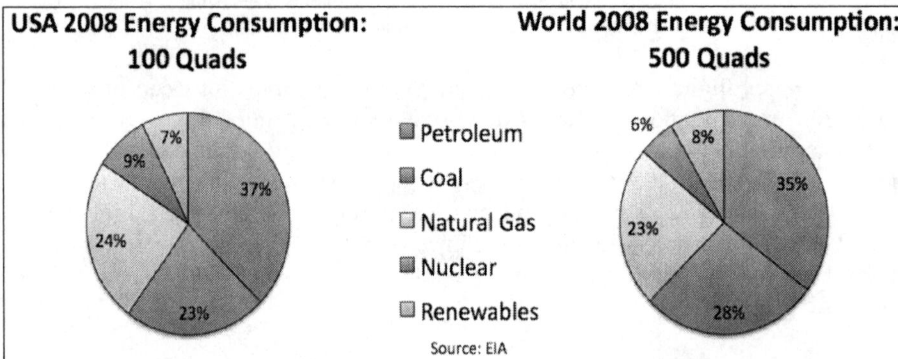

FIGURE 4. The global energy mix is diverse, though fossil fuels satisfy more than 80% of the world's primary energy resources.[38]

TABLE 4. Global water resources are primarily comprised of saltwater (97.5%). Of the 2.5% of the world's water resources that are freshwater, more than two-thirds is in ice or permanent snow cover, with most of the remainder underground (see the bottom part of this table). Relatively little water is easily accessible on the surface. (Table constructed from Figure 2.1 in [40]).

Global Water Resources	Percentage
Saltwater	*97.5%*
Freshwater	*2.5%*
Global Freshwater Resources	**Percentage**
Permafrost	*0.07%*
Rivers, lakes, and swamps	*0.34%*
Groundwater	*30.7%*
Ice and permanent snow cover	*68.7%*

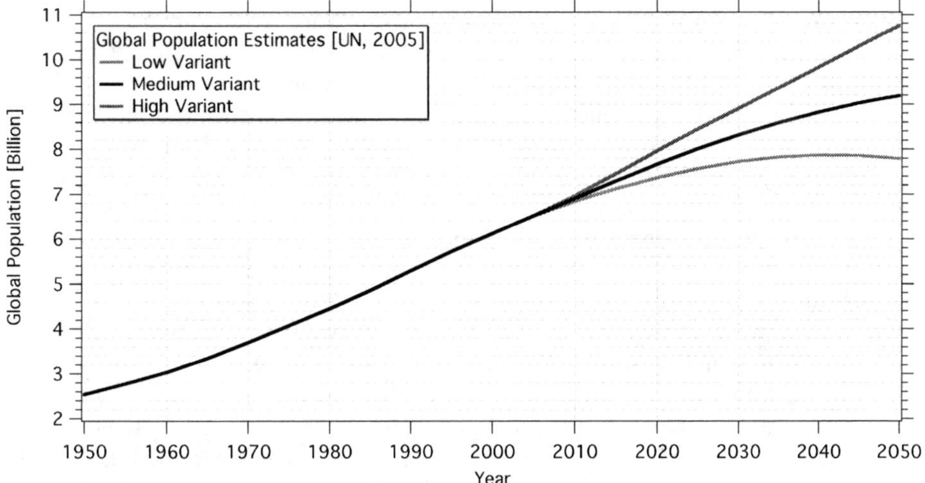

FIGURE 5. Global population projections by the United Nations show growth out to at least 2040 for low-, medium-, and high-growth variants.[42, 43]

Most projections show an increase in demand for energy, predominantly driven by both population (see Figure 5) and economic growth. The United Nations (UN) makes several projections for global population (with low-, medium- and high-growth variants), all showing population growth out to at least 2050, at which time population could potentially decrease.[42, 43] The International Energy Agency (IEA), assumes for its projections that 1) global population will grow 1% per year on average from 6.4 billion in 2004 to 8.1 billion in 2030, and 2) that economic growth will take place at an average of 3.4% per year over the same period.[39] The IEA interprets these growth trends to yield a growth in global primary energy demand of 70% between 2004 and 2030, without a very significant shift in the basic makeup in the fuel mix, despite recent policy prioritization for biofuels and other renewable sources.[39] Similarly, this population growth should lead to increases in global water demand.[12]

Trends Imply Growth In Per Capita Demand for Energy & Water

As the world population grows, the global demand for energy and water are increasing in order to meet the subsistence and lifestyle needs for its inhabitants. On top of this fundamental upward trend in demand is a more alarming trend: the per capita demand for energy and water are also growing. Even though the developed world is looking for ways to conserve energy and water, the developing world is in a phase of rapidly accumulating wealth. One consequence of this affluence is a desire for better transportation, a nicer lifestyle, more meat-intensive diets, and a robust economy. The combined effect is that the demand in developing countries is increasing rapidly for liquid fuels, electricity, and water (for industrial processes, higher-protein diets, and pretty yards and gardens). This statistic means the growth in demand for energy and water is outpacing the growth in population. For a world that already has these resources under strain, accelerating demand might have far-reaching impacts.

Despite population growth estimates of only 19 to 37% [42, 43] from 2004-2030, as noted before, the IEA projects energy demand increases that are much greater, at approximately 70% [39], indicating that annual per capita energy use globally will increase from roughly 1.7 to 2.1 tons of oil equivalent. About half of the growth in demand is projected to be from the power sector [39], revealing that electricity is a preferred form of energy for those who can afford it. Because of the power sector's water intensity, the new demand for electricity will likely lead to increased demand for water withdrawals as well. In its projections, the IEA's *World Energy Outlook 2006* also considers an alternative policy scenario, for which energy growth is lower than the nominal reference projection for energy growth. Even for this alternative case, both the absolute and per-capita energy demand are expected to increase between 2004 and 2030.[39] Even in the United States, where energy-intensive manufacturing has shrunk relatively, the per capita energy use is projected to increase by the EIA.[17]

One of the drivers of increasing energy use per capita is the expectation that people demand better environmental conditions as their incomes rise (that is, they move along the downward-sloping part of the environmental Kuznets curve).[44] This phenomenon is illustrated here for the case of wastewater treatment. The EIA projected in 2006 that the US population would increase about 70 million over the next 25 years, generating a commensurate increase in the amount of wastewater that will need to be treated and POTWs energy use.[32] A US Environmental Protection Agency (EPA) study published in 2002 found that capital needs (collection, pumping and treatment facilities) for wastewater over a twenty year period from 2000 – 2019 would range from $331 billion to $440 billion.[45] Increasingly, POTWs' discharge permits require the removal of specific contaminants (such as nitrogen) not removed by conventional treatment technologies, thus requiring the installation of advanced wastewater treatment technologies.[46] Additionally, concerns about chlorine disinfection byproducts are forcing many POTWs to install new disinfection technology.[47] Advanced wastewater treatment is generally more energy intensive than standard wastewater treatment and so the trend towards these higher treatment standards will likely increase the unit energy needs of wastewater treatment in the future.[47] Because of the growing US population, stricter discharge requirements and aging wastewater infrastructure, the Energy Policy Research Institute (EPRI) projects that national POTWs energy use will increase in the future.[47] However, according to the EIA, it is possible that the introduction of greater energy efficiency at POTWs will offset the expected increases in energy intensity for stricter treatment standards, limiting the projected growth in electricity use at POTWs to approximately 15 percent from 2005 to 2030, which is less that what would have been expected for the nominal 0.8% average annual population growth.[17] The higher per capita energy expenditures for wastewater treatment in order to achieve stricter environmental standards is a scenario likely to be repeated in analogous ways throughout all the societies that are achieving affluence; that is, as nations get richer, they will demand more energy.

While the per capita energy demand is clearly on an upward trend, the situation with water is less clear. At the same time that human population has tripled in the last seventy years, water withdrawals have increased six-fold.[40, 48] Water withdrawals might have increased by a factor of forty in the last three centuries.[40, 49] Furthermore, in the present-day situation, high-income countries have annual per capita freshwater withdrawals in excess of 600 cubic meters, while low-income countries have annual per capita withdrawals of less than half with less than 300 cubic meters and middle-income countries have withdrawals rates that are in between the two.[40] As middle- and low-income countries become wealthier, it is reasonable to expect that their per capita withdrawals will rise to match those of high-income countries, driving up the global per capita freshwater withdrawals. One driving force for increased per capita withdrawals is the expectation that along with economic growth will come growth in meat consumption, which will increase the water demand for fodder production.[12]

It is important to note that not all projections show increasing per capita water withdrawals. For example, water use estimates from a joint project in 1999 between UNESCO and the State Hydrological Institute in St. Petersburg yields a possible global peak in per capita water withdrawals and consumption near 1980, if that water use is normalized by UN population estimates.[41, 42, 50]. However, because those estimates are at low temporal resolution and are not updated with data that include the recent effects of globalization, it is difficult to project their forward-looking implications with much confidence.

Global Climate Change and Trends Will Intensify This Strain

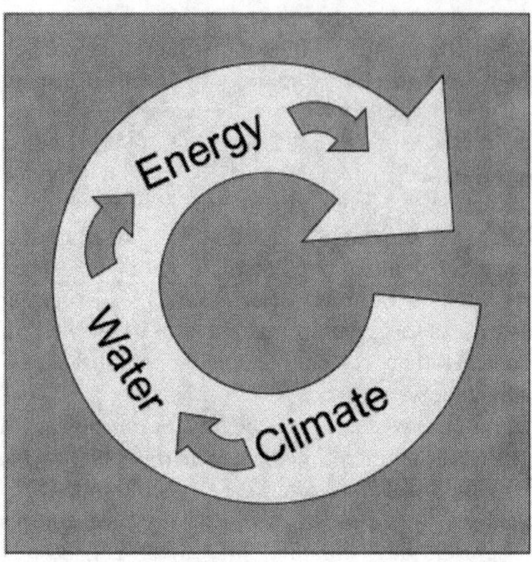

FIGURE 6. The energy-climate-water cycle creates a self-reinforcing challenge. (Image based on suggestion of Jane Long at Lawrence Livermore National Laboratory.)

One important aspect of climate change is that the water systems are likely to be hit hardest and will be a leading indicator of temperature changes. Popular discussion of climate effects often focuses on the risks of rising sea levels, but it is the risk of changes to the hydrological cycles that might be of greater concern. These effects are hard to predict, but it is expected that higher temperatures could induce several consequences, including turning some snowfall into rainfall, moving the snowmelt season earlier (and thereby affected spring water flows), increasing intermittency and intensity of precipitation, affecting water quality and raising the risks of floods and droughts.[12, 61] In addition, the sea level rises can cause contamination of groundwater aquifers with saline water near the coasts, potentially affecting nearly half of the world's population.[12] These challenges can be fixed with greater energy expenditures for mining deeper water, moving it farther, treating water to make it drinkable, or storing it for longer periods of time. With a typical energy mix over the next few decades, these energy expenditures release greenhouse gases, which intensify the hydrological cycle further, compounding the problem in a positive feedback loop (see Figure 4).

POLICY CHOICES ARE MOVING TOWARDS MORE ENERGY-INTENSIVE WATER AND MORE WATER-INTENSIVE ENERGY

On top of the prior three trends is a policy-driven movement towards more energy-intensive water and water-intensive energy.

Trends Imply Growing Energy-Intensity of Water

All the aforementioned trends also indicate a movement towards water-production methods that are increasingly energy-intensive. We are moving towards more energy-intensive water because of a push by many municipalities for new supplies of water from sources that are farther away and lower quality, and thereby require more energy to get them to the right quality and location.

Because of growing environmental concerns, our standards for water treatment get stricter with time, as we expect our water to be ever cleaner, and so the amount of energy we spend per gallon will only increase. And, like water, the treatment standards are getting stricter with time as we wish to treat the wastewater to a higher degree of cleanliness. Prior analysis projects that unit electricity consumption for water treatment has increased at a compound rate of 0.8% per year, with no obvious reason why the trend would stop.[47] In the US, aging wastewater infrastructure will also tend to increase unit electricity use due to age-related losses while other factors (for example, replacing older equipment with more efficient new equipment and processes and larger treatment plants with higher energy economies of scale) will tend to decrease unit energy consumption, but not enough to offset the energy needs of higher level treatment.[47]

In addition to treating water to higher standards of cleanliness, societies are also going to greater lengths to bring freshwater from its sources to dense urban areas. These efforts include digging to ever-deeper underground reservoirs, or by moving water via massive long-haul projects.[79] For example, China is implementing a gargantuan water transport plan named the "South-North Water Transfer Scheme", which is an order of magnitude larger than California's famed aqueduct and will move water from three river basins in the wet southern part of China to the dry northern parts.[13] Two of these routes are more than 1000 km long,[13] therefore representing substantial investments in energy for transport. Similar efforts are happening in Texas, where private investors are proposing a project to move groundwater from the Ogallala Aquifer (one of the world's largest) in the panhandle hundreds of miles across the state of Texas to the thirsty municipalities in the Dallas/Fort Worth metroplex.[62]

On April 12, 1961, President Kennedy said, "If we could ever competitively—at a cheap rate—get fresh water from salt water that would be in the long-range interest of humanity, and would really dwarf any other scientific accomplishment."[63] A few months later, he signed a bill to set the US on a research course to seek a breakthrough in desalination.[64] Since that time, global desalination capacity has enjoyed a decades-long steep upward trend.[65, 66] It's unlikely this trend will end soon given the other trends noted above. While desalination is traditionally associated with the Middle East, where energy resources are plentiful but water is scarce, cities like El

Paso and San Diego are trying to develop desalination plants to get fresh water either from their nearby saline aquifers or coasts. California alone has approximately twenty seawater desalination plants that are proposed.[66] The steep market penetration rates for desalination are particularly relevant for the adoption of new membrane-based technologies.[66, 67] While these membrane-based reverse-osmosis approaches are less energy-intensive than thermal desalination, they still require much more energy than traditional freshwater production from surface sources.

Trends Imply Growing Water-intensity of Energy

At the same time, for a variety of economic, security and environmental reasons, including the desire to produce a higher proportion of our energy from domestic sources and to decarbonize our energy system, many of our preferred energy choices are more water-intensive. For example, nuclear energy is produced domestically, but is also more water-intensive than other forms of power generation. The move towards more water-intensive energy is especially relevant for transportation fuels such as unconventional fossil fuels (oil shale, coal-to-liquids, gas-to-liquids, tar sands), electricity, hydrogen, and biofuels, all of which can require significantly more water to produce than gasoline (depending on how you produce them). It is important to note that the push for renewable electricity also includes solar photovoltaics (PV) and wind power, which require very little water, and so not all future energy choices are worse from a water-perspective.

Almost all unconventional fossil fuels are more water-intensive than domestic, conventional gasoline production. While gasoline might require a few gallons of water for every gallon of fuel that is produced, the unconventional fossil sources are typically a few times more water-intensive. Electricity for plug-in hybrid electric vehicles (PHEVs) or electric vehicles (EVs) are appealing because they are clean at the vehicle's end-use and it's easier to scrub emissions at hundreds of smokestacks millions of tailpipes. However, most powerplants use a lot of cooling water, and consequently electricity can also be about twice as water-intensive than gasoline per mile traveled if the electricity is generated from the standard U.S. grid. If that electricity is generated from wind or other water-free sources, then it will be less water-consumptive than gasoline. Though unconventional fossil fuels and electricity are all potentially more water-intensive than conventional gasoline by a factor of 2-5, biofuels are particularly water-intensive. Growing biofuels consumes approximately 1000 gallons of water for every gallon of fuel that is produced. Sometimes this water is provided naturally from rainfall. However, for a non-trivial and growing proportion of our biofuels production, that water is provided by irrigation.

Today's sustained higher energy prices and emerging political consensus about climate change and energy security have brought fossil fuels into new scrutiny. Consequently, the US in particular is seeking an energy solution that is domestically sourced (addressing some of the national security concerns), abundant (addressing the concerns about resource depletion), and less carbon-intensive (addressing our concerns about climate change). Because the amount of oil imported by the US is approximately the same as what is needed by the transportation sector,[17] and because this sector is a major contributor to carbon emissions, it is on the short-list of targets for change by policymakers, innovators, and entrepreneurs.

Among the options are unconventional fossil fuels (including compressed natural gas, coal-to-liquids, tar sands and oil shale), hydrogen, biofuels and electricity. While these options have their merits, most production methods for these options are more water-intensive than conventional petroleum-based gasoline and diesel (please see Figure 5 or the recent publication by King & Webber for more details).[68] Oil shale and tar sands both are very water-intensive for their production. For example, in-situ oil shale production might use vast amounts of electric power to heat the bitumen underground; that electric power will likely need water cooling. Tar sands are produced through the use of steam injection to reduce the viscosity of the tars. While coal production is not particularly water-intensive, creating liquid fuels from coal using Fischer-Tropsch processes requires water as a process material. Hydrogen can also be very water-intensive if produced via electrolysis.[20] However, if hydrogen is produced from non-irrigated biomass resources or via reforming of fossil fuels, its water-intensity is on par with conventional gasoline production and use.

Electricity is particularly appealing as a transportation fuel for a variety of reasons. By running light-duty vehicles on electricity, it is possible to shift a significant portion of our transportation system away from imported oil to mostly domestic fuels, namely coal, gas, nuclear, wind, solar, and hydroelectric power. Moreover, from an environmental mitigation standpoint, it's easier to manage the emissions from 1500 power plants than from hundreds of millions of tailpipes. It is also convenient that the electrical infrastructure and capacity are already in place. However, as noted before, the power sector is very water intensive, and creating more demand for electricity will only exacerbate those effects.

Biofuels are also very popular because they are grown domestically and they consume CO_2 during photosynthesis. They also hold the potential for displacing fossil fuels, though many have suggested that first generation biofuels (e.g. corn-based ethanol) provide about as much energy as they require for their growth.[69] The real challenge for biofuels is their water intensity, though there are also some important water quality impacts.[81] Recent analysis indicates that irrigated biofuels can require over one hundred gallons of water for every mile traveled in a light-duty vehicle [68, 70], approximately one thousand times more water per mile than conventional gasoline. When scaling up this kind of production to prepare for approximately 140 billion gallons per year of gasoline equivalent energy and trillions of miles traveled, water can become a critical limiting factor. In fact, some municipalities are already in a fight over water resources with the biofuels industry,[71] which might only grow as the industry expands to meet governmental mandates.

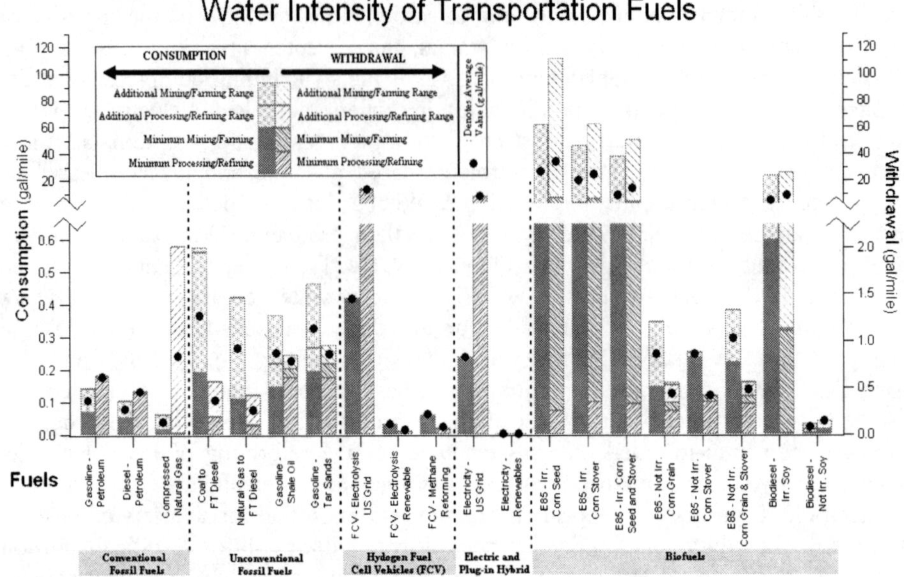

FIGURE 7. The water-intensity of different fuels in gallons of water require per mile traveled, show great variation from irrigated biofuels (at the high end, withdrawing and consuming more than 20 gallons of water per mile) to electricity from wind or solar resources (at the low end, requiring approximately 0 gallons of water per mile). Water consumption (left stacked bars read on left axis) and withdrawal (right stacked bars read on right axis) in gallons of water per mile (gal/mile) for various fuels for light duty vehicles. Water use from mining and farming is designated differently from that used for processing and refining. Where a range of values exists (e.g. different irrigation amounts in different states), a minimum value is listed with an 'additional range'. Otherwise, the values plotted are considered average values. Irr. = irrigated, Not Irr. = not irrigated, FT = Fischer Tropsch, FCV = fuel cell vehicle, US Grid = electricity from average US grid mix, and Renewables = renewable electricity generated without consumption or withdrawal of water (e.g. wind and photovoltaic solar panels).[68]

Recent energy legislation in the US and "energy independence" scenarios envisioned by the US DoE incorporate a rapid shift towards these more water-intensive fuels. For example, the Energy Independence and Security Act of 2007 (EISA 2007) includes requirements for corporate average fuel economy (CAFE) improvements and increased renewable fuel use.[72] These renewable fuels include irrigated biofuels such as corn-based ethanol. For example, the EISA 2007 mandates the production of biofuels increases to 36 billion gallons per year (Bgal/yr) by 2022, of which up to 15 Bgal/yr can be ethanol from corn grain.[72] A recent study by King, *et al* [74], calculated the water consumption and withdrawal that would result from the implementation of EISA 2007.[73] King's study considered light duty vehicle (LDV) travel (in 2005 LDV travel totaled 2.7 trillion miles driven by cars, pickup trucks, vans, and SUVs), and used those two projections of future fuel usage and previously published calculations of water usage rates in gallons per mile (gal H_2O/mile) driven using various fuels.[68] Converting projected fuel usage (units of fuel) into miles driven and then multiplying by water usage rates per mile yielded an estimate for the total water to be consumed and withdrawn for driving LDVs looking forward out to 2030. This analysis shows a considerable increase in water consumption and withdrawal for all analyzed future fuel scenarios. In 2005 it is estimated that 1,440 and 2,800 billion gallons of water were consumed and withdrawn, respectively, for LDV travel. By 2030 it is estimated that for both cases analyzed, approximately 2,600-2,700 billion gallons of water per year, 9% of estimated US fresh water consumption, will be consumed in the farming, mining, and refining of LDV fuels. Water withdrawal for LDVs in 2030 will be 4,700-6,500 Bgal/yr. Agricultural irrigation heavily dominates both water consumption and withdrawal accounting for 80%-85% and 45%-65% of the 2030 total for each category, respectively. Thus, increasing fuel usage, along with a higher diversity of fuels, generally causes an increase in water usage. The amount of this increase depends heavily upon which alternative fuels the US will produce, but the total effect in the US can be an additional few trillion gallons of water consumption annually.

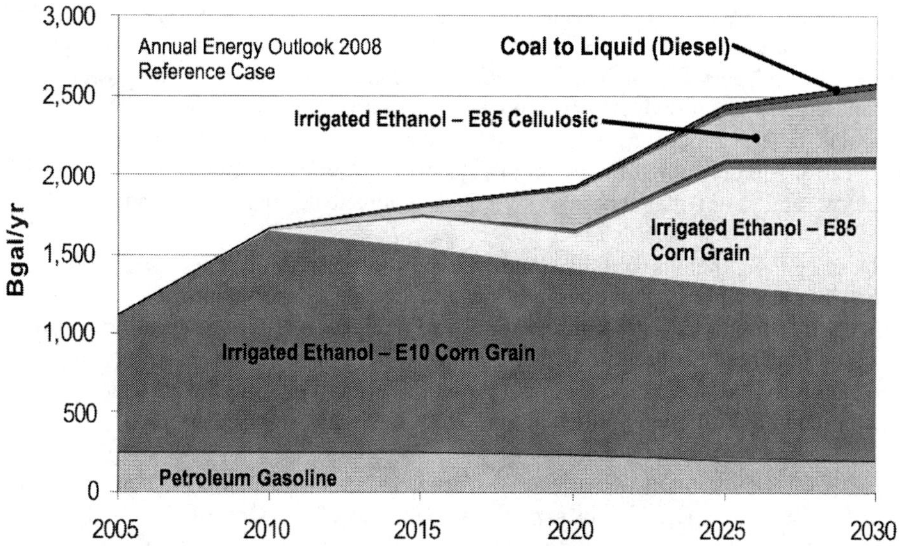

FIGURE 8. The water needs for producing transportation fuels are projected to grow dramatically (by trillions of gallons per year) primarily because of biofuels mandates.[74]

CONCLUSIONS AND RECOMMENDATIONS

The nexus of water and energy is fundamental to society, is intertwined in many ways, is under strain today, and is projected to get worse as worldwide demand for both increases more quickly than population and the effects of climate change manifest themselves. While the energy-water relationship might appear intractable, there are many opportunities to mitigate the worst aspects through new technologies, new concepts for how to reuse water effectively, new markets that put a price on water, and a recognition that conserving water and conserving energy are synonymous. Though the world's water situation appears dire for many reasons, it also presents an opportunity and we have many tools available to tackle the problem. And, in particular, policy engagement is warranted.

Because there are many rivers, watersheds, basins and aquifers that span several states and/or countries, there is a need for federal engagement on energy-water issues. Unfortunately, there are some policy pitfalls at the energy-water nexus. For example, energy and water policymaking are disaggregated. The funding and oversight mechanisms are separate, and there are a multitude of agencies, committees, and so forth, none of which have clear authority. It is not unusual for water planners to assume they have all the energy they need and for energy planners to assume they have the water they need. If their assumptions break down, it could cause significant problems. In addition, the hierarchy of policymaking is dissimilar. Energy policy is formulated in a top-down approach, with powerful federal energy agencies, while water policy is formulated in a bottom-up approach, with powerful local and state water agencies. Furthermore, the data on water quantity are sparse, error-prone, and inconsistent. The United States Geological Survey (USGS) budgets for collecting data on water use have been cut, meaning that their latest published surveys are anywhere from 5 to 15 years out of date. National databases of water use for power plants contain errors, possibly due to differences in the units, format and definitions between state and federal reporting requirements. For example, the definitions for water use, withdrawal and consumption are not always clear. And, water planners in the east use "gallons" and water planners in the west use "acre-feet," introducing additional risk for confusion or mistakes.

Despite the potential pitfalls, there are policy opportunities at the energy-water nexus. For example, water conservation and energy conservation are synonymous. Policies that promote water conservation also achieve energy conservation. Policies that promote energy conservation also achieve water conservation.

Thankfully, the federal government has some effective policy levers at its disposal. I recommend the following policy actions for the energy-water nexus:

1. Collect, maintain and make available accurate, updated and comprehensive water data, possibly through the USGS and EIA. One of the challenges of researching this nexus is the lack of suitable water data that are available. The Department of Energy's Energy Information Administration maintains an extensive database of accurate, up-to-date and comprehensive information on energy production, consumption, trade, and price available with temporal and geographic resolution and standardized units. Unfortunately, there is no equivalent set of data for water. In fact, the National Academy of Sciences complained explicitly that water data and monitoring budgets have been cut.[67] Resuming prior national efforts to collect relevant water data and broaden the scope of monitoring capabilities will enable thoughtful analysis and will illuminate key insights into this problem, after which policymakers can respond. Without good data about water reserves, flows and use, it will be difficult for analysts to assess the situation and for policymakers to respond, and this hurdle remains one of top barriers to effective action.

2. Invest heavily in water-related R&D to match recent increases in energy-related R&D. R&D investments are an excellent policy option for the federal government because state/local governments and industry usually are not in a position to adequately invest in research. Consequently, the amount of R&D in the water sector is much lower than for other sectors such as pharmaceuticals, technology, or energy. Furthermore, since energy-related R&D is expected to go through a surge in funding, it would be appropriate from the perspective of the energy-water nexus to raise water-related R&D in a commensurate way. Topics for R&D include low-energy water treatment, novel approaches to desalination, remote leak detectors for water infrastructure, and air-cooling systems for power plants. In addition, DoE's R&D program for biofuels should emphasize feedstocks such as cellulosic sources or algae that do note require freshwater irrigation.

3. Develop regional water plans that consider increased demands for electricity, and regional energy plans that consider increased demands for water. For example, the rise of biofuels or electricity as fuel substitutes will have very different regional impacts. Biofuels will affect water use in the farm belt, whereas electric vehicles will affect water use at power plants near major population centers.

4. Encourage resource substitution to fuels that have water, emissions and security benefits. Some fuel sources such as natural gas, wind, and solar PV are domestic, need much less water, and reduce emissions of pollutants and carbon.

5. Support the use of reclaimed water for irrigation and process cooling. Using reclaimed water for powerplants, industry, and agriculture can spare a significant amount of energy and cost. However there are financing, regulatory and permitting hurdles in place that restrict this option. Reuse water can also reduce the demand for freshwater, for example by using reclaimed water (e.g. treated wastewater) for power plant cooling or other industrial uses along with irrigation. There is precedent for such action, as "nearly 80% of water used in the industrial sector in Japan is currently recycled."[12] While most cities would refrain from using treated wastewater as a source of drinking water, this avenue is also available and has been implemented in water-scarce Singapore and the International Space Station without ill-effects. Consequently, it is recommended that municipalities affected by water-scarcity should move aggressively towards the use of reclaimed water.

6. Support the use of dry and hybrid wet-dry cooling at powerplants. Not all powerplants need wet cooling all the time. Finding ways to help plants upgrade their cooling to less water-intensive versions can spare significant volumes of water to meet public supply or in-stream flow requirements.

7. Establish strict standards in building codes for water efficiency. Building codes should include revised standards for low-flow appliances, water-heating efficiency, purple-piping for reclaimed water, rain barrels and so forth in order to reduce both water and energy consumption.

8. Invest aggressively in conservation. Water conservation can be a cost-effective way to save energy, and energy conservation can be a cost-effective way to save water. Therefore, conservation has cross-cutting benefits.

Conservation is one of the easiest and most cost-effective approaches to reducing both water and energy use, especially since saving water is synonymous with saving energy, and vice-versa.[28, 77] While conservation will not solve all of our energy and water problems nationally, it will buy us some time while new solutions are developed.

ACKNOWLEDGMENTS

This chapter was written based on a lecture that the author first gave in 2006, and which has evolved dramatically over the years based on the extensive research contributions from a variety of people, especially Dr. Carey King, Mrs. Ashlynn Stillwell, Ms. Kelly Twomey, Dr. Ian Duncan, Ms. Amy Hardberger and Mr. David Hoppock. The author would like to acknowledge them for their significant contributions to this body of work, including prior publications and facts and figures that were incorporated into this manuscript. The research was supported by the Center for International Energy and Environmental Policy at the University of Texas.

REFERENCES

1. *National Geographic*, 1993, 184.
2. *National Geographic*, 1981.
3. *Nature*, 2008, 452, 253-386.
4. *Science*, 2007, 315, 721-896.
5. *Science*, 2006, 313, 1005-1184.
6. M. Hightower and S. A. Pierce, *Nature*, 2008, 452.
7. C. W. King, A. S. Holman and M. E. Webber, *Nature Geoscience*, 2008, 1.
8. H. Fountain, in *New York Times*, New York, NY, Edition edn., 2008.
9. A. Price, in *Austin American-Statesman*, Austin, Texas, Edition edn., 2008.
10. *Subcommittee on Energy and Environment, Committee on Science and Technology*, US House of Representatives, and *Committee on Energy and Natural Resources*, US Senate, Washington, DC, 2007.
11. *Water for Life* ISBN 92 4 156293 5, World Health Organization, 2005.
12. T. Oki and S. Kanae, *Science*, 2006, 313, 1067-1072.
13. R. Stone and H. Jia, *Science*, 2006, 313, 1034-1037.
14. D. Kennedy and B. Hanson, *Science*, 2006, 313.
15. R. P. Schwarzenbach, B. I. Escher, K. Fenner, T. B. Hofstetter, C. A. Johnson, U. v. Gunten and B. Wehrli, *Science*, 2006, 313, 1072-1077.
16. A. Fenwick, *Science*, 2006, 313, 1077-1081.
17. USDOE, *Annual Energy Outlook 2007: With Projections to 2030*, U.S. Department of Energy, Energy Information Administration, 2007.

18. *Lake Mead Could Be Dry by 2021*, http://scrippsnews.ucsd.edu/Releases/?releaseID=876.
19. S. S. Hutson, N. L. Barber, J. F. Kenny, K. S. Linsey, D. S. Lumia and M. A. Maupin, *Estimated Use of Water in the United States in 2000*, U.S. Geological Survey, Reston, VA, 2004.
20. M. E. Webber, *Environmental Research Letters*, 2007, 2, 7.
21. P. Torcellini, N. Long and R. Judkoff, *Consumptive Water Use for U.S. Power Production* NREL/TP-550-33905, National Renewable Energy Laboratory, U.S. Department of Energy, Golden, CO, 2003.
22. *Comparison of Alternate Cooling Technologies for California Power Plants: Economic, Environmental and Other Tradeoffs*, California Energy Commission, 2002.
23. in *The Associated Press*, Editon edn., January 23, 2008.
24. M. Poumadère, C. Mays, S. L. Mer and R. Blong, *Risk Analysis*, 2005, 25, 1483-1494.
25. P. Lagadec, *Journal of Contingencies and Crisis Management*, 2004, 12, 160-169.
26. *Water & Sustainability (Volume 3): U.S. Water Consumption for Power Production - The Next Half Century*, Electric Power Research Institute, 2002.
27. *California's Water-Energy Relationship: Final Staff Report*, California Energy Commission, 2005.
28. R. Cohen, B. Nelson and G. Wolff, *Energy Down the Drain: The Hidden Costs of California's Water Supply*, Natural Resources Defense Council, 2004.
29. USDOE, *End-Use Consumption of Electricity 2001*, U.S. Department of Energy, Energy Information Administration, 2001.
30. A.S. Stillwell, D.C. Hoppock and M.E. Webber, "Energy Recovery from Wastewater Treatment Plants in the United States: A Case Study of the Energy-Water Nexus," *Sustainability* (special issue Energy Policy and Sustainability) 2010, 2(4), 945-962.
31. *Water & Sustainability (Volume 4): U.S. Electricity Consumption for Water Supply & Treatment - The Next Half Century*, Electric Power Research Institute, 2002.
32. USDOE, *Energy Demands on Water Resources: Report to Congress on the Interdependency of Energy and Water*, United States Department of Energy, 2006.
33. M. S. Shannon, P. W. Bohn, M. Elimelech, J. G. Georgiadis, B. J. Marinas and A. M. Mayes, *Nature*, 2008, 452, 301-310.
34. U.S. Environmental Protection Agency Office of Wastewater Management, *Primer for Municipal Wastewater Treatment Systems* EPA 832-R-04-001, U.S. Environmental Protection Agency, Washington D.C., 2004.
35. Water Environment Federation, *Energy Conservation in Wastewater Treatment Facilities Manual of Practice*, Water Environment Federation, Alexandria, VA., 1997.
36. *U.S. Wastewater Treatment Factsheet*, http://css.snre.umich.edu/facts/factsheets.html, Accessed March 9, 2008, 2008.
37. R. B. Seiger and D. Whitlock, CHP and Bioenergy for Landfills and Wastewater Treatment Plants, Salt Lake City, UT, 2005.
38. J. Goldemberg, *Science*, 2007, 315.
39. IEA, *World Energy Outlook 2010*, International Energy Agency, 2010.
40. J. Boberg, *Liquid Assets: How Demographic Changes and Water Management Policies Affect Freshwater Resources*, RAND Corporation, Santa Monica, CA, 2005.
41. I. Shiklomanov, *World Water Resources and Their use*, United Nations Educational Scientific and Cultural Organization and State Hydrological Institute (St. Petersburg), 1999.
42. UN, *World Population Prospects: The 2006 Revision*, Population Division of the Department of Economic and Social Affairs of the United Nations Secretariat, 2006.
43. UN, *World Urbanization Prospects: The 2005 Revision*, Population Division of the Department of Economic and Social Affairs of the United Nations Secretariat, 2005.
44. S. Dasgupta, B. laplante, H. Wang and D. Wheeler, *Journal of Economic Perspectives*, 2002, 16, 147-168.
45. U.S. Enviromental Protection Agency Office of Water, *The Clean Water and Drinking Water Infrastructure Gap Analysis* EPA-816-R-02-020, U.S. Enviromental Protection Agency, Washington D.C., 2002.
46. BASE Energy, *Energy Baseline Study For Municipal Wastewater Treatment Plants*, Pacific Gas & Electric Company, San Francisco, CA., 2006.
47. B. Applebaum, Electric Power Research Institute (EPRI), Palo Alto, CA., 2000.
48. P. H. Gleick, ed., *An Introduction to Global Fresh Water Issues*, Oxford University Press, New York, 1993.
49. J. N. Abramovitz and J. A. Peterson, *Imperiled Waters, Impoverished Future: The Decline of Freshwater Ecosystem*, World Resource Institute, Washington, DC, 1996.
50. D. Zimmer, *Evolution of Water Withdrawals and Consumption Since 2900*, http://www.worldwatercouncil.org/fileadmin/wwc/Water_at_a_glance/Water_withdrawals_and_consumption.ppt, Accessed June 11, 2008.
51. in *Fundacion Proteger*, 2008.
52. P. N. Spotts, in *Christian Science Monitor*, 2008.
53. D. Olinger, in *Denver Post*, Denver, 2008.
54. J. Gertner, in *New York Times Magazine*, 2007.
55. A. Avison, in *The Eagle*, Bryan-College Station, Texas, 2008.

56. L. Mungin, in *Atlanta Journal-Constitution*, Atlanta, Georgia, 2007.
57. K. Kranhold, in *The Wall Street Journal*, New York, New York, 2008.
58. in *BBC*, London, England, 2008.
59. USDOE, *Country Analysis Briefs: Saudi Arabia*, U.S. Department of Energy, Energy Information Administration, 2007.
60. IEA, *World Energy Outlook 2005: Fact Sheet -- Saudi Arabia*, International Energy Agency, 2005.
61. P. H. Gleick, *Water: The Potential Consequences of Climate Variability and Change for the Water Resources of the United States* United States Geological Survey, 2000.
62. S. Berfield, in *BusinessWeek*, 2008.
63. P. H. Gleick, *The World's Water 2000-2001: The Biennial Report on Freshwater Resources*, Island Press, Washington, 2000.
64. *John F. Kennedy Library and Museum, Selected Milestones in the Presidency of John F. Kennedy*, http://www.jfklibrary.org/Historical+Resources/Archives/Reference+Desk/Selected+Milestones+in+the+Presidency+of+John+F.+Kennedy.htm, Accessed June 28, 2008.
65. P. H. Gleick, H. Cooley, D. Katz, E. Lee, J. Morrison, M. Palaniappan, A. Samulon and G. H. Wolff, *The World's Water 2006-2007: The Biennial Report on Freshwater Resources*, Island Press, Washington, 2006.
66. in *The Economist*, 2008, vol. Technology Quarterly.
67. NAS, *Desalination: A National Perspective* 0-309-11924-3, Committee on Advancing Desalination Technology, Water Science and Technology Board, Division on Earth and Life Studies, National Academy of Sciences, Washington, DC, 2008.
68. C.W. King and M.E. Webber, "Water Intensity of Transportation," *Journal of Environmental Science and Technology* 42(21), pp 7866-7872 (2008).
69. A. E. Farrell, R. J. Plevin, B. T. Turner, A. D. Jones, M. O'Hare and D. M. Kammen, *Science*, 2006, 311, 506--508.
70. NAS, *Water Implications of Biofuels Production in the United States* 0-309-11360-1, Committee on Water Implications of Biofuels Production in the United States, National Research Council, National Academy of Sciences, Washington, DC, 2007.
71. J. Paul, in *Journal-Advocate*, Champaign, Illinois, 2006.
72. *Energy Independence and Security Act of 2007, Public Law 110–140*, 2007.
73. K. Kern, P. Balash and B. Schimmoller, *Attaining Energy Security in Liquid Fuels Through Diverse U.S. Energy Alternatives, DOE/NETL-2007/1278*, U.S. Department of Energy, National Energy Technology Laboratory, 2007.
74. C. W. King, M. E. Webber and I. J. Duncan, "The Water Needs for LDV Transportation in the United States," *Energy Policy*, Vol. 38 (2), pp 1157-1167 (2010).
75. USDOE, *Annual Energy Outlook 2008: With Projections to 2030*, U.S. Department of Energy, Energy Information Administration, 2008.
76. E. Marris, *Nature*, 2008, 452, 273-277.
77. A. Hardberger, *From Policy to Reality: Maximizing Urban Water Conservation in Texas*, Environmental Defense Fund, Austin, TX, 2008.
78. A.S. Stillwell, C. W. King, M. E. Webber, I. J. Duncan and A. Hardberger, "The Energy-Water Nexus in Texas," *Ecology and Society (Special Feature: The Energy-Water Nexus: Managing the Links between Energy and Water for a Sustainable Future)* 16 (1): 2 (2011).
79. A.S. Stillwell, C.W. King and M.E. Webber, "Desalination And Long-Haul Water Transfer: A Case Study Of The Energy-Water Nexus In Texas," *Texas Water Journal*, Volume 1, Number 1, Pages 33-41, September 2010.
80. K.M. Twomey and M.W. Webber, "Evaluating The Energy Intensity Of The Us Public Water System," Proceedings of the 5th International Conference on Energy Sustainability, ASME, Washington, DC, 2011.
81. K.M. Twomey, A.S. Stillwell and M.E. Webber, "Nitrate Contamination as a Result of Biofuels Production and Its Unintended Energy Impacts for Treating Drinking Water," *Journal of Environmental Monitoring* (2010), DOI: 10.1039/b913137j.

Energy Efficiency: Transportation and Buildings

Michael S. Lubell[a] and Burton Richter[b]

[a]*Department of Physics*
The City College of CUNY
Convent Ave. & 138th St.
New York, NY 10031

[b]*SLAC National Accelerator Laboratory*
Stanford University
2575 Sand Hill Rd.
Menlo Park, CA 94025

Abstract. We present a condensed version of the American Physical Society's 2008 analysis of energy efficiency in the transportation and buildings sectors in the United States with updated numbers. In addition to presenting technical findings, we include the report's recommendations for policy makers that we believe are in the best interests of the nation.

Keywords: Energy, Efficiency, Transportation, Buildings, Policy, Batteries, Hydrogen, Lighting, Economics
PACS: 88.05.Bc, 88.05.Xj, 88.05.Sv, 88.05.Rt, 88.05.Tg, 88.05.Lg, 88.05.Jk, 88.80.Kg, 88.80.ff, 88.85.jm, 88.85.J-, 88.85.mh

PREFACE

In September 2008, following a yearlong study, the American Physical Society released a report, "Energy Future – Think Efficiency" [1], which is available publicly at www.aps.org/energyefficiencyreport. The report focused on two of the three end-use sectors of the energy economy: transportation and buildings that together account for two-thirds of American energy consumption. It did not address the third sector, industrial usage, because the problems there are industry specific and it was not feasible to assemble a set of generalized findings and recommendations in the time available. The information in this chapter has been excerpted from the 2008 report and updated where appropriate. The authors of this chapter, MSL and BR, were two of the authors of the original report and take full responsibility for all of the edits that appear in this chapter. For brevity this chapter omits all footnotes and endnotes appearing in the original report. Readers should refer to that report for additional information.

As the Executive summary of the report noted, making major gains in energy efficiency is one of the most economical and effective ways our nation can wean itself off its dependence on foreign oil and reduce its emissions of greenhouse gases.

Physics of Sustainable Energy II: Using Energy Efficiently and Producing it Renewably
AIP Conf. Proc. 1401, 107-152 (2011); doi: 10.1063/1.3653848

Transportation and buildings, consume far more than they need to, but even though there are many affordable energy efficient technologies that can save consumers money, market imperfections inhibit their adoption. To overcome the barriers, the federal government must adopt policies that will transform the investments into economic and societal benefit. And the federal government must invest in research and development programs that target energy efficiency. Energy efficiency is one of America's great hidden energy reserves. We should begin tapping it now. In the sections on transportation and buildings that follow, we explore the ways we can use efficiency to reduce energy consumption in the United States, thereby improving our energy security, decreasing our dependence on foreign oil and reducing our green house gas emissions.

TRANSPORTATION

Introduction

Americans driving cars, minivans, sport utility vehicles and pickup trucks burn more than 250,000 gallons of gasoline a minute, dumping carbon dioxide into the atmosphere at a rate of more than 2,000 metric tons per minute. Doing so it isn't cheap. It sends more than half a million dollars per minute to the foreign countries that are supplying the oil from which gasoline is made, and many of those countries do not share our values or world view.

Transportation accounts for 70 percent of the petroleum we use for fuel, and today we import approximately 65 percent of the petroleum we consume, paying other nations about $500 billion a year for the privilege. Transportation's share of the oil bill is about $350 billion. And in terms of total U.S. fossil fuel usage, transportation represents a 28 percent share. It also represents more than 30 percent of U.S. carbon emissions.

These facts leave little doubt our dependence on petroleum for transportation threatens U.S. energy and economic security, as well as the environment. The 1973 OPEC oil embargo made the economic and security risks of that dependence abundantly clear. Yet since 1973 the U.S. transportation sector's thirst for oil has doubled. And so has U.S. reliance on imported oil.

A combination of increased consumption and a steady decline in domestic oil production and reserves has caused our reliance on foreign oil to rise from 33 percent to 65 percent in the intervening four decades. The rest of the world has also changed since 1973. Today the U.S. faces increasing competition for petroleum, as global demand for oil has grown dramatically. Developing countries, especially India and China, are putting unprecedented demands on world supplies as they modernize their economies and rapidly increase the size of their vehicle fleets.

In the ensuing discussion we will focus on the light-duty vehicle sector – cars, pickup trucks, SUVs and minivans – because it represents the largest slice of the transportation pie (See Figure 1), and it is where we believe we can achieve the largest savings in petroleum consumption in the shortest time. We observe, however, that although we can realize significant savings with existing technologies, we believe that fully breaking our addiction to oil will require significant, and in some cases revolutionary, technological advances.

Expanding the use of standard hybrid technologies will unquestionably bring much greater efficiency to the transportation sector. Plug-in hybrids – such as the Chevy Volt and the Fisker Karma – or limited range all electric vehicles – such as the Nissan LEAF, the Ford Focus Electric and the Tesla – will go even further. But our eventual goal should be affordable petroleum-free all-electric vehicles, running either on electricity stored in batteries or generated onboard in a fuel cell, and having a range of 300 miles or more. Unfortunately all-electric and fuel cell vehicles with extended range are proving to be more difficult to develop than first thought. Indeed, it may be years before all-electric vehicles can replace the standard family car, and even longer before practical fuel-cell vehicles may be on the market.

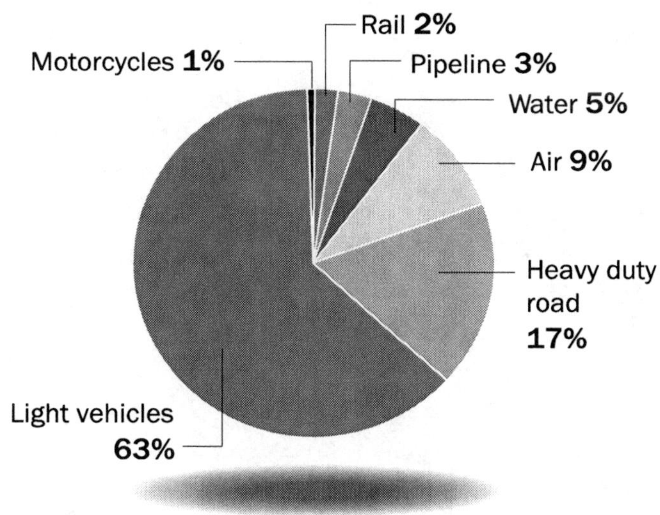

FIGURE 1. Energy usage in the U.S. transportation sector by mode [2].

To appreciate the opportunities for improving fuel economy in gasoline engines, it is instructive to look at fuel-economy trends over the past 30 years for new cars by year of sale, as shown in Figure 2. In 1975, the first year of the federal government's Corporate Average Fuel Economy (CAFE) standards mandating increased fuel economy in the U.S. light-vehicle fleet, the average fuel economy was 14 miles per gallon. Twelve years later, in 1987, the fuel economy of new light vehicles had climbed to 28 miles per gallon for cars and 22 miles per gallon for pickup trucks, minivans and SUVs. But, from 1987 until 2010, there was no additional improvement. New-car fuel economy in 2010 was no better than it was almost a quarter of a century ago.

Before proceeding further, we need to differentiate the terms "fuel economy" and "fuel efficiency." We define the former as vehicle miles per gallon of fuel – how far a car can go on a gallon of fuel. And we define the latter as the fraction of the energy content of fuel used to move the vehicle. In either case, it is important to recognize that different kinds of fuels, such as gasoline, diesel or ethanol, have different energy contents. Any technical analysis must take the differences into account.

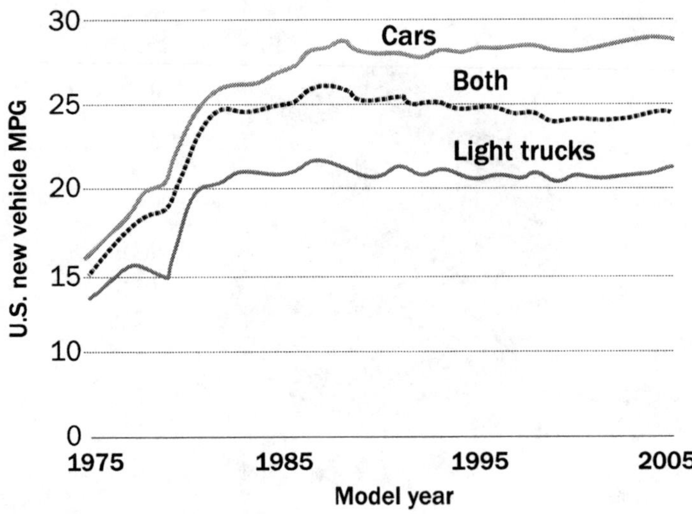

FIGURE 2. Thirty-year trend in fuel economy of cars and light trucks [3].

As Figure 3 illustrates, both fuel efficiency and fuel economy began to improve with the implementation of the CAFE standards program in 1975, but when the government stopped increasing the standards in 1985, fuel economy leveled off, while fuel efficiency continued to improve. Instead of using the technology-driven increases in fuel efficiency of gasoline internal combustion engines to continue to boost fuel economy, the auto industry used the increases to build bigger, more powerful cars, minivans, SUVs and pickup trucks. As a consequence, we have become ever more reliant on foreign sources of oil and, we have stalled in reducing the transportation sector's contribution to CO_2 emissions. Light-duty vehicles now account for nearly half of all U.S. oil consumption and contribute about 20 percent of all CO_2 emissions. To put those emissions in context, we note that in 2005 the U.S. transportation sector, alone, produced nearly 2 billion metric tons of CO_2, more than the **total** CO_2 emissions of every other nation except China.

U.S. fuel economy vs. fuel efficiency

Fuel economy and fuel efficiency for cars and light trucks in the United States for the period 1975 to 2004. (The unit of efficiency in this figure only is ton-miles per gallon. This is the fuel efficiency mentioned in the text multiplied by the weight of the vehicle.)

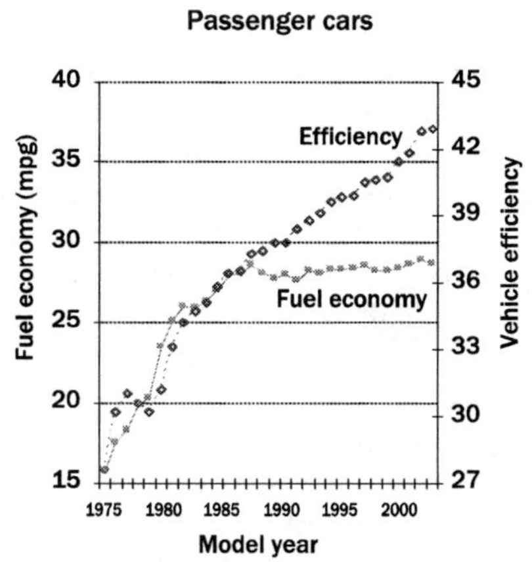

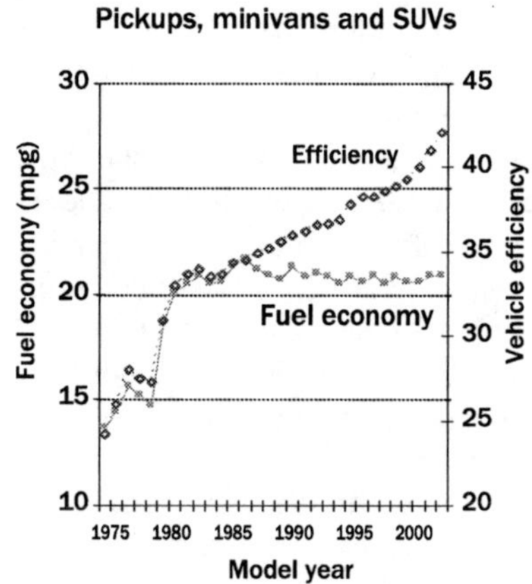

Source: Lutsey and Sperling, 2005

FIGURE 3. Thirty-year trends in fuel economy and fuel efficiency for cars and light tucks, including pickups, minivans and SUVs [4].

The impact of the light-duty transportation sector on greenhouse gas emissions is a serious problem, but its role in deepening our dependence on oil, especially imported oil, undermines our energy security and puts our nation at economic risk. It is not just an extreme event, such as an oil embargo, that can threaten our nation. Oil is a fungible commodity, and its rising world price is a threat to sustained American economic growth. Currently we import between 13 and 14 million barrels of oil per day at a price of more than $100 per barrel, costing our nation more than $1.3 billion dollars per day or $475 billion per year. It is in this context that we present a set of seven findings on transportation and four policy recommendations.

Transportation Finding 1

The fuel economy of conventional gasoline-powered light-duty vehicles can be increased to 35.5 miles per gallon by 2016 – the standard set by the Obama Administration in 2009 – through improvements in internal combustion engines, transmissions, aerodynamics and other technical automotive features using technology that is available today or in the pipeline, with minimal changes in the performance of current vehicles and without the use of hybrid or diesel technology.

Discussion

The 1973 oil embargo showed how vulnerable the economy was to changes in the international oil market. It led to the passage of the Federal Energy Policy and Conservation Act in 1975, which set the CAFE standards for new light-duty-vehicle sales. As Figure 3 shows, substantial efficiency gains in the automobile and light truck fleet occurred over the next decade. In the first few years the standards were in effect, auto manufacturers used both increases in fuel efficiency and significant decreases in the weight of new vehicles (See Figure 4) to achieve increases in fuel economy.

By 1985 fuel economy had risen to 27.5 miles per gallon for cars and 20.7 for light trucks. Efficiency improvements continue to this day due to many incremental improvements, among them the use of front-wheel drive (which reduces drive train losses); enhanced cylinder geometries, fuel injection and computer controls (which lead to more efficient combustion); increased gearing or variable transmissions (which optimize vehicle performance and operation); and low-loss transmissions (which improve energy transfer). But, as Figure 3 illustrates, two years after the CAFE fuel-economy targets were met in 1985, the average combined fuel economy for new cars and light trucks began to decrease, largely due to the increasing fraction of light trucks and SUVs in the automotive fleet. These larger vehicles only had to meet lower CAFE standards as they rolled out of auto plants. The result is that although the fuel economy of both cars and light trucks has increased over the long term, the increased ratio of trucks to cars has caused the combined average fuel economy of the fleet to decrease since the mid-1980s.

In 2002, at the request of Congress the National Academy of Sciences (NAS) evaluated the effectiveness of the 1975 fuel-economy standards and found that not only had the standards produced desired results by the mid 1980s, but that over a 10 to 15-year period beginning in 2002 auto manufacturers could easily employ existing technologies to improve fuel economy by 12 to 27 percent for cars and 25 to 42 percent for light trucks and SUVs, both at a reasonable cost. In 2007, with oil prices climbing dramatically, Congress passed a new CAFE standard for the full light-duty vehicle fleet, and President Bush signed them into law. The Energy Independence and Security Act (EISA) of 2007, Public Law 110-140, Sec. 102, set the new combined fuel-economy standard for cars and light trucks at 35 miles per gallon by 2020 and called for "maximum feasible" increases beyond that date but did not define them. In 2009, President Obama used his executive authority to advance the effective date for the new standard to 2016 with 39 miles per gallon applicable to cars and 30 miles per gallon applicable to light trucks resulting in a combined fleet CAFE standard of 35.5 miles per gallon.

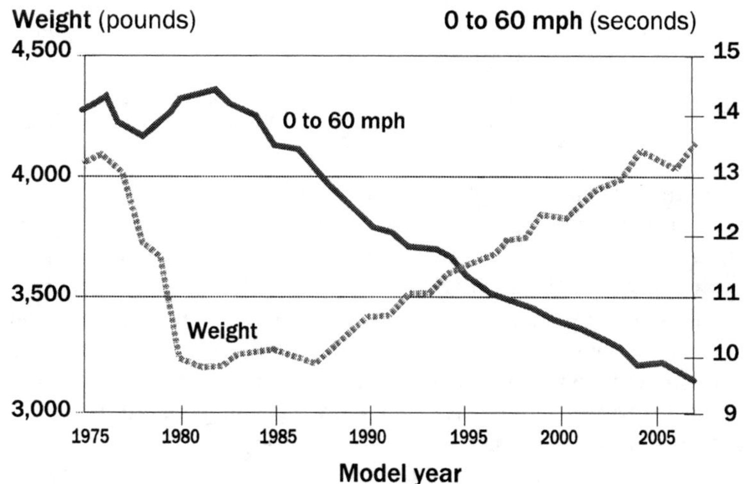

FIGURE 4. Comparison of thirty-year trends in vehicle weight and acceleration [5].

As we noted earlier, fuel economy has remained essentially unchanged constant from 1987 until today while fuel efficiency has increased by more than 20 percent. Had automakers used the gains in fuel efficiency to increase fuel economy rather than increasing the weight and power of vehicles, they would have been able to increase CAFE rating of light-duty vehicles by the same 20 percent. Simply applying that improvement to today's fleet – without assuming any other technological improvements – would increase the average fuel economy from 27.5 to 33 miles per gallon for cars and from 22.2 to 26.6 miles per gallon for light trucks. To meet the new 2016 standards, automakers would have to increase average fuel economies by a further 18 percent for cars and 13 percent for light trucks.

Executives of two major automobile manufacturers have assured us that a 2.5 percent improvement in fuel efficiency per year is possible on a continuing basis for many years to come. Cylinder deactivation, turbo-charging, and improvements in controlling engine valves – all possible with existing technology – can increase operating efficiency, thereby improving fuel economy. Enhancing aerodynamic performance to lower wind resistance and tire performance to reduce rolling resistance can produce further gains.

Transportation Finding 2

Auto makers can increase average fuel economy of new light-duty vehicles to at least 50 miles per gallon by 2030 by (1) further improving the performance of internal combustion engines; (2) decreasing vehicle weight while maintaining vehicle dimensions; and (3) more aggressively deploying hybrid, plug-in hybrid and all electric vehicles in the light-duty vehicle fleet. [We note that the Obama Administration is proposing a light-duty vehicle CAFE standard of 54.5 miles per gallon by 2025. The authors, MSL and BR, consider it a worthy target but difficult to achieve, given consumer behavior.]

Discussion

We call energy-efficiency improvements cost-effective when the value they return exceeds their cost over the life of a vehicle. Since energy-efficient equipment typically comes with a higher initial cost, determining cost effectiveness even in its simplest form involves discounting future benefits. And since future prices of energy are uncertain – as are the costs and performance of future technologies – estimates of cost effectiveness are fraught with uncertainty. Moreover, energy use involves external societal and environmental costs or benefits, such as those related to national security or global climate change, and they are even more difficult to determine.

Given the analytical impediments, we make the following distinctions: for a proven technology, where the value of future energy savings is likely to more than compensate for initial costs, we deem the technology cost effective; but for a future technology, where uncertainties are great, we require only that the technology appear to have the potential to be cost effective, given an appropriate level of research and development. Within the context of these distinctions, we believe that improving fuel economy from the CAFE standard of 35.5 miles per gallon by 2016 to 50 miles per gallon by 2030 is achievable if technological improvements are focused on reducing fuel consumption. The potential for advanced technologies to increase automotive fuel economy by 2030, based on MIT analyses, is shown in Figure 5.

Diesel engines currently have a fuel economy as much as 30 percent greater than gasoline engines, in part because diesel fuel contains 11 percent more energy per liter than gasoline. In Europe 50 percent of new cars sold have diesel engines. (The large market penetration is driven primarily by historically lower diesel-fuel prices, government tax incentives and weaker emission controls on diesel engines than on gasoline engines.) A similar penetration of diesel-powered vehicles in the United States could help achieve the 2030 target. However, the current cost premium of diesel fuel compared to gasoline and American particulate emission standards pose potential barriers.

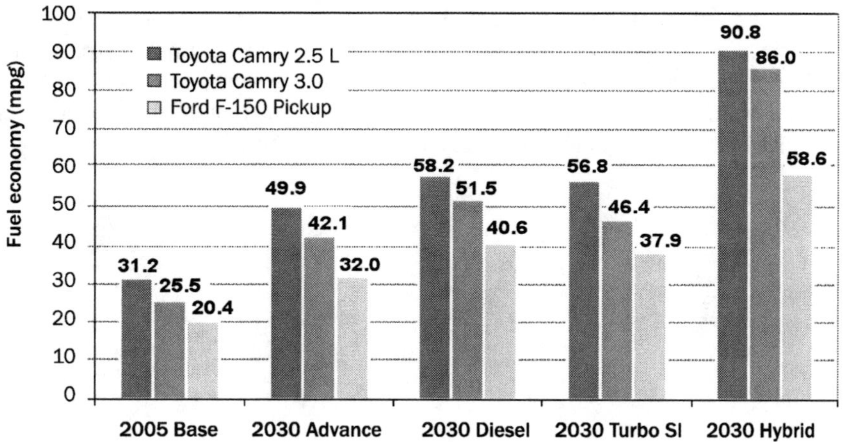

FIGURE 5. Estimates of contributions of technological advances to fuel economy improvements by 2030 [6].

Today, most automobile manufacturers are doing research on homogeneous charge compression ignition (HCCI) a form of combustion that would combine the best of diesel and gasoline engines. An HCCI engine would have the high efficiency of a diesel engine and the relatively low emissions of a gasoline engine. Other technologies under exploration that could come into play in reaching the 2030 goal include engines with variable compression ratios, engines that switch between two and four-stroke operation and engines without camshafts.

Weight reduction, discussed in greater detail below, is another critical part of increasing fuel economy to 50 miles per gallon. Each 10 percent reduction in vehicle weight translates to a 6 or 7 percent increase in fuel economy. The development of strong, lightweight materials, when they become available at an affordable price, could have a dramatic effect on fuel economy while improving overall vehicle safety. Conventional hybrid vehicles, such as the Toyota Prius, the Honda Civic Hybrid and the Ford Fusion Hybrid are currently the most efficient widely available vehicles in the United States, with average fuel-economy ratings between 39 and 50 miles per gallon, depending on the hybrid system and the size of the vehicle. Today hybrids make up almost 3 percent of the U.S. market, and their market share continues to increase (See Figure 6), in part because of rising gasoline prices. But a J.D. Power survey of 4,000 likely new car purchasers conducted in February 2011 [7], projects that U.S. hybrid market share will not exceed 10 percent by 2016 absent reinstatement of federal tax credits or lower vehicle pricing.

The energy efficiency of conventional hybrids can be considerably increased through a number of near-term and long-term improvements, some of which are already in use. Engines could be run at lower revolutions per minute (rpm), and they could be turbocharged. Mechanical pumps and other systems could be replaced with electrical pumps. There are any other improvements based on existing technology that could make relatively efficient hybrid vehicles even more efficient.

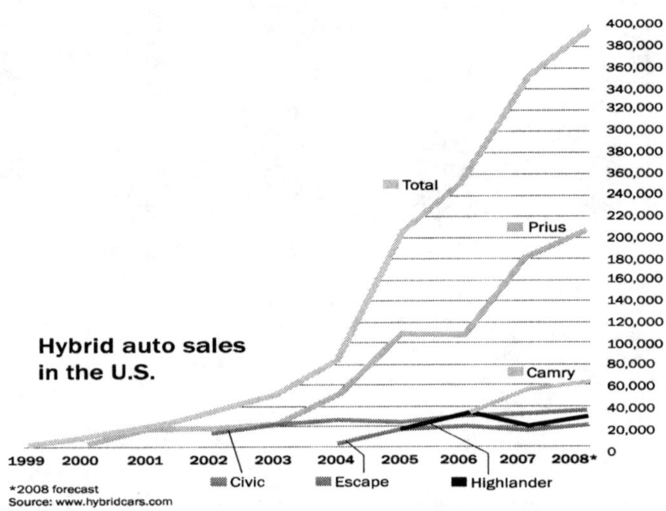

FIGURE 6. Growth in U.S. hybrid auto sales since 1999 [8].

But even with a doubling of today's fuel economy, overall petroleum use and the greenhouse gas emissions associated with it would not decline appreciably if the number of vehicle miles traveled (VMT) by Americans continued to increase as it did almost without exception for the four decades prior to 2007. But as recent Federal Highway Administration data show [9], the growth in VMT has begun to level off and even decline (See Figure 7) as Americans reduce the number of miles they drive in response to rising fuel prices. Although the long-term impact of continued fuel price increases on driving habits is hard to predict accurately, it is extremely likely to depress VMT even further. Limiting urban sprawl by changing the way land use is regulated and financed, developing better models and criteria for transportation infrastructure and factoring vehicle use into urban planning could reduce VMT still more. (See Transportation Finding 7.)

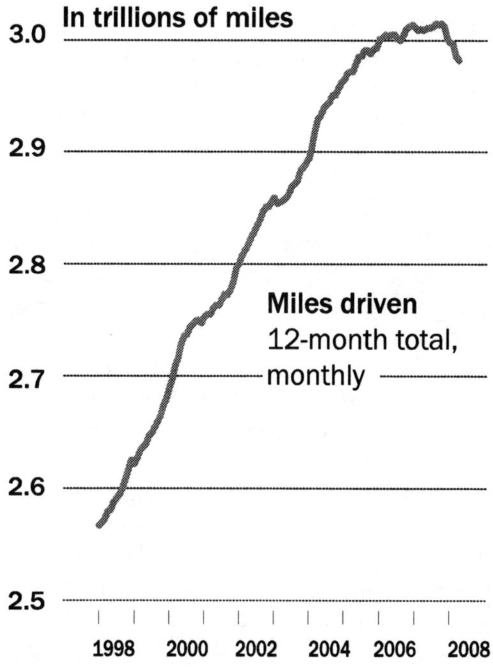

FIGURE 7. Ten-year rend in vehicle miles traveled [10].

Transportation Finding 3

The weight of vehicles can be significantly reduced through design and new materials without compromising safety. Vehicle weight reductions of 20 percent, for example, achieved by greater use of high-strength steel, aluminum and composite materials, would improve fuel economy by approximately 14 percent while reducing traffic injuries and fatalities. Greater reductions in weight, such as the 50 percent goal of the FreedomCAR program (now known as U.S. DRIVE), if achieved by means of advanced lightweight materials, would lead to even greater improvements in fuel economy.

Discussion

Reducing the weight of vehicles by using lighter, stronger materials even while maintaining vehicle size, will significantly improve fuel economy. The new CAFE standards are likely to encourage automobile manufacturers to turn to lighter materials in order to squeeze more miles per gallon out of the same engines. A 10 percent reduction in weight, for example, yields a 6 or 7 percent increase in fuel economy.

A 2002 National Academy of Science study [11] linked the reduction in the weight of a vehicle to a possible increase in fatalities. That view was not unanimous among the NAS panelists, and many experts believe that with advanced materials, vehicles can be made both lighter and safer. This is one of the goals of the U.S. DRIVE (formerly FreedomCAR) program [12].

While it is obvious that in a head-on collision between a very large truck and a car, the passengers in the car will be the losers, that is not necessarily true when the weight differences are not so dramatic. A 2007 International Council on Clean Transportation study (ICCT, an independent organization of transportation experts) [13] looked at the relationship of fuel economy and safety in light vehicles and noted that "the linkages among fuel economy, vehicle size, weight, and safety are manageable and are more a function of smart vehicle design than any other single factor." The report concluded that no trade-off is required between a vehicle's weight and safety. Indeed, other researchers have found that reducing vehicle weight while maintaining the key dimensions of wheelbase and track width could decrease the total number of fatalities [14].

However, there is a crucial difference between making the cars lighter and making them smaller. Smaller cars may indeed be less safe for their occupants than their larger counterparts because of reduced space for "crush zones" in the front and rear; this can be mitigated by proper design. Advanced air-bag technology and smart design mean small cars are safer than they once were, but adequate crush zones are critical for passenger safety. The Insurance Institute for Highway Safety [15], after running crash tests, recently gave its highest safety rating to the diminutive Smart Car, a micro car that weighs only 1800 pounds and is a full three feet shorter and 700 pounds lighter than a Mini Cooper. So it is crush zones and other safety technologies, such as side curtain air bags, electronic stability control, lower bumpers and stronger roofs that ultimately make vehicles safer.

Additional information can be found in Evans 2004, which emphasizes the role of drivers in accidents, and Ahmad and Greene 2005.

Transportation Recommendation 1

Technology is available to move well beyond the 35 mpg CAFE standard mandated in law to be reached by the year 2020 and the 35.5 mpg revised light-duty vehicle standard established by the Obama Administration for 2016. Therefore, Congress should adopt policies to help auto manufacturers achieve the 54.5 mpg standard set by the White House for 2025. The specific policies are beyond the scope of this study but could include more aggressive and even longer-horizon CAFE standards, financial incentives such as "feebates" (fees for not meeting the standard and rebates for surpassing it) and carbon taxes.

Transportation Finding 4

Plug-in hybrid electric vehicles (PHEV), which charge their batteries from the electric grid, could reduce gasoline consumption by 60 percent or more, assuming a range on batteries alone of at least 40 miles. However, plug-in hybrids require more efficient and more durable batteries, able to withstand deep discharges that are not yet in commercial large-scale production. Given the technical difficulties faced in developing the batteries, it cannot be assumed that plug-in hybrids to replace the standard American family car will be available at affordable prices in the near term.

Discussion

Plug-in hybrid electric vehicles (PHEVs), such as the Chevy Volt, differ from conventional hybrid vehicles by being able to travel extended distances not only in a conventional hybrid combined mode but also in a battery-only mode. They charge their batteries from the electricity power grid when the vehicles are not in use. Conventional hybrids, such as the Toyota Prius, have small batteries that do not allow the vehicles to run on them exclusively except in very low-power conditions. Moreover, the battery packs in conventional hybrids cannot be charged from the electric grid because the integrated design of existing systems only allows the battery to be charged within a narrow range.

Conventional hybrid vehicles get all their energy from gasoline, achieving high fuel economy through a combination of methods: recovering some energy from "regenerative braking" (dissipated as waste heat by non-hybrids); using smaller internal-combustion engines that typically operate in their most efficient range; and shutting off the engine when the car is idling.

Plug-in hybrid electric vehicles get some of their energy from gasoline and some from the electricity grid. They have the efficiency advantage of a conventional hybrid and the additional advantage of drawing some of their energy from the grid, rather than gasoline. All-electric battery-powered vehicles, which are discussed later, get all of their energy from the grid.

Figure 8 shows the distribution of daily vehicle miles traveled, using data from the 2001 National Household Travel Survey by the U.S. Department of Transportation [16]. The chart shows, for example, that 30 percent of all miles traveled are associated with vehicles traveling fewer than 40 miles per day. Therefore, if all vehicles could operate solely on batteries charged from the electricity grid for 40 miles per charge, as much as 30 percent of vehicle miles in any one day would not require use of gasoline.

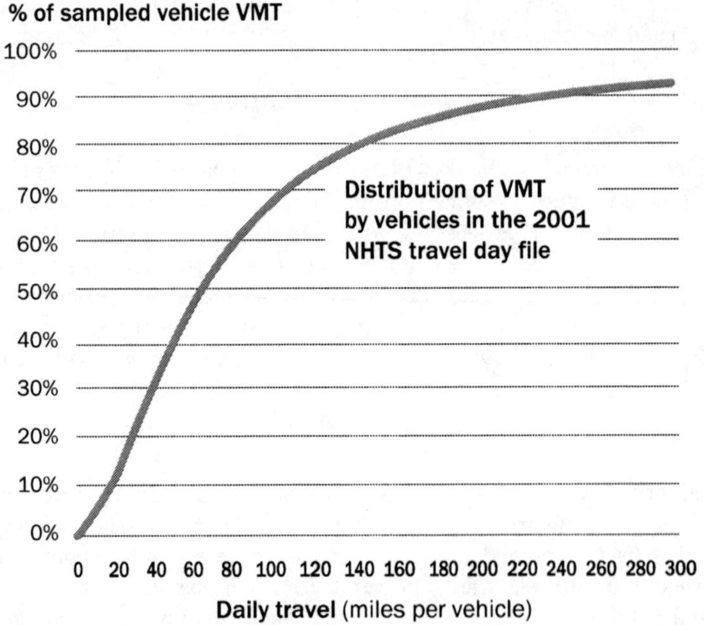

FIGURE 8. Distribution of vehicles mile traveled (VMT) in 2001 as a function of daily miles traveled [17].

Figure 9 shows the percentage of all vehicle miles traveled on electricity as a function of PHEV electric only range. If all vehicles had a 40-mile PHEV electric range, the amount of fuel used in the entire fleet would decrease by 63 percent relative to what was required without the PHEV40 technology. If all trips of 40 miles or less used batteries alone, and if the first 40 miles of longer daily travel relied on batteries alone, the effective fuel economy for a full fleet of PHEV40 vehicles – taking into account only gasoline in the tank (not primary energy) – would be 135 miles per gallon, assuming a 50 mpg fuel economy (as in the case of the Toyota Prius) during gasoline operation [18]. The result would be a 63 percent saving of gasoline per day. The figure is an upper bound, since it assumes the batteries are fully charged each morning and have not deteriorated since purchase. Nonetheless, the impact on both fuel imports and greenhouse gas emission would be dramatic. The reduction in greenhouse gas emission would depend on the energy source of the electricity (coal, natural gas, nuclear, renewables, etc.) used to charge the PHEV batteries.

The key requirement for a PHEV is a battery with large electrical storage capacity and high energy density – the measure of how much energy a battery of a given weight and size can hold. Gasoline stores a remarkable amount of energy for its weight and volume: the energy density of gasoline is 50 times that of a lithium-ion battery by volume, and 100 times that of a lithium-ion battery by weight.

A conventional hybrid vehicle has a relatively small battery that keeps costs low and is still sufficient for storing energy recovered from braking, which is its primary purpose. A Prius, for example, has a battery with a capacity of only about 1.3 kWh with only about 15 percent of it actually available for use in order that the battery have a sufficiently long life.

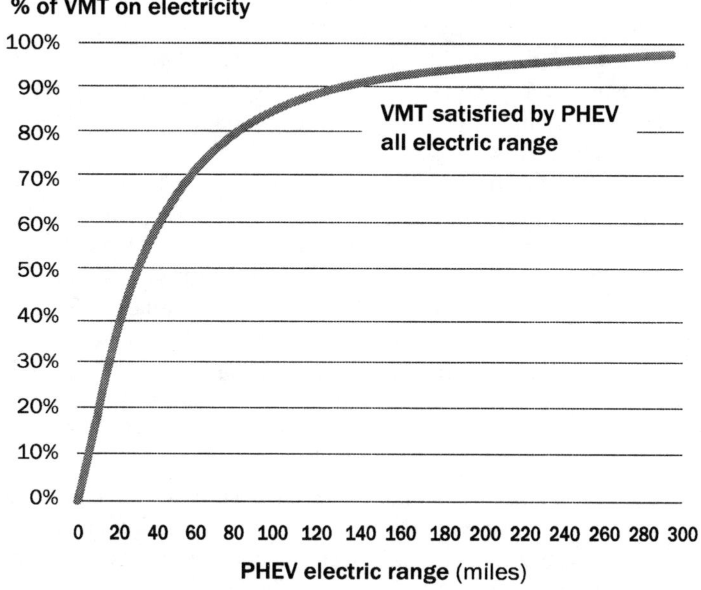

FIGURE 9. Percent of vehicle miles traveled (VMT) on electricity as a function of electric range for plug-in hybrid electric vehicles [19].

A PHEV, by contrast, has a considerably larger battery in order to be able to store significant electrical energy from the grid. A PHEV running on electricity for 40 miles will consume approximately 14 kWh based on the industry standard of 350 Wh/mile. The battery pack has to have a capacity twice that (approximately 28 kWh) to allow for the less-than-complete discharging required for a long battery life. A PHEV battery must also have a reasonable weight, size, cost and recharging time, and, to be commercially viable, it must last many years. At the present cost of about $10,000 for a 28-kWh battery, a PHEV40 will achieve only limited penetration in the market, absent government tax incentives. Battery costs will have to be driven down to make PHEV40s more affordable for the typical consumer.

The recently released Chevrolet Volt, the first PHEV40 intended for the mass market, uses a new type of lithium-ion battery that provides the vehicle with an EPA-rated electric range of 35 miles in urban driving, based on 65 percent discharging of its 16 kWh capacity. General Motors lists the suggested retail price as $41,000, and the car is eligible for $7,500 federal tax credit. Toyota's Prius PHEV, scheduled for release in early 2012, will also use an advanced lithium-ion battery, but one with a smaller capacity, providing an electric range of only 13 miles but allowing the car's anticipated base price to be set at $29,000.

If PHEVs successfully penetrate the consumer market, their widespread use could have major implications for the electrical grid. Charging times for PHEVs are typically several hours, and if a large number of the vehicles are recharged during the day, when electricity demand is already high, the strain on the grid could be significant. By contrast, if the vehicles are charged at night, the impact on the grid should be minimal. A California study concluded that as long as daytime, on-peak charging is avoided, a million vehicles could be charged before a new generation of transmission investments is required [20]. Differential pricing of electricity – cutting the cost for electricity used "off peak" – the study concluded could be used effectively to reduce the daytime demand.

Given the current price of electricity, driving a vehicle powered by electricity would be much cheaper per mile than driving on the power from gasoline. Off-peak electricity costs in California are 9 cents per kWh, corresponding to a cost of about 3 cents per mile for the electric mode in the PHEV. With gas selling for $4.50 per gallon, even a 50-mpg vehicle will cost 8 to 9 cents per mile for fuel. The cost per mile is one-third as much for an electric car as for today's conventional hybrid car. However, major technical and infrastructure issues must be addressed before PHEVs can become a large part of the light vehicle fleet. Among the challenges are the following:

- Present battery costs are too high for an unsubsidized for a commercial market.
- Access to electrical charging stations is obviously required, and many dwellings, such as apartment buildings and condominiums, do not have them.
- Daytime charging will have to be avoided if PHEVs are adopted on a large scale. Note that a car with a 40-mile electric range will only run on electricity for a 20-mile commute one-way. Otherwise daytime charging at the workplace will be required or the return trip will have to rely on gasoline power.

Transportation Finding 5

An all-electric battery-powered vehicle would reduce to zero the use of petroleum as a fuel for light-duty vehicles. However, achieving the same range provided by a gasoline-powered car (300 miles) requires batteries with much larger capacity than is needed for PHEVs. For the standard mid-priced American family vehicle, batteries with the requisite energy storage per unit weight and per unit volume do not exist. A long term R&D program will be required to develop them. A better option might be to increase the range of PHEVs as technology improves. A PHEV with a 100-mile electric range would eliminate 85% of the gasoline requirement for light vehicles.

Discussion

The target range in the Department of Energy's program for an all-electric vehicle is 300 miles, a much more difficult challenge than that for the PHEV. If the battery technology used to make a vehicle with an electric range of 300 miles is the same technology used to make a vehicle with a 40-mile electric range, the 300-mile vehicle would require a battery of 7.5 times the weight, volume and cost of the 40-mile vehicle.

The all-electric vehicle is appealing because it would reduce use of gasoline to zero and reduce consumption of primary energy by roughly 50 percent since electric drive is much more efficient than internal combustion. This reduction includes the electric drive efficiency advantage combined with the average efficiency of 31 percent for electricity delivered from primary energy to the wall plug. The greenhouse gas emission reduction depends on the greening of the electric power generation system. Meeting the all-electric range goal is going to be more difficult than the range goals of PHEVs.

There have been all-electric vehicles before, and there are new ones now. The GM EV-1 and the Toyota RAV-4 were examples using NiMH battery technology. They had relatively short range, but were very popular with their few users. The hardware and software developed for those cars are the starting point for the new generation. Tesla Motors' new two-seater has a range of over 200 miles, very fast acceleration typical of electric propulsion and a price tag of about $100,000. Nissan Motors has introduced its Leaf with a 100-mile range, a top speed of 75 miles per hour and an 8-hour recharge time. They represent a beginning of a journey on an evolutionary road that together with the Toyota and GM PHEV will move the fuel for light-duty vehicles away from petroleum toward electricity generated at central power plants.

Even though PHEVs or all-electric battery-powered vehicles will use little or no gasoline, they are not zero energy vehicles. The electricity they use is made from a primary energy source; for example, fossil, nuclear or renewable fuels. However, the high wheel-to-tank efficiency of an electric vehicle, the existence of the grid, and the potential for freedom from fossil fuels, all favor using electricity to power cars, assuming development of suitable batteries at an acceptable price. In the future, as policies are implemented to reduce greenhouse gas emissions, grid electricity will become a cleaner and cleaner energy source [21].

Transportation Finding 6

Hydrogen fuel cell vehicles (FCVs) are unlikely to be more than a niche product without scientific and engineering breakthroughs in several areas. The main challenges are durability and costs of fuel cells, including their catalysts, cost-effective onboard storage of hydrogen, hydrogen production and deployment of a hydrogen-refueling infrastructure.

Discussion

Beginning in the late 1990s both government and industry began to promote hydrogen fuels and fuel cell vehicle technology heavily. The early promises were not met. The challenge of developing a new technology (fuel cells) and deploying a new fuel supply system proved daunting. A collection of reports by the American Physical Society [22], the DOE Office of Science [23] and the National Academy of Sciences [24] highlighted the challenges. By 2006 it had been recognized that the original plans presented by the automotive companies and a wide range of leaders in the European Union and the United States were not achievable in the near term. A more reasoned view has now emerged, highlighting the breakthroughs that are needed and the longer time to deployment.

Hydrogen vehicles are no longer seen as a short-term or even intermediate-term solution to our oil needs, but as a long-term option requiring fundamental breakthroughs in several areas. The recent NAS report on the FreedomCar and Fuel Partnership program [25] says that even now, "There remain many barriers to achieving the objectives of the Partnership. These barriers include cost and performance at the vehicle, system, and component levels. To be overcome, some of these barriers will require invention, and others will require new understanding of the underlying science." A new NAS report looks at deployment scenarios after the fundamental issues are solved.

There are four principal barriers to large-scale use of fuel-cell vehicles – the fuel cells themselves, onboard hydrogen storage, hydrogen production, and a distribution infrastructure. The fuel cells must be efficient in turning hydrogen into electricity, long lived, and affordable. While considerable progress has been made in the past few years, the necessary cost-effective performance has not yet been achieved. The membranes that are at the heart of fuel cells do not have the durability, permeability, or conductivity to work efficiently in a mass-market vehicle. In addition, a relatively large amount (roughly 60 grams) of the platinum catalyst is presently required to make the chemical reaction run at the necessary rate. This much platinum is too expensive and the material too rare for a mass-market vehicle. Progress in both the membrane and catalyst area is being made. For example, it has recently been shown that control of the atomic structure of the catalyst can, in principle, improve its performance by a factor of 10 or more.

Onboard hydrogen storage remains a significant barrier to development and commercialization of a hydrogen vehicle. Hydrogen has high energy content per molecule, but is a gas at room temperature. Compressed hydrogen systems, the kind typically used on hydrogen demonstration cars, use ultra-high-pressure containers that are heavy, large and typically contain only one-seventh the energy per unit volume of gasoline. Liquid hydrogen has to be stored at lower than -400 °F in special highly insulated containers and 30-40 percent of its energy is lost in the liquefaction-evaporation cycle. An alternative to compressed gas or liquid is highly desirable.

A practical, commercial hydrogen vehicle will most likely have some form of solid-state storage, near atmospheric pressure. In solid-state storage, hydrogen molecules are absorbed onto or chemically bound up in the storage medium. Storage has seen promising new approaches – computer prediction of structure and performance of storage media, and the release of hydrogen from high-density storage media by reaction to structurally different compounds, a process called "destabilization." The latest NAS FreedomCar review notes that finding a solid-state storage material is critical to fulfillment of the vision for the hydrogen economy, and urges that basic and applied research be conducted to establish the necessary technical base. The scientific community has responded to the hydrogen challenge with vigor – the publication rate and activity at meetings on catalysts and membranes for fuel cell reactions and on hydrogen storage media have increased significantly.

Hydrogen production is as important as storage and fuel cells. The United States now produces 90 percent of its hydrogen by reforming natural gas, a process that combines gas and water at high temperature to produce hydrogen and carbon dioxide. If this were to be the source of hydrogen, widespread use of hydrogen cars would simply shift our dependence on oil to a dependence on natural gas with only modest greenhouse gas reduction.

Producing the required hydrogen from coal would be acceptable if carbon capture and storage technology that would eliminate the CO_2 produced in the process were to be commercially successful. More attractive are innovative methods to produce hydrogen with carbon-free sources of energy. These include high-temperature electrolysis in which heat and electricity are both available from solar or nuclear energy, and photo-biological processes that produce hydrogen directly from sunlight. None of these more advanced processes are yet ready for commercialization, but progress is being made.

The last of the four barriers is the hydrogen distribution system. If done by pipeline like natural gas, a new system will be required. It would be far too expensive to transport hydrogen by truck or rail either as a gas or a liquid because of its low energy density. Perhaps one of the solid-state storage solutions will allow the transportation of large amounts of the material.

While large-scale commercialization will not occur any time soon, automakers are planning to produce demonstration fleets. The Honda FCV Clarity is available in limited numbers (200) for lease to customers near hydrogen stations in the Los Angeles area. The vehicle uses a fuel cell in a hybrid-electric vehicle, with hydrogen stored as a pressurized gas at 5,000 pounds per square inch, giving it a range of 270 miles. Each vehicle costs several hundred thousand dollars to produce and receives special servicing from Honda. This demonstration vehicle is a fully functional substitute for gasoline cars, but the cost remains far too high and the life of the fuel cell too short for widespread use in the vehicle fleet now. Honda intends to have 1000

FCVs on the road by 2013; the target for the start of high-volume production is 2015. And this year, General Motors announced that it plans to deliver hydrogen fuel-cell Chevrolet Equinox SUVs to the United States Navy as part of the Hawaii Hydrogen Initiative, which calls for installation of 25 hydrogen fueling stations around Oahu by 2015. Toyota and Daimler also have plans to release hydrogen fuel-cell vehicles in demonstration programs in the near future.

Transportation Recommendation 2

The federal government should ensure that its current R&D program continues to have a broader focus than it did before DOE altered the composition of its portfolio in 2009. A well-balanced portfolio is essential across the full range of potential medium- and long-range advances in automotive technologies, and a sustained research effort is needed in battery technologies suitable for conventional hybrids, plug-in hybrids and battery electric vehicles, and in various types of fuel cells. A well-balanced portfolio should bring significant benefits through the development of a diverse range of efficient modes of transportation and should aid federal agencies in setting successive standards for reduced emissions per mile for vehicles.

Transportation Recommendation 3

"Time of use" electric power metering is needed to make charging of batteries at night the preferred mode. Should large scale testing (currently supported by DOE) reveal deficiencies, improvements in the electric grid will have to be made to allow daytime charging of electric vehicles on a large scale. Grid improvements inevitably will be needed when market penetration of electric vehicles becomes significant.

Transportation Finding 7

There are clearly societal issues that affect fuel use in the transportation sector. Reforms in public policy for land use and urban and transportation infrastructure planning can potentially contribute to energy efficiency by reducing vehicle miles traveled, as can expansions of public transit and various pricing policies. Some could be introduced in the near term, while others, such as changes in land use, would phase in over decades, but might still have significant effects in 10-15 years.

Discussion

It is clear from more than 20 years of research that changes in current policies for urban land use, transportation infrastructure investment, parking and auto insurance can reduce vehicle miles traveled (VMT) and save energy. These are mainly social science issues. Current policies have resulted in growth rates for urban land that exceed the growth rates of population. There has been a similar disproportionate increase in vehicle miles traveled (VMT) since 1973 that cannot be explained by increasing incomes, the cost of driving or the building of more roads – highway congestion has increased since 1973 [26].

Residential density, the availability of public transportation, proximity to jobs, pedestrian friendliness and the mixed-use nature of a community all influence the number of miles people drive. The general conclusions of research on the role of urban planning on vehicle use indicate that to high densities, proximity to reliable public transit, and inclusion of sidewalks and bike lanes correlate with lower household VMT.

An effort to maximize energy efficiency in the transportation sector would require a combination of short-term pricing policies and medium- and longer-term land use and infrastructure investment policies.

A significant obstacle to informing policymakers and the public about these policy options is the lack of an agreed-upon method for quantifying these issues. Different studies frame the questions in different ways, and different sources provide different predictions that are qualitatively in agreement but yield slightly different—or mutually incomparable—predictions.

Transportation Recommendation 4

Federally funded social-science research, currently being questioned by some members of Congress, is vital for determining how land-use and transportation infrastructure can reduce vehicle miles traveled. Studies of consumer behavior as it relates to transportation should be conducted, as should policy and market-force studies on how to reduce VMT. Estimation of the long-term effects of transportation infrastructure on transportation demand should become a required component of the transportation planning process. This program needs a home.

BUILDINGS

Introduction

Americans spend 90 percent of their time indoors, working, living, shopping and entertaining in buildings that consume enormous amounts of energy. In 2006, buildings — more than 118 million residential and commercial structures — were responsible for 39 percent of the nation's primary energy consumption, a level of energy use that has a significant impact on global climate change and potentially on U.S. energy security.

Since most of their energy comes directly or indirectly from fossil fuels, buildings are responsible for large quantities of greenhouse gas (GHG) emissions — about 36 percent of the of CO_2 associated with the nation's total annual energy consumption. Building energy consumption and the resulting GHG emissions, which have been steadily rising, are projected by the Energy Information Administration (EIA) to increase another 30 percent by 2030.

Yet a large fraction of the energy delivered to buildings is wasted because of inefficient building technologies. How much of this energy can ultimately be saved is an open question — as much as 70 percent by the year 2030 in new buildings and perhaps more than 90 percent in the long term if there were pressing reasons to go that far. These energy savings can be made not by reducing the standard of living, but by utilizing more efficient technologies to provide the same, or higher, levels of comfort and convenience we have come to enjoy and appreciate. Some of these technologies are available today; others are beyond our present grasp, but achievable in the future with strong investment in research and development (R&D). Today we can achieve significant energy savings by making cost-effective efficiency improvements in buildings and their equipment, thereby reducing GHG emissions and providing significant economic benefits to consumers.

Buildings consume 72 percent of the nation's electricity, more than 50 percent of which is generated from coal, our nation's most abundant energy resource but one with CO_2 emissions greater than other fossil fuels, according to the EIA. The advantage of electricity is that it is a form of energy that can be fully converted to work and is easy to distribute over the electric grid. Its disadvantage is that it is generated and distributed with 31 percent efficiency — which means 69 percent of the primary energy used to generate electricity is lost as waste heat before reaching the end user.

Building structures pose a more difficult problem than either the equipment they contain or automobiles due to their long lifetimes and slow replacement rates. Whereas vehicles and appliances wear out after a decade or so, buildings typically last for the better part of a century. Most buildings were constructed during the years when energy was cheap, and as a result, they were not designed or built with energy efficiency in mind. The overall number of buildings in the United States is growing by only 1 to 2 percent per year. Hence a major reduction in building energy consumption must involve both improvements in existing buildings and new construction.

Fortunately, widespread use of existing energy efficiency technologies and those that can be developed over the near term would eliminate a sizable portion of the current waste of energy, significantly reducing building energy consumption and greenhouse gas emissions. For the foreseeable future, reducing primary energy consumption through improved efficiency is likely to remain far cheaper than expanding renewable energy production [27].

Residential Buildings

In 2005, residential buildings in the United States consisted of 113 million residences totaling an estimated 180 billion gross square feet, including standalone houses and mobile homes, as well as dwellings located in apartment buildings and other multi-residence units [28].

Data from recent Department of Housing and Urban Development surveys show that the average rate of new construction is about 1.4 percent per year, and when demolition, condemnation, and conversion of residences are factored in, the net growth per year is about 1.2 percent. Once built, a residential building is likely to be usable for about one hundred years [29].

In 2000 (the latest year for which data are available), the average existing residential unit consisted of 1,591 square feet and a household size of 2.7 people [30].

Although the housing market is currently depressed, the trend for at least a half-century has been toward larger residences. The average new single-family home constructed in 2006 was 2,470 square feet, 42 percent larger than in 1980 [31].

Primary energy consumption based on end use for residential buildings in 2005 is summarized in Figure 10. The single largest end use is space heating (32%), followed by air conditioning or space cooling (13%), water heating (13%) and lighting (12%). Note that these four combined account for 70% of the energy consumption.

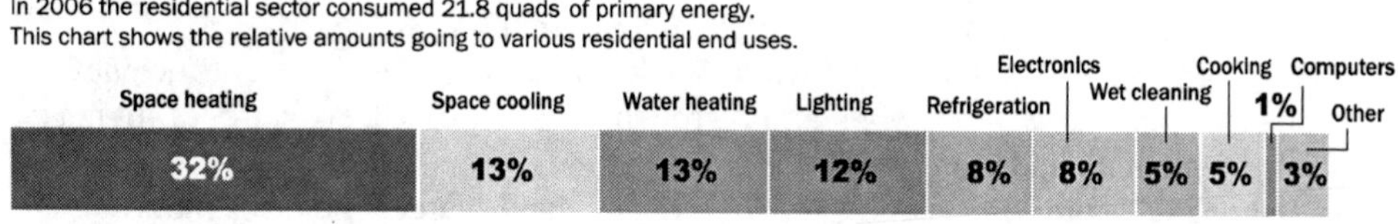

FIGURE 10. Percentage of energy end usage in the residential building energy sector by function [32].

Commercial Buildings

In the United States in 2000, 4.7 million commercial buildings provided 68.5 billion square feet. From 2000 to 2005, the commercial building stock grew by 15 percent to 74.3 billion square feet, double the growth rate of the residential sector [33].

The commercial space breaks down as follows: offices (17%), mercantile (16%), education (14%), warehouse and storage (14%) and lodging (7%), with numerous other functions making up the remaining 32 percent.

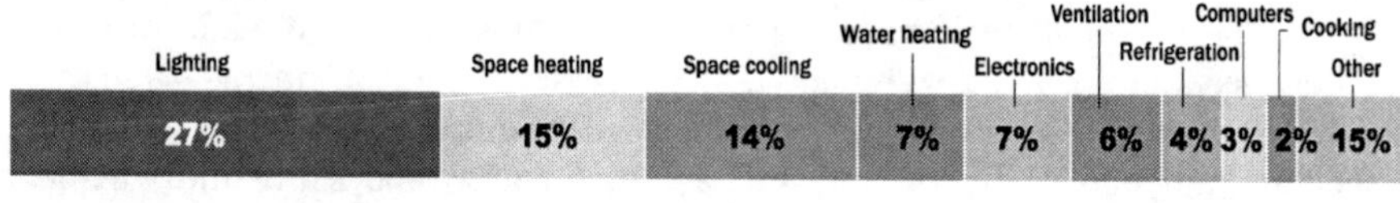

FIGURE 11. Percentage of energy end usage in the commercial building energy sector by function [34].

Primary energy consumption based on end use for commercial buildings for 2005 is summarized in Figure 11. The single largest end use is lighting (27%), followed by space heating (15%), space cooling (14%) and water heating (7%). Together these four end-uses account for 63 percent of primary energy consumption, somewhat lower than the case for the residential sector. Although commercial buildings presently consume less primary energy than residential buildings, the energy use in the commercial sector is experiencing nearly double the growth rate.

Primary Energy

Figure 12 illustrates primary energy used by the residential and commercial sectors from 1950 to the present and projected out to 2030. The graph indicates that energy consumption in the commercial sector is expected to grow faster than that in the residential sector. By 2030 combined primary energy in the two sectors is expected to reach 51 quads, a 30 percent increase over 2006 consumption.

Energy consumption has been growing despite some improvements in efficiency. The main driving forces are population growth and increased standard of living associated with more and more ways to use energy.

As compared with 30 years ago, Americans have larger homes; more air-conditioners, televisions, and computers; and a variety of other devices that use energy. Currently available, cost-effective technologies could significantly reduce the energy consumption of residential and commercial buildings, and the United States is making inadequate use of these measures. But further technologically feasible advances could reduce consumption far more.

Using current and emerging technologies — those already in the pipeline – widespread construction of cost-effective, zero-energy new single-family homes could be achieved in 10 to 15 years, except possibly in hot, humid climates such as those in the Southeast. (By zero energy, we mean buildings that use no fossil fuels. In general, that means reducing a building's energy use by about 70 percent from today's average and fulfilling the remaining power needs with on-site or off-site renewable energy.) Widespread construction of zero-energy commercial buildings will be harder to achieve, but should be possible within 15 to 25 years, with a focused, sustained effort. Achieving 70 percent reductions in energy consumption for new commercial buildings will require both new technologies and greatly expanded use of the concept of integrated design. Such advances are unlikely to occur without greatly expanded research, development and demonstration (RD&D) efforts.

R&D will also be needed to develop more ways to improve energy efficiency in existing buildings through such measures as better wall insulation and windows

But new technology alone will not assure efficiency improvements. Achieving maximum efficiency in our nation's buildings will require expanded use of policy tools such as appliance efficiency standards, building energy codes and utility demand side management programs in order to encourage efficiency.

Clearly, reducing building energy consumption is critical to our nation's future. A first step on the path to limiting greenhouse gas emissions, reducing the national energy bill, avoiding unnecessary construction of power plants and diminishing stresses on fossil energy resources is recognizing that buildings (including factories) as well vehicles now consume vastly more energy than they need to operate efficiently.

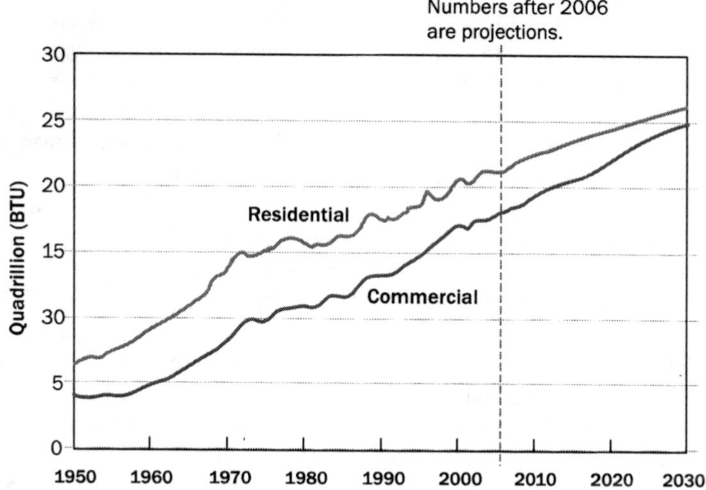

FIGURE 12. Trend in primary energy consumption in residential and commercial buildings energy sectors [35].

Buildings Finding 1

If current and emerging cost-effective energy efficiency measures are employed in new buildings, and in existing buildings as their heating, cooling, lighting and other equipment are replaced, the growth in energy demand by the building sector could be reduced from the projected 30 percent increase to zero between now and 2030.

Discussion

There are a wide variety of technologies and strategies now available that can significantly lower building energy consumption without any loss of service or comfort. Some are appropriate for residential buildings, some for commercial buildings, and some for both. We are not suggesting that all of these items are cost-effective in all cases.

Space heating is the largest **residential** user of energy, and cooling is a close second. Focusing on those two systems, measures for both new construction and renovation that can save significant amounts of energy include the following:

- Increasing insulation in walls, roof, floor and basement to cost-effective levels.
- Using window coatings, chosen based on climate, to reduce the amount of heat gain and loss through thermal transmission.
- Moving heating and cooling ducts into the conditioned space (so that air from leaks is not lost to the outside) for new construction and reducing leakage for new and existing homes.
- Improving heating systems through the use of furnaces that send less than 10% of their heat out the flue, variable-speed and higher efficiency motors/fans for air circulation and efficient ground-source or gas-fired heat pumps.
- Upgrading equipment for cooling to achieve better heat transfer from an air conditioner's evaporator and condenser coils. Using variable-speed drives that allow units to operate efficiently at partial load (rather than turning on and off frequently). In addition to saving energy, this partial load operation also controls humidity more effectively and reduces the internal heat loads on the air conditioner.
- Changing ventilation system installation (mostly for new construction) from the current practice of relying on construction errors and accidental leakage to provide sufficient fresh air to a process that uses the proper amount of mechanical ventilation while sealing the home to nearly airtight standards.
- Controlling ventilation can mitigate problems with indoor air quality and mold, while also recovering energy from the exhaust air stream.
- Expanding use of evaporative cooling, using direct evaporation in arid climates, and evaporation combined with an air-to-air heat exchanger in more humid climates.
- Constructing buildings with "cool" roofs that reflect rather than absorb infrared radiation in warm and hot climates.
- Integrating passive solar heating and cooling into home designs. There are considerable difficulties of custom-designing the orientation and thermal characteristics of individual homes, but when it is done passive solar construction is a very cost-effective measure for saving energy.

The remaining measures focus on the other high-energy end-uses: hot-water heating, lighting, refrigeration, electronics/computers and other appliances.

In the residential sector, water heating uses as much energy as air-cooling. This energy use in all buildings can be cut by utilizing more efficient water heaters, reducing distribution losses in the plumbing system, and reducing the heaviest demands for hot water in the home through water-saving appliances (dishwashers and clothes washers).

Experience indicates a great deal of energy can be saved through increasing the efficiency of appliances. The best example may be refrigerators. Today's refrigerators use one-fifth as much energy as comparable refrigerators did 35 years ago. Also they cost less, after inflation. These energy efficiency improvements have come about at least in part in response to federal regulations that require manufacturers to meet appliance energy efficiency standards that are increasingly strict over time.

Additional equipment that will result in significant energy savings in **commercial** buildings from available technology includes the following:

- More efficient lamps, ballasts and luminaires.
- Improved glazing with lower heat loss and appropriate solar gain.
- Improved controls for air conditioning systems.
- Variable speed fans/drives and pumps.
- Lower-pressure fan systems.
- Occupancy sensors for controlling lights and ventilation.
- Efficient designs for building elevators and escalators.

Although analyses of energy savings stemming from single systems are the easiest to understand, they miss many of the big-picture, cost-effective options that come from integrating systems, such as:

- The use of lighting designs that optimize the distribution of light so that it is brightest where the most light is needed and less intense elsewhere.
- The use of envelope designs that permit daylighting (described in the next section), while controlling solar loads and glare.
- Reduction in size and/or complexity of heating, ventilation, and air-conditioning (HVAC) systems made possible as a consequence of better insulation in walls, roofs, and floors; improved windows; and reduced air leakage.
- The use of separate space conditioning and fresh air ventilation systems that allow occupants to control the systems based on need.
- Separate control of cooling and dehumidification, so that cooling systems can be sized to address cooling alone.

Lighting and window energy-efficiency technologies and strategies are common to **residential** and **commercial** buildings, though some lighting technologies are only appropriate for commercial applications. Lighting presents perhaps the greatest opportunity for immediate, cost-effective energy savings in buildings.

Incandescent lamps, a century-old technology, are the major source of light for **residential** buildings despite converting only 5 percent of their electric energy into light. Electric energy is generated and delivered to end-use sectors with an average of 31 percent efficiency. This means the overall efficiency of converting primary energy into incandescent light is only 1–2 percent. Clearly lighting is an area in which there is great room for improved efficiency. Figure 13 shows the status of lighting technology in the United States as of 2001.

One widely available alternative is the compact fluorescent light (CFL), which uses one-quarter of the energy of an incandescent bulb to deliver the same light intensity and quality. Mercury is an environmental concern in all fluorescent lamps; however, replacing incandescent lamps with CFLs releases less mercury into the environment than traditional light bulbs when the mercury released through the burning of coal for electricity generation is taken into account (at current allowable rates of mercury emissions) [37].

To get a sense of the rough potential of improving lighting efficiency, suppose all incandescent lamps in use in 2001 were replaced by lamps that use one-fourth the energy, such as CFLs. The annual electric savings would be about 240 TWh, corresponding to 2.6 quads of primary energy. (No doubt some of these upgrades have been accomplished since 2001, particularly in the commercial sector.) A more precise analysis of lighting improvements found that upgrading incandescent lamps in residential buildings and upgrading ballasts and lamps in commercial buildings would reduce annual electric energy usage by 120 TWh (1.3 quads primary energy) [38].

Lighting upgrades will accelerate due to the enactment of the federal Energy Independence and Security Act of 2007, which phases in limits on the sale of incandescent bulbs. CFL sales are already booming, with annual sales now at 400 million units compared to 50 million units just 5 years ago. Solid-state lighting now being developed promises to produce lamps that double the energy savings from CFLs.

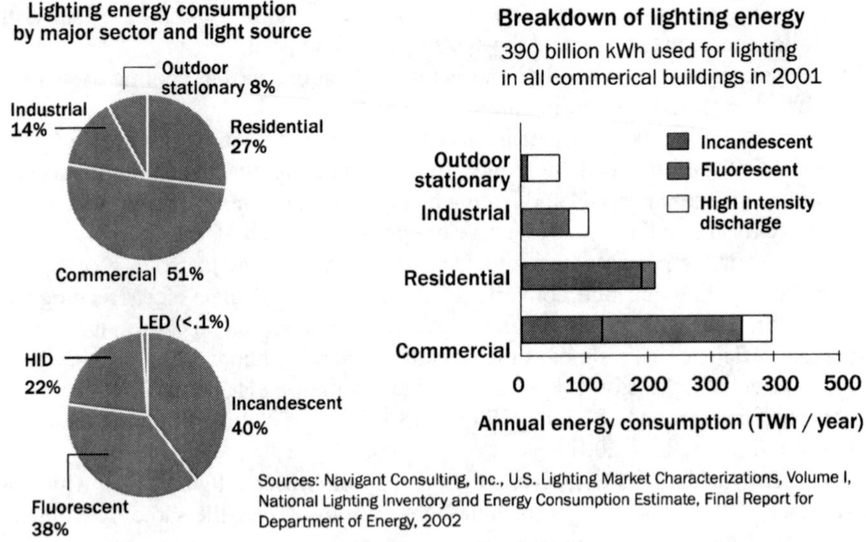
National lighting energy consumption
In 2001, the U.S. consumed 8.2 quads of primary energy (corresponding to 765 TWh of delivered electricity) for incandescent, fluorescent, high-intensity discharge (HID) and light-emitting diode (LED) lighting technologies. Incandescent lamps remain the dominant lighting technology in the residential sector.
Lighting energy consumption by major sector and light source
Outdoor stationary 8%
Industrial 14%
Residential 27%
Commercial 51%
LED (<.1%)
HID 22%
Incandescent 40%
Fluorescent 38%
Breakdown of lighting energy
390 billion kWh used for lighting in all commerical buildings in 2001
Incandescent
Fluorescent
High intensity discharge
Outdoor stationary
Industrial
Residential
Commercial
0
100
200
300
300
500
Annual energy consumption (TWh / year)
Sources: Navigant Consulting, Inc., U.S. Lighting Market Characterizations, Volume I, National Lighting Inventory and Energy Consumption Estimate, Final Report for Department of Energy, 2002

FIGURE 13. Energy consumption in lighting by sector and type of lighting [36].

Expanding the use of natural lighting—so-called "daylighting"—can save an estimated 30–60 percent in lighting energy in many **commercial** buildings [39]. Daylighting uses sensors and controls to adjust artificial lighting in response to changing natural light coming through windows and skylights. Wal-Mart used this approach to upgrade lighting in its 2,100 stores worldwide with energy savings that have a two-year payback in energy costs alone [40].

Lighting energy can also be reduced by making better use of task lighting, combined with sensors and controls, which deliver light at appropriate levels where and when needed. Ironically, commercial buildings use about five times as much energy for lighting (per square foot) as do residential buildings, even though residential buildings are used more at night. One of the primary reasons for the disparity is that residential buildings make better use of natural lighting and task lighting.

The rapid expansion of modern electronic equipment has resulted in homes and businesses containing dozens of smaller electronic loads such as computers, printers, faxes, copiers, microwaves, televisions, VCRs, DVD players and cable boxes. Many of these devices go into a standby mode and continue to use power even when turned off. A recent study estimated that an average California home contained more than forty products constantly drawing power. Together, those products consumed nearly 1000 kWh/year while off or in a low-power mode [41], representing about 8% of the average U.S. household electric energy consumption. Replacing such devices with Energy Star (http://www.energystar.gov/) rated devices would significantly lower energy consumption, particularly in standby or low-power mode.

We pause here to discuss combined heat and power (CHP) because it would enable buildings to make more efficient use of electrical generation plants. However, unlike the technologies mentioned above, CHP would require significant additional R&D to be practical in many cases. Also, CHP is not assumed in reaching the 30 percent energy efficiency improvement cited in Buildings Finding 1.

In addition to gains in end-use energy efficiency in a building, energy supply technologies directly associated with the building — combined heat and power (CHP) — provide a significant opportunity for energy savings, yet one that remains largely unexploited in the United States. The electric power sector discharges roughly two-thirds of its energy — nearly 26 quads annually — to the environment in the form of low-grade heat. That low-grade heat is being lost at the same time residential and commercial buildings are consuming 7.5 quads of natural gas to produce low-grade heat. Clearly a great deal of energy could be saved if waste heat could be delivered to places that need it. It sounds simple, but is very difficult to accomplish with centralized electric power stations. A few power plants do capture this waste heat and distribute it in district heating systems, but those types of plants are more common in Europe.

For U.S. buildings, existing CHP opportunities are mostly limited to large building complexes such as those associated with colleges, universities and hospitals, which provide heating and cooling from a centralized natural gas or coal plant. These plants have the opportunity to produce both electricity and steam, with improved efficiency over plants that just produce heat or electricity. More opportunities could present themselves if communities develop more compact land use patterns, which is desirable from a transportation systems perspective as well.

For individual buildings, CHP has been demonstrated using natural gas micro-turbines and fuel cells, which generate both electricity and heat for space heating and domestic hot water. Balancing the heat and electric demands proves challenging for a single building. For these technologies to achieve widespread use, R&D efforts are needed to bring down the costs of micro-turbines and fuel cells and to address a variety of technical and financial challenges [42].

In determining what efficiency gains are possible with current and emerging technologies, it is useful to start by looking at what is happening under current standard practices. Contractors focused on energy upgrades to existing residential buildings achieve energy efficiency improvements ranging from 15 to 35 percent by installing better and more efficient insulation, windows (in some instances) and lights; by eliminating infiltration and duct leakage; by upgrading furnaces, boilers and air conditioners; by replacing the power supplies that waste electricity when their devices are in standby or low-power mode; and by replacing old appliances with newer, more efficient ones.

Energy service companies (ESCOs) regularly work with larger commercial customers to perform energy audits followed by upgrades in lighting, HVAC equipment and system controls, by which they achieve cost-effective energy savings. We were unable to locate performance data for U.S. ESCOs. In Berlin, Germany, however, ESCOs have improved the energy efficiency of 1,400 buildings by an average of 24 percent at no cost to building owners and a profit to the ESCO that paid for the upgrade [43]. U.S. results are likely to be similar. Generally, it is easier to achieve efficiency gains in new buildings than in existing ones.

Buildings Finding 1 is also based on an analysis conducted in 2000 as part of the Clean Energy Futures study [44] and recently updated to determine the potential for improvements in buildings [45]. The analysis concludes that using currently available technology upgrades as they become cost-effective for current and new buildings would result in a 30 percent decrease in the annual energy consumption by residential and commercial buildings in 2030. It turns out that the reduction erases the projected increase in energy consumption for the buildings sector, so that 2030 consumption by buildings could be the same as it is today.

Far more energy savings are technologically achievable, but they are not cost effective between now and 2030 for the individual consumer. Additional upgrades would be cost-effective if societal costs and benefits were taken into account. As discussed later in this chapter, even the cost-effective energy savings will not be achieved by market forces alone; significant policy tools and incentives will be required. And the policy tools will also likely result in unexpected improvements coming into the marketplace, as has happened in the past.

Buildings Recommendation 1

The federal government has set goals for energy use in its own buildings. There should also be a national goal that the U.S. building sector will use no more primary energy in 2030 than it does in 2008. That goal should be reviewed every 5 years in light of the available technology and revised to reflect even more aggressive goals if justified by technological improvements. Achieving the goal will require that the federal government implement a set of policies and programs such as those discussed later.

Buildings Finding 2

The goal of achieving significant levels of construction of cost-effective new zero-energy **commercial** buildings by 2030 is not obtainable without significant advancement in building technology and without the development and widespread adoption of integrated building design and operation practices.

Discussion

Zero-energy buildings (ZEBs), or "net-zero buildings," are an attractive concept achievable by merging efficient grid-connected buildings with renewable energy generation. The ideal is to use onsite renewable energy sources, typically a photovoltaic (PV) array, to annually generate as much energy as the building uses. A building, at times, buys energy from the grid while at other times, sells energy back to the grid. A ZEB is one that annually sells as much energy as it buys, or more. While ZEBs are being built today, generally they are not yet cost-effective. Indeed, if cost and footprint are not constrained, one can simply add whatever renewable energy sources are necessary to achieve net-zero energy, no matter the efficiency of the building. But widespread construction of ZEBs requires that they be cost-effective and that the renewable energy sources fit into the building footprint. Since efficiency measures are much cheaper per unit energy than on-site renewable energy, both cost and footprint constraint lead to the requirement that such buildings first be made very efficient. Efficiency is also important to reduce energy consumption so that the required renewable energy sources can fit into the building footprint. A 70 percent reduction in energy consumption (as compared with conventional buildings) has been adopted as a consensus target for ZEB — though it is an estimate.

Various organizations, including the U.S. Congress (in the case of federal buildings), the American Institute of Architects (AIA) and the State of California, have called for all new commercial buildings to be ZEB by 2030. The AIA and California have established a 2020 goal for ZEB for all new residential buildings.

Commercial buildings serve a large and widely varying set of occupants and needs. For example, auditoriums and stores may at times be unoccupied, and at other times be crowded with hundreds of people. Some buildings are no larger than small homes while others accommodate 60,000 football fans or 20,000 office workers. And although there are examples of standardized commercial buildings, the largest buildings are often "one-of-a-kind" buildings with specialized criteria. Comfort and health require appropriate ventilation, heating, or more likely, cooling. Design engineers, rightly concerned about liability, commonly design systems for the maximum occupancy, and these systems typically waste enormous amounts of energy when occupancy is low.

There has been growing interest in the construction of green and energy-efficient commercial buildings. The Leadership in Energy and Environmental Design (LEED) certification, introduced in 2000 has rapidly grown in popularity and demand. Despite this growing interest there has been relatively little progress in reducing energy consumption in new commercial buildings.

Information about hundreds of green commercial building projects may be found on the Internet, many with impressive claims about their projected energy consumption. But obtaining actual energy consumption data for green commercial buildings is difficult. There are a growing number of LEED-certified new commercial buildings (552 through 2006), and the public assumes they are energy efficient, but the only study of their energy use is a recent New Buildings Institute review. The Institute obtained energy performance data for only 21 or 22 percent, of the buildings [46]. Of those, only six achieved site energy consumption levels per square foot that were 70 percent below the average for all commercial buildings per square foot. Only three of the buildings achieved that level of savings in primary energy consumption. Still, the New Buildings Institute concluded that the LEED buildings it examined were 25 to 30 percent more efficient than the average new commercial building, but not everyone would reach the same conclusion from the data. Whatever their efficiency, these 121 LEED buildings consume more total energy per square foot (either site or primary) than the average for the entire commercial building stock.

It should be noted that energy efficiency is but one of many criteria for LEED building certification and credits for energy efficiency are awarded based on design simulations, not measured building energy performance. There has been very little work on validating whether projections of performance correspond to actual building performance; that is an area requiring further research. What's needed is a comprehensive system for rating building energy efficiency. More often than not, constructed buildings actually use more energy than predicted by energy simulations performed during the design process [47]. This may be due to flaws in simulation tools; failures in the design, construction or operation of the building; or energy intensive "plug-loads" that were not included in energy simulations. Monthly energy bills cannot distinguish between energy used by building systems (lighting, heating, ventilation, air-conditioning, etc.) and plug-loads. Monthly energy bills for a very efficient hospital are likely to be higher than those for an inefficient elementary school. Neither design energy simulations nor monthly energy bills provide the complete picture of a building's energy efficiency.

Very-low-energy commercial buildings are so rare largely because they are very difficult to design, construct and operate. The biggest barrier is the complexity of the buildings and their HVAC systems, and the important interactions between the various building systems and components. Significant efficiency improvements have been achieved when all of these factors were taken into account — using a process called "integrated design."

Integrated design is a process in which all of the design variables are considered together, and hundreds or even thousands of combinations are analyzed to arrive at the optimal design that meets user requirements and minimizes energy consumption. The usual linear design process simply fails to account for interactions between the various building components — and these can have important energy and cost implications. For instance, the direct energy savings associated with choosing a better window technology may not justify the cost — and the linear design process rejects the upgrade. But the integrated design process goes on to determine that the window upgrade allows a smaller, more efficient HVAC system — with total cost savings that justify the window technology upgrade.

An experimental program, run by Pacific Gas and Electric (PG&E) in the 1990s, showed that 55–65% energy reduction could be accomplished using an integrated design approach [48]. But the process was time-consuming and hard to replicate. The six low-energy LEED buildings offer further proof that 70% reduction in energy use can be accomplished. The challenge is to develop easily replicable design and construction processes that achieve such results cost effectively.

Although it is a crucial component of the solution, integrated design cannot guarantee low-energy commercial building performance. Even the best-designed buildings, with well-thought-out integrated systems, can suffer in their construction by contractors who lack the skills and experience to implement the details faithfully. And facility managers may not know how to operate a new system properly. A $100 home appliance comes with a setup and operating manual; many buildings do not.

Buildings Recommendation 2

To achieve the ZEB goal for commercial buildings by 2030 the federal government should create a research, development, and demonstration program with the goal of making integrated design and operation of buildings standard practice. Such a program should be carried out co-operatively between the federal government, state governments and electric utilities, with funding coming from all three entities.

Since reducing energy consumption and carbon footprint is one of the most important goals for green buildings, any green building rating system, such as LEED, should give energy efficiency the highest priority, based in part on actual energy performance, and require reporting of energy consumption data. The recent revision to the LEED scoring system has moved in this direction.

Buildings Finding 3

The goal of achieving significant levels of construction of cost-effective zero-energy **residential** buildings by 2020 is feasible, except perhaps for hot, humid climates. Most of the required technology to compete with traditional housing is available but inadequately demonstrated. To achieve this goal in hot, humid climates will require increased R&D to develop low-energy dehumidification and cooling technologies and strategies.

Discussion

Cost-effective zero energy homes are not available today, but there has been significant progress in developing efficient single-family homes. Employment of cost-effective efficient technologies has resulted in new, low-budget, single-family homes that use half as much primary energy as comparable conventional homes [49]. And 80 to 90% reduction in energy used for heating (though not total energy) has been achieved by passive solar homes in Germany, Austria, Switzerland, Sweden and France [50].

The U.S. Department of Energy's Building America program directly addresses the fundamental problems of bringing energy efficiency to new residential buildings. The program provides technical support for builders to construct very energy-efficient residential buildings at low or no increased first cost to the consumer. Building America works with builders who are responsible for more than 50 percent of new residential construction in the United States. More than 50,000 competitively priced houses have been constructed under the program, with an average energy use for heating and cooling that is 30 to 40 percent less than that of typical new residences. The DOE's new Builders Challenge sets a more ambitious goal of 30 percent savings in total building energy. Still, this program has a long way to go to meet the ultimate goal of constructing and selling zero-energy houses by 2020.

Building America addresses two basic problems in commercializing zero-energy houses: assuring the cost and energy performance of state-of-the art technologies and acquainting the building industry with the techniques to build such houses. There is an R&D effort associated with this program that supports the need to reduce costs, improve energy performance and address the cooling and dehumidification requirements of hot, humid climates.

The Building America approach is an effective way to create new markets for energy-efficient housing. Funds to support more demonstration activities could speed up the process of commercializing very-low-energy houses. Promoting Building America along with programs that show the value of building energy codes and strict efficiency standards for appliances will produce very large gains in energy savings in new houses.

Buildings Finding 4

The federal government has not been investing sufficient funds in R&D for next-generation building technologies, for training building scientists or for supporting the associated national laboratory, university and private sector research programs although the situation has begun to improve.

Discussion

Federal funding is especially important in the building sector, which is highly fragmented and consists largely of smaller firms that are unable to conduct R&D or have no economic incentive to do so because of an inability to capture the benefits of R&D. Yet during the last two decades funding for energy efficiency R&D for buildings, especially commercial buildings, declined significantly.

In the 1980s, when levels of effort were much higher than they are today, federal R&D on energy efficiency in buildings achieved notable success. A National Academy study [51] estimated the economic benefits from advanced window coatings and electronic fluorescent ballasts to be $23 billion (in 2000 dollars). Both technologies resulted from federally funded energy efficiency R&D efforts that expended far less than $23 billion.

In the 1980s, when levels of effort were much higher than they are today, federal R&D on energy efficiency in buildings achieved notable success. A National Academy study [52] estimated the economic benefits from advanced window coatings and electronic fluorescent ballasts to be $23 billion (in 2000 dollars). Both technologies resulted from federally funded energy efficiency R&D efforts that expended far less than $23 billion.

Examples of research, development and demonstration that could enable the achievement of deep savings for the majority of new commercial buildings include the following:

- **Computer tools:** Improved computer tools are needed to facilitate integrated design by analyzing interactions among building elements that affect energy use. In addition to continued development of complex computer tools such as EnergyPlus, the simulation developed over years by DOE, there is a need for tools that are simpler to use and appropriate during the early stages of design when key decisions are made. The simpler tools need not be crude; indeed, with the low cost of computing, complex programs like EnergyPlus could be made much more user-friendly to meet this need. Such programs could also be used for building labels.
- **Monitoring and control technologies:** Advanced technologies are needed to support diagnostics, fault detection and control in real time for a variety of building energy systems.
- **More efficient building components:** Among the needs are advances in air conditioning and ventilation systems; advances in LED and conventional lighting and their controls; advanced, affordable coatings for windows; envelope systems that optimize air transfer, water transfer and heat transfer together on a climate-sensitive basis; and building-integrated photovoltaic systems.
- **Test facilities:** Controlled experiments for commercial buildings in different climate regions would benefit from the creation of test facilities. These facilities would allow tests of advanced facades (walls, roofs and windows) coupled with innovative HVAC systems and next-generation controls and monitoring. Such facilities are needed in different climate zones: cold winter/hot summer; hot humid summer; and mild winter/summer.
- **Demonstration programs:** Demonstration programs showing that commercial buildings can be built to use 70 percent less energy than current structures would encourage the building industry to pay more attention to integrated design and other energy efficiency practices. Unlike demonstrations for residential buildings, such commercial demonstration programs should be seen as R&D rather than straightforward commercialization of a process.
- **Static insulation:** Nanotechnology developed for direct energy conversion devices can also be applied to create high-performance thermal insulation materials for various thermal systems. Materials with nanometer-sized channels hold the promise of reducing heat transfer, which will open the possibility of a thin, rigid, high R-value (a measure of insulation effectiveness) insulation panel for retrofit of interior surfaces of exterior walls. Such technology could also be applied to improve the performance of foam and fiberglass insulation.
- **Dynamic insulation:** Nanotechnology has the potential to develop switchable insulations in which the thermal conductivity could be varied by an order of

magnitude. For example, the evening by night cooling, insulated in the morning and then used during peak afternoon periods.

- **Lighting:** Solid-state lights can be used to increase lighting efficiency and applied to tailor lighting distribution to specific needs within a commercial building. They are potentially twice as efficient as fluorescent lamps
- **Windows:** Current research is developing windows with high insulation values and selective control of the solar spectrum. Advanced materials for coatings and frames have the potential to produce window systems that achieve net energy gains during the winter and substantially reduced air conditioning loads in the summer.
- **Active building facades:** Long-term R&D could lead to active building facades that can modulate day-lighting, solar gains and ventilation in response to monitoring of interior conditions. For example, application of innovative materials and mirrored systems could distribute daylight much deeper into commercial building interiors and might lead to reductions in lighting energy requirements by 50 percent or more.
- **Advanced air conditioners and heat pump systems:** Today's systems operate at about one-fourth of ideal efficiencies. R&D on systems optimization, heat transfer enhancement and advanced controls can lead to much higher efficiency in space conditioning.
- **Natural ventilation:** Properly designed and operated natural ventilation systems can reduce cooling loads in commercial buildings by 50 percent or more in many U.S. climates. Prediction of airflow and thermal conditions in large, open-plan buildings is needed to assure proper operation under a variety of climatic conditions.
- **Energy performance data and analysis:** Buildings will be increasingly monitored for their energy performance. The creation of these data on a broad scale opens enormous research opportunities to understand energy performance of buildings in the real world. Compilation and analysis of these data is of great importance in informing policy and guiding R&D.
- **Indoor environmental quality, health and productivity:** Concerns exist that very-energy efficient buildings can degrade health and productivity of building occupants. R&D is needed to identify when and if such problems arise from high efficiency and to establish measures to mitigate adverse effects if they occur.

As a means to insuring that R&D on energy use in buildings is able to thrive over the long term, it is essential to train this and future generations of building researchers and leaders among building energy professionals in government and the private sector. For scientists and engineers, graduate programs with opportunities to pursue energy efficiency research need to be established and expanded.

Buildings Recommendation 3

The federal government must maintain strong support of building R&D in order to achieve the ZEB goal of 2030 for commercial buildings and 2020 for residential buildings. In constant dollars the research program of 1980 — which led to important innovations — would be about $250 million today. We recommend that federal funding for building R&D be maintained at that level for the next three years, after which it should be carefully reviewed. The review should determine the level of continuing federal funding needed for the program to reach its goals, including examining what technology is ready to go to market. One use of the additional spending should be the expansion of the existing demonstration program for low-energy construction of residential buildings, along with associated research, as noted in Buildings Finding 3.

Buildings Finding 5

A wide range of market barriers and market failures discourage investment in energy-efficient technologies.

Discussion

If so many energy efficiency measures are cost-effective why are they not adopted? The question has stimulated considerable discussion [53]. Consider some of the barriers that inhibit adoption of cost-effective technologies – barriers faced by consumers, manufacturers, builders, designers and suppliers of efficient products:

- **Not knowing:** The utility customer knows her total bill but not the contribution of the different appliances and the heating and cooling equipment, nor the thermal integrity of the house. Policies such as Energy Star labels and appliance and building standards and labels are essential to overcome this barrier. Even with labels, consumers may not always be aware of highly efficient products on the market or be willing or able to calculate the payback from an initial higher purchase price.
- **Not caring:** For most consumers, energy is a small cost compared with other expenditures. For example, prior to 2002 typical TVs with remote controls used 5 to 7 watts of standby power when turned off to permit the instant-on feature to function. In 2002 TVs were required to reduce standby power to 3 watts or less to qualify for Energy Star. On November 1, 2008, standby power must be reduced to 1 watt or less for new standalone TVs to qualify. For the individual consumer, the reduction from 6 watts to 1 watt represents just a few dollars in savings per TV per year. That sounds trivial, but applied to 300 million televisions across the United States it represents about $1 billion in electric savings. The cost of making the improvement is small, so the manufacturer has a strong incentive to reduce the standby power to 1 watt to qualify for the Energy Star label. But given the overall cost of operating a TV, the consumer is not likely to care about the slight improvement in standby power efficiency.
- **Split incentives:** If a person who does not pay the energy bill owns the energy-using equipment or building, he has little or no incentive to invest in efficiency. Landlords who do not pay for energy, which is typical, are not

likely to gain an advantage from installing energy efficiency measures. In residential buildings, about one-third of all dwellings are occupied by renters. Split incentives can also apply within a single company: often the capital budget for building improvements is under one manager while the operating budget is under the control of another.

- **Stalled demand for innovation:** If manufacturers do not produce energy-efficient products, consumers cannot purchase them. And if consumers do not demand energy efficiency, then producers have little incentive to make their products more efficient. This "chicken and egg" problem applies to appliance manufacturers as well as to builders and building designers. The circle can be broken by policy decisions but is not likely to be resolved by market forces alone.
- **Reluctance to change:** An important barrier to improved efficiency is inertia. For many years, manufacturers produced appliances with little concern about energy efficiency. After appliance standards were implemented, first by California in 1978, and then by the federal government in 1990, electricity consumption by new refrigerators declined over a 30-year period from 1,725 to 498 kWh/yr while increasing considerably in size. The same phenomenon occurred for other appliances, although to a lesser degree. Prior to the standards, energy use had been increasing; for refrigerators it was increasing at 6 percent annually.
- **Utility profits coupled to sales:** Traditionally utilities (typically electric and natural gas companies) have rate structures that connect their profits to energy sales – the more energy they sell, the more money they make. This offers a disincentive for the utility company to help customers become more efficient and use less energy. Yet utility companies are best positioned to assist customers in identifying ways to improve energy efficiency. Establishing rate structures in which utility profits are decoupled from sales removes one of the most important barriers to energy efficiency.

To make the situation even more difficult, the design process itself provides disincentives to incorporate energy efficiency into buildings. For commercial buildings, the lack of coordination between engineers and architects, the payment of design fees that discourage integrated design (which adds to design costs as it later saves in operational costs) and the lack of the required complex knowledge to make the building energy efficient all discourage the use of the best – that is, integrated – approaches to design and construction.

Not only do fragmentation and inefficient design processes provide justification for more federal energy efficiency R&D, they also mean that innovative energy-saving products are unlikely to be produced by manufacturers and thus will not be available to consumers. This problem in the building industry accounts for the inability of the industry to develop first-rate tools for integrated design and operation of buildings.

The example of fluorescent light ballasts makes clear the need for policies to promote energy efficiency. Standard core-coil ballasts were far less efficient than newer ballasts. There was no difference in performance between the two ballasts, and the payback period for the efficient ballast was approximately two years at 1987 electricity prices. In short, the inefficient ballasts made no economic sense. Yet outside of five states that had banned the standard ballasts, inefficient ballasts captured 90 percent of the market in 1987. (The efficient ballasts cost an average of $4.40 more than the inefficient one – $15.40 versus $11.00 – and produced an average savings of $2.15 per year – hence the two-year payback.)

It is worth noting that the largest fraction of purchasers of fluorescent lights are managers of commercial buildings, who might be expected to make purchases with high paybacks and be familiar with technology as simple as fluorescent ballasts. But it took the passage of a federal ban through a 1988 amendment to the National Appliance Energy Conservation Act of 1987 to move the market away from the inefficient ballasts. These barriers are not unique to the United States. They occur throughout the world. Even developing economies and centrally planned economies are subject to the same failures.

Experience has shown that particularly in the case of buildings, even the best cost-effective technologies are not readily adopted without policies to pull them into the market place. This may be especially true for the buildings sector, where unnecessary energy costs that may make little difference to the individual consumer can have large cumulative effects. Below we discuss several policy tools that we believe should be part of a portfolio of efforts to promote energy efficiency in buildings. The detail about how to apply these tools is beyond the scope of this study, and this is not meant to be a comprehensive list. For example, we do not discuss electric rate decoupling, which would enable utilities to make money from reducing consumption, as mentioned above. Our main point is to emphasize yet again the absolute need for both research and policy to make progress in energy efficiency.

Buildings Finding 6

Among the most effective tools for increasing energy efficiency in buildings are building energy codes, labeling, audit programs and tax and other incentives for the purchase of efficient technology. For appliances, heating and cooling equipment and lighting, both mandatory efficiency standards (standards for appliances), voluntary standards (industry consensus guidelines for lighting usage) and energy labels (the Energy Star label developed and promoted by the Environmental Protection Agency and DOE) have been effective. Utility demand-side management (DSM) programs that provide incentives for energy efficiency have been very successful.

Discussion

We limit our discussion to appliance standards, building energy codes, and utility DSM programs, as those have been especially effective in the United States. Figure 14 shows the impact of the three programs in California, calculated conservatively by the California Energy Commission. Since the mid-1970s electric energy use per capita nationally has risen steadily while for California it has remained relatively flat. Today Californians use about 5,000 kWh per person per year less than the average American. Appliance standards, building energy codes, and utility DSM programs are estimated to be responsible for one-fourth [54] to one-third [55] of the difference.

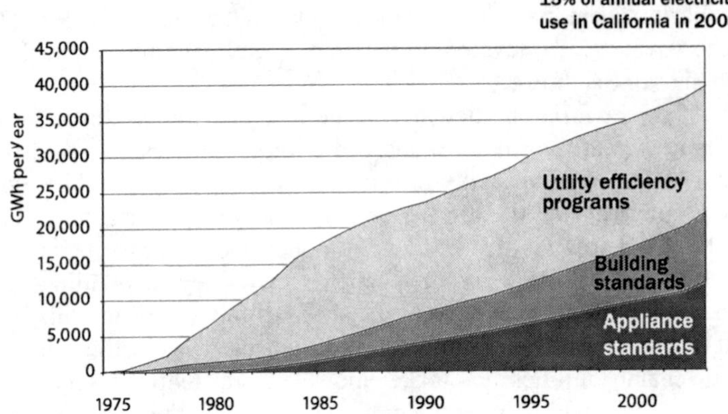

FIGURE 14. Trend in California electrical energy savings from appliance and building energy standards and demand side management (DSM) programs in place since 1975 [56].

Appliance Standards

In 2000, appliance standards reduced U.S. electricity use by approximately 88 billion kWh, 2.5% of total U.S. electricity use. That same year, the standards reduced peak generating needs by approximately 21 GW (roughly equivalent to 21 large power plants). Over the 1990–2000 period, standards have reduced consumer energy bills by approximately $50 billion, with benefits being more than three times the cost of meeting the standards [57].

By 2020, existing appliance standards are expected to cut annual U.S. electricity use by 483 billion kWh or 11 percent of projected electricity use. Peak electricity savings are estimated to total 158 GW in 2020 and annual carbon dioxide savings, 375 Mt. The net savings in expenditures resulting from the standards should approach $300 billion [58], with standards adopted after 2008 and implemented in mid-2011 providing even greater savings. The American Council for an Energy Efficient Economy (ACEEE), for example, estimates that by 2030 the new standards have the potential to add another 190 billion kWh in yearly energy savings, to reduce peak demand by an additional 80 GW and to cut annual carbon dioxide emissions by another 165 Mt. Figure 15 shows the effect of the appliance standards on the efficiency of three major appliances.

Impact of standards on efficiency of 3 household appliances

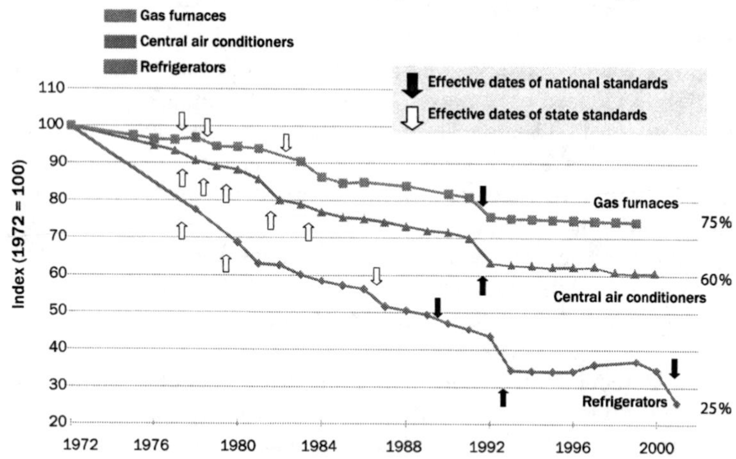

Source: A. Rosenfeld, California Energy Commission; S. Nadel, ACEEE, in ECEEE 2003 Summer Study, www.eceee.org

FIGURE 15. Illustration of the influence of appliance standards on efficiency using 1972 baseline [59].

Utility Demand-Side Management Directed at Customer Energy Efficiency

Demand-side management (DSM) programs are programs in which some central agency, often an electric or natural gas utility, invests money to assist customers in becoming more energy efficient. The investment may be in education programs or customer rebates to encourage purchase of more efficient appliances, or the agent may pay for the bulk of the efficiency upgrade, as in weatherization programs for low-income customers.

Such DSM programs involving customer energy efficiency have reduced growth in electricity sales in the short run by providing financial incentives for energy efficiency purchases by consumers. Utility DSM has also served to transform markets by aiding the commercialization of new energy efficient products.

Analysis of specific DSM programs has shown benefits greater than costs. For the nation, total annual utility expenditures on customer energy efficiency from 1995 through 2006 have varied from a low of $880 million in 1998 to a high of $1,700 million in 1995. Levels are expected to continue to increase for the foreseeable future.

Building Energy Codes

Energy codes are adopted at the state or occasionally local level in the United States and are enforced by local code officials at the city or county level. Most states follow national models established by the two nonprofit organizations that write model codes, the International Code Council and the American Society of Heating, Refrigerating and Air-Conditioning Engineers (ASHRAE).

California is one of the states that has not followed these models; it has been a leader in building energy standards that it develops itself. Energy codes adopted since 1975 in California reduced peak power demand in 2003 by 5.75 GW while reducing electric energy use by 11 TWh/yr. The economic value of energy savings is more than $30 billion or more than $2000 per household. The electric energy needed to cool a new home in California has declined by two-thirds (about 2400 kWh/yr to 800) from 1970 to 2006, despite the fact that today's new home is about 50 percent bigger and is in a warmer climate as new development occurs farther from the coast. The California energy code was revised in 2002, 2005, and 2008; each revision cut energy use by 10-15 percent compared to the previous iteration, an annual rate of improvement of about 4 percent.

There is little federal involvement in establishing building energy codes. The federal government, through the Department of Housing and Urban Development, sets standards for manufactured housing and DOE provides modest technical assistance to the model codes organizations.

While energy codes are often thought of in a context limited to new construction, they also save energy in existing buildings. When a new tenant moves into a space in a commercial building and replaces the lights or the HVAC system, that action triggers the energy code requirements. When a home is remodeled, the systems affected must meet energy code: Thus, a kitchen remodel requiring changes to the electrical system in California triggers the need to meet the lighting efficacy standards. A few localities also require retrofits at time of sale for both commercial and residential properties.

Energy codes typically offer two methods of compliance: a prescriptive checklist approach and a performance-based approach that relies on simulated energy performance of the proposed building compared to a comparable reference building. Builders in states where a usable method of calculating and displaying performance is available overwhelmingly prefer the performance approach, because it allows the builder to meet the energy goal at the lowest first cost. Calculation software for use by architectural and engineering firms and consulting companies that provide technical expertise for meeting codes is widely available throughout the United States for homes, but only in California is it widely available for commercial buildings. By contrast, in the European Union (EU), which has mandated that an energy label be developed for all new buildings and an energy evaluation take place whenever a building is sold, easily used software is expected to be available throughout the member states. Efforts to harmonize software across member states and with the residential system used in the United States are in progress.

Buildings Recommendation 4

The DOE should promulgate appliance efficiency standards at levels that are cost-effective and technically achievable, as required by the federal legislation enabling the standards. The DOE should promulgate standards for all products for which it has been granted authority to do so, including those appliances for which there is not a specific congressional mandate. A streamlined procedure is needed to avoid delays in releasing the standards.

Buildings Recommendation 5

Considering the cost effectiveness of utility DSM to date, and the fact that many states have hesitated in creating such programs, the federal government should encourage states to initiate DSM programs through their utilities. The federal role could be to provide rewards to states that have significant and effective DSM programs and disincentives to those that do not.

Buildings Recommendation 6

Building energy standards, such as those promulgated in California, should be implemented nationwide. States should be strongly encouraged to set standards for residential buildings and require localities to enforce them. For commercial buildings, performance-based standards that rely on computer software to compare a building design with a reference building are implemented only in California. The federal government should develop a computer software tool much like that used in California to enable states to adopt performance standards for commercial buildings. States should set standards that are tight enough to spur innovation in their building industries.

ACKNOWLEDGMENTS

We wish to acknowledge the study group members and authors of the original report, “Energy Future – Think Efficiency,” George Crabtree, Leon Glicksman, David Goldstein, David Goldston, David Greene, Dan Kammen, Mark Levine, Maxine Savitz and Daniel Sperling; the study group research and editorial staff, Fred Schlachter, John Scofield and James Dawson; and the administrative, composition and graphics assistance of Jeanette Russo and Kerry Johnson.

REFERENCES

1. B. Richter, *et al.*, Rev. Mod. Phys. 80, S1 (2008)
2. S.C. Davis and S.W. Diegel, *Transportation Energy Data Book*, edition 26, ORNL-6978, Oak Ridge National Laboratory, Oak Ridge, Tennessee, http://cta.ornl.gov/data/index.shtml
3. Environmental Protection Agency, *Light-Duty Automotive Technology and Fuel-Economy Trends: 1975 through 2007*, http://www.epa.gov/otaq/fetrends.htm#2 (2007); National Highway Traffic Safety Administration.
4. N. Lutsey and D. Sperling, "Energy Efficiency, Fuel Economy, and Policy Implications," *Transportation Research Record* **1941**, 8 (2005).
5. Environmental Protection Agency, *Light-Duty Automotive Technology and Fuel-Economy Trends: 1975 through 2007*, http://www.epa.gov/otaq/fetrends.htm#2 (2007).
6. M.A. Kromer and J.B. Heywood, "Electric Powertrains: Opportunities and Challenges in the U.S. Light-duty Fleet," *LFEE* 2007-03 RP, MIT, Cambridge, MA (May 2007); E.P. Kasseris and J.B. Heywood, *Comparative Analysis of Powertrain Choices for the Near to Mid-term Future*, Master's thesis, MIT, Cambridge, MA (2006).
7. www.jdpower.com/news/pressRelease.aspx?ID=2011039
8. www.hybridcars.com
9. www.fhwa.dot.gov/policyinformation/travel/tvt/history/
10. Federal Highway Administration, Traffic Volume Trend, FHWA, Washington, DC (2008), http://www.fhwa.dot.gov/ohim/tvtw/tvtpage.htm
11. NAS, 2002: National Academy of Sciences, *Effectiveness and Impact of Corporate Average Fuel Economy (CAFE) Standards*, NAS Washington, DC (2002).
12. EERE, 2005: U.S. Department of Energy, *Energy Efficiency and Renewable Energy, FreedomCAR and Fuel Technical Partnership Technical Goals*, (2005), http://www1.eere.energy.gov/vehicles andfuels/about/partnerships/freedomcar/fc_goals.html.
13. D. Gordon, D.L. Greene, M.H. Ross and T.P. Wenzel, *Increasing Vehicle Fuel Economy without Sacrificing Safety*, International Council on Clean Transportation, Washington, DC (2006).
14. R.M. Van Auken and J.W. Zellner, "An Assessment of the Effects of Vehicle Weight and Size on Fatality Risk in 1985 to 1998 Model Year Passenger Cars and 1985 to 1997 Model Year Light Trucks and Vans," *SAE Transactions* **114** (6) 1354 (2004).
15. Insurance Institute for Highway Safety, press release May 14, 2008, http://www.iihs.org/news/rss/pr051408.html
16. Cited by Santini and Vyas 2008; Vyas and Santini, 2008 — see references 17 and 19
17. A. Vyas and D. Santini, "Use of National Survey for Estimating 'Full' PHEV Potential for Oil-use Reduction," presented at PLUG-IN 2008 Conference in San Jose, CA, July 2008.
18. Vyas and Santini, 2008; Santini and Vyas, 2008; E.D. Tate, General Motors, private communication.
19. D. Santini and A. Vyas, "How to Use Life-cycle Analysis Comparisons of PHEVs to Competing Powertrains," presented at 8th International Advanced Automotive Battery and Ultracapacitor Conference May 12-16, 2008, Tampa, Florida; D. Santini and A. Vyas, "More Complications in Estimation of Oil Savings via Electrification of Light-duty Vehicles," presented at PLUG-IN 2008 Conference in San Jose, CA, July 2008.
20. D.M. Lemoine, D.M. Kammen and A.E. Farrell, "An Innovative and Policy Agenda for Commercially Competitive Plug-In Hybrid Electric Vehicles," *Environ. Res. Lett.* **3**, 014003 (2008).
21. *Energy Information Administration, Energy Market and Economic Impacts of S. 2192, the Lieberman-Warner Climate Security Act of 2007*, SR/OIAF/2008-01, U.S. Department of Energy, Washington, DC, April 2008.
22. C. Davis, B. Edelstein, B. Evenson, A. Brecher and D. Cox, *Hydrogen Fuel-Cell-Vehicle Study*, report prepared for the Panel on Public Affairs, American Physical Society, New York (2003).

23. http://www.sc.doe.gov/bes/hydrogen.pdf
24. NAS 2005: National Academy of Sciences, *Review of the Research Program of the FreedomCAR and Fuel Partnership: First Report*, NAS, Washington, DC (2005), http://books.nap.edu/openbook.php?isbn=0309097304
25. NAS 2008b: National Academy of Sciences, *Review of the Research Program of the FreedomCAR and Fuel Partnership: Second Report*, NAS, Washington, DC (2008).
26. A.M. Eaken and D.B. Goldstein; "Quantifying the Third Leg: The Potential for Smart Growth to Reduce Greenhouse Gas Emissions," in Proceedings of the 2008 ACEEE Summer Study on Energy Efficiency in Building, in press; M.G. Boarnet and R. Crane, "Travel by Design: The Influence of Urban Form on Travel," Oxford University Press, New York (2001); R. Ewing, K. Bartholomew, S. Winkelman, J. Walters and D. Chen, *Growing Cooler: The Evidence on Urban Development and Climate Change*, Urban Land Institute, Chicago (2007); S. Handy, X. Cao and P.L. Mokhtarian, "Self-Selection in the Relationship between the Built Environment and Walking," *Journal of the American Planning Association* **72** (1), 55-74 (2006); R.A. Johnston, *Review of U.S. and European Regional Modeling Studies of Policies Intended to Reduce Transportation Greenhouse Gas Emissions*, Institute of Transportation Studies, University of California, Davis, research report UCD-ITS-RR-08-12.; S.L. Handy, L. Weston and P.L. Mokhtarian "Driving by Choice or Necessity?" *Transportation Research Part A* **39** (3), 183-203 (2005).
27. L.R. Glicksman, "Energy Efficiency in the Built Environment," *Physics Today*, July 2008, p.35.
28. *2007 Buildings Energy Data Book*, prepared for the Buildings Technologies Program and Office of Planning, Budget, and Analysis, Energy Efficiency and Renewable Energy, U.S. Department of Energy by D&R International, Ltd., September 2007, p. 2-1
29. I. Johnstone, *"Energy and Mass Flows of Housing: Estimating Mortality," Building and Environment*, v.36, pp. 43-51 (2001).
30. *2007 Buildings Energy Data Book, op. sit.*, p. 2-1
31. *2007 Buildings Energy Data Book, op. sit* , p. 2-3
32. *2007 Buildings Energy Data Book, op. sit*
33. *2007 Building Energy Data Book, op. sit*, p. 2-5
34. *2007 Buildings Energy Data Book, op. sit*
35. EIA 2008 Annual Energy Outlook
36. Navigant Consulting, Inc., U.S. Lighting Market Characterizations, Volume I, National Lighting Inventory and Energy Consumption Estimate, Final Report for Department of Energy, 2002.
37. U.S. Environmental Protection Agency, Energy Star Program, *Frequently Asked Questions: Information on Compact Fluorescent Light Bulbs (CFLs) and Mercury*, http://www.energystar.gov/ia/partners/promotions/change_light/downloads/Fact_Sheet_Mercury.pdf , June 2008.
38. J. McMahon, "Buildings: Energy Efficiency Options," presentation at the Haas School, University of California, Berkeley, Nov. 27, 2007.
39. V. Loftness, "Improving Building Energy Efficiency in the U.S.: Technologies and Policies for 2010 to 2050," in Proceedings of the Workshop, *The 10-50 Solution: Technologies and Policies for a Low-Carbon Future*, sponsored by the Pew Center on Global Climate Change and the National Commission on Energy Policy (2004).
40. C. Zimmerman, Wal-Mart, "The Continuing Evolution of Sustainable Facilities at Wal-Mart," private communication, (2008).
41. A. Meier, Lawrence Berkeley National Laboratory, private communication (2008).
42. C. Marnay, G. Venkataramanan, M. Stadler, A.S. Siddiqui, R. Firestone and B. Chandran, "Optimal Technology Selection and Operation of Commercial-Building Microgrids," *IEEE Transactions on Power Systems* (2007), and references therein.
43. C40 Cities Climate Leadership Group, http://www.c40cities.org/bestpractices/buildings/berlin_efficiency.jsp, 2008
44. M.A. Brown, M.D. Levine, W. Short and J.G. Koomey, "Scenarios for a Clean Energy Future," *Energy Policy* **29**, 1179-1196 (2001).
45. R. Brown, S. Borgeson and J. Koomey, *Building-Sector Energy Efficiency Potential Based on the Clean Energy Futures Study*, 2008, to be published.
46. C. Turner and M. Frankel, *Energy Performance of LEEP for New Construction Buildings, Final*

Report, New Buildings Institute, White Salmon, WA, 2008.

47. J.L. Barrientos Sacari, U. Bhattacharjee, T. Martinez and J.J. Duffy, "Green Buildings in Massachusetts: Comparison between Actual and Predicted Energy Performance," *Proceedings of the American Solar Energy Society*, Cleveland, OH, July 9-13, 2007.
48. G.J. Brohard *et al., Advanced Customer Technology Test for Maximum Energy Efficiency (ACT^2) Project: The Final Report*, Pacific Gas & electric, 1997.
49. P. Norton *et al., Affordable High-Performance Homes: A Cold-Climate Case Study*, NREL/TP-550-31650, National Renewable Energy Laboratory, Golden, CO, April 2005; J. Christian, *Affordable and Demand Responsive Net Zero House*, draft report, private communication (2008).
50. J. Schnieders, "CEPHEUS – Measurement Results from More than 100 Dwelling Units in Passive Houses," posted at http://www.cepheus.de/eng/index.html.
51. Committee on Benefits of DOE R&D on Energy Efficiency and Fossil Energy, *Energy Research at DOE: Was It Worth It? Energy Efficiency and Fossil Energy Research 1978 to 2000,* National Academy Press, Washington, DC (2001).
52. *IBID*
53. International Energy Agency, *Mind the Gap: Quantifying Principal-Agent Problems in Energy Efficiency*, IEA, Paris, 2007; Panel on Policy Implications of Greenhouse Warming, *Policy Implications of Greenhouse Warming: Mitigation, Adaptation, and the Science Base*, National Academy Press, Washington, DC (1992); R. Cavanagh, *Energy Efficiency in Buildings and Equipment: Remedies for Pervasive Market Failures*, National Commission on Energy Policy, 2004; D.B. Goldstein, *Saving Energy, Growing Jobs*, Bay Tree Publishing, Berkeley, CA (2007).
54. A. Sudarshan and J. Sweeney, "Deconstructing the 'Rosenfeld Curve'," submitted to *Energy Journal* (2008).
55. A. Rosenfeld, California Energy Commission, private communication (2008).
56. *IBID*
57. American Council for an Energy Efficient Economy (ACEEE), http://www.aceee.org/energy/applstnd.htm (2008).
58. S. Nadel *et al., Leading the Way: Continued Opportunities for New State Appliance and Equipment Standards*, ACEEE, Washington DC (2006); also see Energy Bill Savings Estimates as Passed by the Senate, ACEEE, Washington DC (December 14, 2007), posted at http://www.aceee.org/energy/index.htm.
59. S. Nadel, A.M. Shipley and R.N. Elliott, *The Technical, Economic and Achievable Potential for Energy Efficiency in the United States: A Meta-Analysis of Recent Studies*, American Council for an Energy Efficient Economy, Washington, DC (2004).

California State Policy on Sustainable Energy

Dian M. Grueneich

Former Commissioner
California Public Utilities Commission
Partner, Morrison & Foerster

Abstract. California has set an ambitious goal of pursuing all cost-effective energy efficiency and increasing the percent of electrical power generated by renewable energy sources to 33% by 2020. Through a large mixture of projects, many overseen by the California Public Utilities Commission, the state is aiming to greatly increase its reliance on sustainable energy.

ACHIEVING CALIFORNIA'S GREENHOUSE GAS REDUCTION GOALS

This is an exciting time to be involved with energy policy at the state level, and particularly in California. Beginning with the creation of the California Energy Commission (CEC) in 1974 and continuing through today, California has led the nation in energy efficiency. It is also leading the effort to produce more power from renewable energy sources. Many of the technical advances facilitating this effort have come from California universities and from national laboratories located in the state. Given California's leadership and the exciting opportunities in sustainable energy, I encourage more students and post-docs in the sciences to consider careers in the energy area.

From 2005–2010, I served as a Commissioner on the California Public Utilities Commission (CPUC), which is particularly focused on renewable energy sources, including distributed generation, and energy efficiency. Pursuit of these energy resources is an essential component of California's goal of reducing emissions of greenhouse gases. Under the mandate of California's climate change law (Assembly Bill (AB) 32), the state must reduce greenhouse gas emissions to 1990 levels by 2020. This target translates to emissions of 427 million metric tons of CO_2 equivalent by 2020. In the energy sector, energy efficiency and renewable power development, including local distributed generation, are the major policy tools being used to reduce emissions.

California has formulated concrete policies to address the AB 32 mandate for reducing greenhouse-gas emission. As summarized in Figure 1, the projected total reduction is 174 million metric tons (MMT) per year of carbon-dioxide equivalent by 2020. About 29% of this decrease, or 50 MMT, is expected to come from the energy sector. Another 33% would come from the transportation sector, 20% from cap-and-

Physics of Sustainable Energy II: Using Energy Efficiently and Producing it Renewably
AIP Conf. Proc. 1401, 153-161 (2011); doi: 10.1063/1.3653849

trade schemes (which allow emitters to buy and sell credits for activities beyond those otherwise mandated) and 33% from other efforts, such as addressing greenhouse gases other than CO_2. Of the energy-sector reductions, about half of are expected to result from enhanced end-use efficiency in appliances and buildings, and the other half from the producing electricity from renewable energy sources.

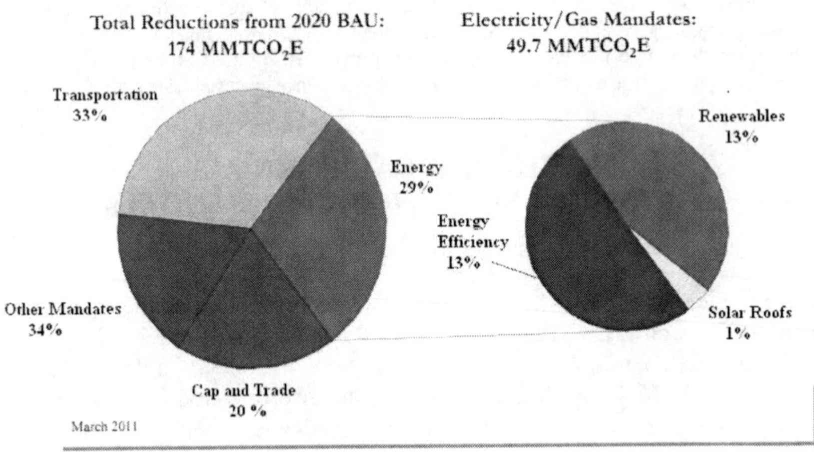

FIGURE 1. Projected reductions in carbon dioxide emissions for all programs (left) and energy programs (right).

CALIFORNIA'S PREFERRED RESOURCES

In 2003, California formulated an Energy Action Plan [1] under the guidance of the CPUC, the CEC, and the Consumer Power and Conservation Financing Authority. It was updated in February 2008. The Energy Action Plan introduced the concept of a "loading order," or a prioritized list of how the state and its utilities should invest to meet California's energy needs. The loading order (codified in California Public Utilities Code Section 454.5) is as follows:

- **Energy Efficiency.** Efficiency has been a mainstay to reduce California's energy needs and continues to play the leading role. As an example, the energy requirements for refrigerators have been reduced to one-quarter of their 1975 levels, while their volumes have doubled.

- **Demand response.** This term describes the management of customer consumption of electricity in response to supply conditions. For example, the price of peak power can be adjusted to encourage users to shift their consumption to off-peak hours. Reducing peak demand can avoid the need to build very expensive "peaking plants" that run few hours per year. It can also save money for utilities because electricity is usually sold at an average price, although its cost is at least a factor of 2 higher at peak power compared to off-peak power. The advent of smart meters can play an important role in facilitating demand response by allowing real time price feedback to consumers and businesses.

- **Distributed generation.** This term refers to the generation of electricity from small sources, which are often located closer to the point of use. These sources are increasingly renewable-energy sources, such as solar photovoltaics on commercial and residential roofs.

- **Renewable generation.** This term refers to large-scale solar, wind, biomass, and geothermal projects. In the category of solar energy, solar thermal power, which produces steam that drives turbines to generate electricity, has been the favored approach, although solar photovoltaic is gaining in interest as prices have been dropping dramatically.

- **Cleanest available fossil resources.** Only after squeezing all the energy gains from the sources listed above will California consider adding more new fossil-fuel plants. Utilities are prohibited by law from entering into long-term contracts for fossil-fuel plants unless they are state-of-the-art natural gas plants.

The bottom line is that California's policy focus is clear: Reduce demand first, look to fill the remaining need with renewable sources, and pursue new fossil plants, using the cleanest technology, as the final resort.

CALIFORNIA'S LONG-TERM ENERGY EFFICIENCY STRATEGIC PLAN

During a 2007 trip to meet with energy leaders in Europe, I learned that the European Union (EU) was requiring each member nation to develop a national energy efficiency action plan. Following the EU example, the CPUC decided to develop an energy efficiency action roadmap for California and particularly the investor-owned utilities regulated by the Commission. The CPUC worked with a wide-range of stakeholders, including other state and federal agencies, utilities, national labs and universities, consumers, and businesses. The strategic plan was finished in 2008 and updated in 2010.[2] The Plan is the most comprehensive roadmap for deep, sustained energy efficiency and it encompasses all of California's main economic sectors—residential, commercial, industrial, and agricultural. For each sector, the roadmap has a chapter with very specific goals and strategies.

The roadmap also deals with seven crosscutting areas, which are as follows:

- heating, ventilation and air-conditioning
- codes and standards
- demand-side management integration
- workforce education and training
- marketing, education and outreach
- research and technology
- local governments
- lighting.

One key theme in the Strategic Plan is "market transformation," or how to bring in new technologies, test them, lower their costs, get them into the marketplace, and then move their use into codes and standards and/or general public use. For such market transformation, it is critical for government policy makers to understand not just what the private sector is doing and how that sector is investing its capital but also what researchers in academic institutions are doing.

The California Strategic Plan is being vigorously implemented. Between 2010 and 2012, $3.8 billion will be spent on programs funded by California's investor-owned utility ratepayers to improve energy efficiency. That is the largest such program in the United States and, except possibly for China, the world. The bulk of the money, or $3.13 billion, is being spent for general energy efficiency programs and the remaining $750 million is earmarked for appliances and energy retrofits in low-income homes. The estimated three-year savings potential is 7000 gigawatt-hours (GWh). These energy savings will avoid the emission of 3 million metric tons of carbon dioxide equivalent and avoid the building of three large power plants, one per year for three years.

The economic impact of this program should be very significant. It is estimated[3] that the $3.8 billion investment will generate 15–20,000 new jobs. Some of the program funds will be spent on work-force training ($122 million) and some will go to local government programs ($260 million).

Implementation includes the following elements:

- Consolidation of programs. Already, 200 statewide programs have been merged into 12 programs.
- A statewide education campaign to enhance energy efficiency. A major program is looking at behavior changes because scientific studies are increasingly finding that motivating people is key to achieving energy efficiency.
- Creation of a web portal for people interested in efficiency to share information.[4]
- Continuous energy improvement programs for industry in which industries continually seek to achieve "best practices" as new technologies and processes evolve.
- Investment in advanced lighting technologies.

• Review of the best practices for measurement and verification.

RENEWABLE ENERGY SOURCES

California is the nation's leader in development of renewable resources. California relies upon a "Renewable Portfolio Standard" (RPS) which sets mandated target levels for the state's utilities and other power providers. The state's 20% RPS by 2010 was expanded in 2011 by legislation (SB XI 2) that sets a 33% RPS requirement by 2020.

The impact of California's RPS is shown Figure 2, which charts the growth in renewable capacity since 2003. One can see the rapid rise in renewable capacity over the past five years. Also note that 78% of the projects and 59% of the generating capacity in MW have been added from in-state sources. These figures reflect the abundance of renewable resources in California. Those numbers are expected to rise significantly with the addition of new solar projects, many using photovoltaic technologies, and wind developments.

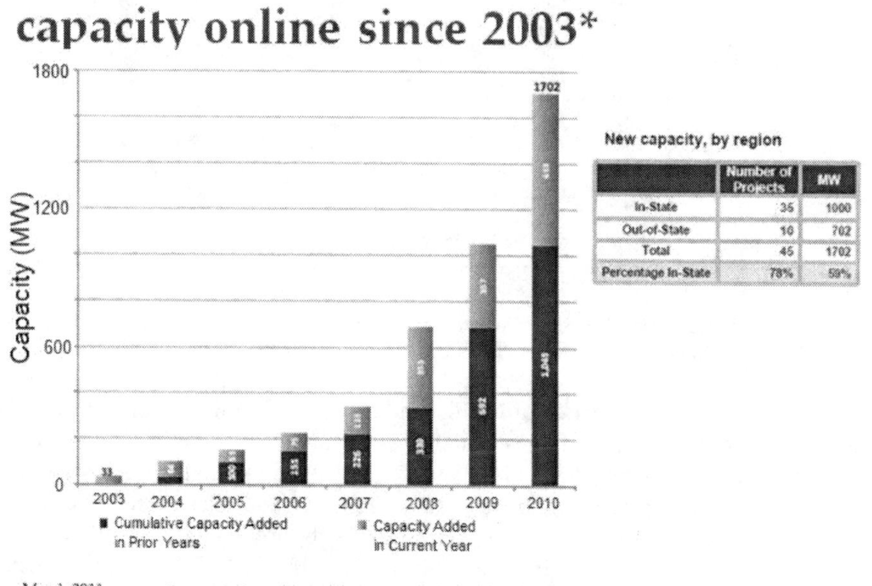

FIGURE 2. Renewable energy capacity added in California since 2003. *Additional out-of-state projects—over 1,000 MW—have come online after 2005 under short-term contracts.

Figure 3 charts the respective investments in renewable energy sources by the three large California investor-owned utilities: Pacific Gas & Electric (PG&E), Southern California Edison (SCE), and San Diego Gas and Electric (SDG&E). The utilities are forecast to reach the 20% goal in 2011-12. Since 2003, the CPUC has approved more than 180 contracts for more than 14,000 MW of new and existing eligible renewable energy capacity. Contracts for an additional 4,000 MW of mostly new capacity are under review. The 2009 RPS solicitation resulted in 100,000 GWh of bids. Those bids alone would meet half of the 33% target.

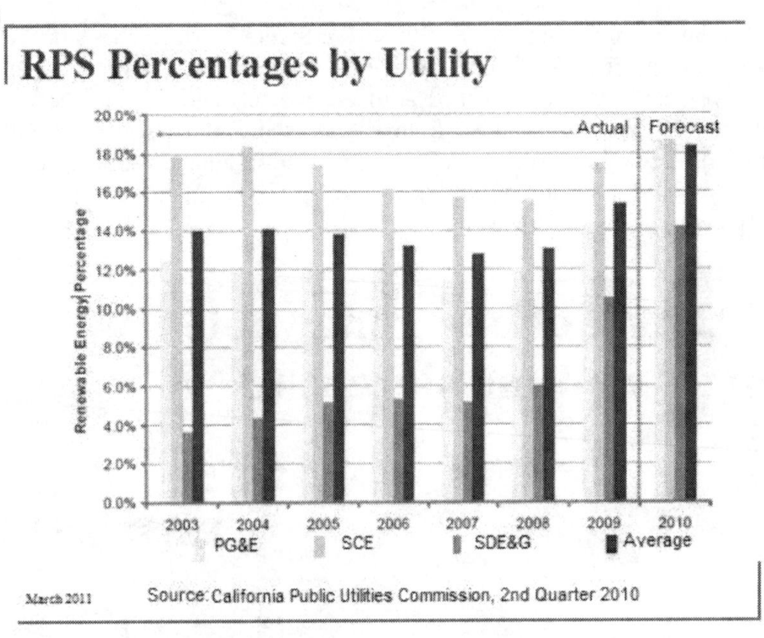

FIGURE 3. Renewable Portfolio Standard for Pacific Gas and Electric (PG&E), Southern California Edison (SCE), and San Diego Gas and Electric (SDG&E) utilities.

In the past six years the State of California has issued permits for the construction of three new large transmission lines, which are now under construction at cost of over $6 billion. The CPUC determined these lines were needed to provide transmission for new renewable development, while also helping improve the reliability of California's electric grid and, in some cases, lowering costs by relieving transmission line congestion. Permitting and building new transmission lines is often difficult, because major lines can cause significant environmental impacts and few people voluntarily seek to have new lines built near where they live. But the new California transmission lines were approved after full review under California's stringent environmental laws

ABOUT THE CALIFORNIA SOLAR INITIATIVE

A central ingredient in California's renewable program is solar energy. Recent RPS bids have shifted toward solar in the past few years, as seen in Figure 4. The move to solar is encouraged by the CPUC's California Solar Initiative (CSI) program, which provides incentives for solar system installations to customers of the state's three investor-owned utilities (IOUs): PG&E, SCE and SDG&E. The CSI Program provides upfront incentives for solar systems installed on existing residential homes, as well as existing and new commercial, industrial, government, non-profit, and agricultural properties within the service territories of the IOUs.

The CSI Program, also known as the "Million Solar Roofs," was authorized by the CPUC through a number of regulatory decisions throughout 2006. When it was launched in January 2007, the CSI Program built upon nearly 10 years of state support for solar, including other incentive programs such as the Emerging Renewables Program and the Self-Generation Incentive Program.

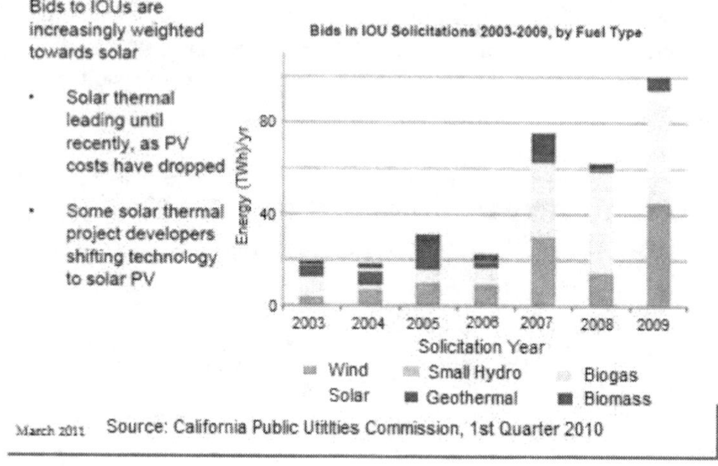

FIGURE 4. The number of bids for renewable energy projects received by investor-owned utilities (IOUs) as a function of the energy source. Note the trend toward a larger percentage of the bids based on solar energy.

The CSI Program has a budget of $2.167 billion over 10 years, and the goal is to reach 1,940 MW of installed solar capacity by the end of 2016. The goal includes 1,750 MW of capacity from the general market program, as well as 190 MW of capacity from the low-income programs. The general market program is the main incentive component of the CSI and is administered through three Program Administrators: PG&E, SCE, and the California Center for Sustainable Energy in

SDG&E territory.

In addition to the general market program, the CSI Program has four other program components, each with their own program administrator and 10-year budgets:

- A research and development program
- The Single-family Solar Affordable Solar Housing program which provides solar incentives to single-family low income housing.
- The Multifamily Affordable Solar Housing program.
- The Solar Water Heating Pilot.

Current information about the CSI and other solar efforts in California can be found on the CPUC and CEC websites.

DISTRIBUTED GENERATION

Another important ingredient of California's renewable energy program is smaller-scale, distributed generation of electricity. California Governor Jerry Brown, the CPUC, developers, and the utilities have increasingly focused on tapping small-scale renewable energy sources (under 20 MW). These projects can come on line faster, require less financing, and have fewer or no new transmission interconnection requirements.

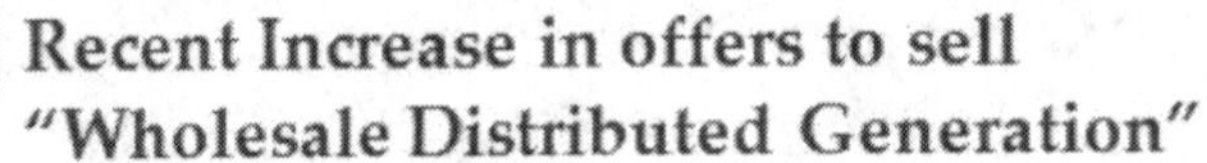

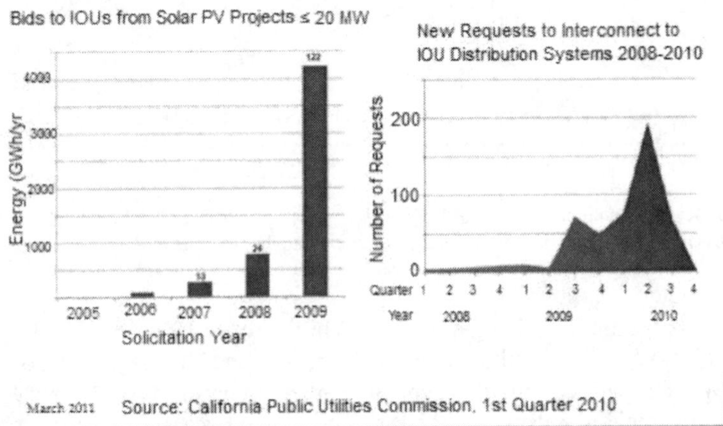

FIGURE 5. Distributed-solar power bids at less than 20 MW are increasing, with 122 in 2009.

The interest in smaller-scale distributed energy is reflected in Figure 5, which shows both the total energy represented by RPS bids for solar projects under 20 MW and the number of requests to interconnect small projects to the investor-owned utility distribution system. A major part of the distributed-energy program is the effort to install solar photovoltaics on rooftops of large commercial buildings such as "big box" stores.

California is also encouraging the development of new technology in such areas as wind power and fuel cells. All these programs should ensure that the state meet its ambitious goal of increasing the percentage of electrical energy generated by renewable to 33% by 2020.

CONCLUSION

California has been a leader in sustainable energy policy for over three decades and its efforts continue unabated. It has launched innovative programs and policies in energy efficiency, demand response, large renewable power development, and distributed generation. These efforts require a complex set of actions by the public and private sectors, and continuing introduction of new technologies and approaches. California's academic institutions and the national laboratories have provided key analysis, R&D, and policy support on many levels and must continue to do so in the future. California's success as the leader in sustainable energy development requires the involvement and support of all stakeholders.

REFERENCES

1. Energy Action Plan 2003 and Update 2008 (www.cpuc.ca.gov/NR/rdonlyres/58ADCD6A-7FE6-4B32-8C70-7C85CB31EBE7/0/2008_EAP_UPDATE.PDF)

2. The California Long-Term Energy Efficiency Strategic Plan (www.californiaenergyeffciency.com)

3. The job benefit calculation resulting from California's Long Term Energy Efficiency Strategic Plan is based on the Council of Economic Advisor's May 2009 publication of "Estimates of Job Creation by the American Recovery and Reinvestment Act of 2009."

4. For the energy-efficiency web portal, see www.engage360.com.

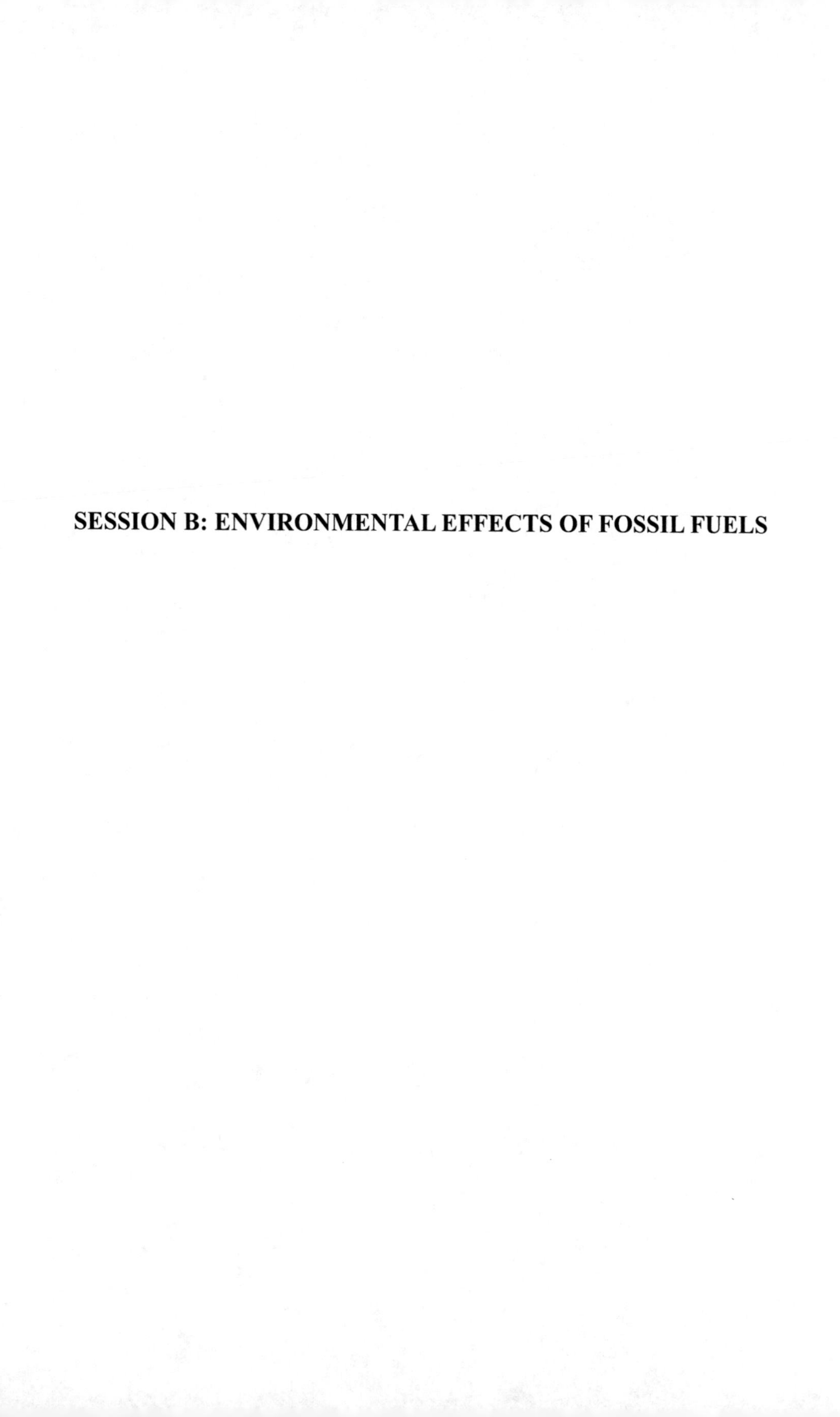

SESSION B: ENVIRONMENTAL EFFECTS OF FOSSIL FUELS

Hidden Costs of Energy: Unpriced Consequences of Energy Production and Use*

Committee on Health, Environmental, and Other External Costs and Benefits of Energy Production and Consumption
National Research Council

National Research Council
500 5th St., NW
Washington, DC 20001

Abstract. The U.S. Congress directed the U.S. Department of the Treasury to arrange for a review by the National Academy of Sciences to define and evaluate the health, environmental, security, and infrastructural external costs and benefits associated with the production and consumption of energy—costs and benefits that are not or may not be fully incorporated into the market price of energy, into the federal tax or fee, or into other applicable revenue measures related to production and consumption of energy. In response, the National Research Council established the Committee on Health, Environmental, and Other External Costs and Benefits of Energy Production and Consumption, which prepared the report summarized in this chapter.

The report estimates dollar values for several major components of these costs. The damages the committee was able to quantify were an estimated $120 billion in the U.S. in 2005, a number that reflects primarily health damages from air pollution associated with electricity generation and motor vehicle transportation. The figure does not include damages from climate change, harm to ecosystems, effects of some air pollutants such as mercury, and risks to national security, which the report examines but does not monetize.

SUMMARY

Modern civilization is heavily dependent on energy from sources such as coal, petroleum, and natural gas. Yet, despite energy's many benefits, most of which are reflected in energy market prices, the production, distribution, and use of energy also cause negative effects. Beneficial or negative effects that are not reflected in energy market prices are termed "external effects" by economists. In the absence of

---

* Reprinted with permission from the National Research Corporation from *Hidden Costs of Energy: Unpriced Consequences of Energy Production and Use*, Summary, pages 3-20, Copyright 2010. The report authors are Jared L. Cohon (chair), Maureen L. Cropper (vice chair), Mark R. Cullen, Elisabeth M. Drake, Mary R. English, Christopher B. Field, Daniel S. Greenbaum, James K. Hammitt, Rogene F. Henderson, Catherine L. Kling, Alan J. Krupnick, Russell Lee, H. Scott Matthews, Thomas E. McKone, Gilbert E. Metcalf, Richard G. Newell, Richard L. Revesz, Ian Sue Wing, Terrance G. Surles.

Physics of Sustainable Energy II: Using Energy Efficiently and Producing it Renewably
AIP Conf. Proc. 1401, 165-182 (2011); doi: 10.1063/1.3653850
2011 American Institute of Physics 978-0-7354-0972-9/$30.00

government intervention, external effects associated with energy production and use are generally not taken into account in decision making.

When prices do not adequately reflect them, the monetary value as-signed to benefits or adverse effects (referred to as damages) are "hidden" in the sense that government and other decision makers, such as electric utility managers, may not recognize the full costs of their actions. When market failures like this occur, there may be a case for government interventions in the form of regulations, taxes, fees, tradable permits, or other instruments that will motivate such recognition.

Recognizing the significance of the external effects of energy, Congress requested this study in the Energy Policy Act of 2005 and later directed the Department of the Treasury to fund it under the Consolidated Appropriations Act of 2008. The National Research Council committee formed to carry out the study was asked to define and evaluate key external costs and benefits—related to health, environment, security, and infrastructure—that are associated with the production, distribution, and use of energy but not reflected in market prices or fully addressed by current government policy. The committee was not asked, however, to recommend specific strategies for addressing such costs because policy judgments that transcend scientific and technological considerations—and exceed the committee's mandate—would necessarily be involved. The committee studied energy technologies that constitute the largest portion of the U.S. energy system or that represent energy sources showing substantial increases (>20%) in consumption over the past several years.

We evaluated each of these technologies over their entire life cycles—from fuel extraction to energy production, distribution, and use to disposal of waste products—and considered the external effects at each stage. Estimating the damages associated with external effects was a multi-step process, with most steps entailing assumptions and their associated uncertainties. Our method, based on the "damage function approach," started with estimates of burdens (such as air-pollutant emissions and water-pollutant discharges). Using mathematical models, we then estimated these burdens' resultant ambient concentrations as well the ensuing exposures. The exposures were then associated with consequent effects, to which we attached monetary values in order to produce damage estimates. One of the ways economists assign monetary values to energy-related adverse effects is to study people's preferences for reducing those effects. The process of placing monetary values on these impacts is analogous to determining the price people are willing to pay for commercial products. We applied these methods to a year close to the present (2005) for which data were available and also to a future year (2030) to gauge the impacts of possible changes in technology.

A key requisite to applying our methods was determining which policy-relevant effects are truly external, as defined by economists. For example, increased food prices caused by the conversion of agricultural land from food to biofuel production, are not considered to represent an external cost, as they result from (presumably properly functioning) markets. Higher food prices may of course raise important social concerns and may thus be an issue for policy makers, but because they do not constitute an external cost they were not included in the study.

Based on the results of external-cost studies published in the 1990s, we focused especially on air pollution. In particular, we evaluated effects related to emissions of particulate matter (PM), sulfur dioxide (SO_2), and oxides of nitrogen (NO_x), which form criteria air pollutants.[1] We monetized effects of those pollutants on human health, grain crop and timber yields, building materials, recreation, and visibility of outdoor vistas. Health damages, which include premature mortality and morbidity (such as chronic bronchitis and asthma), constituted the vast majority of monetized damages, with premature mortality being the single largest health-damage category.

Some external effects could only be discussed in qualitative terms in this report. Although we were able to quantify and then monetize a wide range of burdens and damages, many other external effects could not ultimately be monetized because of insufficient data or other reasons. In particular, the committee did not monetize impacts of criteria air pollutants on ecosystem services or nongrain agricultural crops, or effects attributable to emissions of hazardous air pollutants.[2] In any case, it is important to keep in mind that the individual estimates presented in this report, even when quantifiable, can have large uncertainties. In addition to its external effects in the present, the use of fossil fuels for energy creates external effects in the future through its emissions of atmospheric greenhouse gases (GHGs)[3] that cause climate change, subsequently resulting in damages to ecosystems and society. This report estimates GHG emissions from a variety of energy uses, and then, based on previous studies, provides ranges of potential damages. The committee determined that attempting to estimate a single value for climate-change damages would have been inconsistent with the dynamic and unfolding insights into climate change itself and with the extremely large uncertainties associated with effects and range of damages. Because of these uncertainties and the long time frame for climate change, our report discusses climate-change damages separately from damages not related to climate change.

OVERALL CONCLUSIONS AND IMPLICATIONS

Electricity

Although the committee considered electricity produced from coal, natural gas, nuclear power, wind, solar energy, and biomass, it focused mainly on coal and natural gas—which together account for nearly 70% of the nation's electricity—and on monetizing effects related to the air pollution from these sources. From previous studies, it appeared that the electricity-generation activities accounted for the majority of such external effects, with other activities in the electricity cycle, such as mining and drilling, playing a lesser role.

Coal

Coal, a nonrenewable fossil fuel, accounts for nearly half of all electricity produced in the United States. We monetized effects associated with emissions from 406 coal-fired power plants, excluding Alaska and Hawaii, during 2005. These facilities represented 95% of the country's electricity from coal. Although coal-fired electricity generation from the 406 sources resulted in large amounts of pollution overall, a plant-by-plant breakdown showed that the bulk of the damages were from a relatively small number of them. In other words, specific comparisons showed that the source-and-effect landscape was more complicated than the averages would suggest.

Damages Unrelated to Climate Change. The aggregate damages associated with emissions of SO_2, NO_x, and PM from these coal-fired facilities in 2005 were approximately $62 billion, or $156 million on average per plant.[4] However, the differences among plants were wide—the 5th and 95th percentiles of the distribution were $8.7 million and $575 million, respectively. After ranking all the plants according to their damages, we found that the 50% of plants with the lowest damages together produced 25% of the net generation of electricity but accounted for only 12% of the damages. On the other hand, the 10% of plants with the highest damages, which also produced 25% of net generation, accounted for 43% of the damages. Figure S-1 shows the distribution of damages among coal-fired plants.

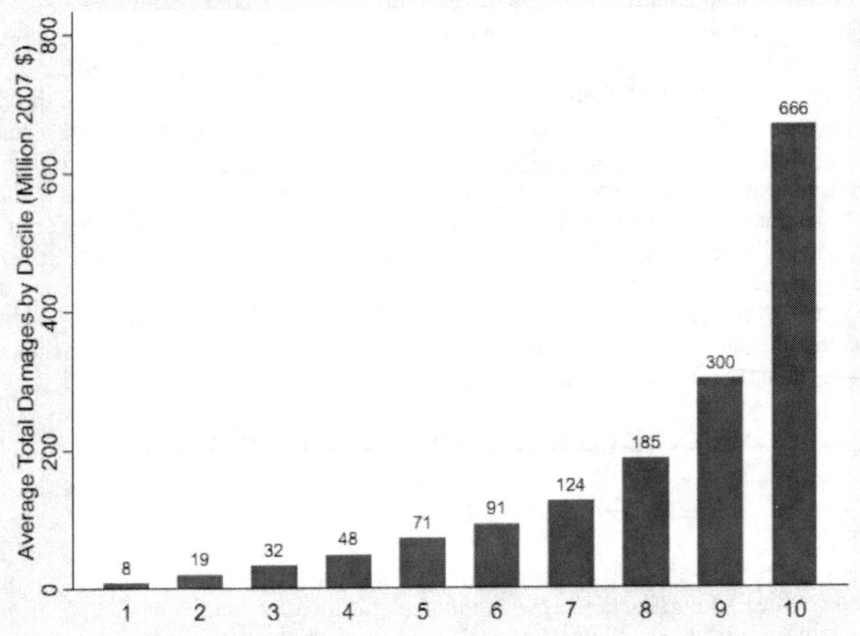

FIGURE S-1. Average Total Damages by Decile (Million 2007 $). Distribution of aggregate damages among the 406 coal-fired power plants analyzed in this study. In computing this chart, plants were sorted from smallest to largest based on damages associated with each plant. The lowest decile (10% increment) represents the 40 plants with the smallest damages per plant (far left). The decile of plants that produced the most damages is on the far right. The figure on the top of each bar is the average damage across all plants of damages associated with sulfur dioxide, oxides of nitrogen, and particulate matter. Damages related to climate-change effects are not included.

Some of the variation in damages among plants occurred because those that generated more electricity tended to produce greater damages; hence, we also reported damages per kilowatt hour (kWh) of electricity produced. If plants are weighted by the amount of electricity they generate, the mean damage is 3.2 cents per kWh. For the plants examined, variation in damages per kWh is primarily due to variation in pollution intensity (emissions per kWh) among plants, rather than variation in damages per ton of pollutant. Variations in emissions per kWh mainly reflected the sulfur content of the coal burned; the adoption, or not, of control technologies (such as scrubbers); and the vintage of the plant—newer plants were subject to more stringent pollution-control requirements. As a result, the distribution of damages per kWh was highly skewed: There were many coal-fired power plants with modest damages per kWh as well as a small number of plants with large damages. The 5th percentile of damages per kWh is less than half a cent, and the 95th percentile of damages is over 12 cents.[5]

The estimated air-pollution damages associated with electricity generation from coal in 2030 will depend on many factors. For example, damages per kWh are a function of the emissions intensity of electricity generation from coal (for example, pounds [lb] of SO_2 per megawatt hour [MWh]), which in turn depends on future regulation of power-plant emissions. Based on government estimates, net power generation from coal in 2030 is expected to be 20% higher on average than in 2005. Despite projected increases in damages per ton of pollutant resulting mainly from population and income growth—average damages per kWh from coal plants (weighted by electricity generation) are estimated to be 1.7 cents per kWh in 2030 as compared with 3.2 cents per kWh in 2005. This decrease derives from the assumption that SO_2 emissions per MWh will fall by 64% and that NO_x and PM emissions per MWh will each fall by approximately 50%.

Natural Gas

An approach similar to that used for coal allowed the committee to estimate criteria-pollutant-related damages for 498 facilities in 2005 that generated electricity from natural gas in the contiguous 48 states. These facilities represented 71% of the country's electricity from natural gas. Again, as with coal, the overall averages masked some major differences among plants, which varied widely in terms of pollution generation.

Damages Unrelated to Climate Change. Damages from gas-fueled plants tend to be much lower than those from coal plants. The sample of 498 gas facilities produced $740 million in aggregate damages from emissions of SO_2, NO_x, and PM. Average annual damages per plant were $1.49 million, which reflected not only lower damages per kWh at gas plants but smaller plant sizes as well; net generation at the median coal plant was more than six times larger than that of the median gas facility. After sorting the gas plants according to damages, we found that the 50% with the lowest damages accounted for only 4% of aggregate damages. By contrast, the 10% of plants with the largest damages produced 65% of the air-pollution damages from all 498 plants (see Figure S-2). Each group of plants accounted for approximately one-quarter of the sample's net generation of electricity.

Mean damages per kWh were 0.16 cents when natural-gas-fired plants were weighted by the amount of electricity they generated. However, the distribution of damages per kWh had a large variance and was highly skewed. The 5th percentile of damages per kWh is less than 5/100 of a cent, and the 95th percentile of damages is about 1 cent.[6] Although overall electricity production from natural gas in 2030 is predicted to increase by 9% from 2005 levels, the average pollution intensity for natural-gas facilities is expected to decrease, though not as dramatically as for coal plants. Pounds of NO_x emitted per MWh are estimated to fall, on average, by 19%, and emissions of PM per MWh are estimated to fall by about 32%. The expected net effect of these changes is a decrease in the aggregate damages related to the 498 gas facilities from $740 million in 2005 to $650 million in 2030. Their average damage per kWh is expected to fall from 0.16 cents to 0.11 cents over that same period.

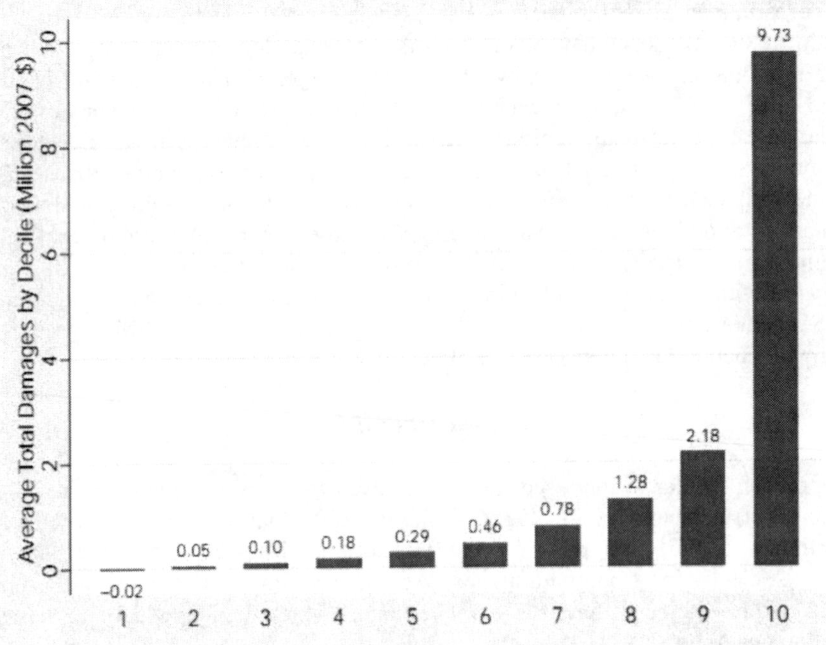

FIGURE S-2. Average Total Damages by Decile (Million 2007 $). Distribution of aggregate damages among the 498 natural-gas-fired power plants analyzed in this study. In computing this chart, plants were sorted from smallest to largest based on damages associated with each plant. The lowest decile (10% increment) represents the 50 plants with the smallest damages per plant (far left). The decile of plants that produced the most damages is on the far right. The figure on the top of each bar is the average damage across all plants of damages associated with sulfur dioxide, oxides of nitrogen, and particulate matter. Damages related to climate-change effects are not included.

Nuclear

The 104 U.S. nuclear reactors currently account for almost 20% of the nation's electrical generation. Overall, other studies have found that damages associated with the normal operation of nuclear power plants (excluding the possibility of damages in the remote future from the disposal of spent fuel) are quite low compared with those of fossil-fuel-based power plants.[7] However, the life cycle of nuclear power does pose some risks. If uranium mining activities contaminate ground or surface water, people could potentially be exposed to radon or other radionuclides through ingestion. Because the United States mines only about 5% of the world's uranium supply, such risks are mostly experienced in other countries.

Low-level nuclear waste is stored until it decays to background levels and currently does not pose an immediate environmental, health, or safety hazard. However, regarding spent nuclear fuel, development of full-cycle, closed-fuel processes that recycle waste and enhance security could further lower risks.

A permanent repository for spent fuel and other high-level nuclear wastes is perhaps the most contentious nuclear-energy issue, and considerably more study of the external cost of such a repository is warranted.

Renewable Energy Sources

Wind power currently provides just over 1% of U.S. electricity, but it has large growth potential. Because no fuel is involved in electricity generation, neither gases nor other contaminants are released during the operation of a wind turbine. Its effects do include potentially adverse visual and noise effects, and the killing of birds and bats. In most cases, wind-energy plants currently do not kill enough birds to cause population-level problems, except perhaps locally and mainly with respect to raptors. The tallies of bats killed and the population consequences of those deaths have not been quantified but could be significant. If the number of wind-energy facilities continues to grow as fast as it has recently, bat and perhaps bird deaths could become more significant.

Although the committee did not evaluate in detail the effects of solar and biomass generation of electricity, it has seen no evidence that they currently produce adverse effects comparable in aggregate to those of larger sources of electricity. However, as technology improves and penetration into the U.S. energy market grows, the external costs of these sources will need to be reevaluated.

Greenhouse Gas Emissions and Electricity Generation

Emissions of carbon dioxide (CO_2) from coal-fired power plants are the largest single source of GHGs in the United States. CO_2 emissions vary; their average is about 1 ton of CO_2 per MWh generated, having 5th-to-95th-percentile range of 0.95-1.5 tons. The main factors affecting these differences are the technology used to generate the power and the age of the plant. Emissions of CO_2 from gas-fired power plants also are significant, having an average of about 0.5 ton of CO_2 per MWh generated and a 5th-to-95th-percentile range of 0.3-1.1 tons. Life-cycle CO_2 emissions from nuclear, wind, biomass, and solar appear so small as to be negligible compared with those from fossil fuels.

Heating

The production of heat as an end use accounts for about 30% of U.S. primary energy demand, the vast majority of which derives from the combustion of natural gas or the application of electricity. External effects associated with heat production come from all sectors of the economy, including residential and commercial (largely for the heating of living or work spaces) and industrial (for manufacturing processes).

Damages Unrelated to Climate Change. As with its combustion for electricity, combustion of natural gas for heat results in lower emissions than from coal, which is the main energy source for electricity generation. Therefore health and environmental dam-ages related to obtaining heat directly from natural-gas combustion are much less than damages from the use of electricity for heat. Aggregate damages from the combustion of natural gas for direct heat are estimated to be about $1.4 billion per year, assuming that the magnitude of external effects resulting from heat production for industrial activities is comparable to that of residential and commercial uses.[8] The median estimated damages attributable to natural-gas combustion for heat in residential and commercial buildings are approximately 11 cents per thousand cubic feet. These damages do not vary much across regions when considered on a per-unit basis, although some counties have considerably higher external costs than others. In 2007, natural-gas use for heating in the industrial sector, excluding its employment as a process feedstock, was about 25% less than natural-gas use in the residential and commercial building sectors. Damages associated with energy for heat in 2030 are likely to be about the same as those that exist today, assuming that the effects of additional sources to meet demand are offset by lower-emitting sources. Reduction in damages would only result from more significant changes—largely in the electricity-generating sector, as emissions from natural gas are relatively small and well controlled. However, the greatest potential for reducing damages associated with the use of energy for heat lies in greater attention to improving efficiency. Results from the recent National Research Council report America's Energy Future: Technology and Transformation suggest a possible improvement of energy efficiency in the buildings and industrial sectors by 25% or more between now and 2030. Increased damages would also be possible, however, if new domestic energy development resulted in higher emissions or if additional imports of liquefied natural gas, which would increase emissions from the production and international transport of the fuel, were needed.

Greenhouse Gas Emissions

The combustion of a thousand cubic feet of gas generates about 120 lb (0.06 tons) of CO_2. Methane, the major component of natural gas, is a GHG itself and has a global-warming potential about 25 times that of CO_2. Methane enters the atmosphere through leakage, but the U.S. Energy Information Administration estimates that such leakage amounted to less than 3% of total U.S. CO_2-equivalent (CO2-eq) emissions[9] (excluding water vapor) in 2007. Thus, in the near term, where domestic natural gas remains the dominant source for heating, the average emissions factor is likely to be about 140 lb CO_2-eq per thousand cubic feet (including upstream methane emissions); in the longer term—assuming increased levels of liquefied natural gas or shale gas as part of the mix—the emissions factor could be 150 lb CO_2-eq per thousand cubic feet.

Transportation

Transportation, which today is almost completely reliant on petroleum, accounts for nearly 30% of U.S. energy consumption. The majority of transportation-related emissions come from fossil-fuel combustion—whether from petroleum consumed during conventional-vehicle operation, coal or natural gas used to produce electricity to power electric or hybrid vehicles, petroleum or natural gas consumed in cultivating biomass fields for ethanol, or electricity used during vehicle manufacture.

The committee focused on both the nonclimate-change damages and the GHG emissions associated with light-duty and heavy-duty on-road vehicles, as they account for more than 75% of transportation energy consumption in the United States. Although damages from nonroad vehicles (for example, aircraft, locomotives, and ships) are not insignificant, the committee emphasized the much larger highway component.

Damages Unrelated to Climate Change. In 2005, the vehicle sector produced $56 billion in health and other nonclimate-change damages, with $36 billion from light-duty vehicles and $20 billion from heavy-duty vehicles. Across the range of light-duty technology and fuel combinations considered, damages expressed per vehicle miles traveled (VMT) ranged from 1.2 cents to 1.7 cents (with a few combinations having higher damage estimates).[10]

The committee evaluated motor-vehicle damages over four life-cycle stages:

(1) vehicle operation, which results in tailpipe emissions and evaporative emissions;
(2) production of feedstock, including the extraction of the resource (oil for gasoline, biomass for ethanol, or fossil fuels for electricity) and its transportation to the refinery;
(3) refining or conversion of the feedstock into usable fuel and its transportation to the dispenser; and
(4) manufacturing and production of the vehicle.

It is important that, in most cases, vehicle operation accounted for less than one-third of total damages; other components of the life cycle contributed the rest. Life-cycle stages 1, 2, and 3 were somewhat proportional to actual fuel use, while stage 4 (which is a significant source of life-cycle emissions that form criteria pollutants) was not.

The estimates of damage per VMT among different combinations of fuels and vehicle technologies were remarkably similar (see Figure S-3). Because these assessments were so close, it is essential to be cautious when interpreting small differences between combinations. The damage estimates for 2005 and 2030 also were very close, despite an expected rise in population. This result is attributable to the expected national implementation of the recently revised "corporate average fuel economy" (CAFE) standards, which require the new light-duty fleet to have an average fuel economy of 35.5 miles per gallon by 2016 (although an increase in VMT could offset this improvement somewhat).

Despite the general overall similarity, some fuel and technology combinations were associated with greater nonclimate damages than others. For example, corn ethanol, when used in E85 (fuel that is 85% ethanol and 15% gasoline), showed estimated damages per VMT similar to or slightly higher than those of gasoline, both for 2005

a

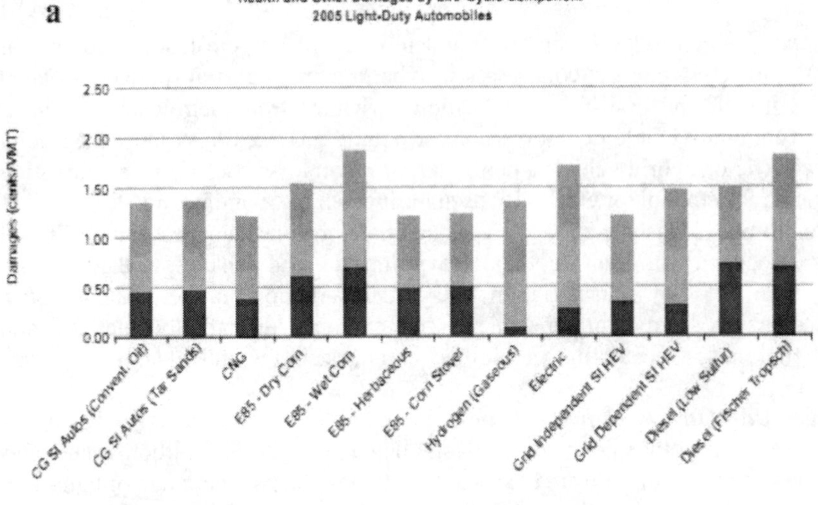

Health and Other Damages by Life-Cycle Component
2005 Light-Duty Automobiles
Damages (cents/VMT)
2.50
2.00
1.50
1.00
0.50
0.00
CG SI Autos (Convent. Oil)
CG SI Autos (Tar Sands)
CNG
E85 - Dry Corn
E85 - Wet Corn
E85 - Herbaceous
E85 - Corn Stover
Hydrogen (Gaseous)
Electric
Grid Independent SI HEV
Grid Dependent SI HEV
Diesel (Low Sulfur)
Diesel (Fischer Tropsch)
Operation
Feedstock
Fuel
Vehicle

b

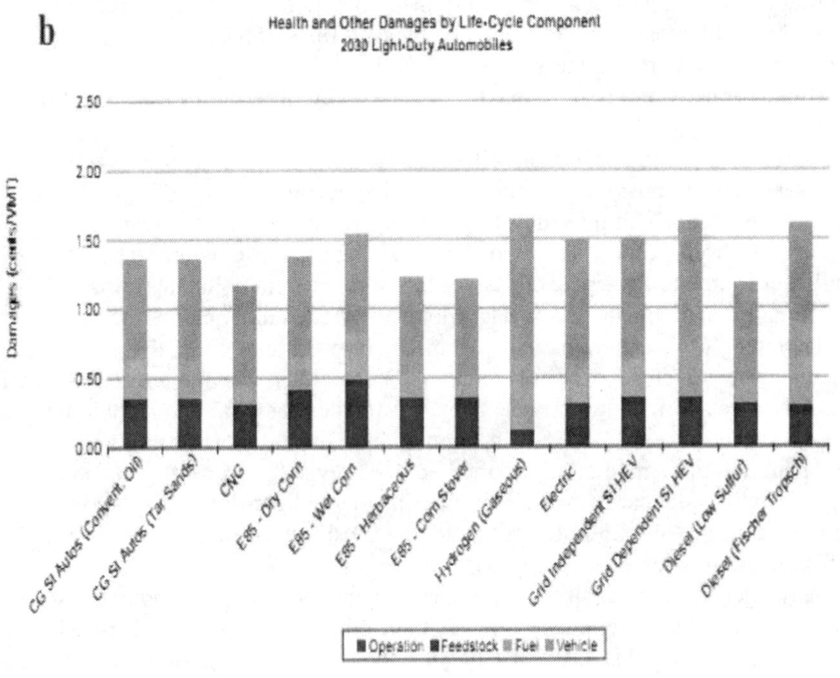

Health and Other Damages by Life-Cycle Component
2030 Light-Duty Automobiles
Damages (cents/VMT)
2.50
2.00
1.50
1.00
0.50
0.00
CG SI Autos (Convent. Oil)
CG SI Autos (Tar Sands)
CNG
E85 - Dry Corn
E85 - Wet Corn
E85 - Herbaceous
E85 - Corn Stover
Hydrogen (Gaseous)
Electric
Grid Independent SI HEV
Grid Dependent SI HEV
Diesel (Low Sulfur)
Diesel (Fischer Tropsch)
Operation
Feedstock
Fuel
Vehicle

FIGURE S-3. Health and Other Damages by Life-Cycle Component. Health effects and other nonclimate damages are presented by life-cycle component for different combinations of fuels and light-duty automobiles in 2005 (a) and 2030 (b). Damages are expressed in cents per VMT (2007 U.S. dollars). Going from bottom to top of each bar, damages are shown for life-cycle stages as follows: vehicle operation, feedstock production, fuel refining or conversion, and vehicle manufacturing. Damages related to climate change are not included. ABBREVIATIONS: VMT, vehicle miles traveled; CG SI, conventional gasoline spark ignition; CNG, compressed natural gas; E85, 85% ethanol fuel; HEV, hybrid electric vehicle.

and 2030, because of the energy required to produce the biofuel feedstock and convert it to fuel. Yet cellulosic (nonfood biomass) ethanol made from herbaceous plants or corn stover had lower damages than most other options when used in E85. The reason for this contrast is that the feedstock chosen and growing practices used influence the overall damages from biomass-based fuels. We did not quantify water use and indirect land use for biofuels.[11] Electric vehicles and grid-dependent hybrid vehicles showed somewhat higher damages than many other technologies for both 2005 and 2030.

Although operation of the vehicles produces few or no emissions, electricity production at present relies mainly on fossil fuels and, based on current emission control requirements, emissions from this stage of the life cycle are expected to still rely primarily on those fuels by 2030, albeit at significantly lower emission rates. In addition, battery and electric motor production—being energy- and material-intensive—added up to 20% to the damages from manufacturing.

Compressed natural gas had lower damages than other options, as the technology's operation and fuel produce very few emissions.

Although diesel had some of the highest damages in 2005, it is expected to have some of the lowest in 2030, assuming full implementation of the Tier 2 vehicle emission standards of the U.S. Environmental Protection Agency (EPA). This regulation, which requires the use of low-sulfur diesel, is expected to significantly reduce PM and NOx emissions as well. Heavy-duty vehicles have much higher damages per VMT than light-duty vehicles because they carry more cargo or people and, therefore, have lower fuel economies. However, between 2005 and 2030, these damages are expected to drop significantly, assuming the full implementation of the EPA Heavy-Duty Highway Vehicle Rule.

Greenhouse Gas Emissions

Most vehicle and fuel combinations had similar levels of GHG emissions in 2005 (see Figure S-4). Because vehicle operation is a substantial source of life-cycle GHGs, enforcement of the new CAFE standards will have a greater impact on lowering GHG emissions than on lowering life-cycle emissions of other pollutants. By 2030, with improvements among virtually all light-duty-vehicle types, the committee estimates that there will be even fewer differences in the GHG emissions of the various technologies than there were in 2005. However, in the absence of additional fuel-efficiency requirements, heavy-duty vehicle GHG emissions are expected to change little between 2005 and 2030, except from a slight increase in fuel economy in response to market conditions.

For both 2005 and 2030, vehicles using gasoline made from petroleum extracted from tar sands and diesel derived from Fischer-Tropsch fuels[12] have the highest life-cycle GHG emissions among all fuel and vehicle combinations considered. Vehicles using celluosic E85 from herbaceous feedstock or corn stover have some of the lowest GHG emissions because of the feedstock's ability to store CO_2 in the soil. Those using compressed natural gas also had comparatively low GHG emissions.

Future Reductions

Substantially reducing nonclimate damages related to transportation would require major technical breakthroughs, such as cost-effective conversion of cellulosic biofuels, cost-effective carbon capture, and storage for coal-fired power plants, or a vast increase in renewable energy capacity or other forms of electricity generation with lower emissions.[13] Further enhancements in fuel economy will also help, especially for emissions from vehicle operations, although they are only about one-third of the total life-cycle picture and two other components are proportional to fuel use. In any case, better understanding of potential external costs at the earliest stage of vehicle research should help developers minimize those costs as the technology evolves.

ESTIMATING CLIMATE-CHANGE DAMAGES

Energy production and use continue to be major sources of GHG emissions, principally CO_2 and methane. Damages from these emissions will result as their increased atmospheric concentrations affect climate, which in turn will affect such things as weather, freshwater supply, sea level, bio-diversity, and human society and health.[14]

Estimating these damages is another matter, as the prediction of climate-change effects, which necessarily involves detailed modeling and analysis, is an intricate and uncertain process. It requires aggregation of potential effects and damages that could occur at different times (extending centuries into the future) and among different populations across the globe. Thus, rather than attempt such an undertaking itself, especially given the constraints on its time and resources, the committee focused its efforts on a review of existing integrated assessment models (IAMs) and the associated climate-change literature.

We reviewed IAMs in particular, which combine simplified global-climate models with economic models that are used to (1) estimate the economic impacts of climate change, and (2) identify emissions regimes that balance the economic impacts with the costs of reducing GHG emissions. Because IAM simulations usually report their results in terms of mean values, this approach does not adequately capture some possibilities of catastrophic outcomes. Although a number of the possible outcomes have been studied—such as release of methane from permafrost that could rapidly accelerate warming and collapse of the West Antarctic or Greenland ice sheets, which could raise sea level by several meters—the damages associated with these events and their probabilities are very poorly understood. Some analysts nevertheless believe that the expected value of total damages may be more sensitive to the possibility of low-probability catastrophic events than to the most likely or best-estimate values.

In any case, IAMs are the best tools currently available. An important factor in using them (or virtually any other model that accounts for monetary impacts over time) is the "discount rate," which converts costs and benefits projected to occur in the future into amounts ("present values") that are compatible with present-day costs and benefits. Because the choice of a discount rate for the long periods associated with climate change is not well-established, the committee did not choose a particular discount rate for assessing the value of climate change's effects; instead, we considered a range of discount-rate values.

Under current best practice, estimates of global damages associated with a particular climate-change scenario at a particular future time are translated by researchers into an estimate of damages per ton of emissions (referred to as marginal damages) by valuating the linkage between current GHG emissions and future climate-change effects. Marginal damages are usually expressed as the net present value of the damages expected to occur over many future years as the result of an additional ton of CO_2-eq emitted into the atmosphere. Estimating these marginal damages depends on the temperature increase in response to a unit increase in CO_2-eq emissions, the additional climate-related effects that result, the values of these future dam-ages relative to the present, and how far into the future one looks. Because of uncertainties at each step of the analysis, a given set of possible future conditions may yield widely differing estimates of marginal damages.

Given the preliminary nature of the climate-damage literature, the committee found that only rough order-of-magnitude estimates of marginal damages were possible at this time. Depending on the extent of projected future damages and the discount rate used for weighting them, the range of estimates of marginal damages spanned two orders of magnitude, from about $1 to $100 per ton of CO_2-eq, based on current emissions. Approximately one order of magnitude in difference was attributed to discount-rate assumptions and another order of magnitude to assumptions about future damages from emissions used in the various IAMs. The damage estimates at the higher end of the range were associated only with emission paths without significant GHG controls. Estimates of the damages specifically to the United States would be a fraction of the levels in the range of estimates, because this country represents only about one-quarter of the world's economy, and the proportionate impacts it would suffer are generally thought to be lower than for the world as a whole.

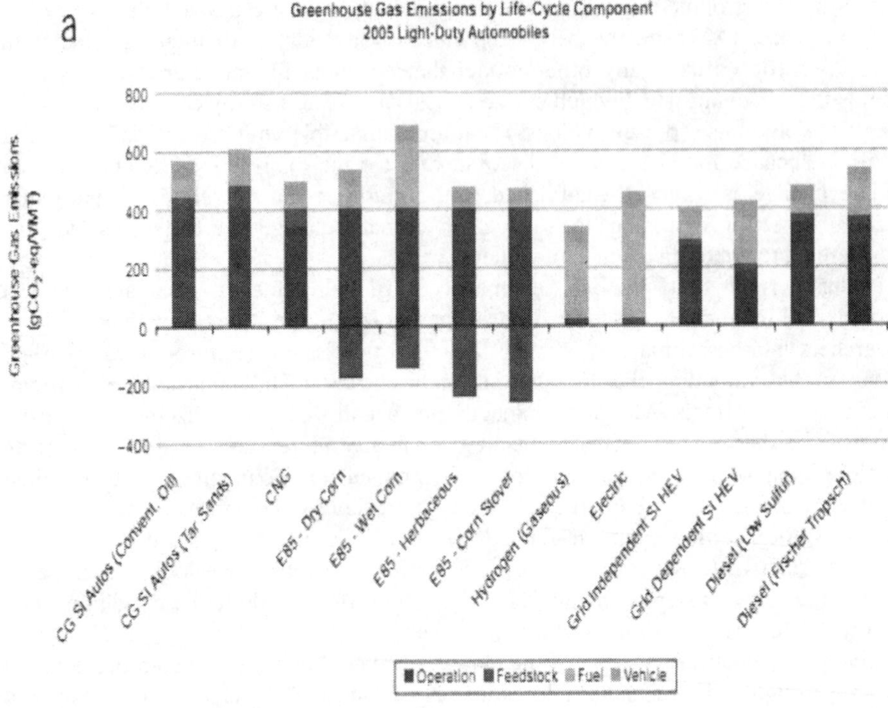
a
Greenhouse Gas Emissions by Life-Cycle Component
2005 Light-Duty Automobiles
Greenhouse Gas Emissions
(gCO_2-eq/VMT)
800
600
400
200
0
-200
-400
CG SI Autos (Convent. Oil)
CG SI Autos (Tar Sands)
CNG
E85 - Dry Corn
E85 - Wet Corn
E85 - Herbaceous
E85 - Corn Stover
Hydrogen (Gaseous)
Electric
Grid Independent SI HEV
Grid Dependent SI HEV
Diesel (Low Sulfur)
Diesel (Fischer Tropsch)
Operation
Feedstock
Fuel
Vehicle

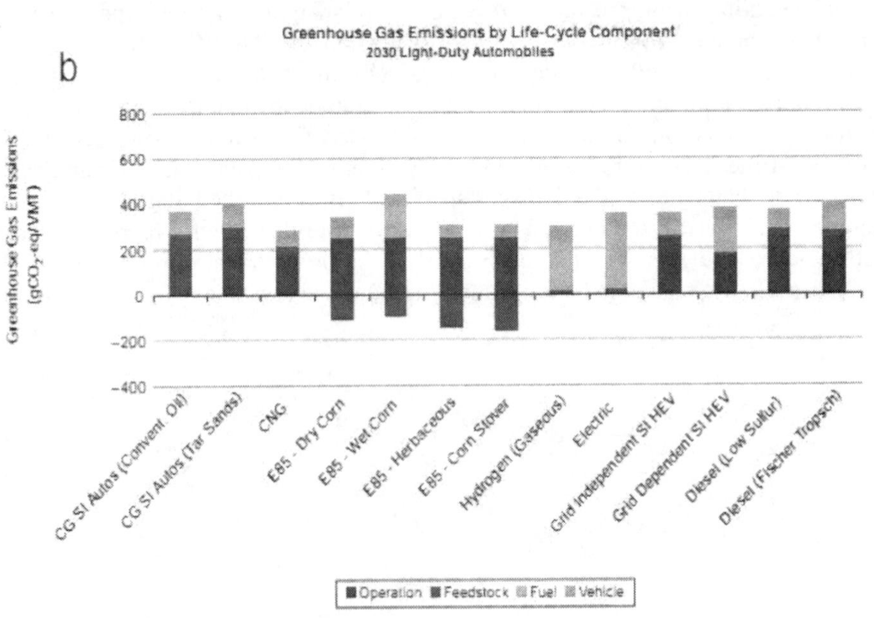
b
Greenhouse Gas Emissions by Life-Cycle Component
2030 Light-Duty Automobiles
Greenhouse Gas Emissions
(gCO_2-eq/VMT)
800
600
400
200
0
-200
-400
CG SI Autos (Convent. Oil)
CG SI Autos (Tar Sands)
CNG
E85 - Dry Corn
E85 - Wet Corn
E85 - Herbaceous
E85 - Corn Stover
Hydrogen (Gaseous)
Electric
Grid Independent SI HEV
Grid Dependent SI HEV
Diesel (Low Sulfur)
Diesel (Fischer Tropsch)
Operation
Feedstock
Fuel
Vehicle

FIGURE S-4. Greenhouse Gas Emissions by Life-Cycle. Greenhouse gas emissions (grams CO_2-eq)/VMT by life-cycle component for different combinations of fuels and light-duty automobiles in 2005 (a) and 2030 (b). Going from bottom to top of each bar, damages are shown for life-cycle stages as follows: vehicle operation, feedstock production, fuel refining or conversion, and vehicle manufacturing. One exception is ethanol fuels for which feedstock production exhibits negative values because of CO_2 uptake. The amount of CO_2 consumed should be subtracted from the positive value to arrive at a net value. ABBREVIATIONS: g CO_2-eq, grams CO2-equivalent; VMT, vehicle mile traveled; CG SI, conventional gasoline spark ignition; CNG, compressed natural gas; E85, 85% ethanol fuel; HEV, hybrid electric vehicle.

Comparing Climate and Nonclimate Damage Estimates

Comparing nonclimate damages to climate-related damages is extremely difficult. The two measures differ significantly in their time dimensions, spatial scales, varieties of impacts, and degrees of confidence with which they can be estimated. For 2005, determining which type of external effect caused higher damages depended on the energy technology being considered and the marginal damage value selected from the range of \$1 to \$100 per ton of CO_2-eq emitted. For example, coal-fired electricity plants were estimated to emit an average of about 1 ton of CO_2 per MWh (or 2 lb/kWh). When multiplying that emission rate by an assumed marginal dam-age value of \$30/ton CO_2-eq, climate-related damages equal 3 cents/kWh, comparable to the 3.2 cents/kWh estimated for nonclimate damages. It is important to keep in mind that the value of \$30/ton CO_2-eq is provided for illustrative purposes and is not a recommendation of the committee.

Natural Gas: The climate-related damages were higher than the nonclimate damages from natural-gas-fired power plants, as well as from combustion of natural gas for producing heat, regardless of the marginal damage estimate. Because natural gas is characterized by low emissions that form criteria pollutants, the nonclimate damages were about an order of magnitude lower than the climate damages estimated by the models, if the marginal climate damage were assumed to be \$30/ton CO_2-eq.

Coal: The climate-related damages from coal-fired power plants were estimated to be higher than the nonclimate damages when the assumed marginal climate damage was greater than \$30/ton CO_2-eq. If the marginal climate damage was less than \$30/ton CO_2-eq, the climate-related damages were lower than the nonclimate damages.

Transportation: As with coal, the transportation sector's climate-change damages were higher than the nonclimate damages only if the marginal damage for climate was higher than \$30/ton CO2-eq.

Overall: All of the model results available to the committee estimated that the climate-related damages per ton of CO_2-eq would be 50-80% worse in 2030 than in 2005. Even if annual GHG emissions were to remain steady between now and 2030, the damages per ton of CO_2-eq emissions would be substantially higher in 2030 than at present. As a result, the climate-related damages in that year from coal-fired power plants and transportation are likely to be greater than their nonclimate damages.

Infrastructure Risks and Security

The committee also considered external effects and costs associated with disruptions in the electricity-transmission grid, energy facilities' vulnerability to accidents and possible attack, oil-supply disruptions, and other national security issues. We concluded as follows:

• The nation's electric grid is subject to periodic failures because of transmission congestion and the lack of adequate reserve capacity. These failures are considered an external effect, as individual consumers of electricity do not take into account the impact of their consumption on aggregate load. The associated and possibly significant damages of grid failure underscore the importance of carefully analyzing the costs and benefits of investing in a modernized grid—one that takes advantage of new smart technology and that is better able to handle intermittent renewable-power sources.

• The external costs of accidents at energy facilities are largely taken into account by their owners and, at least in the case of our nation's oil and gas transmission networks, are of negligible magnitude per barrel of oil or thousand cubic feet of gas shipped.

• Because the United States is such a large consumer of oil, policies to reduce domestic demand can also reduce the world oil price, thereby benefiting the nation through lower prices on the remaining oil it imports. Government action may thus be a desirable countervailing force to monopoly or cartel-producer power. However, the committee does not consider this influence of a large single buyer (known as monopsony power) to be a benefit that is external to the market price of oil. It was therefore deemed to be outside the scope of this report.

• Although sharp and unexpected increases in oil prices adversely affect the U.S. economy, the macroeconomic disruptions they cause do not fall into the category of external effects and damages. Estimates in the literature of the macroeconomic costs of disruptions and adjustments range from $2 to $8 per barrel.

• Dependence on imported oil has well-recognized implications for foreign policy, and although we find that some of the effects can be viewed as external costs, it is currently impossible to quantify them. For example, the role of the military in safeguarding foreign supplies of oil is often identified as a relevant factor. However, the energy-related reasons for a military presence in certain areas of the world cannot readily be disentangled from the nonenergy-related reasons. Moreover, much of the military cost is likely to be fixed in nature. For example, even a 20% reduction in oil consumption, we believe, would probably have little impact on the positioning of U.S. military forces throughout the world.

• Nuclear waste raises important security issues and poses tough policy challenges. The extent to which associated external effects exist is hard to assess, and even when identified they are very difficult to quantify. Thus, although we do not present numerical values in this report, we recognize the importance of studying these issues further.

CONCLUSION

In aggregate, the damage estimates presented in this report for various external effects are substantial. Just the damages from external effects the committee was able to quantify add up to more than $120 billion for the year 2005.[15] Although large uncertainties are associated with the committee's estimates, there is little doubt that this aggregate total substantially underestimates the damages, because it does not include many other kinds of damages that could not be quantified for reasons explained in the report, such as damages related to some pollutants, climate change, ecosystems, infrastructure, and security. In many cases, we have identified those omissions, within the chapters of this report, with the hope that they will be evaluated in future studies.

Even if complete, our various damage estimates would not automatically offer a guide to policy. From the perspective of economic efficiency, theory suggests that damages should not be reduced to zero but only to the point where the cost of reducing another ton of emissions (or other type of burden) equals the marginal damages avoided—that is, the degree to which a burden should be reduced depends on its current level and the cost of lowering it. The solution cannot be determined from the amount of damage alone. Economic efficiency, however, is only one of several potentially valid policy goals that need to be considered in managing pollutant emissions and other burdens. For example, even within the same location, there is compelling evidence that some members of the population are more vulnerable than others to a particular external effect.

Although not a comprehensive guide to policy, our analysis does indicate that regulatory actions can significantly affect energy-related damages. For example, the full implementation of the federal diesel-emission rules would result in a sizeable decrease in nonclimate damages from diesel vehicles between 2005 and 2030. Similarly, major initiatives to further reduce other emissions, improve energy efficiency, or shift to a cleaner electricity-generating mix (for example, renewables, natural gas, and nuclear) could substantially reduce the damages of external effects, including those from grid-dependent hybrid and electric vehicles.

It is thus our hope that this information will be useful to government policy makers, even in the earliest stages of research and development on energy technologies, as an understanding of their external effects and dam-ages could help to minimize the technologies' adverse consequences.

ENDNOTES

1. Criteria pollutants, also known as "common pollutants" are identified by the U.S. Environmental Protection Agency (EPA), pursuant to the Clean Air Act, as ambient pollutants that come from numerous and diverse sources and that are considered to be harmful to public health and the environment and to cause property damage.
2. Hazardous air pollutants, also known as toxic air pollutants, are those pollutants that are known or suspected to cause cancer or other serious health effects, such as reproductive effects and birth defects, or adverse environmental effects.
3. Greenhouse gases absorb heat from the earth's surface and lower atmosphere, resulting in much of the energy being radiated back toward the surface rather than into space. These gases include water vapor, CO_2, ozone, methane, and nitrous oxide.
4. Costs are reported in 2007 dollars.
5. When damages per kWh are weighted by electricity generation, the 5th and 95th percentiles are 0.19 and 12 cents; the unweighted figures are .53 and 13.2 cents per kWh.
6. When damages per kWh are weighted by electricity generation, the 5th and 95th percentiles are 0.001 and 0.55 cents; the unweighted figures are .0044 and 1.7 cents per kWh.
7. The committee did not quantify damages associated with nuclear power. Such an analysis would have involved power-plant risk modeling and spent-fuel transportation modeling that would have required far greater resources and time than were available for this study.
8. Insufficient data were available to conduct a parallel analysis of industrial activities that generate useful heat as a side benefit.
9. CO_2-eq expresses the global-warming potential of a given stream of GHGs, such as methane, in terms of CO_2 quantities.
10. The committee also estimated damages on a per-gallon basis, with a range of 23 to 38 cents per gallon (with gasoline vehicles at 29 cents per gallon). Interpretation of the results is complicated, however, by the fact that fuel and technology combinations with higher fuel efficiency appear to have markedly higher damages per gallon than those with lower efficiency solely due to the higher number of miles driven per gallon.
11. Indirect land use refers to geographical changes occurring indirectly as a result of biofuels policy in the United States and the effects of such changes on GHG emissions.
12. The Fischer-Tropsch reaction converts a mixture of hydrogen and carbon monoxide—de-rived from coal, methane, or biomass—into liquid fuel. In its analysis, the committee considered only the use of methane for the production of Fischer-Tropsch diesel fuel.
13. The latter two changes are needed to reduce the life-cycle damages of grid-dependent vehicles.
14. In response to a request from Congress, the National Research Council has launched America's Climate Choices, a suite of studies designed to inform and guide responses to climate change across the nation.
15. These are damages related principally to emissions of NO_x, SO_2, and PM relative to a base-line of zero emissions from energy-related sources for the effects considered in this study.

Studying the Causes of Recent Climate Change

Benjamin D. Santer

Program for Climate Model Diagnosis and Intercomparison
Lawrence Livermore National Laboratory
Livermore, CA 94550

Abstract. This chapter describes progress in the field of "detection and attribution" (D&A) research, which seeks to identify certain "fingerprints," or patterns of climate change, and to correlate them with possible human factors influencing the climate. Such studies contributed to the scientific confidence with which the Fourth Assessment Report of the Intergovernmental Panel on Climate Change was able to assert that anthropogenic greenhouse gases had had a discernible effect on global warming since the mid-20th century. D&A methods have greatly improved to incorporate many more climate variables and to include increasingly finer variations in space and time. The chapter also describes the intercomparison of global climate models and the comprehensive data base of model simulations now available to anyone free of charge.

The following is the testimony given by Benjamin Santer to the U.S. House of Representative Committee on Science and Technology, Subcommittee on Energy and Environment, on November 17, 2010. It is adapted from a chapter that Tom Wigley and Benjamin Santer published in a book edited by the late Stephen Schneider [1] and from previous testimony given by Dr. Santer to the House Select Committee on Energy Independence and Global Warming.[2]

INTRODUCTION

In 1988, the Intergovernmental Panel on Climate Change (IPCC) was jointly established by the World Meteorological Organization and the United Nations Environment Programme. The goals of this panel were threefold: to assess available scientific information on climate change, to evaluate the environmental and societal impacts of climate change, and to formulate response strategies. The IPCC's first major scientific assessment, published in 1990, concluded that "unequivocal detection of the enhanced greenhouse effect from observations is not likely for a decade or more."[3].

In 1996, the IPCC's second scientific assessment made a more definitive statement regarding human impacts on climate, and concluded that "the balance of evidence suggests a discernible human influence on global climate."[4] This cautious sentence marked a paradigm shift in our scientific understanding of the causes of recent climate change. The shift arose for a variety of reasons. Chief amongst these was the realization that the cooling effects of sulfate aerosol particles (which are produced by

Physics of Sustainable Energy II: Using Energy Efficiently and Producing it Renewably
AIP Conf. Proc. 1401, 183-197 (2011); doi: 10.1063/1.3653851

burning fossil fuels that contain sulfates) had partially masked the warming signal arising from increasing atmospheric concentrations of greenhouse gases.[5]

A further major area of progress was the increasing use of "fingerprint" studies.[6-8] The strategy in this type of research is to search for a "fingerprint" (the climate change pattern predicted by a computer model) in observed climate records. The underlying assumption in fingerprinting is that each "forcing" of climate – such as changes in the Sun's energy output, volcanic dust, sulfate aerosols, or greenhouse gas concentrations – has a unique pattern of climate response (see Figure 1). Fingerprint studies apply signal processing techniques very similar to those used in electrical engineering.[6] They allow researchers to make rigorous tests of competing hypotheses regarding the causes of recent climate change.

The third IPCC assessment was published in 2001, and went one step further than its predecessor. The third assessment reported on the magnitude of the human effect on climate. It found that "There is new and stronger evidence that most of the warming observed over the last 50 years is attributable to human activities."[9] This conclusion was based on improved estimates of natural climate variability, better reconstructions of temperature fluctuations over the last millennium, continued warming of the climate system, refinements in fingerprint methods, and the use of results from more (and improved) climate models, driven by more accurate and complete estimates of the human and natural "forcings" of climate.

This gradual strengthening of scientific confidence in the reality of human influences on global climate continued in the IPCC AR4 report, which stated that "warming of the climate system is unequivocal", and that "most of the observed increase in global average temperatures since the mid-20th century is very likely due to the observed increase in anthropogenic greenhouse gas concentrations" [10] (where "very likely" signified >90% probability that the statement is correct). The AR4 report justified this increase in scientific confidence on the basis of "…longer and improved records, an expanded range of observations and improvements in the simulation of many aspects of climate and its variability."[10] In its contribution to the AR4, IPCC Working Group II concluded that anthropogenic warming has had a discernible influence not only on the physical climate system, but also on a wide range of biological systems which respond to climate.[11]

Extraordinary claims require extraordinary proof.[13] The IPCC's extraordinary claim that human activities significantly altered both the chemical composition of Earth's atmosphere and the climate system <u>has</u> received extraordinary scrutiny. This claim has been independently corroborated by the U.S. National Academy of Sciences [14], the Science Academies of eleven nations[15], and the Synthesis and Assessment Products of the U.S. Climate Change Science Plan.[16] Many of our professional scientific organizations have also affirmed the reality of a human influence on global climate.[17]

Despite the overwhelming evidence of pronounced anthropogenic effects on climate, important uncertainties remain in our ability to quantify the human influence. The experiment that we are performing with the Earth's atmosphere lacks a suitable control: we do not have a convenient "undisturbed Earth", which would provide a reference against which we could measure the anthropogenic contribution to climate change. We must therefore rely on numerical models and paleoclimate evidence.[18-

20] to estimate how the Earth's climate might have evolved in the absence of any human intervention. Such sources of information will always have significant uncertainties.

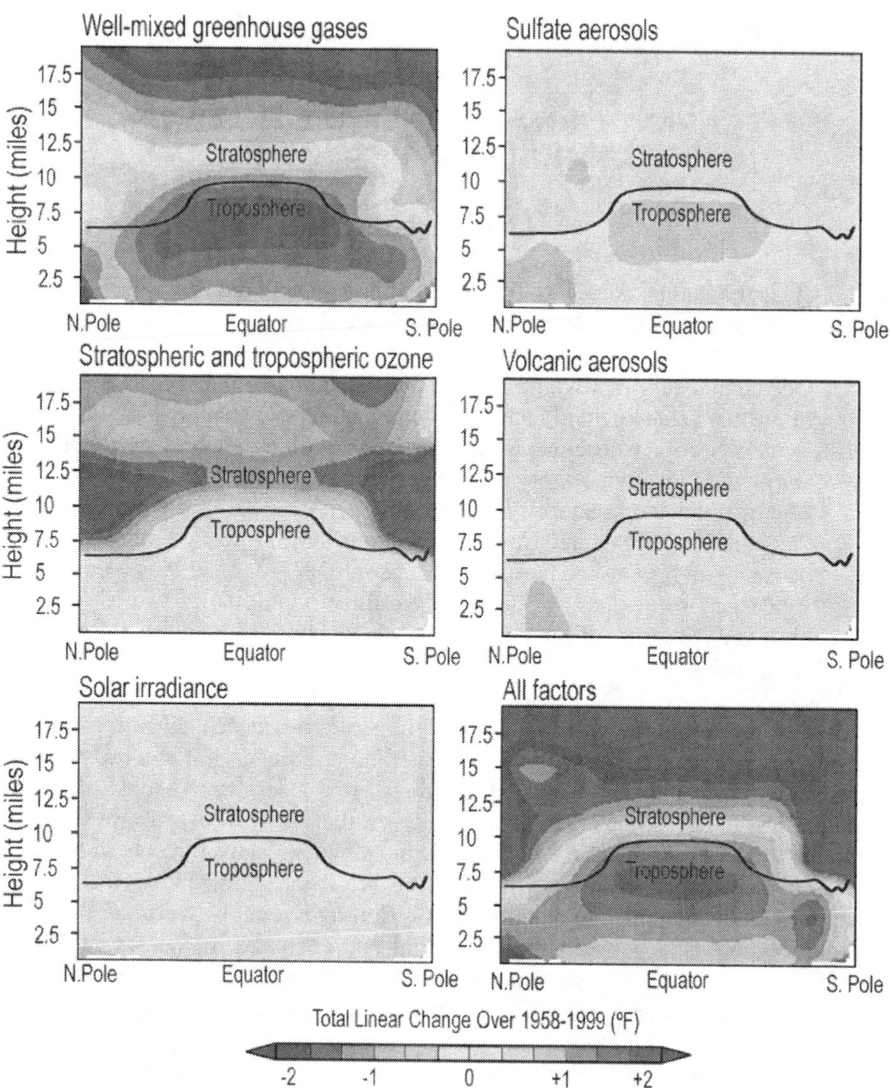

FIGURE 1. Climate simulations of the vertical profile of temperature change due to five different factors, and the effect due to all factors taken together. The panels above represent a cross-section of the atmosphere from the North Pole to the South Pole, and from the surface up into the statosphere. The black lines show the approximate location of the tropopause, the boundary between the lower atmosphere (the troposphere) and the statosphere. This Figure is reproduced from Karl *et al.*[12]

In the following testimony, I provide a personal perspective on recent developments in the field of detection and attribution ("D&A") research. Such research is directed towards detecting significant climate change, and then attributing some portion of the detected change to a specific cause or causes [21-24]. I also make some brief remarks about openness and data sharing in the climate modeling community, and accommodation of "alternative" views in the IPCC.

RECENT PROGRESS IN DETECTION AND ATTRIBUTION RESEARCH

Fingerprinting

The IPCC and National Academy findings that human activities are affecting global-scale climate are based on multiple lines of evidence:

1. Our continually-improving physical understanding of the climate system, and of the human and natural factors that cause climate to change;
2. Evidence from paleoclimate reconstructions, which enables us to place the warming of the 20th century in a longer-term context [25, 26];
3. The qualitative consistency between observed changes in different aspects of the climate system and model predictions of the changes that should be occurring in response to human influences [10, 27];
4. Evidence from rigorous quantitative fingerprint studies, which compare observed patterns of climate change with results from computer model simulations.

Most of my testimony will focus on the fingerprint evidence, since this is within my own area of scientific expertise. As noted above, fingerprint studies search for some pattern of climate change (the "fingerprint") in observational data. The fingerprint can be estimated in different ways, but is typically obtained from a computer model experiment in which one or more human factors are varied according to the best-available estimates of their historical changes. Different statistical techniques are then applied to quantify the level of agreement between the fingerprint and observations and between the fingerprint and estimates of the natural internal variability of climate. This enables researchers to make rigorous tests of competing hypotheses [28] regarding the possible causes of recent climate change.[21-24]

While early fingerprint work dealt almost exclusively with changes in near-surface or atmospheric temperature, more recent studies have applied fingerprint methods to a range of different variables, such as changes in ocean heat content [29, 30], Atlantic salinity [31], sea-level pressure [32], tropopause height [33], rainfall patterns [34, 35], surface humidity [36], atmospheric moisture [37, 38], continental river runoff [39], and Arctic sea ice extent [40]. The general conclusion is that for each of these variables, natural causes alone cannot explain the observed climate changes over the second half of the 20th century. The best statistical explanation of the observed climate changes invariably involves a large human contribution.

These fingerprint results are robust to the processing choices made by different groups, and show a high level of physical consistency across different climate variables. For example, observed atmospheric water vapor increases [41] are physically consistent with increases in ocean heat content [42, 43] and near-surface temperature.[44, 45]

There are a number of popular misconceptions about fingerprint evidence. One misconception is that fingerprint studies consider global-mean temperatures only, and thus provide a very poor constraint on the relative contributions of human and natural factors to observed changes [46]. In fact, fingerprint studies rely on information about the detailed spatial structure (and often the combined space and time structure) of observed and simulated climate changes. Complex patterns provide much stronger constraints on the possible contributions of different factors to observed climate changes.[47-49]

Another misconception is that computer model estimates of natural internal climate variability ("climate noise") are accepted uncritically in fingerprint studies, and are never tested against observations.[50] This is demonstrably untrue. Many fingerprint studies test whether model estimates of climate noise are realistic. Such tests are routinely performed on year-to-year and decade-to-decade timescales, where observational data are of sufficient length to obtain reliable estimates of observed climate variability.[51-54]

Because regional-scale climate changes will determine societal impacts, fingerprint studies are increasingly shifting their focus from global to regional scales.[55] Such regional studies face a number of challenges. One problem is that the noise of natural internal climate variability typically becomes larger when averaged over increasingly finer scales [56], so that identifying regional and local climate signals becomes more difficult.

Another problem relates to the climate "forcings" used in computer model simulations of historical climate change. As scientific attention shifts to ever smaller spatial scales, it becomes more important to obtain reliable information about these forcings. Some forcings are both uncertain and highly variable in space and time.[57, 58] Examples include human-induced changes in land surface properties [59] or in the concentrations of carbon-containing aerosols.[60, 61] Neglect or inaccurate specification of these factors complicates D&A studies.

Despite these problems, numerous researchers have now shown that the climate signals of greenhouse gases and sulfate aerosols are identifiable at continental and sub-continental scales in many different regions around the globe.[62-65] Related work [66, 67] suggests that a human-caused climate signal has already emerged from the background noise at spatial scales at or below 500 km [68], and may be contributing to regional changes in the distributions of plant and animal species.[69]

In summarizing this section of my testimony, I note that the focus of fingerprint research has evolved over time. Its initial emphasis was on global-scale changes in Earth's surface temperature. Subsequent research demonstrated that human fingerprints were identifiable in many different aspects of the climate system – not in surface temperature only. We are now on the verge of detecting human effects on climate at much finer regional scales of direct relevance to policymakers, and in variables tightly linked to climate change impacts.[70-74]

Assessing Risks of Changes in Extreme Events

We are now capable of making informed scientific statements regarding the influence of human activities on the likelihood of extreme events.[75-77] As noted previously, computer models can be used to perform the control experiment (no human effects on climate) that we cannot perform in the real world. Using the "unforced" climate variability from a multi-century control run, it is possible to determine how many times an extreme event of a given magnitude should have been observed in the absence of human interference. The probability of obtaining the same extreme event is then calculated in a perturbed climate – for example, in a model experiment with historical or future increases in greenhouse gases, or under some specified change in mean climate.[78] Comparison of the frequencies of extremes in the control and perturbed experiments allows climate scientists to make probabilistic statements about how human-induced climate change may have altered the likelihood of the extreme event.[53, 78, 79] This is sometimes referred to as an assessment of "fractional attributable risk."[78]

Recently, a "fractional attributable risk" study of the 2003 European summer heat wave concluded that "*there is a greater than 90% chance that over half the risk of European summer temperatures exceeding a threshold of 1.6 K is attributable to human influence on climate.*"[78]

This study (and related work) illustrates that the "D&A" community has moved beyond analysis of changes in the mean state of the climate. We now apply rigorous statistical methods to the problem of estimating how human activities may alter the probability of occurrence extreme events. The demonstration of human culpability in changing these risks is likely to have significant implications for the debate on policy responses to climate change.

SUMMARY OF DETECTION AND ATTRIBUTION EVIDENCE

In evaluating how well a novel has been crafted, it is important to look at the internal consistency of the plot. Critical readers examine whether the individual storylines are neatly woven together, and whether the internal logic makes sense.

We can ask similar questions about the "story" contained in observational records of climate change. The evidence from numerous sources (paleoclimate data, rigorous fingerprint studies, and qualitative comparisons of modeled and observed climate changes) shows that the climate system is telling us an internally consistent story about the causes of recent climate change.

Over the last century, we have observed large and coherent changes in many different aspects of Earth's climate. The oceans and land surface have warmed.[29, 30, 42-45, 80, 81] Atmospheric moisture has increased.[36-38, 41] Rainfall patterns have changed.[34, 35] Glaciers have retreated over most of the globe.[82-84] The Greenland Ice Sheet has lost some of its mass.[85] Sea level has risen [86]. Snow and sea-ice extent have decreased in the Northern Hemisphere.[40, 87-89] The stratosphere has cooled [90], and there are now reliable indications that the troposphere has warmed.[16, 91-100] The height of the tropopause has increased.[33] Individually, all of these changes are consistent with our scientific understanding of how the climate system should be responding to anthropogenic forcing. Collectively, this behavior is inconsistent with the changes that we would expect to occur due to natural variability alone.

There is now compelling scientific evidence that human activity has had a discernible influence on global climate. However, there are still significant uncertainties in our estimates of the size and geographical distribution of the climate changes projected to occur over the 21st century.[10] These uncertainties make it difficult for us to assess the magnitude of the mitigation and adaptation problem that faces us and our descendants. The dilemma that confronts us, as citizens and stewards of this planet, is how to act in the face of both hard scientific evidence that our actions are altering global climate and continuing uncertainty in the magnitude of the planetary warming that faces us.

OPENNESS AND DATA SHARING IN THE CLIMATE MODELING COMMUNITY

Recently, concerns have been expressed about ease of access to the information produced by computer models of the climate system. "Climate modeling" is sometimes portrayed as a secretive endeavor. This is not the case.

In the 1970s and 1980s, the evaluation and intercomparison of climate models was largely a qualitative endeavor, mostly performed by modelers themselves. It often involved purely visual examination of maps from a single model and observations (or from several different models). There were no standard benchmark experiments, and there was little or no community involvement in model diagnosis. It was difficult to track changes in model performance over time.[101]

This situation changed dramatically with the start of the Atmospheric Model Intercomparison Project (AMIP) in the early 1990s. AMIP involved running different Atmospheric General Circulation Models (AGCMs) with observed sea-surface temperatures and sea-ice changes over 1979 to 1988. Approximately 30 modeling groups from 10 different countries participated in the design and diagnosis of the AGCM simulations. Subsequent "revisits" of AMIP enabled the climate community to track changes in model performance over time.[102]

The next major Model Intercomparison Project ("MIP") began in the mid-1990s. In phase 1 of the Coupled Model Intercomparison Project (CMIP-1), over a dozen fully-coupled Atmosphere/Ocean General Circulation Models (A/OGCMs) were used to study the response of the climate system to an idealized climate-change scenario – a 1% per year (compound interest) increase in levels of atmospheric CO_2.[103] The key aspect here was that each modeling group performed the same benchmark simulation, allowing scientists to focus their attention on the task of quantifying (and understanding) uncertainties in computer model projections of future climate change. AMIP and CMIP have spawned literally dozens of other international Model Intercomparison Projects. "MIPs" are now a *de facto* standard in the climate science community. They have allowed climate scientists to:

- Identify systematic errors common to many different models;
- Track changes in model performance over time (in individual models and collectively);
- Make informed statements about the relative quality of different models;
- Quantify uncertainties in model projections of future climate change.

Full community involvement in "MIPs" has led to more thorough model diagnosis, and to improved climate models.

Perhaps the best-known model intercomparison is phase 3 of CMIP. The CMIP-3 project was a valuable resource for the Fourth Assessment Report (FAR) of the IPCC.[10] In the course of CMIP-3, simulation output was collected from 25 different A/OGCMs. The models used in these simulations were from 17 modeling centers and 13 countries. Twelve different types of simulation were performed with each model. The simulations included so-called "climate of the 20th century" experiments (with estimated historical changes in greenhouse gases, various aerosol particles, volcanic dust, solar irradiance, *etc.*), pre-industrial control runs (with no changes in human or natural climate forcings), and scenarios of future changes in greenhouse gases. All of the simulation output was stored at LLNL's PCMDI.

At present, 35 Terabytes of CMIP-3 data are archived at PCMDI, and nearly 1 Petabyte of model output (1 Petabyte = 10^{15} bytes) has been distributed to over 4,300 users in several dozen countries. The CMIP-3 multi-model archive has transformed the world of climate science. As of November 2010, over 560 peer-reviewed publications used CMIP-3 data. These publications formed the scientific backbone of the IPCC FAR. The CMIP-3 archive provided the basis for roughly 75% of the figures in Chapters 8-11 of the Fourth Assessment Report, and for 4 of the 7 figures in the IPCC "Summary for Policymakers."[10]

The CMIP-3 database can be used by anyone, free of charge. It is one of the most successful data-sharing models in any scientific community – not just the climate science community.

ACCOMMODATION OF "ALTERNATE" VIEWS OF THE IPCC

Some parties critical of the IPCC have claimed that it does not accommodate the full range of scientific views on the subject of the nature and causes of climate change. In my opinion, such claims are specious. I would contend that all four previous IPCC Assessments [3, 4, 9, 10] have dealt with "alternative viewpoints" in a thorough and comprehensive way. The IPCC reports have devoted extraordinary scientific attention to a number of highly-publicized (and incorrect) claims.

Examples include the claim that the tropical lower troposphere cooled over the satellite era; that the water vapor feedback is zero or negative; that variations in the Sun's energy output explain all observed climate change. The climate science community has not dismissed these claims out of hand. Scientists have done the research necessary to determine whether these "alternative viewpoints" are scientifically credible, and have shown that they are not.

CONCLUDING THOUGHTS

My job is to evaluate climate models and improve our scientific understanding of the nature and causes of climate change. I chose this profession because of a deep and abiding curiosity about the world in which we live. The same intellectual curiosity motivates virtually all climate scientists I know.

As my testimony indicates, the scientific evidence is compelling. We know, beyond a shadow of a doubt, that human activities have changed the composition of Earth's atmosphere. And we know that these human-caused changes in the levels of greenhouse gases make it easier for the atmosphere to trap heat. This is simple, basic physics. While there is legitimate debate in the scientific community about the size of the human effect on climate, there is really no serious scientific debate about the scientific finding that our planet warmed over the last century, and that human activities are implicated in this warming.

REFERENCES AND NOTES

1. Santer, B.D., and T.M.L. Wigley, 2010: Detection and attribution. In: Climate Change Science and Policy: [Schneider, S.H., A. Rosencranz, M.D. Mastrandrea, and K. Kuntz-Duriseti (eds.)]. Island Press, Washington D.C., pp. 28-43.
2. This testimony was given on May 20, 2010.
3. [1]Houghton, J.T., *et al.*, 1990: *Climate Change. The IPCC Scientific Assessment.* Cambridge University Press, Cambridge, U.K., page xxix.
4. Houghton, J.T., *et al.*, 1996: *Climate Change 1995: The Science of Climate Change.* Cambridge University Press, Cambridge, U.K., page 4.
5. Wigley, T.M.L., 1989: Possible climatic change due to SO_2-derived cloud condensation nuclei. *Nature*, **339**, 365-367.
6. Hasselmann, K., 1979: On the signal-to-noise problem in atmospheric response studies. In: *Meteorology of Tropical Oceans* (Ed. D.B. Shaw). Royal Meteorological Society of London, London, U.K., pp. 251-259.
7. Hasselmann, K., 1993: Optimal fingerprints for the detection of time dependent climate change. *Journal of Climate*, **6**, 1957-1971.
8. North, G.R., K.Y. Kim, S.S.P Shen, and J.W. Hardin, 1995: Detection of forced climate signals. Part I: Filter theory. *Journal of Climate*, **8**, 401-408.
9. Houghton, J.T., et al., 2001: *Climate Change 2001: The Scientific Basis.* Cambridge University Press, Cambridge, U.K., page 4.
10. IPCC, 2007: Summary for Policymakers. In: *Climate Change 2007: The Physical Science Basis.* Contribution of Working Group I to the Fourth Assessment Report of the Intergovernmental Panel on Climate Change [Solomon, S., D. Qin, M. Manning, Z. Chen, M. Marquis, K.B. Averyt, M. Tignor, and H.L. Miller (eds.)]. Cambridge University Press, Cambridge, United Kingdom and New York, NY, USA.
11. IPCC, 2007: Summary for Policymakers. In: *Climate Change 2007: Impacts, Adaptation and Vulnerability.* Contribution of Working Group II to the Fourth Assessment Report of the

Intergovernmental Panel on Climate Change [Parry, M. et al. (eds.)]. Cambridge University Press, Cambridge, United Kingdom and New York, NY, USA.

12. Karl, T.R., J.M. Melillo, and T.C. Peterson, 2009: Global Climate Change Impacts in the United States. Cambridge University Press, 189 pages.
13. This phrase is often attributed to the late sociologist Marcello Truzzi http://en.wikipedia.org/wiki/Marcello_Truzzi
14. NRC (National Research Council), 2001: *Climate Change Science. An Analysis of Some Key Questions.* Board on Atmospheric Sciences and Climate, National Academy Press, Washington D.C., 29 pp.
15. Prior to the Gleneagles G8 summit in July 2005, the Science Academies of 11 nations issued a joint statement on climate change (http://www.nasonline.org/site). The statement affirmed the IPCC finding that "most of the warming observed over the last 50 years is attributable to human activities" (ref. 10). The signatories were from the Academia Brasiliera de Ciências, the Royal Society of Canada, the Chinese Academy of Sciences, the Academié des Sciences, France, the Deutsche Akademie der Naturforscher, the Indian National Science Academy, the Accademia dei Lincei, Italy, the Science Council of Japan, the Russian Academy of Sciences, the United Kingdom Royal Society, and the U.S. National Academy of Sciences.
16. Karl, T.R., S.J. Hassol, C.D. Miller, and W.L. Murray (eds.), 2006: *Temperature Trends in the Lower Atmosphere: Steps for Understanding and Reconciling Differences.* A Report by the U.S. Climate Change Science Program and the Subcommittee on Global Change Research. National Oceanic and Atmospheric Administration, National Climatic Data Center, Asheville, NC, USA, 164 pp.
17. See, for example, the position statements on climate change issued by the American Geophysical Union (AGU), the American Meteorological Society (AGU), and the American Statistical Association (ASA). These can be found at: http://www.agu.org/sci_pol/positions/climate_change2008.shtml (AGU); http://www.ametsoc.org/amsnews/2007climatechangerelease.pdf (AMS); and http://www.amstat.org/news/climatechange.cfm (ASA).
18. Mann, M.E., and P.D. Jones, 2003: Global surface temperatures over the past two millenia. *Geophysical Research Letters*, **30**, 1820, doi:10.1029/2003GL017814.
19. Mann, M.E., Z. Zhang, M.K. Hughes, R.S. Bradley, S.K. Miller, S. Rutherford, and F. Ni, 2008: Proxy-based reconstructions of hemispheric and global surface temperature variations over the past two millennia. *Proceedings of the National Academy of Sciences*, **105**, 13252-13257.
20. Chapman, D.S., and M.G. Davis, 2010: Climate change: Past, present, and future. *Eos*, **91**, 325-326.
21. Mitchell, J.F.B. *et al.*, 2001: Detection of climate change and attribution of causes. In: *Climate Change 2001: The Scientific Basis.* Contribution of Working Group I to the Third Assessment Report of the Intergovernmental Panel on Climate Change [Houghton, J.T. *et al.*, (eds.)]. Cambridge University Press, Cambridge, United Kingdom and New York, NY, USA, pp. 695-738.
22. IDAG (International Detection and Attribution Group), 2005: Detecting and attributing external influences on the climate system: A review of recent advances. *Journal of Climate*, **18**, 1291-1314.
23. Santer, B.D., J.E. Penner, and P.W. Thorne, 2006: How well can the observed vertical temperature changes be reconciled with our understanding of the causes of these changes? In: *Temperature Trends in the Lower Atmosphere: Steps for Understanding and Reconciling Differences.* A Report by the U.S. Climate Change Science Program and the Subcommittee on Global Change Research [Karl, T.R., S.J. Hassol, C.D. Miller, and W.L. Murray (eds.)]. National Oceanic and Atmospheric Administration, National Climatic Data Center, Asheville, NC, USA, pp. 89-108.
24. Hegerl, G.C., F.W. Zwiers, P. Braconnot, N.P. Gillett, Y. Luo, J.A. Marengo Orsini, J.E. Penner and P.A. Stott, 2007: Understanding and Attributing Climate Change. In: *Climate Change*

2007: The Physical Science Basis. Contribution of Working Group I to the Fourth Assessment Report of the Intergovernmental Panel on Climate Change [Solomon, S., D. Qin, M. Manning, Z. Chen, M. Marquis, K.B. Averyt, M. Tignor, and H.L. Miller (eds.)]. Cambridge University Press, Cambridge, United Kingdom and New York, NY, USA, pp. 663-745.

25. A recent assessment of the U.S. National Academy of Sciences concluded that "It can be said with a high level of confidence that global mean surface temperature was higher during the last few decades of the 20th century than during any comparable period during the preceding four centuries" (ref. 26, page 3). The same study also found "it plausible that the Northern Hemisphere was warmer during the last few decades of the 20th century than during any comparable period over the preceding millennium" (ref. 26, pages 3-4).
26. National Research Council, 2006: *Surface Temperature Reconstructions for the Last 2,000 Years*. National Academies Press, Washington D.C., 196 pp.
27. Examples include increases in surface and tropospheric temperature, increases in atmospheric water vapor and ocean heat content, sea-level rise, widespread retreat of glaciers, *etc*.
28. An example includes testing the null hypothesis that there has been no external forcing of the climate system against the alternative hypothesis that there has been significant external forcing. Currently, all such hypothesis tests rely on model-based estimates of "unforced" climate variability (also known as natural internal variability). This is the variability that arises solely from processes internal to the climate system, such as interactions between the atmosphere and ocean. The El Niño phenomenon is a well-known example of internal climate noise.
29. Barnett, T.P. *et al.*, 2005: Penetration of human-induced warming into the world's oceans. *Science*, **309**, 284-287.
30. Pierce, D.W. *et al.*, 2006: Anthropogenic warming of the oceans: Observations and model results. *Journal of Climate*, **19**, 1873-1900.
31. Stott, P.A., R.T. Sutton, and D.M. Smith, 2008: Detection and attribution of Atlantic salinity changes. *Geophysical Research Letters*, **35**, L21702, doi:10.1029/2008GL035874.
32. Gillett, N.P., F.W. Zwiers, A.J. Weaver, and P.A. Stott, 2003: Detection of human influence on sea level pressure. *Nature*, **422**, 292-294.
33. Santer, B.D. *et al.*, 2003: Contributions of anthropogenic and natural forcing to recent tropopause height changes. *Science*, **301**, 479-483.
34. Zhang, X. *et al.*, 2007: Detection of human influence on 20th century precipitation trends. *Nature*, **448**, 461-465.
35. Min, S.-K., X. Zhang, and F. Zwiers, 2008: Human-induced Arctic moistening. *Science*, **320**, 518-520.
36. Willett, K.M., N.P. Gillett, P.D. Jones, and P.W. Thorne, 2007: Attribution of observed surface humidity changes to human influence. *Nature*, **449**, doi:10.1038/nature06207.
37. Santer, B.D., *et al.*, 2007: Identification of human-induced changes in atmospheric moisture content. *Proceedings of the National Academy of Sciences*, **104**, 15248-15253.
38. Santer, B.D., *et al.*, 2009: Incorporating model quality information in climate change detection and attribution studies. *Proceedings of the National Academy of Sciences*, **106**, 14778-14783.
39. Gedney, N., P.M. Cox, R.A. Betts, O. Boucher, C. Huntingford, and P.A. Stott, 2006: Detection of a direct carbon dioxide effect in continental river runoff records. *Nature*, **439**, 835-838.
40. Min, S.-K., X. Zhang, F.W. Zwiers, and T. Agnew, 2008: Human influence on Arctic sea ice detectable from early 1990s onwards. *Geophysical Research Letters*, **35**, L21701, doi:10.1029/2008GL035725.
41. Trenberth, K.E., J. Fasullo, and L. Smith, 2005: Trends and variability in column-integrated atmospheric water vapor. *Climate Dynamics*, **24**, doi:10.1007/s00382-005-0017-4.

42. Levitus, S., J.I. Antonov, and T.P. Boyer, 2005: Warming of the world ocean, 1955-2003. *Geophysical Research Letters*, **32**, L02604, doi:10.1029/2004GL021592.
43. Domingues, C.M., *et al.*, 2008: Rapid upper-ocean warming helps explain multi-decadal sea-level rise, *Nature*, **453**, 1090-1093.
44. Jones, P.D., M. New, D.E. Parker, S. Martin, and I.G. Rigor, 1999: Surface air temperature and its changes over the past 150 years. *Reviews of Geophysics*, **37**, 173-199.
45. Brohan, P., J.J. Kennedy, I. Harris, S.F.B. Tett, and P.D. Jones, 2006: Uncertainty estimates in regional and global observed temperature changes: A new dataset from 1850. *Journal of Geophysical Research*, **111**, D12106, doi:10.1029/2005JD006548.
46. The argument here is that some anthropogenic "forcings" of climate (particularly the so-called indirect forcing caused by the effects of anthropogenic aerosols on cloud properties) are highly uncertain, so that many different combinations of these factors could yield the same global-mean changes. While this is a valid concern for global-mean temperature changes, it is highly unlikely that different combinations of forcing factors could produce the same complex space-time patterns of climate change (see Figure 1).
47. Some researchers have argued that most of the observed near-surface warming over the 20th century is attributable to an overall increase in the Sun's energy output. The effect of such an increase would be to warm most of the atmosphere (from the Earth's surface through the stratosphere; see Figure 1, lower left panel). Such behavior is not seen in observations. While temperature measurements from satellites and weather balloons do show warming of the troposphere, they also indicate that the stratosphere has cooled over the past 2-4 decades (ref. 16). Stratospheric cooling is fundamentally inconsistent with a 'solar forcing only' hypothesis of observed climate change, but is consistent with simulations of the response to anthropogenic greenhouse gas increases and ozone decreases (see Figures 1, top left and middle left panels). The possibility of a large solar forcing effect has been further weakened by recent research indicating that changes in solar luminosity on multi-decadal timescales are likely to be significantly smaller than previously thought (refs. 48, 49).
48. Foukal, P., G. North, and T.M.L. Wigley, 2004: A stellar view on solar variations and climate. *Science*, **306**, 68-69.
49. Foukal, P., C. Fröhlich, H. Spruit, and T.M.L. Wigley, 2006: Physical mechanisms of solar luminosity variation, and its effect on climate. *Nature*, **443**, 161-166.
50. In order to assess whether observed climate changes over the past century are truly unusual, we require information on the amplitude and structure of climate noise on timescales of a century or longer. Unfortunately, direct instrumental measurements are of insufficient length to provide such information. This means that detection and attribution studies must rely on decadal- to century-timescale noise estimates from computer model control runs.
51. Allen, M.R., and S.F.B. Tett, 1999: Checking for model consistency in optimal fingerprinting. *Climate Dynamics*, **15**, 419-434.
52. Thorne, P.W. *et al.*, 2003: Probable causes of late twentieth century tropospheric temperature trends. *Climate Dynamics*, **21**, 573-591.
53. Santer, B.D. *et al.*, 2006: Causes of ocean surface temperature changes in Atlantic and Pacific tropical cyclogenesis regions. *Proceedings of the National Academy of Sciences*, **103**, 13905-13910.
54. AchutaRao, K.M., M. Ishii, B.D. Santer, P.J. Gleckler, K.E. Taylor, T.P. Barnett, D.W. Pierce, R.J. Stouffer, and T.M.L. Wigley, 2007: Simulated and observed variability in ocean temperature and heat content. *Proceedings of the National Academy of Sciences*, **104**, 10768-10773.
55. Stott, P.A. *et al.*, 2010: Detection and attribution of climate change: A regional perspective. *Wiley Interdisciplinary Reviews*, doi: 10.1002/WCC.34.
56. Wigley, T.M.L., and P.D. Jones, 1981: Detecting CO_2-induced climatic change. *Nature*, **292**, 205-208.

57. Ramaswamy, V., *et al.*, 2001: Radiative forcing of climate change. In: *Climate Change 2001: The Scientific Basis*. Contribution of Working Group I to the Third Assessment Report of the Intergovernmental Panel on Climate Change [Houghton, J.T. *et al.*, (eds.)]. Cambridge University Press, Cambridge, United Kingdom and New York, NY, USA, pp. 349-416.
58. NRC (National Research Council), 2005: *Radiative Forcing of Climate Change: Expanding the Concept and Addressing Uncertainties*. Board on Atmospheric Sciences and Climate, National Academy Press, Washington D.C., 168 pp.
59. Feddema, J. *et al.*, 2005: A comparison of a GCM response to historical anthropogenic land cover change and model sensitivity to uncertainty in present-day land cover representations. *Climate Dynamics*, **25**, 581-609.
60. Penner, J.E. *et al.*, 2001: Aerosols, their direct and indirect effects. In: *Climate Change 2001: The Scientific Basis*. Contribution of Working Group I to the Third Assessment Report of the Intergovernmental Panel on Climate Change [Houghton, J.T. *et al.* (eds.)]. Cambridge University Press, Cambridge, United Kingdom and New York, NY, USA, pp. 289-348.
61. Menon, S., J. Hansen, L. Nazarenko, and Y.F. Luo, 2002: Climate effects of black carbon aerosols in China and India. *Science*, **297**, 2250-2253.
62. Stott, P.A., 2003: Attribution of regional-scale temperature changes to anthropogenic and natural causes. *Geophysical Research Letters*, **30**, doi: 10.1029/2003GL017324.
63. Zwiers, F.W., and X. Zhang, 2003: Towards regional-scale climate change detection. *Journal of Climate*, **16**, 793-797.
64. Karoly, D.J. *et al.*, 2003: Detection of a human influence on North American climate. *Science*, **302**, 1200-1203.
65. Min, S.-K., A. Hense, and W.-T. Kwon, 2005: Regional-scale climate change detection using a Bayesian detection method. *Geophysical Research Letters*, **32**, L03706, doi: 11/12/2010 7:00 PM11/12/2010 7:00 PM10.1029/2004GL021028.
66. Karoly, D.J., and Q. Wu, 2005: Detection of regional surface temperature trends. *Journal of Climate*, **18**, 4337-4343.
67. Knutson, T.R. *et al.*, 2006: Assessment of twentieth-century regional surface temperature trends using the GFDL CM2 coupled models. *Journal of Climate*, **19**, 1624-1651.
68. Knutson *et al.* (ref. 67) state that their "*regional results provide evidence for an emergent anthropogenic warming signal over many, if not most, regions of the globe*".
69. Root, T.L., D.P. MacMynowski, M.D. Mastrandrea, and S.H. Schneider, 2005: Human-modified temperatures induce species changes: Joint attribution. *Proceedings of the National Academy of Sciences*, **102**, 7465-7469.
70. Examples include snowpack depth (refs. 71, 72), maximum and minimum temperatures in mountainous regions of the western U.S. (refs. 71, 73), and the timing of streamflow in major river basins (refs. 71, 74).
71. Barnett, T.P., *et al.*, 2008: Human-induced changes in the hydrology of the western United States. *Science*, **319**, 1080-1083.
72. Pierce, D.W., *et al.*, 2008: Attribution of declining western U.S. snowpack to human effects. *Journal of Climate*, **21**, 6425-6444.
73. Bonfils, C., *et al.*, 2008: Detection and attribution of temperature changes in the mountainous Western United States. *Journal of Climate*, **21**, 6404-6424.
74. Hidalgo, H., *et al.*, 2009: Detection and attribution of streamflow timing changes to climate change in the western United States. *Journal of Climate*, **22**, 3838-3855.
75. Allen, M.R., 2003: Liability for climate change. *Nature*, **421**, 891-892.
76. Wigley, T.M.L., 1988: The effect of changing climate on the frequency of absolute extreme events. *Climate Monitor*, **17**, 44-55.
77. Meehl, G.A., and C. Tebaldi, 2004: More intense, more frequent, and longer lasting heat waves in the 21st century. *Science*, **305**, 994-997.
78. Stott, P.A., D.A. Stone, and M.R. Allen, 2004: Human contribution to the European heatwave of 2003. *Nature*, **423**, 61-614.

79. Tebaldi, C., K. Hayhoe, J.M. Arblaster, and G.A. Meehl, 2006: Going to the extremes: An intercomparison of model-simulated historical and future changes in extreme events. *Climatic Change*, **79**, 185-211.
80. Jones, P.D, and A. Moberg, 2003: Hemispheric and large scale surface air temperature variations: an extensive revision and an update to 2001. *Journal of Climate*, **16**, 206-223.
81. Gillett, N.P., D.A. Stone, P.A. Stott, T. Nozawa, A.Y. Karpechko, G.C. Hegerl, M.F. Wehner, and P.D. Jones, 2008: Attribution of polar warming to human influence. *Nature Geoscience*, **1**, 750-754.
82. Arendt, A.A. *et al.*, 2002: Rapid wastage of Alaska glaciers and their contribution to rising sea level. *Science*, **297**, 382-386.
83. Paul, F., A. Kaab, M. Maisch, T. Kellenberger, and W. Haeberli, 2004: Rapid disintegration of Alpine glaciers observed with satellite data. *Geophysical Research Letters*, **31**, L21402, doi:10.1029/2004GL020816.
84. Meier, M.F., *et al.*, 2007: Glaciers dominate eustatic sea-level rise in the 21st century. *Science*, **317**, 1064-1067.
85. Luthcke, S.B. *et al.*, 2006: Recent Greenland ice mass loss by drainage system from satellite gravity observations. *Science*, **314**, 1286-1289.
86. Cazenave, A., and R.S. Nerem, 2004: Present-day sea level change: Observations and causes. *Reviews of Geophysics*, **42**, RG3001, doi:10.1029/2003RG000139.
87. Vinnikov, K.Y. *et al.*, 1999: Global warming and Northern Hemisphere sea ice extent. *Science*, **286**, 1934-1937.
88. Stroeve, J., *et al.*, 2008: Arctic sea ice plummets in 2007. *EOS*, **89**, 2, 13-14.
89. Stroeve, J., M.M. Holland, W. Meier, T. Scambos, M. Serreze, 2007: Arctic sea ice decline: Faster than forecast. *Geophysical Research Letters*, **34**, L09501, doi: 10.1029/2007GL029703.
90. Ramaswamy, V. *et al.*, 2006: Anthropogenic and natural influences in the evolution of lower stratospheric cooling. *Science*, **311**, 1138-1141.
91. Trenberth K.E., *et al.*, 2007: Observations: Surface and atmospheric climate change. In: *Climate Change 2007: The Physical Science Basis.* Contribution of Working Group I to the Fourth Assessment Report of the Intergovernmental Panel on Climate Change. [Solomon S., Qin D., Manning M., Chen Z., Marquis M., Averyt K.B., Tignor M., Miller H.L. (eds.)]. Cambridge University Press, Cambridge, United Kingdom and New York, NY, USA, pp. 235-336
92. Wentz, F.J., and M. Schabel, 1998: Effects of orbital decay on satellite-derived lower-tropospheric temperature trends. *Nature*, **394**, 661-664.
93. Mears, C.A., M.C. Schabel, and F.W. Wentz, 2003: A reanalysis of the MSU channel 2 tropospheric temperature record. *Journal of Climate*, **16**, 3650-3664.
94. Fu, Q., C.M. Johanson, S.G. Warren, and D.J. Seidel, 2004: Contribution of stratospheric cooling to satellite-inferred tropospheric temperature trends. *Nature*, **429**, 55-58.
95. Zou, C.-Z., M. Gao, and M.D. Goldberg, 2009: Error structure and atmospheric temperature trends in observations from the Microwave Sounding Unit. *Journal of Climate*, **22**, 1661-1681.
96. Sherwood, S.C., J. Lanzante, and C. Meyer, 2005: Radiosonde daytime biases and late 20th century warming. *Science*, **309**, 1556-1559.
97. Mears, C.A., and F.W. Wentz, 2005: The effect of diurnal correction on satellite-derived lower tropospheric temperature. *Science*, **309**, 1548-1551.
98. Allen, R.J., and S.C. Sherwood, 2008: Warming maximum in the tropical upper troposphere deduced from thermal winds. *Nature Geoscience*, **1**, 399-403.
99. Sherwood, S.C., C.L. Meyer, and R.J. Allen, 2008: Robust tropospheric warming revealed by iteratively homogenized radiosonde data. *Journal of Climate*, **21**, 5336-5350.
100. Titchner, H.A., P.W. Thorne, M.P. McCarthy, S.F.B. Tett, L. Haimberger, and D.E. Parker, 2008: Critically reassessing tropospheric temperature trends from radiosondes using realistic validation experiments. *Journal of Climate*, **22**, 465-485.

101. Gates, W.L., 1992: AMIP: The Atmospheric Model Intercomparison Project. *Bulletin of the American Meteorological Society*, **73**, 1962-1970.
102. Gates, W.L. *et al.*, 1999: An overview of the Atmospheric Model Intercomparison Project (AMIP). *Bulletin of the American Meteorological Society*, **80**, 29-55.
103. Meehl, G.A. *et al.*, 2007: The WCRP CMIP-3 multi-model dataset: A new era in climate change research. *Bulletin of the American Meteorological Society*, **88**, 1383-1394.

Methane: A Menace Surfaces*

Katey Walter Anthony

Water and Environmental Research Center
University of Alaska at Fairbanks
Fairbanks, Alaska 99775

Abstract: The arctic permafrost is thawing, releasing organic matter that was frozen in the ground into the bottoms of lakes. This organic matter feeds microbes that produce methane, which in turn escapes to the atmosphere. Permafrost, a rich source of organic carbon, covers 20% of the earth's land surface, and one third to one half of permafrost is now within 1.0° C to 3° C of thawing. New estimates indicate that by 2100, thawing permafrost could boost emissions of methane—a greenhouse gas that's 25 times more potent than carbon dioxide—by 20 to 40 percent beyond what would be produced by all natural and man-made sources. As a result, the earth's mean annual temperature could rise by an additional 0.32° C, further upsetting weather patterns and sea level.

INTRODUCTION

Touchdown on the gravel runway at Cherskii in remote northeastern Siberia sent the steel toe of a rubber boot into my buttocks. The shoe had sprung free from gear stuffed between me and my three colleagues packed into a tiny prop plane. This was the last leg of my research team's five-day journey from the University of Alaska Fairbanks across Russia to the Northeast Science Station in the land of a million lakes, which we were revisiting as part of our ongoing efforts to monitor a stirring giant that could greatly speed up global warming.

These expeditions help us to understand how much of the perennially frozen ground, known as permafrost, in Siberia and across the Arctic is thawing, or close to thawing, and how much methane the process could generate. The question grips us—and many scientists and policy makers—because methane is a potent green house gas, packing 25 times more heating power, molecule for molecule, than carbon dioxide. If the permafrost thaws rapidly because of global warming worldwide, the planet could get hotter more quickly than most models now predict. Our data, combined with

 The research described in this article and the photos credited to K. W. Anthony were funded by National Science Foundation, NASA, and the Department of Energy.

Physics of Sustainable Energy II: Using Energy Efficiently and Producing it Renewably
AIP Conf. Proc. 1401, 198-210 (2011); doi: 10.1063/1.3653852
2011 American Institute of Physics 978-0-7354-0972-9/$30.00

complementary analyses by others, are revealing troubling trends.

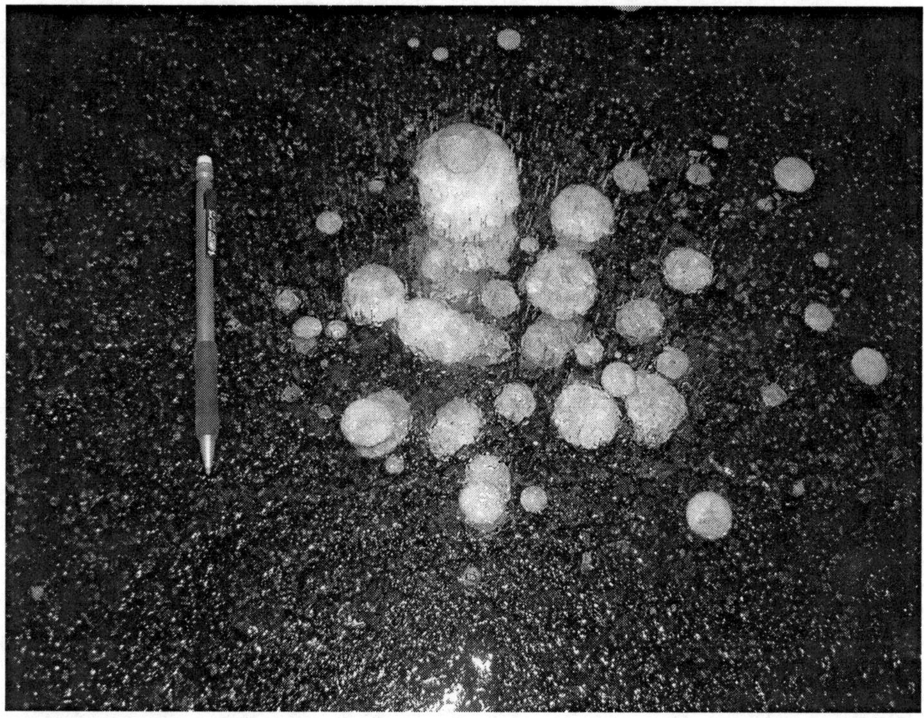

FIGURE 1. Methane bubbles. Methane gas (white) rising from an Arctic lake bottom is frozen into ice that has formed across the surface. (Photo courtesy of K. W. Anthony, UAF, Institute of Northern Engineering (INE.)

Leaving the Freezer Door Open

Changes in permafrost are so worrisome because the frozen ground, which covers 20 percent of the earth's land surface, stores roughly 1670 billion tons of carbon in the top three to tens of meters. (More permafrost can extend downward hundreds of meters.) This carbon, in the form of dead plant and animal remains, has accumulated over tens of thousands of years. As long as it stays frozen beneath and between the many lakes, it is safely sequestered from the air.

But when permafrost thaws, the carbon previously locked away is made available to microbes, which rapidly degrade it, producing gases. The same process happens if a freezer door is left open; given long enough, food thaws and begins to rot. Oxygen stimulates bacteria and fungi to aerobically decompose organic matter, producing carbon dioxide. But oxygen is depleted in soil that is waterlogged, such as in lake-bottom sediments; in these conditions, anaerobic decomposition occurs, which releases methane (in addition to some carbon dioxide). Under lakes, the methane gas molecules form bubbles that escape up through the water column, burst at the surface and enter the atmosphere.

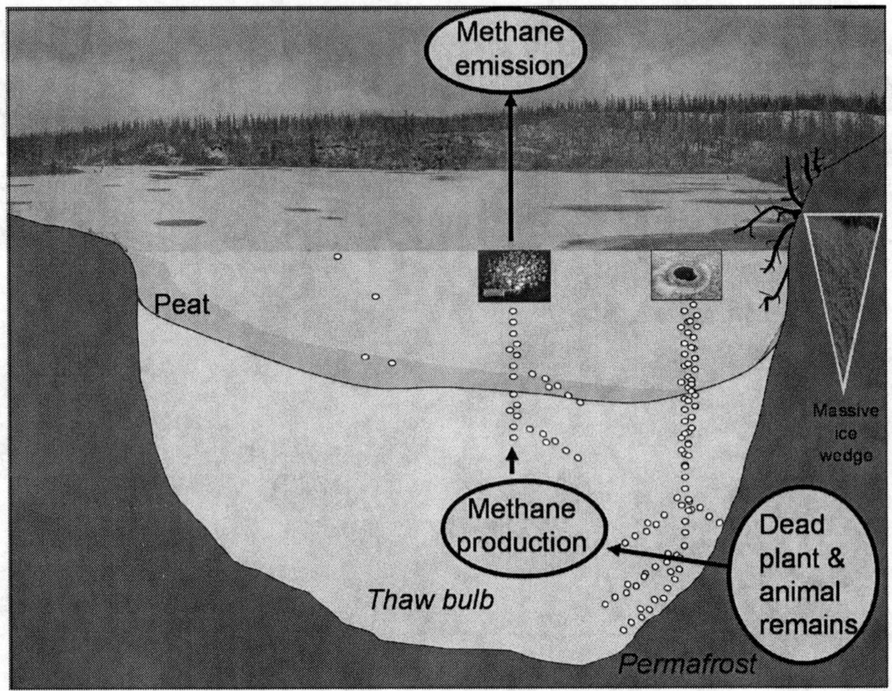

FIGURE 2. Process of methane release. In the cold Arctic environment, dead plant and animal matter lies frozen in ancient permafrost below a thin layer of modern soil. As the atmosphere warms, the ground thaws, forming a thaw bulb, and anaerobic bacteria convert the organic matter there to methane, which bubbles to the surface. In hotspots, where methane production is especially high (as sketched near the left side of the lake), the bubbles can maintain open holes in the lake ice throughout the winter. [Adapted from K. Walter Anthony et al., *Phil. Trans. Royal Soc. A,* **365**, No. 1856, pages 1657-1676 (2007).]

Anaerobic decomposition is the primary source of methane in the Arctic. Melting ice in permafrost causes the ground surface to subside, a process known as thermokarst. Runoff water readily fills the depressions, creating many small, newly formed lakes, which begin to spew vast quantities of methane as the permafrost that now lines their bottom thaws much more extensively. Scars left behind reveal that this process has been going on for the past 10,000 years, since the earth entered the most recent interglacial warm period. Satellite recordings made during recent decades suggest, however, that permafrost thaw may be accelerating.

Those recordings are consistent with observations made at numerous field-monitoring sites across Alaska and Siberia maintained by my Fairbanks colleague Vladimir E. Romanovsky and others. Romanovsky notes that permafrost temperature

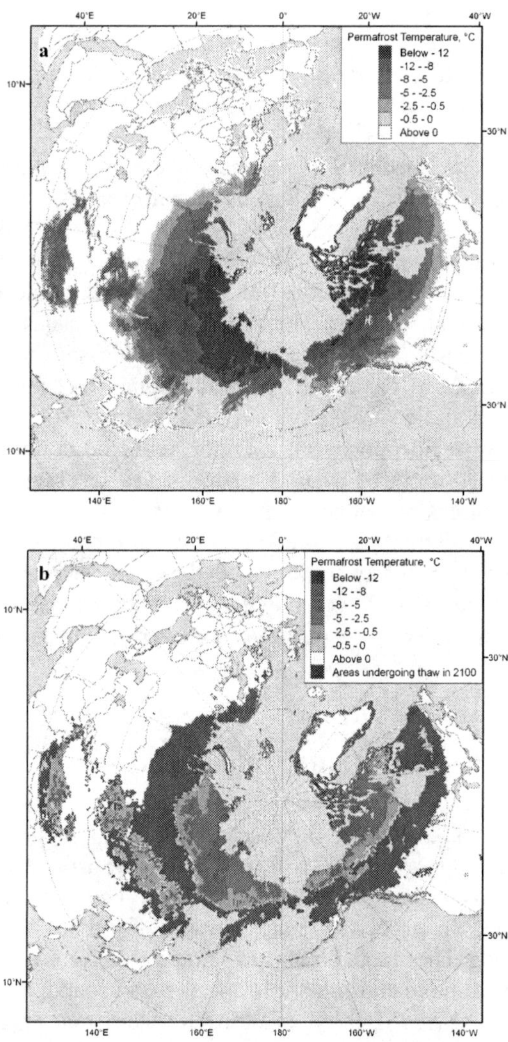

FIGURE 3. Permafrost temperatures predicted for two time periods: (a) 1990-2006 and (b) 2080-2099. Significant areas of the permafrost are already within 0.5° C of thawing (lightest blue shading). Considerable warming and thawing (red) are expected by the end of the century, as predicted by the Spatially Distributed Permafrost Dynamics Model[**] of the Geophysical Institute Permafrost Lab (GIPL) at the University of Alaska, Fairbanks (UAF). (Figure courtesy of Sergey Marchenko and Vladimir Romanovsky, GIPL, UAF.)

[**] The model (S. Marchenko, V. Romanovsky, *EOS Trans.* **88,** 52, 2007) was run for the Northern Hemisphere using a climate scenario for a gradual doubling of CO_2 by the end of the century. The model was derived from the MIT-2D integrated global system model developed at MIT. That model is a two-dimensional (zonally averaged) atmospheric model coupled with a diffusive ocean model that simulates the surface climate over the land and ocean for 23 latitudinal bands globally (A. P. Sokolov, P. H. Stone, *Clim. Dyn.* **14**, 291, 1998). Snow data for the period 2000-2100 were derived from the terrestrial ecosystem model (E. S. Euskirchen et al., *Global Change Bio.* **12**, 731, 2006).

at the sites has been rising since the early 1970s. Based on those measurements, he calculates that one third to one half of permafrost in Alaska is now within one degree to one and a half degrees Celsius of thawing; in some places worldwide, it is already crossing that critical zero degrees C threshold.

Ongoing observations, made by my research team during trips to Cherskii and numerous other sites and by our colleagues, reinforce the sense that thawing is accelerating and indicate that the emissions could be much greater than anticipated. My group's latest estimates are that under current warming rates, by 2100 permafrost thawing could boost methane emissions far beyond what would be produced by all other natural and man-made sources. The added green house gas, along with the extra carbon dioxide that exposed, thawing ground would release, together could raise the mean annual temperature of the earth by an additional 0.32 degree C, according to Vladimir Alexeev, also at the University of Alaska at Fairbanks.

That increase may sound minor, but it is not; it would contribute significantly to global warming-induced upset of weather patterns, sea level, agriculture and disease dispersal. If deeper sources of methane were to escape—such as that stored in material known as methane hydrates [see section below: Future Methane Sources from Hydrates]—the temperature rise could be as high as several degrees. Therefore, humankind has more reason than ever to aggressively slow the current rate of warming so that we do not push large regions of the Arctic over the threshold.

The Mother Lode in Siberia

Probing regions such as Cherskii is key to verifying—or revising—our estimations. Walking along a Siberian riverbank with my colleague from the Northeast Science Station, Sergei A. Zimov, I am careful where I step. The skin of the earth is only a half-meter thick, made up largely of muddy, mossy peat that sits loosely atop ice that is 40 to 80 meters deep. The stunted trees are slanted at various angles in this "drunken forest" because they cannot send roots into the frozen ground, and cycles of summer thaws generate large heaves. Behind me, one drunken tree crashes to the ground; through the torn blanket of forest floor we see the shiny black surface of solid ice and catch the musty scent of decomposing organic matter. It is also hard not to stub one's toe on the plethora of scattered bones: woolly rhinoceros, mammoth, Pleistocene lion, bear and horse.

To Zimov, this region is a goldmine—and not because of the tusks and skulls of extinct fauna. In 1989, spurred by an interest in the amount of carbon locked in the ground, he led a group of young scientists that set up the isolated Northeast Science Station to monitor permafrost in tundra and taiga year-round. The researchers traveled the great Russian rivers in small skiffs and scaled cliffs of permafrost without ropes to measure carbon content, the harbinger of methane release. With army tanks and bulldozers, they simulated disturbances that remove surface soil in the way that severe wildfires do. Their experiments proved the size and importance of the permafrost carbon pool to the world.

FIGURE 4. Yedoma ice complex. Exposed in the cliff face is a type of permafrost called yedoma, rich in ice and carbon—both central to the methane story. Massive wedges of ice 10 to 80 meters high and smaller lenses constitute up to 90 percent of the ground volume; the remainder is columns of organic rich soil, a cornucopia of the remains of Pleistocene mammals and the grasses they once ate. (Photo courtesy of K. W. Anthony, UAF, INE.)

But why did Zimov—and my group later—concentrate studies here, in a region known previously only for its Soviet gulags? Because not all permafrost is the same. Any ground where the mean annual temperature is below zero degrees C for at least two consecutive years is classified as permafrost, whether ice is present or not. This vast part of Siberia contains a distinct type of permafrost called yedoma, rich in ice and carbon—both central to the methane story. Massive wedges of ice 10 to 80 meters high and smaller lenses constitute up to 90 percent of the ground volume; the remainder is columns of organic rich soil, a cornucopia of the remains of Pleistocene mammals and the grasses they once ate.

Yedoma formed over roughly 1.8 million square kilometers in Siberia and in a few pockets of North America during the end of the last Ice Age. The organic matter froze in place before microbes could decompose it. A huge storehouse of food was being locked away until conditions would change, leaving the freezer door open.

A warmer climate recently has helped melt the yedoma ice, creating lakes. Vegetation collapses into the edges as the ground thaws and subsides, a process known as thermokarst. Today lakes cover up to 30 percent of Siberia. Further melting makes them larger and deeper, coalescing into broad methane-producing water bodies.

Blown Away by Bubbles

During the 1990s researchers at the Northeast Science Station observed that methane was bubbling out of the bottoms of lakes year-round but they did not know how important the lakes might be globally. Hence, my rough landing by plane in Cherskii this past August, for my ninth expedition of wading into voraciously expanding thermokarst lakes, to measure changes in permafrost and the release of methane.

My quest had begun as a Ph.D. research project in 2000. At the time, scientists knew that levels of methane—the third most abundant green house gas in the atmosphere after carbon dioxide and water vapor—were rising. The amount and the rate of increased emissions were unprecedented during the previous 650,000 years. Evidence indicated that in bygone eras the methane concentration in the atmosphere fluctuated by 50 percent in association with natural climate variations over thousands of years. But that change was slim by comparison with the nearly 160 percent increase that had occurred since the mid-1700s, rising from 700 parts per billion (ppb) before the industrial revolution to almost 1,800 ppb when I started my project.

Scientists also knew that agriculture, industry, landfills and other human activities were clearly involved in the recent rise, yet roughly half of the methane entering the atmosphere every year was coming from natural sources. No one, however, had determined what the bulk of those sources were.

From 2001 to 2004 I split my time between my cabin in Fairbanks and working with Zimov and others in Cherskii, living with the few local Russian families. In the attic library above our little, yellow wooden research station I spent long nights cobbling together plastic floats that I could place on the lakes to capture bubbles of methane. I dropped the bubble traps by leaning over the side of abandoned boats that I claimed, and I checked them daily to record the volume of gas collected under their large jellyfish-like skirts. In the beginning I did not capture much methane.

Winter comes early, and one October morning when the black ice was barely thick enough to support my weight I walked out onto the shiny surface and exclaimed, "Aha!" It was as if I was looking at the night sky. Brilliant clusters of white bubbles were trapped in the thin black ice, scattered across the surface, in effect showing me a map of the bubbling point sources, or seeps, in the lake bed below. I stabbed an iron spear into one big white pocket and a wind rushed upward. I struck a match, which ignited a flame that shot up five meters high, knocking me back ward, burning my face and singeing my eye brows. Methane!

FIGURE 5. Burning methane from a hotspot. The author has ignited the escaping methane with a match. Unfortunately, harvesting methane from the millions of lakes scattered across vast regions is not economically viable because the seeps are too diffuse. [Photo courtesy of Todd Paris, UAF; reprinted with permission from K. Walter Anthony et al., *Limnol. Oceanogr.:Methods*, **8**, 592 (2010).]

All winter I ventured across frozen lakes to set more traps above these seeps. More than once I stepped unknowingly on a bubbling hotspot and plunged into ice-cold water. Methane hotspots in lake beds can emit so much gas that the convection caused by bubbling can prevent all but a thin skin of ice from forming above, leaving brittle openings the size of manhole covers even when the air temperature reaches –50 degrees C in the dark Siberian winter. I caught as much as 25 liters (eight gallons) of methane each day from individual seeps, much more than scientists usually find. I kept maps of the hotspots and tallies of their emissions across numerous lakes. The strongest bubbling occurred near the margins of lakes where permafrost was most actively thawing. The radiocarbon age of the gas, up to 43,000 years old in some places, pointed to yedoma carbon as the culprit.

From 2002 to 2009 I conducted methane seep surveys on 60 lakes of different types and sizes in Siberia and Alaska. What scientists were not expecting was that the increase in methane emissions across the study region was disproportional to the increase in lake area over that same region. It was nearly 45 percent greater. It was accelerating.

Extrapolated to lakes across the Arctic, my preliminary estimate indicated that 14 million to 35 million metric tons of methane a year were being released. Evidence from polar ice-core records and radiocarbon dating of ancient drained lake basins has revealed that 10,000 to 11,000 years ago thermokarst lakes contributed substantially to abrupt climate warming—up to 87 percent of the Northern Hemisphere methane that helped to end the Ice Age. This outpouring tells us that under the right conditions, permafrost thaw and methane release can pick up speed, creating a positive feedback loop: Pleistocene-age carbon is released as methane, contributing to atmospheric warming, which triggers more thawing and more methane release. Now man-made warming threatens to once again trigger large feedbacks.

How fast might these feedbacks occur? In 2007 global climate models reported by the Intergovernmental Panel on Climate Change (IPCC) projected the strongest future warming in the high latitudes, with some models predicting a rise of seven to eight degrees C by the end of the 21st century. Based on numerous analyses, my colleagues and I predict that at least 50 billion tons of methane will escape from thermokarst lakes in Siberia as yedoma thaws during the next decades to centuries. This amount is 10 times all the methane currently in the atmosphere.

FIGURE 6. Newly forming lakes. Lakes form across Siberia as warming air thaws formerly frozen ground. (Photo courtesy of K. W. Anthony, UAF, INE.)

Future Methane Sources from Hydrates: Deeper Trouble

Permafrost is not the world's only methane concern. Vast quantities of the gas lie trapped in ice cages hundreds of meters down in the ground and below ocean bottoms. If these "methane hydrates" were to somehow melt and release their gas to the atmosphere, they would almost certainly trigger abrupt climate change. Evidence in seafloor sediments suggests that this very event, spurred by rapidly rising ocean temperatures, may have occurred 55 million years ago.

Some Russian scientists claim that more than 1,000 billion tons of methane lie beneath the Siberian shelf—submerged land extending seaward from the coastline that eventually drops to the deep ocean. If even 10 percent escaped—100 billion tons—it would be twice the 50 billion tons we project could be released by permafrost thaw. Warming of the deep ocean is unlikely in the near future. But high concentrations of methane in shallow waters along the shelf have recently been observed; continuing research there should determine whether the source is hydrates or (more likely) decomposing organic matter in permafrost thawing in the shallow seafloor.

On land, if lake-bed thawing extended like fingers deeper into the earth below, it could conceivably break into hydrate deposits and give them a channel to bubble upward to and through the water and into the atmosphere. My group is collaborating with U.S. Geological Survey scientists Carolyn Ruppel and John Pohlman to evaluate this possibility.

If hydrates prove to be a threat, the effect might be counteracted a bit by extracting the methane as a fuel before it is released. The methane in global hydrates would produce more energy than all the natural gas, oil and coal deposits on earth combined. Very little of it would ever be economically recoverable, however, because it is too dispersed in geologic strata, making exploration and extraction too expensive, even if oil was $100 a barrel. In a few places, mining concentrated hydrates might prove more affordable, and countries such as Japan, South Korea and China, eager to reduce fossil-fuel imports, are investing in technology to possibly extract those deposits. ConocoPhillips and British Petroleum are assessing the commercial feasibility of certain hydrates in the U.S.

Tapping hydrates is controversial. If enough evidence suggested an imminent, uncontrolled release of methane from destabilized hydrates, then capturing the gas instead would help mitigate climate warming. No proof of large hydrate releases exists yet, however, so commercial extraction would simply exasperate fossil-fuel driven climate changes. From a global-warming point of view, we are better off leaving those hydrates deep underground.

FINE-TUNING THE MODELS

Even with our best efforts, our current estimates beg more sophisticated modeling as well as consideration of potential negative feedbacks, which could serve as breaks on the system. For instance, in Alaska, a record number of thermokarst lakes are draining. Lakes formed in upland areas grow until they hit a slope. Then the water flows downhill, causing erosion and further drainage, sending melted sediment into rivers and eventually the ocean. Drained basins fill in with new vegetation, often becoming wetlands. Although they produce methane when they are unfrozen in summer, their total annual emissions are often less than those of lakes.

There are also negative feedbacks in the thermokarst lake cycle. Peat accumulates in drained lake basins.

FIGURE 7. Possible negative feedback. As temperatures warm and more permafrost thaws, the ground subsides and the resulting thermokarst lakes grow. These lakes grow until they hit a slope and begin to drain. Then wetlands form. The net production of methane can be less than that of lakes. (Figure courtesy of K. W. Anthony, UAF, INE.)

It is hard to say whether such potential processes would lessen methane release by a sizable amount or just a few percentage points. Two projects of mine, with my Fairbanks colleague Guido Grosse, Lawrence Plug of Dalhousie University in Nova Scotia, Mary Edwards of the University of Southampton in England and others, began in 2008 to improve the first-order approximations of positive and negative feedbacks. A key step is to produce maps and a classification of thermokarst lakes and carbon cycling for regions of Siberia and Alaska. The cross-disciplinary research links ecological and emissions measurements, geophysics, remote sensing, laboratory incubation of thawed permafrost soils and lake sediments, and other disciplines. The goal is to inform a quantitative model of methane and carbon dioxide emissions from thermokarst lakes from the Last Glacial Maximum (21,000 years ago) to the present and to forecast climate-warming feedbacks of methane from lakes for the up-coming decades to centuries.

To help predict how future warming could affect thermokarst lakes, Plug and a postdoctoral researcher working with us, Mark Kessler, are developing two computer models. The first, a single-lake model, will simulate the dynamics of a lake basin. The second, a landscape model, includes hill-slope processes, surface-water movement and landscape-scale permafrost changes. The models will first be validated by comparison with landscapes we are already studying, then against data from sediment cores going back 15,000 years in Siberia and Alaska, and then against other climate simulations from 21,000 years ago. The final step will be to couple the thermokarst-lake models with the vast Hadley Center Coupled Model that describes the circulation of oceans and atmosphere—one of the major models used in IPCC assessment reports. The result, we hope, will be a master program that can fully model the extent and effects of permafrost thaw, allowing us to calculate a future rate of methane release and assess how that would drive global temperatures.

More fieldwork, of course, will continue to refine the data going into such models. In August 2011, with the help of a hovercraft, we will investigate lakes along nearly 1,000 miles of Siberian rivers and Arctic coast. A huge expedition will also retrieve sediment cores from lakes dating back millennia. Field data, together with remote sensing, will ultimately be used in the Hadley Center program to model climate change drivers from the Last Glacial Maximum to 200 years into the future.

SOLUTIONS

If, as all indicators suggest, arctic methane emissions from thawing permafrost are accelerating, a key question becomes: Can anything be done to prevent methane release? One response would be to extract the gas as a relatively clean fuel before it escapes. But harvesting methane from the millions of lakes scattered across vast regions is not economically viable, because the seeps are too diffuse. Small communities that are close to strong seeps might tap the methane as an energy source, however.

Zimov and his son, Nikita, have devised an intriguing plan to help keep the permafrost in Siberia frozen. They are creating a grassland ecosystem maintained by large northern herbivores similar to those that existed in Siberia more than 10,000 years ago. They have introduced horses, moose, bears and wolves to "Pleistocene Park," a 160-square-kilometer scientific reserve in northeastern Siberia. They brought back muskox from Wrangell Island and plan on bison from North America, depending on funding, which comes from independent sources, the Russian government and U.S. agencies.

These grazing animals, along with mammoths, maintained a steppe-grassland ecosystem years ago. The bright grassland biome is much more efficient in reflecting incoming solar radiation than the dark boreal forest that has currently replaced it, helping to keep the under lying permafrost frozen. Furthermore, in winter the grazers trample and excavate the snowpack to forage, which allows the bitter cold to more readily chill the permafrost.

One man and his family have taken on a mammoth effort to save the world from climate change by building Pleistocene Park. Yet a global response is needed, in which every person, organization and nation takes responsibility to reduce their carbon footprint. Slowing emissions of carbon dioxide is the only way humankind can avoid amplifying the feedback loop of greater warming causing more permafrost thaw, which causes further warming. We predict that if carbon emissions increase at their current projected rate, northern lakes will release 100 million to 200 million tons of methane a year by 2100, much more than the 14 million to 35 million tons they emit annually today. Total emissions from all sources worldwide is about 550 million tons a year, so permafrost thaw, if it remains unchecked, would add another 20 to 40 percent, driving the additional 0.32 degree C rise in the earth's mean annual temperature noted earlier. The world can ill afford to make climate change that much worse. To reduce atmospheric carbon dioxide and thereby slow permafrost thaw, we all must confront the elephant in the room: people burning fossil fuels.

REFERENCES

1. K.M. Walter, "Methane Bubbling from Siberian Thaw Lakes as a Positive Feedback to Climate Warming," *Nature,* Vol. 443, pages 71–75; September 7, 2006.
2. K.M. Walter, "Thermokarst Lakes as a Source of Atmospheric CH_4 during the Last Deglaciation," *Science*, Vol. 318, pages 633–636; October 26, 2007.
3. G. Grosse, K. Walter and V. E. Romanovsky, *Assessing the Spatial and Temporal Dynamics of Thermokarst, Methane Emissions, and Related Carbon Cycling in Siberia and Alaska*, NASA Carbon Cycle Sciences Project, April 2008–March 2011.
4. K. Walter, G. Grosse, M. Edwards, L. Plug and L. Slater, *Understanding the Impacts of Icy Permafrost Degradation and Thermokarst-Lake Dynamics in the Arctic on Carbon Cycling, CO_2 and CH_4 emissions, and Feedbacks to Climate Change*, Project 0732735 for National Science Foundation/International Polar Year, July 2008–June 2011
5. V. W. Romanovsky et al., "*Thermal State and Fate of Permafrost in Russia:* First Results of IPY, Ninth International Conference on Permafrost, Vol. VII, pp. 1511-1518 (2008). http://www.blue-europa.org/nicop_proceedings/6%20Vol%202%20(1281-1530).pdf
6. E. A. G. Schuur, et al., "The effect of permafrost thaw on old carbon release and net carbon exchange from tundra," *Nature* **459**, pages 556-559; May 28, 2009.
7. C. Tarnocai, et al. "Soil organic carbon pools in the northern circumpolar permafrost region," *Global Biogeochem. Cycles* **23,** GB2023, 2009.
8. S. A. Zimov, E. A. G. Schuur, F. S. Chapin."Permafrost and the global carbon budget," *Science*, **312,** pages 1612-1613; June 16, 2006.
9. To learn more and to get involved in collecting data, see the Pan-Arctic Lake-Ice Methane Monitoring network (http://ine.uaf.edu/werc/palimmn/).

Climatic Consequences of Afforestation

Inez Fung[a] and Abigail Swann[b]

[a]*Department of Earth and Planetary Science*
University of California, Berkeley
Berkeley, CA 94720

[b]*Department of Organismic and Evolutionary Biology*
Harvard University
Cambridge, MA 02138

Abstract. One strategy proposed to mitigate the climate change associated with increased greenhouse gases is to plant more trees, which absorb carbon dioxide in the process of photosynthesis. We have used a climate model to explore the possible impact of a large-scale afforestation, both in the Arctic and at mid-latitudes in the northern hemisphere. The global climate responds in surprising ways, largely because of the water vapor released by plant transpiration. The experiments illustrate the importance of exploring the entire spectrum of consequences from any proposed action to deal with climate change.

INTRODUCTION

Global climate models are the principal tool for projecting global and regional climate change in response to different anthropogenic activities. They solve the equations for the conservation of mass, angular momentum, and energy for the atmosphere and the oceans. The models include also prognostic equations for the distribution of water vapor in the atmosphere and of salinity in the oceans. Typically the time histories of atmospheric composition and land cover are specified, and the model is integrated forward in time, to yield the transient or equilibrium climate response. A new generation of climate models predicts, rather than specifies, the abundance of CO_2 and other trace constituents in the atmosphere. For CO_2, the trajectory of fossil-fuel emissions is specified, and the atmospheric CO_2 increment is the residual after climate-sensitive interactions with the land biosphere and the oceans are taken into effect. [1, 2]

Physics of Sustainable Energy II: Using Energy Efficiently and Producing it Renewably
AIP Conf. Proc. 1401, 211-219 (2011); doi: 10.1063/1.3653853

ESTIMATES OF CLIMATE FORCING

The main driver of climate change is the radiative forcing (in watts per square meter, W/m^2) contributed by various agents external to the climate system. The forcing associated with each greenhouse gas is determined by the change in the abundance of the gases since the pre-industrial era together with its absorption spectrum. Figure 1 summarizes the changes in forcings that the Fourth Assessment Report of the Intergovernmental Panel on Climate Change found to be associated with the changes in atmospheric composition since the preindustrial era [3]. Note that the radiative forcing for greenhouse gases is not in proportion to their abundance in the atmosphere, as the molecules with more vibrational and rotational degrees of freedom are more potent absorbers. Thus, the forcing associated with each additional molecule of chlorofluorocarbon (CFC) is about 10,000 times greater than that associated with a molecule of CO_2, and the radiative forcing by CFC is 20-25% that of CO_2, even though the abundance of CFC is several orders of magnitude smaller than that of CO_2.

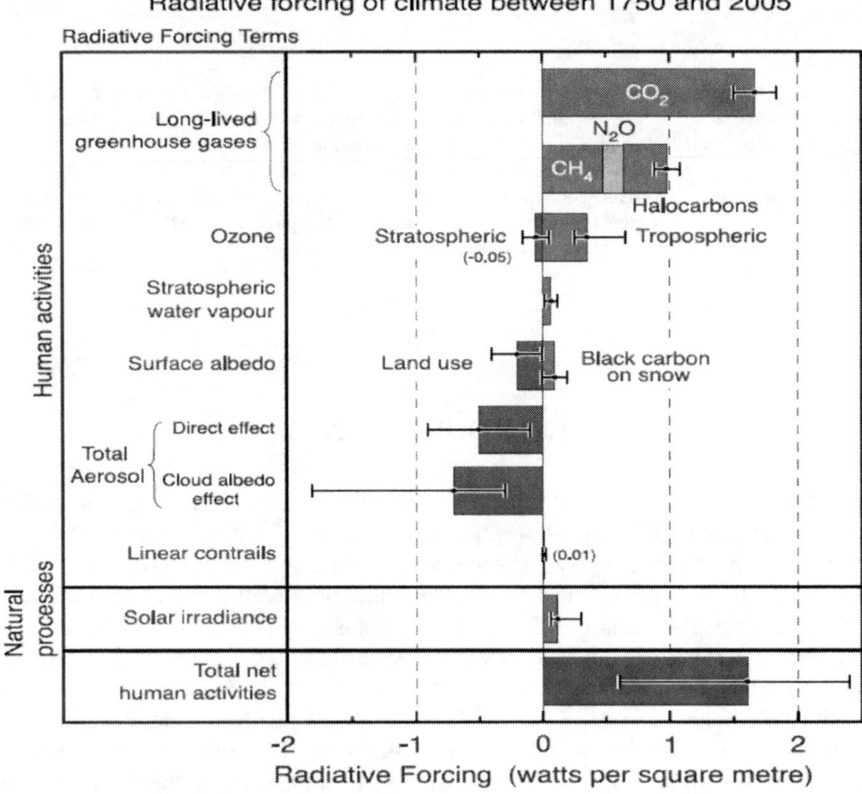

FIGURE 1. Changes in Radiative Forcing. Estimates done by the Intergovernmental Panel on Climate Changes for its 2007 report.[1] The table lists the changes in forcings (in watts per square meter) calculated for various changes in atmospheric constituents caused by man's activities since the preindustrial age.

Figure 1 also includes both the direct and indirect forcing associated with aerosols. While most aerosols reflect sunlight, some (e.g. mineral aerosols, soot) can also absorb in the visible and/or the infrared parts of the spectrum. Hence, the direct forcing can be either positive or negative, but the net impact of the increase in aerosols is estimated to be a negative forcing. The magnitude of the forcing depends on the size of the aerosols as well as their concentrations. For the same aerosol mass, the number density and hence total reflective area are inversely proportional to aerosol radius, and so small droplets are more reflective than large ones. The indirect effect of aerosols results from their ability to act as cloud condensation nuclei and influence cloud microphysics and cloud dynamics. There is large uncertainty associated with the cloud effect, as reflected in the large error bars shown.

The net sum of all the global mean forcings from 1750 to 2005, as summarized in Figure 1, is a radiative forcing of 2.9 +/- 0.3 W/m^2 from greenhouse gases, countered by the negative forcing from aerosols, with a median of -1.3 W/m^2 and a range of -2.2 to -0.5 W/m^2 90% confidence range. The combined forcing is thus 1.6 W/m^2 with a 0.6 to 2.4 W/m^2 90% confidence range. Greenhouse gases are typically long-lived, with lifetimes of weeks to centuries and longer, while aerosols have much shorter lifetimes, typically days to weeks. Hence climate change would be greater without the effects of mostly polluting aerosols.

CLIMATE FEEDBACKS

The radiative forcing can be amplified or diminished by feedbacks internal to the climate system. The major feedbacks are associated with the three phases of water are summarized in Figure 2. As the climate warms, increased evaporation from the oceans leads to an increase in atmospheric water vapor, itself a greenhouse gas, thus leading to more heating that magnifies the initial warming, i.e. a positive feedback. The vapor exceeding the saturation threshold condenses and forms clouds which reflect sunlight, leading to a reduction in the warming, i.e. a negative feedback. Finally, warming causes the melting of the snowpack and sea ice, leading to a decrease of surface albedo, enhanced solar absorption and greater warming -- a positive feedback.

Although the net feedback associated with cloud changes is expected to be negative, the uncertainties are large. The impact of a cloud on climate depends greatly on its horizontal and vertical extent, among other things, and hence its impact on both the solar and infrared radiation. Contrast, for example, tall cumulonimbus clouds in the tropics with maritime stratus clouds off the coast of California. Cumulonimbus towers are narrow, with tops at high altitudes. Because temperature decreases with height, high clouds radiate at a colder temperature than low clouds, and so could have a net warming effect because of the reduced energy loss to space. Maritime stratus clouds, which are low and have a large areal coverage, reflect more sunlight and have a cooling effect on climate. This is an active area of research.

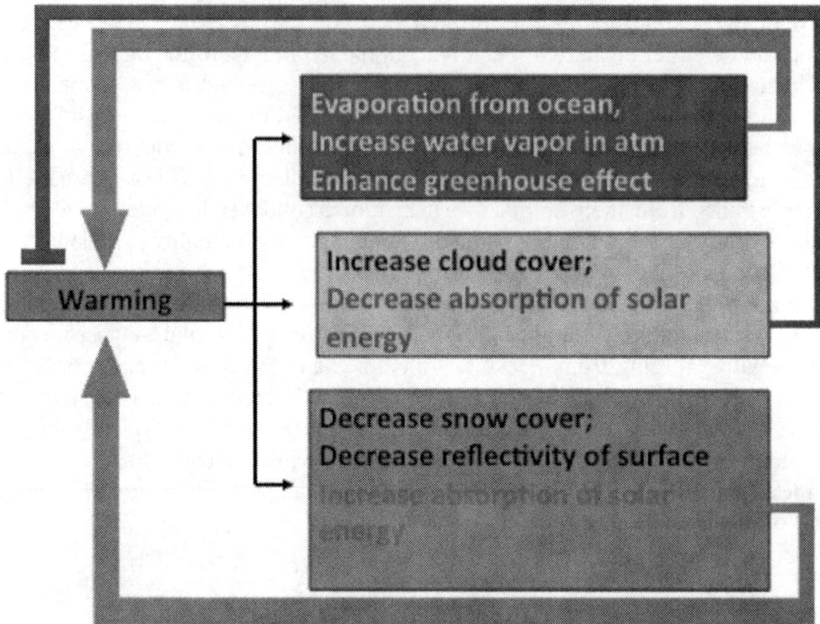

FIGURE 2. Climate Feedbacks associated with changes in the phase of water.

AFFORESTATION AND CLIMATE CHANGE

Large-scale planting of trees has been proposed as a climate-mitigation strategy, as carbon is removed from the atmosphere via photosynthesis and stored in the forests. Such large-scale modification of the land surface also impacts the climate directly. Vegetation influences the surface energy budget in several ways. First, albedo is determined by the type of landcover. Trees growing in a previously snow-covered area would lead to a decrease in albedo, with the decrease much less for the brighter deciduous trees than for the darker evergreen trees. A second impact on the energy balance is the change in net longwave radiation, which is determined by the vertical profiles of temperature and greenhouse gases (especially water vapor) in the atmospheric column. The net radiation (net solar minus net longwave radiation) is countered by latent and sensible heat fluxes, with the residual energy heating the surface. Latent heat flux is dominated by transpiration from plants and evaporation from the soil, both of which contribute water vapor to the atmosphere. Deciduous trees transpire more than evergreen trees. Sensible heating directly warms the air above the surface.

We carried out two sets of afforestation experiments with the carbon-climate model of the National Center for Atmospheric Research [4] to explore the impact of afforestation on climate. We were interested in the resultant equilibrium climate, and so the atmospheric general circulation model was coupled to a simplified ocean. Each set of experiments comprises four model runs, two with the oceans represented by a simple "slab" ocean, with interactive thermodynamics but prescribed heat transports, and two with prescribed sea surface temperature and sea ice conditions, so that the oceans do not play a role in the resultant climate change. For each ocean configuration, the vegetation is either left as is (the "control run") or modified (described below). Each run was integrated to equilibrium.

In the first set of experiments, we replaced bare ground in the Arctic with deciduous trees in the model[5]. The experiment is motivated by suggestions that broad-leaf trees may invade warming tundra more readily than evergreen trees would [6]. Furthermore, evidence for such vegetation has been seen in paleoclimate data for warm periods in the past.[7]

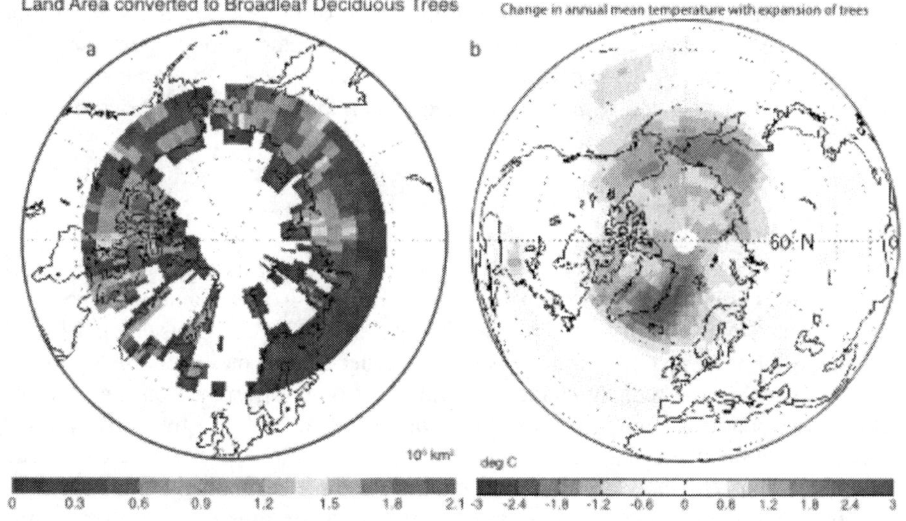

FIGURE 3. Arctic Afforestation Experiment. (a) Bare ground area converted to deciduous forests in the climate model. (b) Near-Surface temperature anomalies (Kelvin). From Swann *et al.* (2010).

The results of our study are illustrated in Figure 3. In the model runs with interactive oceans, we found, to our surprise, widespread temperature increases across the Arctic, exceeding 2°C in places, even away from the afforested areas. In particular, we found that the top-of-atmosphere radiative imbalance from enhanced transpiration (associated with the expanded forest cover) is up to 1.5 times larger than the forcing due to albedo change from the forest. Furthermore, the greenhouse warming by the additional water vapor transpired melts sea-ice and triggers a positive feedback through changes in ocean albedo and evaporation. Land surface albedo change has been considered to be the dominant mechanism by which trees directly modify climate at high-latitudes, but our experiments suggest an additional mechanism through transpiration of water vapor and feedbacks from the ocean and sea ice. The amplification of the warming is not found in the runs with prescribed ocean.

To understand these results, we employed a 1-D (vertical) radiative-convective model, in which the impacts of a single perturbation can be assessed, with all other variables held at the values of the control run. We estimated the changes in the forcing at the top of the atmosphere caused only by a change in albedo and only by an increase in water vapor due to transpiration in the runs with fixed ocean conditions. The two forcings are comparable, ~1 W/m^2 averaged poleward of 60°N. Because water vapor is rapidly mixed in the Arctic atmosphere, the increased water vapor greenhouse effect over the Arctic ocean in the run with an interactive ocean leads to an enhanced ice-albedo feedback, associated with melting sea ice, increased evaporation from the open ocean.

MID–LATITUDE AFFORESTATION

In our second set of experiments, we replaced mid-latitude grasslands and croplands with deciduous forests.[8] Again, in the model runs with an interactive ocean, we found widespread warming, extending well beyond the afforested areas (Figure 4). To start, the northern hemisphere surface is darkened, leading to increased absorption of solar radiation. The subsequent changes are however quite different from those in the Arctic: transpiration is limited by the finite amount of moisture in the soils, and the mid-latitude atmosphere has a greater moisture capacity than the cold Arctic atmosphere. Thus enhanced transpiration contributes only a relatively small increase in water vapor in the mid-latitudes and the water vapor greenhouse effect, while positive, is weak. Ice-albedo feedback is also weak, due to distance from the forcing. The change in net radiation is no longer balanced by changes in latent heat loss, but by changes in sensible heat loss from the surface, which directly warms the atmosphere.

An unexpected consequence of the warmer northern hemisphere is a shift in the position of the intertropical convergence zone (ITCZ), the rain belt that demarcates the Northern and Southern Hadley cells (Figure 5). The Southern Hemisphere Hadley cell expands northwards to enhance heat transport from the warmer to the cooler hemisphere: a northward shift in the rain belt results. Such changes in global circulation lead to changes in net primary production worldwide: net primary production decreases in the southern Amazon, because of the drying (Figure 6).

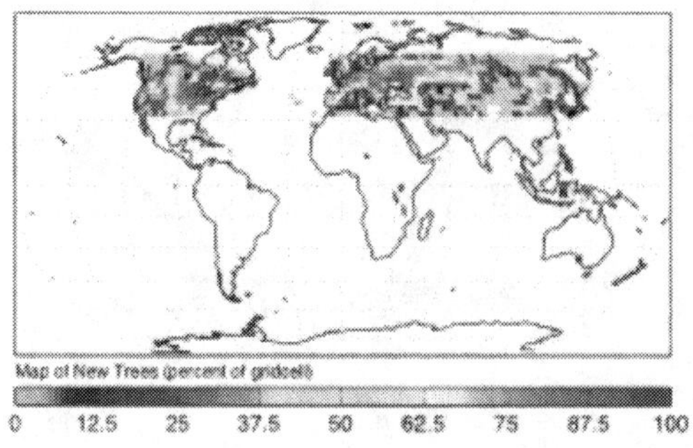

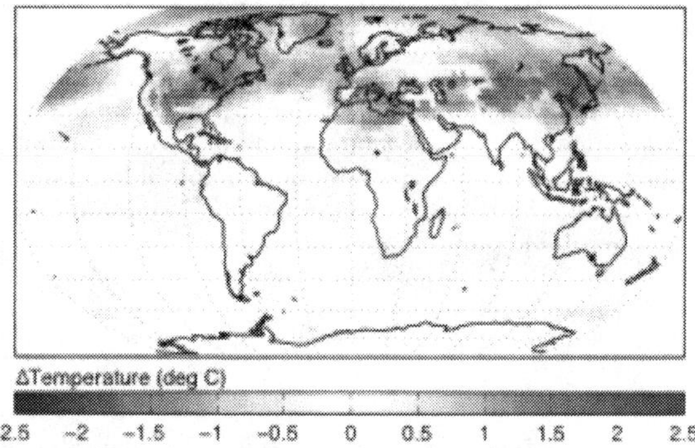

FIGURE 4. Mid-Latitude Afforestation Experiment. (a) Areas in which deciduous trees were assumed to replace grassland or cropland in the climate model. (b) The resulting temperature changes. Note that temperatures were affected even well beyond the afforested regions. From Swann et al. (2011).

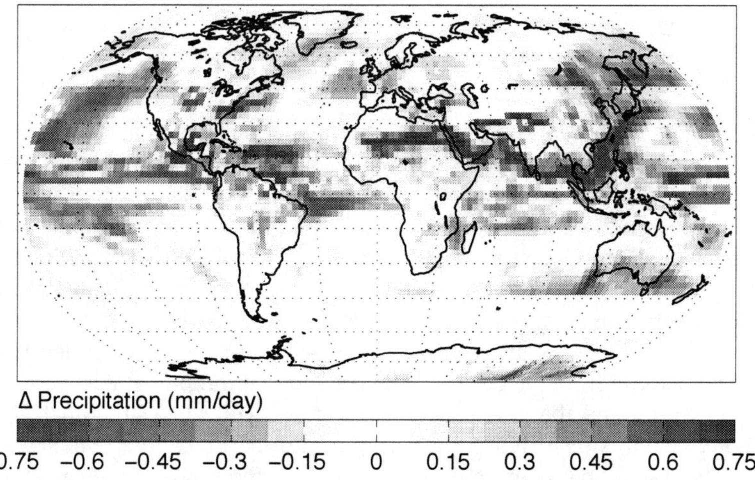

FIGURE 5. Teleconnections. The northern hemisphere warming caused by the hypothetical planting of mid-latitude trees led to changes in atmospheric circulation and changes in precipitation patterns. From Swann *et al.* (2011)

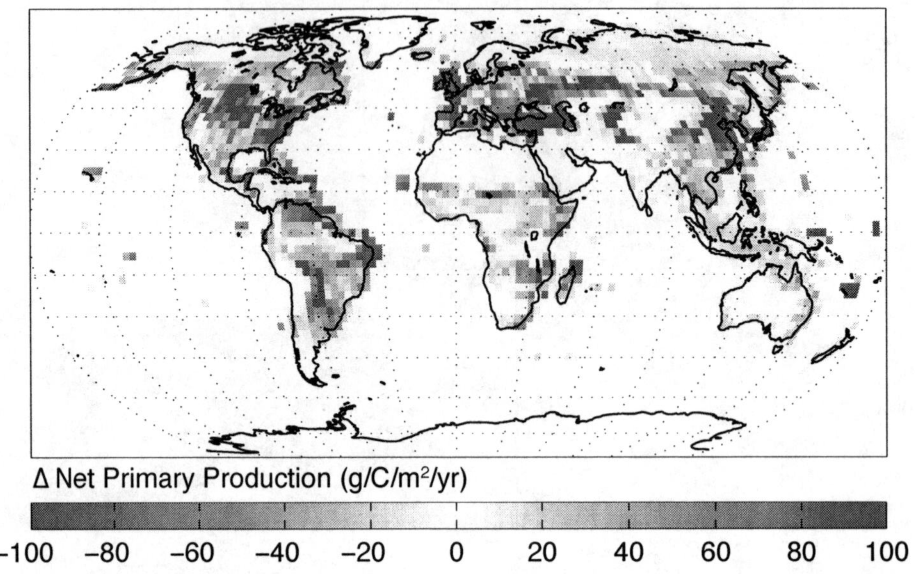

FIGURE 6. Changes in Net Primary Productivity (NPP). The changed precipitation patterns caused by the warming of the northern hemisphere had implications for the growth of vegetation throughout the world. From Swann *et al.* (2011).

CONCLUDING REMARKS

The afforestation experiments with the climate model prescribed landcover changes on a scale much larger than any afforestation project under consideration. Still, the results serve to underscore the kinds of unexpected consequences that might ensue and hence inform climate modification strategies. Our mid-latitude afforestation experiment resulted in an additional storage of 230 petagrams carbon (1 Pg = 10^{15} g) carbon in the terrestrial biosphere. The removal of this much carbon from the atmosphere would produce a global cooling of 1–2.2 °C. However, a decrease in the atmospheric carbon alters the gradient in CO_2 partial pressure across the air-sea interface and leads to an outgassing of carbon from the oceans. The resultant net drop in atmospheric carbon is only 30–50 Pg C, and the global cooling is small while the latitudinal gradient in temperature, which is the driving force for changes in circulation, remains (Swann et al. 2011). Such a global perspective obscures substantial regional changes in ocean, atmosphere and ecosystems.

REFERENCES

1. Fung, I., "Challenges of climate modeling," *Discrete and Continuous Dynamical Systems – Series B,* 7, 543-551 (2007).
2. Friedlingstein, P., P. Cox, R. Betts, L. Bopp, W. Von Bloh, V. Brovkin, P. Cadule, S. Doney, M. Eby, I. Fung, G. Bala, J. John, C. Jones, F. Joos, J.T. Kato, M. Kawamiya, W. Knorr, K. Lindsay, H. D. Matthews, T. Raddatz, P. Rayner, C. Reick, E. Roeckner, K.G. Schnitzler, R. Schnur, K. Strassmann, A.J. Weaver, C. Yoshikawa and N. Zeng, "Climate carbon cycle feedback analysis: Results from the C4MIP Model Intercomparison," *J. Climate* 19, 3337-3353 (2006).
3. "Climate Change 2007: The Physical Basis" (The Contribution of Working Group I to the Fourth Assessment Report of the Intergovernmental Panel on Climate Change), S. D. Solomon, D. Qin, M. Manning, Z. Chen, M. Marquis, K. B. Averyt, M. Tignor, H.L. Miller (eds.), Cambridge University Press (Cambridge, UK, 2007).
4. CollinsW, *et al*, "The formulation and atmospheric simulation of the community atmosphere model version 3 (CAM3)," J. *Climate* 19, 2144–2161 (2006).
5. A.L. Swann, I.Y. Fung, S. Levis, G.B. Bonan and S.C. Doney, "Changes in Arctic vegetation amplify high-latitude through the greenhouse effect," *Proc. Nat'l. Acad. Sci.* 107, 1295-1300 (2010). www.pnas.org/cgi/doi/10.1073/pnas.0913846107
6. Rupp T, F. Chapin and A. Starfield, "Response of subarctic vegetation to transient climatic change on the Seward Peninsula in north-west Alaska," Global Change Biol. 6, 541–555 (2000).
7. Edwards M, L. Brubaker, A. Lozhkin and P. Anderson, "Structurally novel biomes: A response to past warming in Beringia," *Ecology* 86, 1696–1703 (2005).
8. Swann, A.L, I.Y. Fung and J.C.H. Chiang, "Mid-latitude afforestation shifts general circulation and tropical precipitation," submitted to *Proceedings Proc. Nat. Acad. Sci., USA* (2011).

The Thinning of Arctic Ice*

Ronald Kwok[a] and Norbert Untersteiner[b]

[a]*Jet Propulsion Laboratory*
California Institute of Technology
Pasadena, California.

[b]*Department of Atmospheric Sciences and Geophysics*
University of Washington
Seattle, Washington

Abstract. The surplus heat needed to explain the loss of Arctic sea ice during the past few decades is on the order of 1 W/m^2. Observing, attributing, and predicting such a small amount of energy remain daunting problems.

INTRODUCTION

During the first half of the 20th century, the Arctic sea-ice cover was thought to be in a near-steady seasonal cycle, reaching an area of roughly 15 million km^2 each March and retreating to 7 million km^2 each September. Ice thick enough to survive the melt season, termed perennial or multiyear ice (MYI), adds to the ice cover. A large fraction of MYI typically remained in the Arctic Basin for several years and grew to an equilibrium thickness of about 3.5 m—melting half a meter at the surface from June through August and growing by about half a meter at the bottom from October through March. In the late 1970s, MYI occupied more than two-thirds of the surface area of the Arctic Basin, with first year ice (FYI) covering the remaining one-third. FYI is the thinner, seasonal ice that fills cracks in the ice cover and grows on the open ocean with the southward advance of the ice edge at the end of each summer.

That picture began to change significantly in the latter part of the century. Since 1979, passive microwave measurements by satellite, which can distinguish between the brightness signatures of ice and water, have established a more accurate account of the seasonal cycle of ice extent. The satellite record reveals that over the past 30 years the average September ice extent has been declining at an astonishing rate of more than 11% per decade (see figure 1 and the article by Josefino Comiso and Claire Parkinson in PHYSICS TODAY, August 2004, page 38).

---

* Reprinted with permission from Ronald Kwok and Norbert Untersteiner, *Physics Today* Vol. 64, April 2011, pages 36-41,

Physics of Sustainable Energy II: Using Energy Efficiently and Producing it Renewably
AIP Conf. Proc. 1401, 220-231 (2011); doi: 10.1063/1.3653854
2011 American Institute of Physics 978-0-7354-0972-9/$30.00

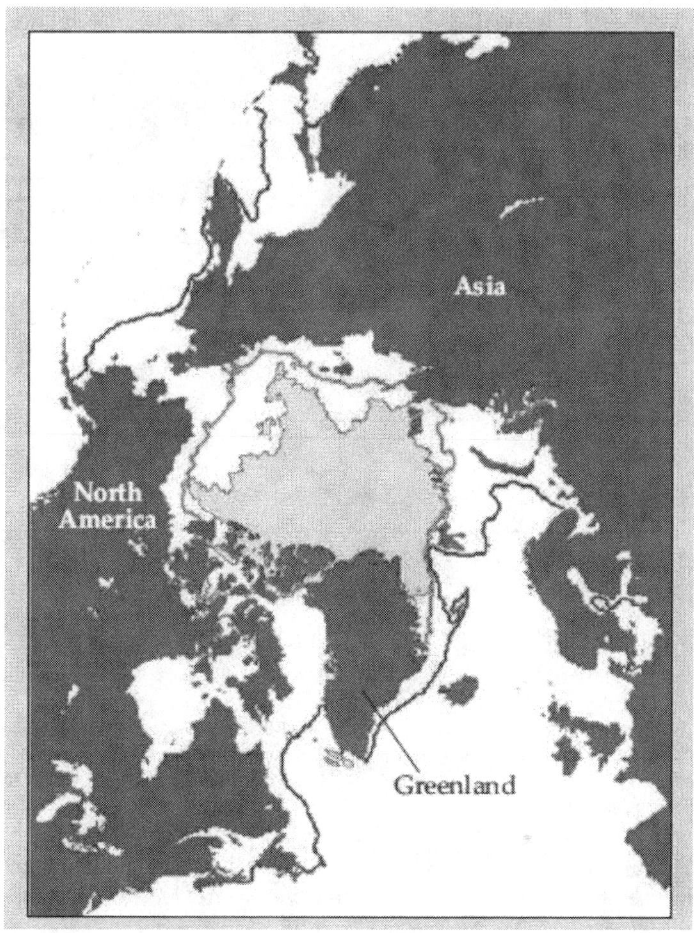

FIGURE 1. The extent of sea ice covering the Arctic Ocean expands and recedes seasonally. The median ice edges in March (blue) and in September (red) illustrate the extremes in area over the period 1979–2000. The yellow area shows the sea-ice extent at the end of summer 2010. (Data courtesy of the National Snow and Ice Data Center.)

As a consequence, FYI has replaced much of the MYI in the Arctic Ocean. Satellite-borne radar scatterometers have made it possible during the past decade to identify and directly map the two primary ice types. As ice ages, brine drains from it, leaving air pockets behind; the older, less saline MYI is more than twice as reflective as seasonal ice. From that complementary satellite record, scientists witnessed a dramatic loss of MYI during the past decade, as illustrated in figure 2. Between 2004 and 2008, the winter cover of MYI shrank by 1.5 million km^2—more than twice the size of Texas—and now covers only one-third of the Arctic Basin.

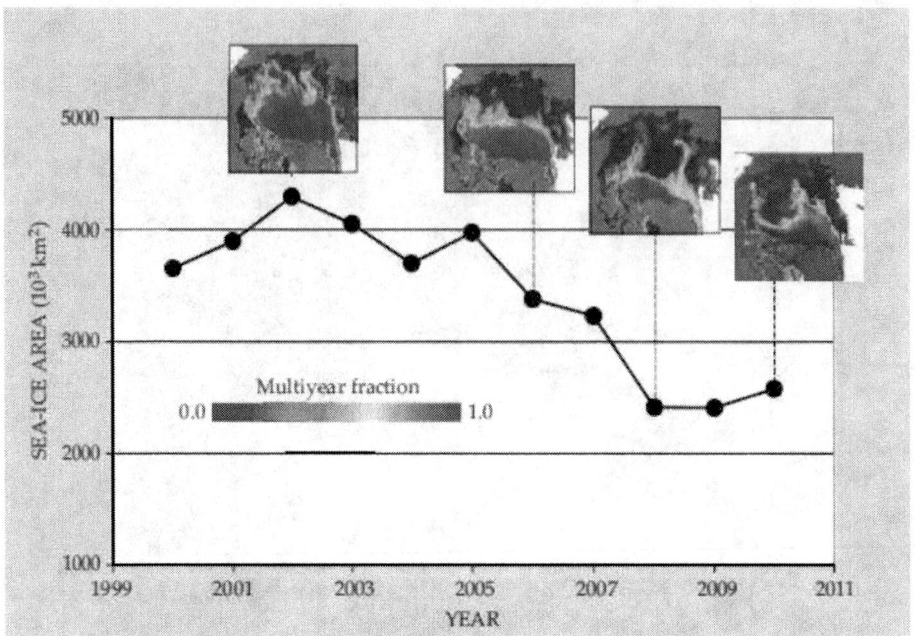

FIGURE 2. The decline of winter's multiyear sea-ice coverage is evident from an analysis of data taken over the years 2000–10 by NASA's *Quick Scatterometer* satellite and the European Space Agency's *Advanced Scatterometer* satellite. The electromagnetic scattering properties of first-year ice and multiyear ice—that which survives more than one summer melt season—differ in salinity, surface roughness, and volume inclusions (that is, air pockets) that develop as sea ice ages. Those differences alter reflectivity and thus distinguish the two ice types in radar backscatter measurements.

To determine the volume of melted sea ice and the associated changes in the heat lost or gained by the ice cover, one must make a basin-wide sampling not only of the ice's area but also of its thickness. The latter is the technically more difficult measurement. Cross-Arctic estimates of thickness came with the first under-ice crossing of the Arctic Basin by the nuclear submarine USS *Nautilus* in 1958. Since the 1960s the US Navy has periodically declassified measurements of ice draft—the depth of the submerged portion of the floating ice observed by upward-looking sonars on submarines—for scientific analysis. The ice draft is converted to thickness using Archimedes's principle and the densities of ice and seawater. Since 1979, researchers have been able to infer changes in the sea-ice thickness of the central Arctic using available submarine profiles.[1] The launch of the *Ice, Cloud, and Land Elevation Satellite* (*ICESat*) in 2003 made possible near-basin-scale mapping of ice thickness from space. The satellite's light detection-and-ranging (lidar) altimeter took readings of sea ice freeboard—that part of the ice above the ocean surface—and thicknesses could then be deduced from those freeboard measurements just as they are from the ice draft.

The combined submarine and *ICESat* records, plotted in figure 3, show that the average sea-ice thickness of the central Arctic during winter has decreased from 3.5 m to less than 2 m over the past three decades.[2] Along with the observed decrease in sea-ice extent, there is a parallel thinning of the ice cover. If those rates persist, we are likely to eventually experience a seasonally ice-free Arctic Ocean (see PHYSICS TODAY, September 2009, page 19).

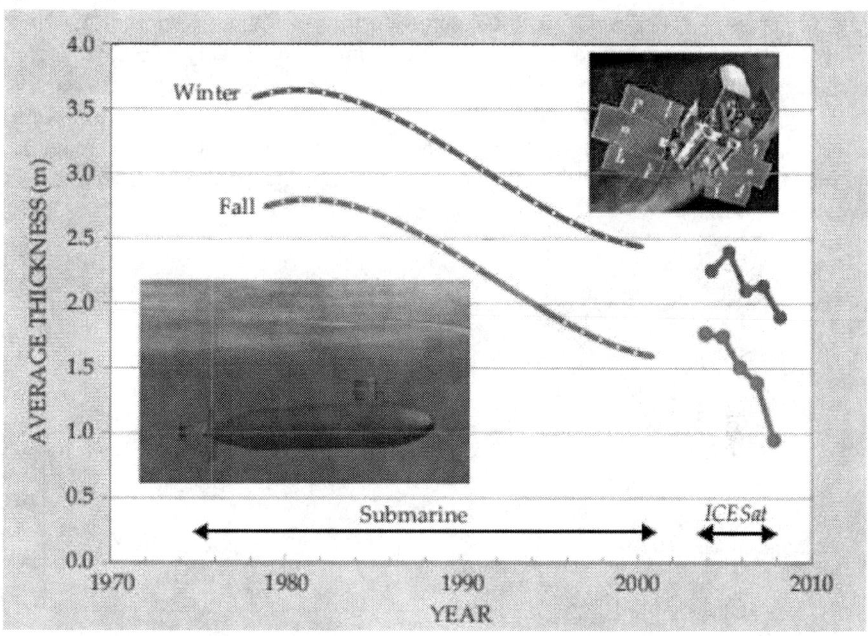

FIGURE 3. The thinning of the central Arctic sea-ice cover from 1978 to 2008 is evident from upward-looking sonar data recorded by US Navy submarines and by altimetry from NASA's *Ice, Cloud, and Land Elevation Satellite* (*ICESat*), launched in 2003. The overall mean winter thickness of 3.64 m in 1980 can be compared with a 1.89-m mean during the last winter of the *ICESat* record—an astonishing decrease of 1.75 m in thickness. Between 1975 and 2000 the steepest rate of change was −0.08 m/yr in 1990. During *ICESat's* Recent five-year run through 2008, it recorded a still higher rate of −0.10 to −0.20 m/yr. (Adapted from ref. 2.)

That possibility has received increased public attention because the presence or absence of Arctic sea ice is a striking, important, and leading indicator of climate change. The shrinking ice cover has far-reaching consequences. Shifts in local climate affect marine ecosystems, endanger survival of birds and mammals, and pose a threat to the livelihood of indigenous communities around the Arctic Basin. Moreover, an ice-free ocean raises a plethora of issues concerning commercial shipping and resource extraction, all with long-term geopolitical and economic implications.

Changes in Arctic sea ice also influence deep convection in the marginal waters such as the Greenland and Labrador Seas. Those seas are sources of North Atlantic Deep Water, which contributes to the meridional overturning circulation (sometimes referred to as the conveyor belt), a global system of surface and deep currents that transports large amounts of water, heat, salt, carbon, nutrients, and other substances around the major oceans. That global circulation connects the ocean surface and atmosphere with the huge reservoir of the deep sea. Changes in the rate of production of North Atlantic Deep Water in the Arctic marginal seas have been shown to affect the Gulf Stream and hence climate, particularly that of Europe.

The observed rates of shrinking and thinning of sea ice in the Arctic Basin during the past three decades were greatly underestimated by the 2007 Intergovernmental Panel on Climate Change Fourth Assessment Report (IPCC–AR4) climate models;[3] indeed, none of the models can quantitatively explain the trends experienced in the Arctic. But the ice-free summers widely forecast in press reports as impending have not yet occurred. As long as some of the FYI is thick enough to survive the summer, and as long as the annual export of ice out of the Arctic Basin continues to be no more than the current annual average, a precipitous decline of the ice cover is not likely.

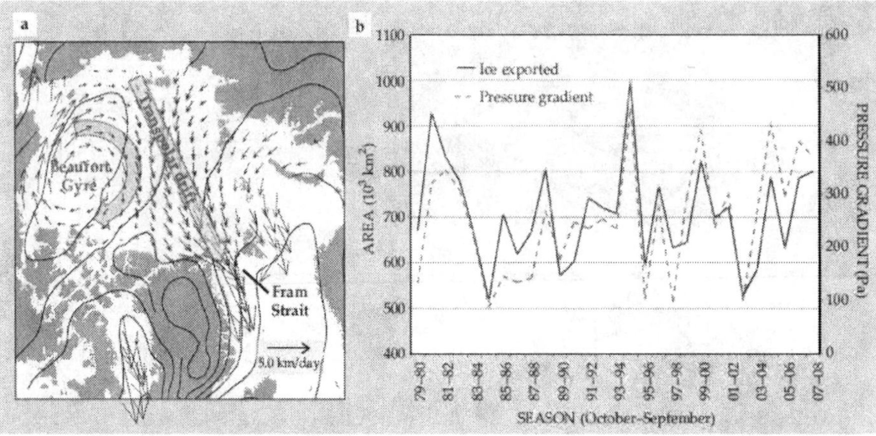

FIGURE 4. Sea-ice circulation. **(a)** The two prominent features in the circulation of sea ice in the Arctic Ocean are the clockwise drift in the western Arctic's Beaufort Gyre, which shoves sea ice against Greenland and the Canadian archipelago, and the Transpolar Drift Stream, which transports sea ice from the Siberian sector of the Arctic Basin out through the Fram Strait into the Greenland Sea. Ice drift is, on average, parallel to the atmospheric-pressure isobars (black lines). **(b)** The record of how much ice (blue) was annually transported through the Fram Strait between 1979 and 2008 correlates well with the atmospheric pressure gradient across the strait (red) at sea level. Every year about 10% of the Arctic Basin's area is exported into the Greenland Sea. (Adapted from ref. 4.)

Thus the questions remain as to what actually caused the dramatic loss of ice and why the climate models have so underestimated its rate. Here we offer a perspective on the quality of the observational record, the gaps in our present understanding of the physical processes involved in maintaining and altering the sea ice.

SEA-ICE DYNAMICS

The dynamics of the ice cover is attributable to the wind and, to a lesser degree, the ocean currents. Due to the counterbalancing action of the atmospheric pressure gradients and the Coriolis effect, sea ice drifts roughly parallel to the frictionless wind above the surface, at about 1% of its speed. During winter, when the ice concentration—the fraction of the surface covered by ice—is near 100% and the mechanical strength of the ice is high, the surface stresses are propagated over distances comparable to the length scale of atmospheric weather systems. Fracture of the ice cover due to the gradients of the external stress results in the formation of ubiquitous welts of compressed ice blocks, known as pressure ridges, and openings in the ice caused by either diverging stresses or shear along jagged boundaries.

The approximate circulation pattern of sea ice has been known for more than a century, but it took the development of suitable satellite technology; automatic, drifting data buoys; and sophisticated methods of data transmission to develop a more detailed picture. Twenty institutions from nine different countries currently support the International Arctic Buoy Programme. Satellites have provided observations of ice motion on many different length scales. Generally, the circulation of sea ice is highly variable on weekly to monthly time scales but is dominated, on average, by a clockwise motion pattern in the western Arctic and by a persistent southward flow—the Transpolar Drift Stream—that exports approximately 10% of the area of the Arctic Basin through the Fram Strait every year. Figure 4a shows the average drift pattern and velocity of Arctic sea ice. An animation of the combined expression of the dynamic and thermodynamic processes—the drift of the ice and its seasonal expansion and regression during the years 1979–2009—is available at http://iabp.apl.washington.edu/data_movie.html.

From a mass-balance perspective, the Arctic Ocean loses ice volume by melt and by export—hence the interest in southward transport of ice through the Fram Strait. The annual record of areal ice loss by export, based on satellite data of ice motion, can be seen in figure 4b.[4] Several authors have studied its anomalies and trends; remarkably, the data show no decadal trend. Much less can be said about a possible decadal trend in volume export—a more definitive measure of mass balance—due to the lack of an extended record of the thickness of ice floes that are exported through the Fram Strait. Although a recent study quite clearly shows that MYI loss in the Arctic Basin has occurred by melting during the past decade,[5] the relative contributions of melt and export to the loss remain uncertain.

Because of the system's complexity, projections of sea-ice decline using global climate simulations are also problematic. Present-day sea-ice models include variations in the ice thickness distributions that capture the interactions between dynamics and thermodynamics.[6] As ice thickens, it both becomes mechanically stronger and conducts less heat. The models compute ice velocities from the balance of forces acting on the ice: external stresses exerted by wind and ocean currents, and internal stresses that are due to the mechanical response of the ice cover. This response depends on the ice's strength and thus its thickness distribution. During winter, the alternating diverging and converging motions of the ice cover modify the extremes of that distribution: Open water is exposed from cracks in some areas of the ice and pressure ridges develop in others. During summer, divergence of the ice controls the abundance of open water and alters the albedo feedback.

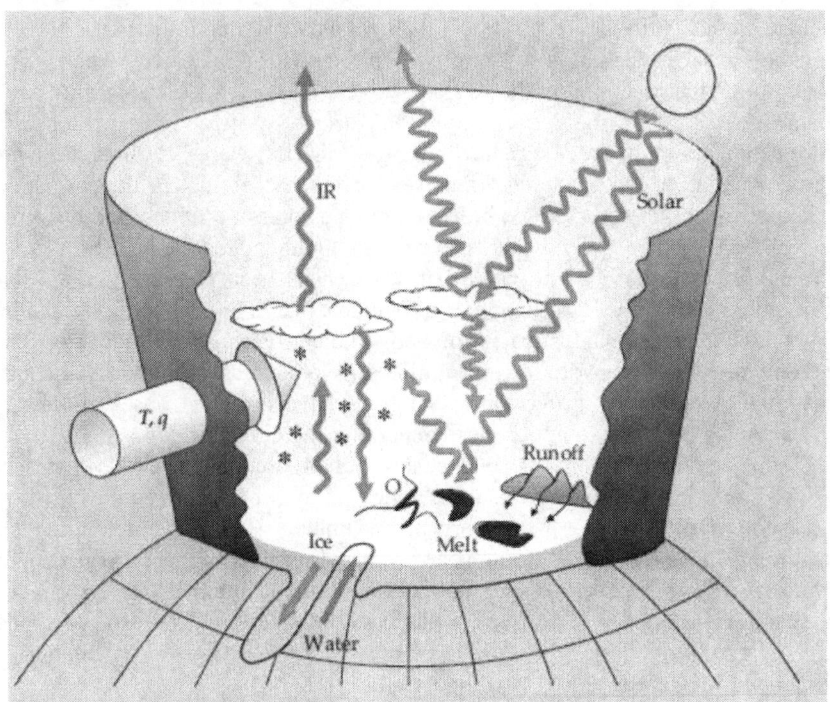

FIGURE 5. The heat and mass balance of the Arctic Basin. Incoming solar radiation (blue) is partially reflected, absorbed, and transmitted by clouds. The radiation reaching the surface is then partially reflected and absorbed in amounts that depend on the albedo of bare ice, open water (O), and numerous melt ponds formed during summer. River runoffs from surrounding continents feed the Arctic Ocean with fresh water. Infrared radiation (red) is emitted and absorbed by the clouds and the surface. Some of the atmospheric water vapor condenses and falls as snow, adding to the mass of the ice. The general circulation of the atmosphere results in a net influx of sensible and latent heat (T and q) from lower latitudes. The outflow of ice is primarily through the Fram Strait.

Projected September ice coverages from the global climate models used in the IPCC–AR4 range from ice-free conditions in September by 2060 to considerably more ice than is observed today. None of the models or their averages predicts the trends of the past three decades. Although a majority of the global climate models include simulations of ice dynamics, proper interpretation of their results is confounded by uncertainties in the simulated atmospheric and oceanic forcing of the ice cover. For instance, two IPCC models with sophisticated ice dynamics and one with no ice-motion component all predict a near-zero September ice cover by 2060. With such discrepancies, it is difficult to identify the actual role of sea-ice dynamics in the projected ice behavior.

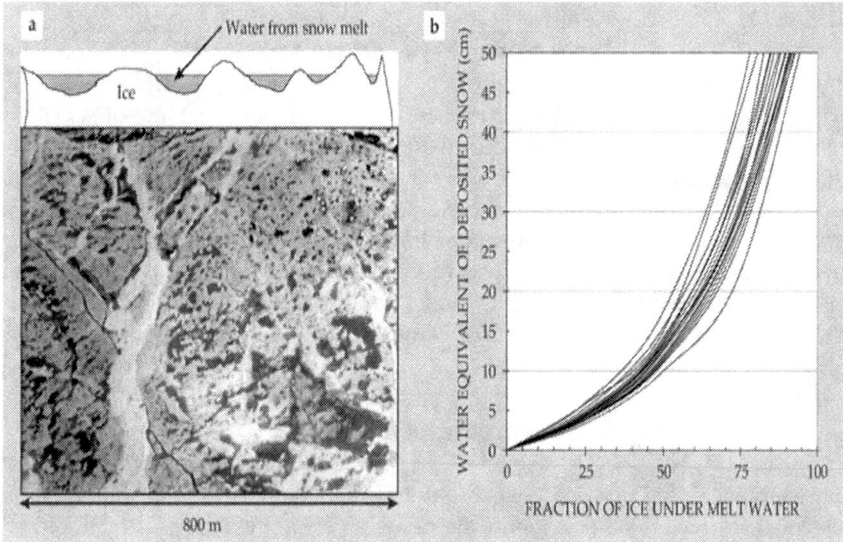

FIGURE 6. (a) During summer, the Arctic Ocean's ice is covered with residual patches of snow and melt ponds of variable depth and darkness—and thus variable albedo—as seen in this photograph from early July 1972; the white dots in the upper right are the huts of a research camp. Natural variations in surface topography create depressions that fill with water from snow melt during summer. (b) A 170-km elevation profile from an airborne laser altimeter taken northeast of Greenland—and parsed into 17 equal sections—shows the dependence of pond coverage on the amount of water available from snow melt. Even a modest increase in melt water can strongly influence the fraction of pond coverage and hence the surface albedo. (Data courtesy of Josefino Comiso.)

THERMODYNAMICS

In the heat-energy balance, which describes the gain or loss of heat in the system, sketched in figure 5, the solar and atmospheric radiation terms dominate. Smaller in magnitude are the latent and sensible heat transported across the Arctic boundaries by atmospheric circulation and the sensible heat carried into the basin by the warm West Spitsbergen Current and by the Pacific inflow through the Bering Strait.

The surplus flux of thermodynamic energy needed to cause the observed thinning of the ice during the past half-century is about 1 W/m^2. (That flux is equivalent to a reduction in ice thickness of approximately 0.1 m/year.) In this section, we summarize the gaps in our understanding of the atmospheric and oceanic processes that are behind the surplus.

Ice–atmosphere interactions

Radiative energy fluxes from the atmosphere and the annual advection of sensible and latent energy from lower latitudes are two orders of magnitude larger than 1 W/m^2. (Figure 5 illustrates those and other components of the mass and heat balance in the Arctic Basin.) At the moment, uncertainties in the heat-balance measurements observed at manned drifting stations and in the meridional heat transport calculated from radiosonde (balloon-based) observations around the Arctic perimeter prevent researchers from resolving those heat fluxes to an accuracy required to attribute the surplus of heat to any particular source or mechanism that explains the observed ice loss.[7]

Ice–ocean heat storage

During summer, when the ice concentration is less than 100% and numerous melt ponds cover the ice, some fraction of the radiative energy is temporarily stored in the exposed water and delays the onset of freezing in autumn. In areas of low ice concentration and low albedo, the energy causes melting at the bottom and laterally around the perimeter of the ice floes. Recent observations[8] have found rates of ice-bottom melting as high as 1 m/month. Field observations of lateral melt are logistically difficult and laborious; repeated measurements of the same ice floe over the melt season are needed to characterize the process. The few reported measurements suggest that even thick ice floes can melt laterally up to several meters during summer, but the contribution of that process to the loss of MYI is not known.

One prospective approach for learning more about lateral melting is to mine the 1-m-resolution images collected by intelligence satellites—the so-called National Technical Means—and released to the public at http://gfl.usgs.gov. The fixed-location images acquired since the summer of 1999 reveal telling features of the ice surface. But studying the details of processes such as lateral melt will require sequential images of the same ensemble of ice floes to trace the history of surface changes during the melt season. Samples of such acquisitions have recently been released on the above mentioned website.

Ice–Ocean Heat Flux

The rate of basal ice growth or melt is proportional to the difference between vertical heat conduction in the ice and turbulent heat flux in the ocean. One-dimensional thermodynamic ice models show that provided the surface energy balance is kept constant, the ocean heat flux derived from warmer-than-freezing water affects the equilibrium ice thickness most sensitively when the ice is thick. With that same constant-energy provision, an increase in the ocean heat flux from 1 to 2 W/m^2 thins the equilibrium ice by 1 m/yr and an increase from 3 to 4 W/m^2 thins it by only 0.5 m/yr. The heat carried into the Arctic Basin by the West Spitsbergen Current[9] and the Bering Strait inflow[10] has been documented by oceanographic moorings. But the mixing of those flows inside the Arctic Basin and the processes by which the imported warm water gives up its heat to the ice remain a subject of research. The only certainty is that a small change in ocean heat flux can have a large effect on ice thickness.[11]

SNOW AND MELT WATER

Owing to its low thermal conductivity, a blanket of snow slows the growth of underlying ice during the cold season. On the other hand, the onset of melting in early summer darkens the snow and, by albedo feedback, accelerates its own melting and that of the underlying ice. According to the thermodynamic model by Gary Maykut and one of us (Untersteiner),[12] an average snow depth of less than 1 m has little effect on the equilibrium ice thickness so long as, again, the energy balance at the surface is held constant. The less snow, the less time it takes to melt; the more the underlying ice melts, the thinner the ice is at the end of summer and the faster it grows during the following winter.

Over the Arctic Ocean, most of the snow falls in September and October. Thus new ice grown early in the season has the thickest snow cover. In contrast, calculations suggest, ice that starts to grow later in the cold of autumn or early winter in dynamically opened leads—areas of exposed water amid the pack ice—grows very quickly. That newly formed ice can thus overtake the older seasonal ice in thickness.

After the snow falls it is redistributed by the wind, which produces snow drifts behind pressure ridges; sweeps clean areas of flat, young ice; and blows snow into open leads. However, the clear and cold weather that usually follows a snowstorm induces a steep temperature gradient in the snow, which causes the snow to sublime and the vapor to diffuse upward toward surface layers. The process petrifies the snow within a day or two; rendered stiff, the snow remains in place for the balance of the winter. The overall impact of snow depth and its relationship to the underlying ice topography are not well understood, though.

When the snow melts, however, melt-water ponds, which begin forming in June in the lowest or thinnest places on the surface, can exert a profound thermodynamic impact. The average snow cover on Arctic sea ice is about 33 cm, the equivalent of 11 cm of water, with an annual variability of 2 cm.[13] According to data shown in fig. 6 an increase in the snow depth from 5 to 15 cm water equivalent would nearly double the area covered by ponds, where the melt rate is about 2.5 times that of bare ice. Given the variability in the spatial density of cracks and leads in different types of ice—thick, thin, young, old—it's not known how far the melt water can travel over the surface to fill in the low places. There's no doubt that surface topography and even small changes in available melt water substantially influence the ice loss during the melt season. But in the absence of measurements of surface relief and snow depth, it is difficult to quantitatively account for that influence.

ICESat stopped taking data in 2009, and its successor, second-generation *ICESat-2*, is not scheduled to launch until early 2016. NASA's IceBridge mission, the largest airborne survey of Earth's polar ice cover ever flown, was launched in 2009 to fill the gap and ensure a continuous series of measurements. The hope is that high-resolution ice- and snow-depth profiles collected by the radars and lidars on IceBridge flights, along with imagery taken by reconnaissance National Technical Means satellites, may provide new insights into the problem. The European Space Agency's ice mission known as *CryoSat-2*, launched last year, is also tasked with measuring changes in sea-ice thickness.

OUTLOOK

As stated above, the net heat required to account for the average loss of ice during the past three decades is of similar magnitude to a 1-W/m^2 global heat surplus.[14] Assuming that the surplus continues, and assuming that the global system does not undergo fundamental shifts, the share of heat received by the Arctic can be attributed to a host of variables and processes, including the cloudiness of Arctic skies; the distribution in the types of clouds; the temperature at the base of those clouds; changes in ocean-surface albedo; variations in the meridional transport of heat by the atmosphere, ice, and ocean; and the effect of greenhouse gases on all those factors. Gaps in our understanding of the processes that affect each factor represent a significant challenge to researchers attempting to assign specific causes for the thinning and loss of MYI or to project more detail than a general trend toward less Arctic ice in the future.

The loss of ice in the Arctic has made the region a crossroads of research, where the interests of science, environmental conservation and protection, resource development, and public policy meet. To produce useful ice forecasts that support societal needs, we see the following prospects. On time scales of days to weeks, forecasting the state of the ice cover can be expected to proceed along traditional lines based mainly on meteorological methods and satellite observations.

On time scales of years to decades, reliable projections face the problems of forecasting winds, cloudiness, surface albedo, and oceanic heat advection—all confounded by a plethora of climate-system feedbacks. Because sea ice is extremely sensitive to the least well-modeled and simulated part of the climate system—radiative heating from the clouds (see the article by Raymond T. Pierrehumbert in PHYSICS TODAY, January 2011, page 33)—it seems difficult to predict more than the fact that Arctic sea ice is likely to diminish.

Year-round field programs and repeated airborne surveys by aircraft are operationally limited and expensive. The best prospects for supporting and improving seasonal ice prediction may well come from initializing model ensembles with the most current atmospheric, ocean, and ice analyses. Input to those analyses would come from satellite surveys capable of providing near real-time observations of key ice, ocean, and atmospheric parameters. Instead of the project-based, sporadic deployments of oceanographic moorings now common, a sustained international program to deploy and maintain such instruments at strategic locations will be especially useful.

Perhaps equally useful as a predictive tool for long-term behavior are simplified, low-order models of the physical processes described in this article. Those models may provide insight regarding quantitative changes one might expect on multiple time scales.

Satellite altimetry and imagery used to track the changes of ice properties within ice parcels on the scale of a meter will remain crucial for understanding the physical processes that control the Arctic's evolution, particularly when the observations are supplemented with occasional short-term field studies. We believe that a greater degree of coordination between those field studies and the use of civilian and intelligence satellites is essential.

ACKNOWLEDGEMENT

We thank John Wettlaufer, Ian Eisenman, and Steve Warren for their careful reading and constructive criticism of the manuscript.

REFERENCES

1. D. A. Rothrock, D. B. Percival, M. Wensnahan, *J. Geophys. Res.* 113, C05003 (2008), doi:10.1029/2007JC004252.
2. R. Kwok, D. A. Rothrock, *Geophys. Res. Lett.* 36, L15501 (2009), doi:10.1029/2009GL039035.
3. J. Stroeve et al., *Geophys. Res. Lett.* 34, L09501 (2007), doi:10.1029/2007GL029703.
4. R. Kwok, *J. Clim.* 22, 2438 (2009), doi: 10.1175/2008JCLI2819.1.
5. R. Kwok, G. F. Cunningham, *Geophys. Res. Lett.* 37, L20501(2010), doi:10.1029/2010GL044678.
6. A. S. Thorndike, *J. Geophys. Res.* 97, 9401 (1992), doi:10.1029/92JC00695.
7. For an examination of the Arctic's large-scale energy budget compared with reanalysis by the National Centers for Environmental Prediction and the National Center for Atmospheric Research, see M. C. Serreze et al., *J. Geophys. Res.* 112, D11122 (2007), doi:10.1029/2006JD008230.
8. D. K. Perovich et al., *J. Geophys. Res.* 108, 8050 (2003), doi:10.1029/2001JC001079.
9. E. Fahrbach et al., *Polar Res.* **20**, 217 (2001).
10. R. A. Woodgate, T. Weingartner, R. Lindsay, *Geophys. Res. Lett.* 37, L01602 (2010), doi:10.1029/2009GL041621.
11. I. V. Polyakov et al., *J. Phys. Oceanogr.* **40**, 2743 (2010).
12. G. A. Maykut, N. Untersteiner, *J. Geophys. Res.* 76, 1550 (1971), doi:10.1029/JC076i006p01550.
13. S. G. Warren et al., *J. Clim.* 12, 1814 (1999).
14. K. E. Trenberth, *Curr. Opin. Environ. Sustain.* 1, 19 (2009).

Infrared Radiation and Planetary Temperature

Raymond T. Pierrehumbert*

Department of Geophysical Sciences
University of Chicago
Chicago, Illinois

Abstract. Infrared radiative transfer theory, one of the most productive physical theories of the past century, has unlocked myriad secrets of the universe including that of planetary temperature and the connection between global warming and greenhouse gases.

INTRODUCTION

In a single second, Earth absorbs 1.22×10^{17} joules of energy from the Sun. Distributed uniformly over the mass of the planet, the absorbed energy would raise Earth's temperature to nearly 800,000 K after a billion years, if Earth had no way of getting rid of it. For a planet sitting in the near-vacuum of outer space, the only way to lose energy at a significant rate is through emission of electromagnetic radiation, which occurs primarily in the subrange of the IR spectrum with wavelengths of 5–50 μm for planets with temperatures between about 50 K and 1000 K. For purposes of this article, that subrange is called the thermal IR. The key role of the energy balance between short-wave solar absorption and long-wave IR emission was first recognized in 1827 by Joseph Fourier,[1,2] about a quarter century after IR radiation was discovered by William Herschel. As Fourier also recognized, the rate at which electromagnetic radiation escapes to space is strongly affected by the intervening atmosphere. With those insights, Fourier set in motion a program in planetary climate that would take more than a century to bring to fruition.

Radiative transfer is the theory that enables the above to be made precise. It is a remarkably productive theory that builds on two centuries of work by many of the leading lights of physics. Apart from its role in the energy balance of planets and stars, it lies at the heart of all forms of remote sensing and astronomy used to observe planets, stars, and the universe as a whole. It is woven through a vast range of devices that are part of modern life, from microwave ovens to heat-seeking missiles. This article focuses on thermal IR radiative transfer in planetary atmospheres and its consequences for planetary temperature. Those aspects of the theory are of particular current interest, because they are central to the calculations predicting that global climate disruption arises from anthropogenic emission of carbon dioxide and other radiatively active gases.

* Reprinted with permission from Raymond T. Pierrehumbert, *Physics Today*, Vol. 64, January 2011, pages 33-38, Copyright 2011, American Institute of Physics.

Physics of Sustainable Energy II: Using Energy Efficiently and Producing it Renewably
AIP Conf. Proc. 1401, 232-244 (2011); doi: 10.1063/1.3653855
2011 American Institute of Physics 978-0-7354-0972-9/$30.00

An atmosphere is a mixed gas of matter and photons. Radiative transfer deals with the nonequilibrium thermodynamics of a radiation field interacting with matter and the transport of energy by the photon component of the atmosphere. Except in the tenuous outer reaches of atmospheres, the matter can generally be divided into parcels containing enough molecules for thermodynamics to apply but small enough to be regarded as isothermal and hence in local thermodynamic equilibrium (LTE).

The local radiation field need not be in thermodynamic equilibrium with matter at the local temperature. Nonetheless, the equations predict that the radiation field comes into thermodynamic equilibrium in the limiting case in which it interacts very strongly with the matter. For such blackbody radiation, the distribution of energy flux over frequency is given by a universal expression known as the Planck function B(ν,T), where ν is the frequency and T is the temperature.

Integrating the Planck function over all directions and frequencies yields the Stefan–Boltzmann law for the flux F exiting from the surface of a blackbody, $F = \sigma T^4$, where $\sigma = 2\pi^5 k_B^4/(15c^2 2h^3) \approx 5.67 \times 10^{-8}$ W m^{-2} K^{-4}. Here, k_B is the Boltzmann thermodynamic constant, c is the speed of light, and h is Planck's constant. The fourth-power increase of flux with temperature is the main feedback allowing planets or stars to come into equilibrium with their energy source. Since such bodies are not actually isothermal, there is a question as to which T to use in computing the flux escaping to space. Radiative transfer is the tool that provides the answer.

The appearance of h and c in the Stefan–Boltzmann constant means that relativity and quantization—the two non-classical aspects of the universe—are manifest macroscopically in things as basic as the temperatures of planets and stars. It is intriguing to note that one can construct a universe that is classical with regard to quantization but nonetheless is well behaved with regard to the thermodynamics of radiation only if one also makes the universe classical with regard to relativity. That is, σ remains fixed if we let $h \rightarrow 0$ but also let c tend to infinity as $h^{-3/2}$.

A FEW FUNDAMENTALS

At planetary energy densities, photons do not significantly interact with each other; their distribution evolves only through interaction with matter. The momentum of atmospheric photons is too small to allow any significant portion of their energy to go directly into translational kinetic energy of the molecules that absorb them. Instead, it goes into changing the internal quantum states of the molecules. A photon with frequency ν has energy hν, so for a photon to be absorbed or emitted, the molecule involved must have a transition between energy levels differing by that amount.

Coupled vibrational and rotational states are the key players in IR absorption. An IR photon absorbed by a molecule knocks the molecule into a higher-energy quantum state. Those states have very long lifetimes, characterized by the spectroscopically measurable Einstein A coefficient. For example, for the CO_2 transitions that are most significant in the thermal IR, the lifetimes tend to range from a few milli-seconds to a few tenths of a second. In contrast, the typical time between collisions for, say, a nitrogen-dominated atmosphere at a pressure of 10^4 Pa and temperature of 250 K is well under 10^{-7} s. Therefore, the energy of the photon will almost always be assimilated by collisions into the general energy pool of the matter and establish a new Maxwell–Boltzmann distribution at a slightly higher temperature. That is how radiation heats matter in the LTE limit.

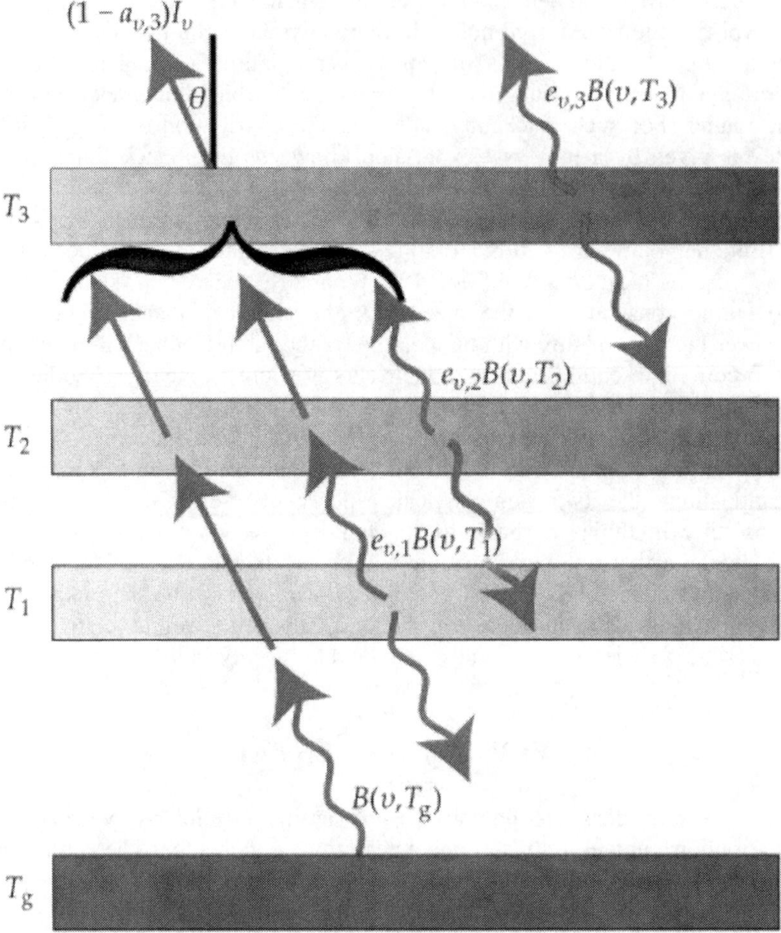

FIGURE 1. Three isothermal layers model the atmosphere in this illustration of upward-moving electromagnetic radiation with frequency ν. The radiation, assumed not to scatter, propagates at an angle θ with respect to the vertical and emerges from layer 3, the topmost atmospheric slice. The ground below the atmosphere emits as an ideal blackbody, characterized by the Planck function B. Each layer, at its own temperature T, emits with its own emissivity e_ν and, by Kirchhoff's law, absorbs a proportion $a_\nu = e_\nu$ of the incident radiation. The radiation flux distribution incident on layer 3 is I_ν. It is the sum of the thermal emission from the ground, layer 1, and layer 2, attenuated by absorption in the intervening layers 1 and 2. Squiggly arrows indicate thermal emission; straight arrows indicate transmitted radiation.

According to the equipartition principle, molecular collisions maintain an equilibrium distribution of molecules in higher vibrational and rotational states. Many molecules occupy those higher-energy states, so even though the lifetime of the excited states is long, over a moderately small stretch of time a large number of molecules will decay by emitting photons. If that radiation escapes without being reabsorbed, the higher-energy states are depopulated and the system is thrown out of thermodynamic equilibrium. Molecular collisions repopulate the states and establish a new thermodynamic equilibrium at a slightly cooler temperature. That is how thermal emission of radiation cools matter in the LTE limit.

Now consider a column of atmosphere sliced into thin horizontal slabs, each of which has matter in LTE. Thermal IR does not significantly scatter off atmospheric molecules or the strongly absorbing materials such as those that make up Earth's water and ice clouds. In the absence of scattering, each direction is decoupled from the others, and the linearity of the electromagnetic interactions means that each frequency can also be considered in isolation. If a radiation flux distribution I_ν in a given propagation direction θ impinges on a slab from below, a fraction a_ν will be absorbed, with $a_\nu \ll 1$ by assumption. The slab may be too thin to emit like a black-body. Without loss of generality, though, one can write the emission in the form $e_\nu B(\nu, T)$; here $e_\nu \ll 1$ is the emissivity of the slab (see figure 1). Both a_ν and e_ν are proportional to the number of absorber–emitter molecules in the slab.

The most fundamental relation underpinning radiative transfer in the LTE limit is Kirchhoff's law, which states that $a_\nu = e_\nu$. Gustav Kirchhoff first formulated the law as an empirical description of his pioneering experiments on the interaction of radiation with matter, which led directly to the concept of blackbody radiation. It can be derived as a consequence of the second law of thermodynamics by requiring, as Kirchhoff did, that radiative transfer act to relax matter in a closed system toward an isothermal state. If Kirchhoff's law were violated, isolated isothermal matter could spontaneously generate temperature inhomogeneities through interaction with the internal radiation field.

Given Kirchhoff's law, the change in the flux distribution across a slab is $\Delta I_\nu = e_\nu [-I_\nu + B(\nu, T)]$, assuming $e_\nu \ll 1$. The radiation decays exponentially with rate e_ν, but it is resupplied by a source $e_\nu B$. The stable equilibrium solution to the flux-change iteration is $I_\nu = B(\nu, T)$, which implies that within a sufficiently extensive isothermal region the solution is the Planck function appropriate to a blackbody. The recovery of blackbody radiation in that limit is one of the chief implications of Kirchhoff's law, and it applies separately for each frequency.

In the limit of infinitesimal slabs, the iteration reduces to a linear first-order ordinary differential equation for I_ν. Or, as illustrated in figure 1, one can sum the contributions from each layer, suitably attenuated by absorption in the intervening layers. The resulting radiative transfer equations entered 20th-century science through the work of Karl Schwarzschild (of black hole fame) and Edward Milne, who were interested in astrophysical applications; Siméon Poisson published a nearly identical formulation of radiative transfer[3] in 1835, but his equations languished for nearly 100 years without application.

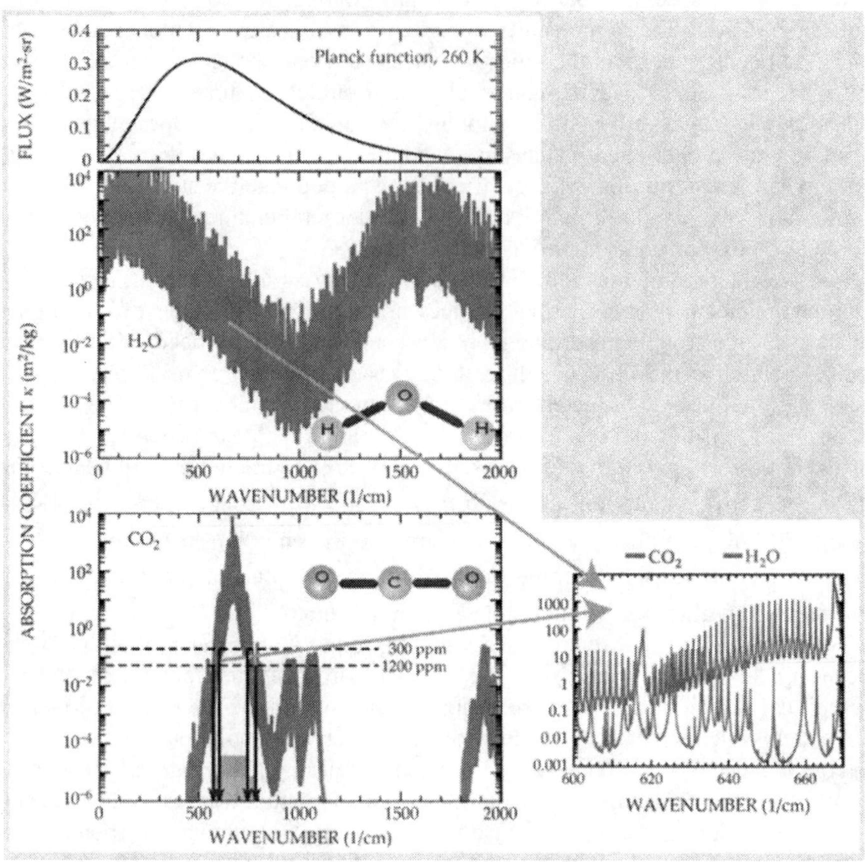

FIGURE 2. Absorption coefficients for water vapor and carbon dioxide as a function of wavenumber are synthesized here from spectral line data in the HITRAN database. The upper panel gives the Planck function B(ν,T) for a 260-K surface, which indicates the spectral regions that are important for planetary energy balance. The wavenumber, defined as the reciprocal of the wavelength, is proportional to frequency. If a layer of atmosphere contains M kilograms of absorber for each square meter at the base of the layer, then light is attenuated by a factor $\exp(-\kappa M)$ when crossing the layer, where κ is the absorption coefficient. The horizontal dashed lines on the CO_2 plot give the value of absorption coefficient above which the atmosphere becomes very strongly absorbing for CO_2 concentrations of 300 ppm and 1200 ppm; the green rectangle shows the portion of the spectrum in which the atmosphere is optically thick for the lower concentration, and the orange rectangle indicates how the optically thick region expands as the concentration increases. The inset shows fine structure due to rotational levels.

SPECROSCOPY OF GREENHOUSE GASES

Because of its numerous uses throughout science and technology, gaseous spectroscopy is a highly developed subject. The application of gaseous spectroscopy to atmospheric constituents began with John Tyndall, who discovered in 1863 that most of the IR opacity of Earth's atmosphere was attributable to two minor constituents—CO_2 and water vapor. All spectral absorption lines acquire a finite width by virtue of a number of processes that allow a molecule to absorb a photon even if the energy is slightly detuned from that of an exact transition. For reasonably dense atmospheres, the most important of those processes is collisional broadening, which borrows some kinetic energy from recent collisions to make up the difference between the absorbed photon's energy and a transition.

Databases of spectral-line properties lie at the foundations of all calculations of IR radiative transfer in gases. The HITRAN database,[4] culled from thousands of meticulously cross-validated, published spectroscopic studies, provides line properties for 39 molecules; it has been extensively used for applications across engineering and atmospheric sciences. The database is freely available at http://www.cfa.harvard .edu/hitran. A simple, flexible Python-language interface to HITRAN is included in the online software supplement to reference 5, available at http://geosci.uchicago.edu/~rtp1/PrinciplesPlanetaryClimate.

Measurements of absorption cross sections allow one to relate the absorption–emission properties of a layer of atmosphere to its composition. Figure 2, for example, shows absorption cross sections for CO_2 and water vapor as a function of wavenumber, which is proportional to frequency; the thermal IR ranges from 200 cm^{-1} to 2000 cm^{-1}. The spectra there are computed from the HITRAN database with lines broadened by collision with air at a pressure of 5×10^4 Pa, which corresponds to about the middle of Earth's atmosphere by mass. The CO_2 molecule has four main groups of absorption features in the thermal IR, of which the most important for Earthlike conditions is the one with the wavenumber near 667 cm^{-1}. (The corresponding wavelength is 15 μm.) The feature arises from vibrational bending modes of the linear triatomic molecule, which are given a fine structure by mixing with rotational transitions; the inset to figure 2 shows the corresponding lines. Water vapor is a polar molecule, and its richer set of vibrational and rotational modes allows it to absorb effectively over a much broader range of frequencies than CO_2.

Gases exhibit continuum absorption, as does the condensed matter making up clouds of all kinds. In some cases continua result from the overlap of nearby lines, but in other cases continua appear where no lines are in the vicinity. Loosely speaking, those continua arise because, over the finite duration of a collision, a pair of colliding molecules acts somewhat like a single, more complex molecule with transitions of its own. Equivalently, they result from the overlap of the tails of remote collisionally broadened lines. The statistical mechanics governing the far regions of long line tails is not at all well understood;[6] nonetheless, the continua have been quite well characterized, at least for those cases relevant to radiative transfer in Earth's atmosphere. For present-day Earth, the only important continuum is the water vapor continuum in the window around 1000 cm^{-1}. Carbon dioxide continua are unimportant for conditions that have prevailed on Earth during the past several billion years, but they are important for plugging the gaps in the line spectra for the dense CO_2 atmospheres of Venus and early Mars. Diatomic homoatomic molecules like N_2, which are transparent to IR in Earthlike conditions, have collisional continua that become important in cold, dense atmospheres. For example, the continuum makes N_2 one of the most important greenhouse gases on Saturn's largest moon, Titan.

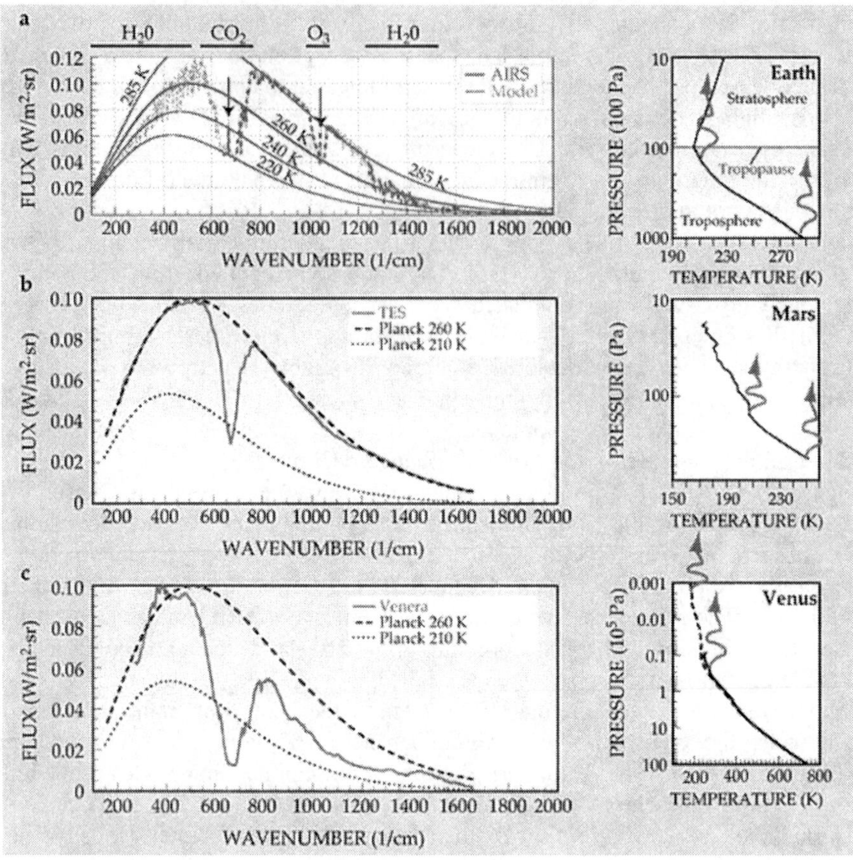

FIGURE 3. Satellite measurements of emission spectra are not limited to Earth. (a) The left panel compares a computed global-mean, annual-mean emission spectrum for Earth (blue) with observations from the satellite-borne AIRS instrument (red); both are superimposed over a series of Planck distributions. Two arrows point to absorption spikes discussed in the text. The temperature profile to the right, also an annual and global average, is based on in situ measurements. (b) The panel to the left shows a summer-afternoon emission spectrum for Mars observed by the TES instrument on the Mars Global Surveyor. Its accompanying temperature profile was obtained from radio-occultation measurements corresponding to similar conditions. (c) The panels here show a Venusian equatorial night thermal spectrum as measured by the *Venera 15* orbiter[14] together with a typical temperature profile for the planet. The upper portion (dashed curve) of the temperature sounding is based on radio-occultation observations from the Magellan mission; the lower portion (solid curve) was observed by a Pioneer Venus descender probe. For all three planets, squiggly arrows on the temperature profiles indicate the range of altitudes from which IR escapes to space.

The intricate variation of absorption with frequency makes it difficult to efficiently solve the radiative transfer equations. In line-by-line models, the equations are solved separately on a grid of millions of frequencies and the results are summed to obtain net fluxes. Climate models, however, require greater computational efficiency; one needs to compute the frequency-averaged radiation flux at each of several thousand model time steps for each of several thousand grid boxes covering a planet's atmosphere. Modelers use various approximations to represent the aggregate effects of spectral lines averaged over bands about 50 cm^{-1} wide. Such approximations are validated against line-by-line codes that have, in turn, been validated against laboratory and atmospheric observations. When averaged over a broad band, radiative flux decays algebraically rather than exponentially with distance traversed, because the progressive depletion of flux at strongly absorbed frequencies leaves behind flux at frequencies that are more weakly absorbed.[5]

CONFIRMED BY OBSERVED SPECTRA

The Sun radiates approximately like a blackbody having a temperature of 6000 K, even though the temperature of the solar interior is many millions of degrees. That's because the visible-wavelength and IR photons that predominate in solar radiation can escape from only the cooler outer layers of the Sun. Similarly, the 2.7-K cosmic microwave background radiation gives the temperature of the radiating layer of the very early universe, redshifted down from its original, much higher temperature.

The radiating layer of a planet is the IR equivalent of the Sun's photosphere. When a planet is viewed from above, the emission seen at a given frequency originates in the deepest layer that is optically thin enough for significant numbers of photons to escape. The effective emission temperature for that frequency is a suitably weighted average temperature of that layer. If the atmospheric temperature varies with height, variations of the absorption coefficients of atmospheric constituents with frequency show up in planetary emission spectra as variations of emission temperature; the more transparent the atmosphere is, the deeper one can probe.

For atmospheres heated partly from below—either as a consequence of solar absorption at the ground as in the case of Earth, Mars, and Venus, or due to internal absorption and escaping interior heat as with Jupiter and Saturn—the lower layers of the atmosphere are stirred by convection and other fluid motions, and the constant lifting and adiabatic cooling establish a region whose temperature decline with height approximates that of an adiabat. That region is the troposphere. At higher altitudes, heat transfer is dominated by radiative transfer instead of fluid motions; the corresponding region is the stratosphere. Stratospheric temperature is constant or gently decaying with height for pure IR radiative equilibrium, but in situ absorption of solar radiation can make the stratospheric temperature increase with height. Ozone facilitates such absorption on Earth, and organic hazes have a similar effect on Titan. Typical temperature profiles for Earth, daytime Mars, and Venus are shown in the right-hand column of figure 3.

The top panel of figure 3 compares global-mean, annual-mean, clear-sky spectra of Earth observed by the Atmospheric Infrared Sounder (AIRS) satellite instrument with spectra calculated after the radiative transfer equations were applied to output of a climate model driven by observed surface temperatures.[7] The agreement between the two is nearly perfect, which confirms the validity of the radiative transfer theory, the spectroscopy used to implement it, and the physics of the climate model. The AIRS instrument covers only wavenumbers above 650 cm^{-1}, but the theory and spectroscopic data sources used for radiative transfer at lower wavenumbers do not differ in any significant way from those used in the wavenumber range probed by AIRS. Numerous observations—notably, downward-looking radiation measurements from high-altitude aircraft—have confirmed the validity of radiative transfer models in the low-wavenumber water-vapor region.[8]

In the window region from roughly 800 to 1300 cm^{-1}, Earth radiates to space at very nearly the mean temperature of the ground, except for a dip due to ozone near 1050 cm^{-1}. At higher wavenumbers, one can see the reduction of radiating temperature due to water-vapor opacity. The main CO_2 absorption group leads to a pronounced reduction of radiating temperature in a broad region centered on 667 cm^{-1}. The emission spike at the center of the feature arises because CO_2 absorbs so strongly that the radiating level is in the upper stratosphere, which is considerably warmer than the tropopause; the ozone feature exhibits a similar spike.** The spectrum thus reveals the presence of CO_2, water vapor, ozone, and other gases not discussed here. We can infer that the planet has a stratosphere in which temperature increases with height, indicating the presence of an upper-level solar absorber. We can determine that temperatures of the atmosphere and ground range at least from 220 K to 285 K. But absent additional information, we cannot tell that the high end of that range actually comes from the ground.

Climate scientists routinely use spectral inferences such as those discussed above to monitor the state of Earth's atmosphere from space. Every time you see an IR weather satellite image, you are seeing radiative transfer in action. Earth's liquid or frozen water clouds act essentially as blackbodies. They emit at the cloud-top temperature, which is cold if the clouds are deep. On an IR satellite image, clouds appear as regions of weak emission, though by convention IR weather satellite images are usually presented with an inverted gray scale that makes clouds look white, as one expects from everyday experience. Weather forecasting centers worldwide use such images many times every day, as they show cloud patterns even on Earth's night side and, unlike visible-light images, allow forecasters to determine the height of cloud tops. Observations in selected IR and microwave bands are routinely used to retrieve temperature profiles and patterns of atmospheric constituents such as water vapor and CO_2.

Figure 3 also shows emission spectra for Mars and Venus. The Martian spectrum, obtained on a summer afternoon, mainly takes the form of blackbody emission from a 260-K surface, but as with Earth's spectrum, it has a region centered on the main CO_2 absorption band where the radiating temperature is much colder. As far as one can tell from its IR spectra, nighttime Venus looks about as cold as daytime Mars. However, based on microwave emissions (to which the atmosphere is largely transparent), Venera landers, and Pioneer descenders, we now know that Venus has an extremely hot surface, a nearly pure CO_2 atmosphere, and a surface pressure of nearly 100 Earth atmospheres. Because of the thick atmosphere, essentially all the IR escaping from Venus originates in the top region of the atmosphere, where the pressure is less than 2.5×10^4 Pa. The highest-temperature radiating surface in that layer is primarily attributable to CO_2 continuum absorption, which fills in the transparent regions of the line spectrum shown in figure 2. Sulfuric-acid clouds and trace amounts of water

** In reality, the spike in the ozone feature arises because ozone has a minimum opacity at that point in the spectrum, and the radiation leading to that feature comes from the ground and not the stratosphere.

vapor also contribute to plugging the gaps.

SATURATION FALLACIES

The path to the present understanding of the effect of carbon dioxide on climate was not without its missteps. Notably, in 1900 Knut Ångström (son of Anders Ångström, whose name graces a unit of length widely used among spectroscopists) argued in opposition to his fellow Swedish scientist Svante Arrhenius that increasing CO_2 could not affect Earth's climate. Ångström claimed that IR absorption by CO_2 was saturated in the sense that, for those wavelengths CO_2 could absorb at all, the CO_2 already present in Earth's atmosphere was absorbing essentially all of the IR. With regard to Earthlike atmospheres, Ångström was doubly wrong. First, modern spectroscopy shows that CO_2 is nowhere near being saturated. Ångström's laboratory experiments were simply too inaccurate to show the additional absorption in the wings of the 667-cm^{-1} CO_2 feature that follows upon increasing CO_2. But even if CO_2 were saturated in Ångström's sense—as indeed it is on Venus—his argument would nonetheless be fallacious. The Venusian atmosphere as a whole may be saturated with regard to IR absorption, but the radiation only escapes from the thin upper portions of the atmosphere that are not saturated. Hot as Venus is, it would become still hotter if one added CO_2 to its atmosphere.

A related saturation fallacy, also popularized by Ångström, is that CO_2 could have no influence on radiation balance because water vapor already absorbs all the IR that CO_2 would absorb. Earth's very moist, near-surface tropical atmosphere is nearly saturated in that sense, but the flaw in Ångström's argument is that radiation in the portion of the spectrum affected by CO_2 escapes to space from the cold, dry upper portions of the atmosphere, not from the warm, moist lower portions. Also, as displayed in the inset to figure 2, the individual water-vapor and CO_2 spectral lines interleave but do not totally overlap. That structure limits the competition between CO_2 and water vapor.

ENERGY BALANCE AND SURFACE TEMPERATURE

The same considerations used in the interpretation of spectra also determine the IR cooling rate of a planet and hence its surface temperature. An atmospheric greenhouse gas enables a planet to radiate at a temperature lower than the ground's, if there is cold air aloft. It therefore causes the surface temperature in balance with a given amount of absorbed solar radiation to be higher than would be the case if the atmosphere were transparent to IR. Adding more greenhouse gas to the atmosphere makes higher, more tenuous, formerly transparent portions of the atmosphere opaque to IR and thus increases the difference between the ground temperature and the radiating temperature. The result, once the system comes into equilibrium, is surface warming. The effect is particularly spectacular for Venus, whose ground temperature is 730 K. If the planet were a blackbody in equilibrium with the solar radiation received by the planet, the ground temperature would be a mere 231 K.

The greenhouse effect of CO_2 on Earth and Mars is visually manifest as the ditch carved out of the Planck spectrum near 667 cm^{-1}. That dip represents energy that would have escaped to space were it not for the opacity of CO_2. On Venus, the CO_2 greenhouse effect extends well beyond the ditch, owing to the opacity of the continuum associated with so much CO_2. In the Earth spectrum, one can also see a broad region in which water vapor has reduced the radiating temperature to a value well below the surface temperature.

For Earth and Mars, the width of the CO_2 ditch corresponds approximately to the width of the spectral region over which the atmosphere is nearly opaque to IR. Increasing atmospheric CO_2 increases the width of the ditch and hence increases the CO_2 greenhouse effect. But the increase occurs in the wings of the absorption feature rather than at the center (see figure 2). That limitation is the origin of the logarithmic relation between CO_2 concentration and the resulting perturbation in Earth's energy budget. It has been a feature of every climate model since that of Svante Arrhenius in 1896. Per square meter of surface, Mars has nearly 70 times as much CO_2 in its atmosphere as Earth, but the low Martian atmospheric pressure results in narrower spectral lines. That weakens absorption so much that the Martian CO_2 ditch has a width somewhat less than Earth's.

The planetary warming resulting from the greenhouse effect is consistent with the second law of thermodynamics because a planet is not a closed system. It exchanges heat with a high-temperature bath by absorbing radiation from the photosphere of its star and with a cold bath by emitting IR into the essentially zero-temperature reservoir of space. It therefore reaches equilibrium at a temperature intermediate between the two. The greenhouse effect shifts the planet's surface temperature toward the photospheric temperature by reducing the rate at which the planet loses energy at a given surface temperature. The way that works is really no different from the way adding fiberglass insulation or low-emissivity windows to your home increases its temperature without requiring more energy input from the furnace. The temperature of your house is intermediate between the temperature of the flame in your furnace and the temperature of the outdoors, and adding insulation shifts it toward the former by reducing the rate at which the house loses energy to the outdoors. As Fourier already understood, when it comes to relating temperature to the principles of energy balance, it matters little whether the heat-loss mechanism is purely radiative, as in the case of a planet, or a mix of radiation and turbulent convection, as in the case of a house—or a green-house. Carbon dioxide is just planetary insulation.

For present Earth conditions, CO_2 accounts for about a third of the clear-sky greenhouse effect in the tropics and for a somewhat greater portion in the drier, colder extratropics (see reference 9, figure 12.1); the remainder is mostly due to water vapor. The contribution of CO_2 to the greenhouse effect, considerable though it is, understates the central role of the gas as a controller of climate. The atmosphere, if CO_2 were removed from it, would cool enough that much of the water vapor would rain out. That precipitation, in turn, would cause further cooling and ultimately spiral Earth into a globally glaciated snowball state.[10] It is only the presence of CO_2 that keeps Earth's atmosphere warm enough to contain much water vapor. Conversely, increasing CO_2 would warm the atmosphere and ultimately result in greater water-vapor content—a now well understood situation known as water-vapor feedback.[9,11]

Though the first calculation of the warming of Earth due to CO_2 increase was carried out by Arrhenius in 1896, accurate CO_2 and water-vapor spectroscopy and a fully correct formulation of planetary energy balance did not come together until the work of Syukuro Manabe and Richard Wetherald in 1967.[2,12] With that development, the theory was brought to its modern state of understanding. It has withstood all subsequent challenges and without question represents one of the great triumphs of 20th-century physics.

PLANETS FAR AND NEAR

The foundations of radiative transfer were laid by some of the greatest physicists of the 19th and 20th centuries—Fourier, Tyndall, Arrhenius, Kirchhoff, Ludwig Boltzmann, Max Planck, Albert Einstein, Schwarzschild, Arthur Eddington, Milne, and Subrahmanyan Chandrasekhar—plus many more whose names are not well known, even among physicists, but probably deserve to be. The subject has had a century of triumphs (and, as the section above, ENERGY BALANCE AND SURFACE TEMPERATURE describes, some wrong turns)[13] and is about to go into high gear because of the dawning era of extrasolar planet discovery. What kind of atmospheres would render a planet in the potentially habitable zone of its star actually habitable,[13] and how would astronomers detect it? If they see a high-albedo object with CO_2 in its atmosphere, how will they determine if it is a snowball or a large Venus-like rocky planet?

Whatever the future holds for newly discovered planets, interest remains intense in maintaining the habitability of the planet likely to be our only home for some time to come. The contributions of fundamental physics to achieving that aim are clear. The CO_2 greenhouse effect is directly visible in satellite observations of the bite taken out of the IR spectrum near 667 cm^{-1}, a feature whose details agree precisely with results of calculations based on first-principles radiative transfer calculations. Laboratory spectroscopy demonstrates that the width of the bite will increase as CO_2 increases, and warming inevitably follows as a consequence of well-established energy-balance principles. The precise magnitude of the resulting warming depends on the fairly well-known amount of amplification by water-vapor feedbacks and on the less-known amount of cloud feedback. There are indeed uncertainties in the magnitude and impact of anthropogenic global warming, but the basic radiative physics of the anthropogenic greenhouse effect is unassailable.

ACKNOWLEDGMENT

I am grateful to Yi Huang for providing me with AIRS spectra, to David Crisp for providing Venera digital data and for many illuminating discussions on the subject of radiative transfer over the years, and to Joachim Pelkowski for pointing out Poisson's work on radiative transfer.

REFERENCES

1. R. T. Pierrehumbert, *Nature* 432, 677 (2004).
2. D. A. Archer, R. T. Pierrehumbert, eds., *The Warming Papers: The Scientific Foundation for the Climate Change Forecast,* Wiley-Blackwell, Hoboken, NJ (in press).
3. S. D. Poisson, *Théorie mathématique de la chaleur*, Bachelier, Paris (1835).
4. L. S. Rothman et al., *J. Quant. Spectrosc. Radiat. Transfer* 110, 533 (2009).
5. R. T. Pierrehumbert, *Principles of Planetary Climate*, Cambridge U. Press, New York (2010).
6. I. Halevy, R. T. Pierrehumbert, D. P. Schrag, *J. Geophys. Res.* 114, D18112 (2009), doi:10.1029/2009JD011915.
7. Y. Huang et al., *Geophys. Res. Lett.* 34, L24707 (2007), doi:10.1029/2007GL031409.
8. See, for example, D. Marsden, F. P. J. Valero, *J. Atmos. Sci.* 61, 745 (2004).
9. R. T. Pierrehumbert, H. Brogniez, R. Roca, in *The Global Circulation of the Atmosphere*, T. Schneider, A. Sobel, eds., Princeton U. Press, Princeton, NJ (2007), p. 143.
10. A. Voigt, J. Marotzke, *Clim. Dyn.* 35, 887 (2010).
11. A. E. Dessler, S. C. Sherwood, *Science* 323, 1020 (2009).
12. S. Weart, *The Discovery of Global Warming*, Harvard U. Press, Cambridge, MA (2008), and online material at http://www.aip.org/history/climate/index.htm.
13. R. T. Pierrehumbert, *Ap. J. Lett.* (in press).
14. V. I. Moroz et al., *Appl. Opt.* 25, 1710 (1986).

Touring the Atmosphere Aboard the A-Train*

Tristan S. L'Ecuyer[a] and Jonathan H. Jiang[b]

[a]*Department of Atmospheric Science*
Colorado State University
*Fort Collins, Colorado***

[b]*Jet Propulsion Laboratory*
California Institute of Technology
Pasadena, California

Abstract. A convoy of satellites orbiting Earth measures cloud properties, greenhouse gas concentrations, and more to provide a multifaceted perspective on the processes that affect climate.

INTRODUCTION

Growing evidence indicates that human activity is altering the climate in significant and potentially hazardous ways. The most recent assessment from the Intergovernmental Panel on Climate Change asserts that global temperature may rise by 2–5 °C (4–9 °F) during the next 100 years in response to rising greenhouse gas concentrations.[1] Current predictions also suggest that regional climates may experience significant changes in the frequency and intensity of precipitation, shifts in surface vegetation and soil fertility, and rises in global sea level, to give some examples. Indeed, some changes are already evident, including the dramatic reduction in size of many glaciers, the rapid shrinking of the summertime Arctic ice cap, and a 20-cm rise in sea level since preindustrial times. Predictions of future climate, however, are predicated on model simulations. Of necessity, such models approximate climate scientists' often incomplete knowledge of the fundamental physical processes that govern the evolution of the climate system. Consequently, significant

---

* Reprinted with the permission of Tristan S. L'Ecuyer and Jonathan H. Jiang, *Physics Today*, Vol. 63, July 2010, pages 36-43. Copyright, American Institute of Physics.

** Now at the Department of Atmospheric and Oceanic Sciences, University of Wisconsin – Madison, Madison, Wisconsin.

Physics of Sustainable Energy II: Using Energy Efficiently and Producing it Renewably
AIP Conf. Proc. 1401, 245-256 (2011); doi: 10.1063/1.3653856
2011 American Institute of Physics 978-0-7354-0972-9/$30.00

uncertainties remain in current climate-change projections, particularly at the regional level.[2]

Central to climate modeling is the challenge of accurately representing both the water cycle, which governs the distribution of water around the planet, and the exchange of heat between atmosphere, surface, and space. The global energy and water cycles, in turn, are intimately coupled to the large-scale atmospheric and oceanic circulation patterns that redistribute the surplus of radiative energy received in the tropics to higher-latitude regions that radiate away more energy than they receive from the Sun. Those large-scale atmospheric circulations are also strongly coupled to clouds and rainfall that can influence regional circulations by redistributing energy in the atmosphere. Indeed, the largest source of uncertainty in current projections of future climate is from incomplete knowledge of the feedbacks through which clouds can either amplify or diminish temperature changes induced by greenhouse gases.[3]

To better understand the climate system, climate scientists need to quantify the complex relationships that connect water in all three phases to heat exchanges between the surface, atmosphere, and space; to aerosols; and to trace gases. That is a daunting task, given the sheer number and diversity of measurements and parameters involved, but a one-of-a-kind constellation of satellites collectively known as the A-Train is helping scientists to meet the challenge.

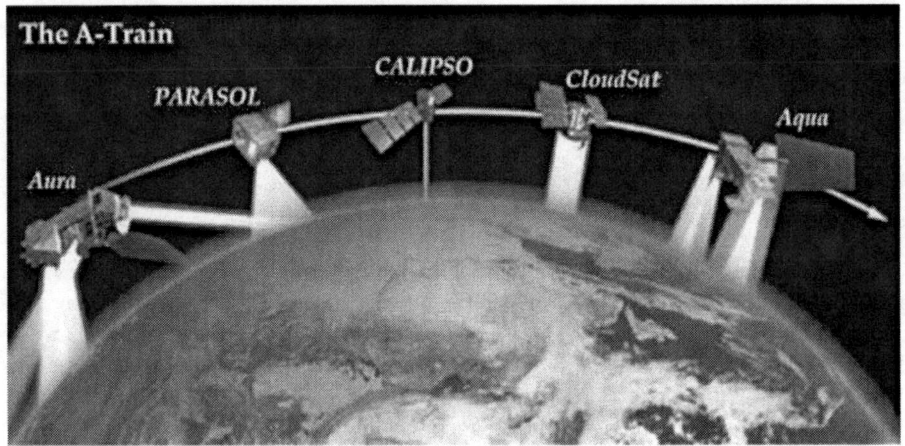

FIGURE 1. The A-Train constellation included five satellites and more than a dozen instruments during the period 2006–09. *PARASOL* has since dropped out, but at least one new satellite will join the train within the next couple of years. Some of the A-Train's instruments observe wide swaths of Earth's surface and atmosphere; others observe narrower regions in greater detail.

A LITTLE HISTORY

The idea behind the A-Train emerged in the mid-1990s, as engineers and scientists were developing the *Aura* mission, then called *EOS Chem*. The *Aura* satellite had to be Sun synchronous, meaning that it must always cross the equator at the same local time, and in order for it to measure solar backscatter, the crossing time had to be within 1.5 hours of local noon. Otherwise, the orbit was unconstrained. Since the infrastructure of the *Aura* spacecraft was identical to that of its older sister *Aqua*, which was dedicated to water- and energy-cycle measurements, the scientists and engineers decided that *Aura* would follow its older sibling at an altitude of 705 km and an inclination of 98.2°. That way, the *Aqua* launch computations could simply be updated and applied to *Aura*. Due to limitations in data transmission rates, however, mission engineers decided that *Aura* should fly 15 minutes (6300 km) behind *Aqua*.

Meanwhile, NASA was developing a new mission that would, for the first time, combine spaceborne radar and lidar (light detection and ranging) to simultaneously measure the vertical structure of clouds and aerosol layers in the atmosphere. Due to budget constraints, the mission was split into two separate proposals—*CALIPSO* (*Cloud-Aerosol Lidar and Infrared Pathfinder Satellite Observation*), which was focused on profiling aerosols and thin clouds with lidar, and *CloudSat*, directed toward profiling thicker clouds using radar. The two missions competed against one another and several others to be part of NASA's Earth System Science Pathfinder program, and both were selected for further development. Unaware that *Aura* was already planned to follow *Aqua*, the *CALIPSO* and *CloudSat* teams also requested orbits close behind the *Aqua* spacecraft to take advantage of its cloud and humidity measurements. Across the Atlantic, France's CNES had plans to launch a small satellite called *PARASOL* (*Polarization and Anisotropy of Reflectances for Atmospheric Sciences Coupled with Observations from a Lidar*), whose imaging polarimeter could also benefit greatly from coordination with *Aqua*'s higher spatial resolution measurements.

As scientists and engineers refined their mission plans, they began to fully appreciate the potential advantages of formation flying. A single platform could not accommodate the mass and power demands of all the missions' instruments. Moreover, if they were all crowded together on a single craft, the sensors would get in each others' way and interfere electronically. Carefully coordinating the orbits of five individual satellites, however, would enable researchers to benefit from a unique multisensor perspective of our planet. Figure 1 shows the resulting convoy of satellites as it was configured in 2006–09. During that period, it comprised *CloudSat*, *CALIPSO*, and *PARASOL*, bracketed by *Aqua* and *Aura*, two of the cornerstones of NASA's Earth Observing System program. The unique satellite configuration led *Aura* project scientist Mark Schoeberl to coin the name A-Train, after the famous 1930s jazz piece composed by Billy Strayhorn and popularized by the Duke Ellington Orchestra.

Within a few years after the launch of *Aura* in 2004, data transmission rates had improved sufficiently. Thus, over the course of a year ending in May 2008, *Aura* was gradually moved to within 7 minutes of *Aqua*. The closer proximity enabled better coordination between the two satellites. In particular, since clouds change very little in 7 minutes, the move meant that *Aqua* cloud observations could be used to improve *Aura* trace-gas measurements. Launched in 2006, *CloudSat* and *CALIPSO* were placed in a tight formation, with a separation of 12.5 s or 93.8 km. The satellites are so close that *CloudSat* must make regular orbit maneuvers to compensate for the different atmospheric drag it experiences. Since the launch of the two satellites, *CALIPSO*'s lidar beam and *Cloud-Sat*'s radar beam have coincided at Earth's surface more than 90% of the time; that remarkable pointing precision has allowed data from the two spacecraft to be used in tandem for many applications. Launched in 2004, *PARASOL* flew an average of 30 s behind *CALIPSO* until decoupled from the A-Train on 2 December 2009 because it no longer had enough fuel to match the orbital maneuvers of the other satellites.

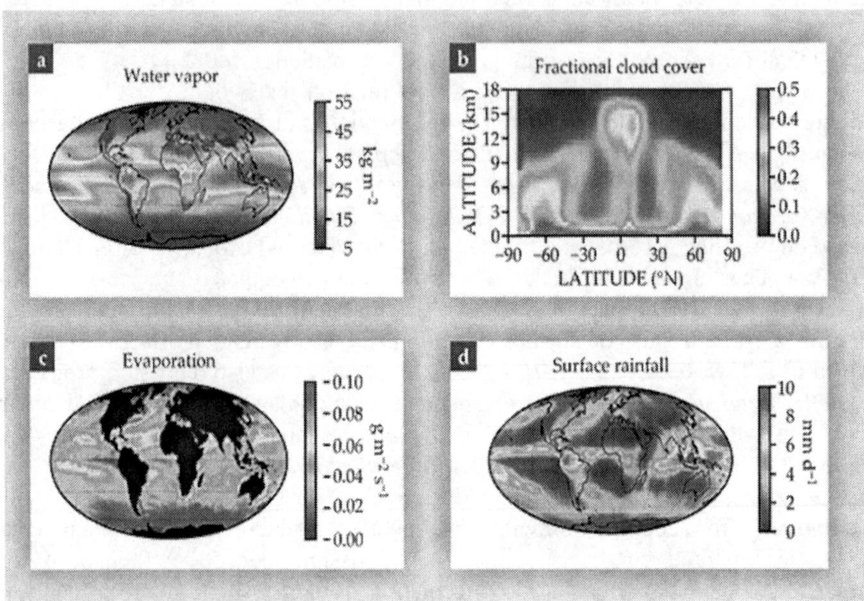

FIGURE 2. The global water cycle is the set of processes in which water evaporates from Earth's surface, moistens the atmosphere, is redistributed horizontally and vertically by atmospheric circulations, condenses into clouds, and falls back to Earth as precipitation. As these measurements from the A-Train satellites show, water vapor concentrations tend to be largest in the tropics (a), where clouds form at the highest altitudes (b). The horizontal and vertical distributions seen in panels a and b can be explained by the enhanced evaporation from warmer waters in the tropics (c) and atmospheric circulations that transport water vapor toward the equator. The convergence of water vapor in the tropics leads to a band of enhanced precipitation near the equator, called the Intertropical Convergence Zone, and to a large area of intense precipitation in the western Pacific Ocean extending southeast, known as the South Pacific Convergence Zone (d). Also evident in panel d are the mid-latitude storm tracks, paths that storms often follow off the coasts of North America and Asia, and the widespread precipitation band between 45° S and 60° S. Sailors refer to those latitudes as the roaring 40s and furious 50s because of the persistent westerly winds and stormy weather there. In panel a, the water vapor is given as a column density. In all panels, data are averages over the year 2007. (Panel c courtesy of Carol Anne Clayson and the GEWEX SeaFlux project.)

THE A-TRAIN PERSPECTIVE

The satellites in the A-Train carry both active instruments that transmit and receive signals and passive sensors that only receive them. Together the satellites view Earth from the UV to the microwave—a wavelength span of four orders of magnitude.[4]

That wavelength diversity, coupled with the distinct viewing geometries and scanning patterns of the instruments aboard the A-Train, provides composite information about a wide variety of climate parameters. At the front of the train, *Aqua* carries several instruments that obtain, for example, profiles of temperature, water vapor, and surface rainfall—important components of the atmospheric branch of the global water cycle. *Aqua* also measures cloud properties, aerosol concentrations, and radiative fluxes at the top of the atmosphere—all key quantities related to the global radiation budget.

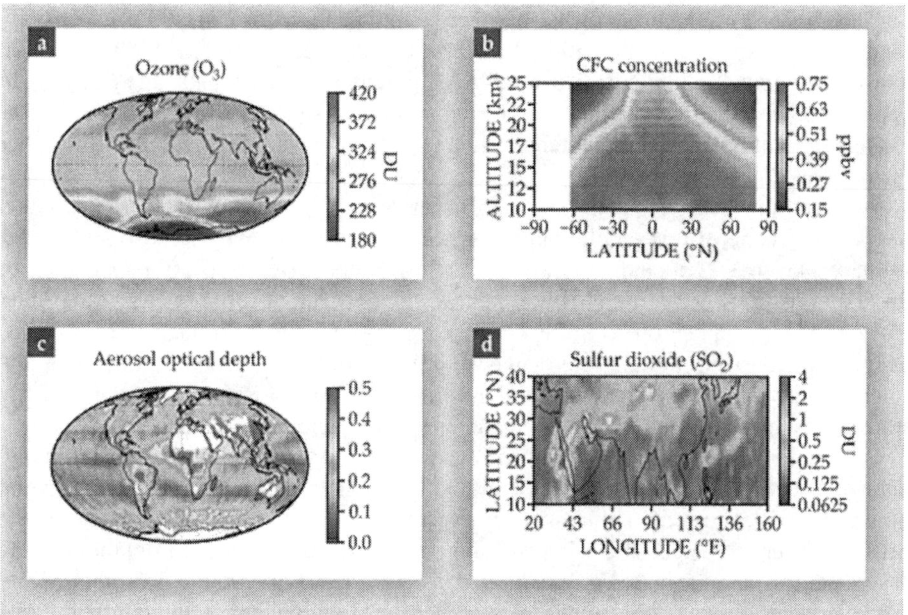

FIGURE 3. Trace gases and aerosols affect public health and reveal surface activity on Earth. (a) Measurements of the ozone distribution taken in September 2007 from the A-Train satellites clearly reveal the ozone hole over the Antarctic. Thickness is measured in Dobson units (DU): 100 DU corresponds to a quantity of ozone that would give a 1-mm-thick layer at a temperature of 0 °C and a pressure of 1 atmosphere. (b) At the same time, the A-Train also measured the concentrations of two long-lived chlorofluorocarbons (CFCs) that catalyze ozone depletion. The depletion reactions occur on the surfaces of ice particles that make up the polar stratospheric clouds that often exist over Antarctica. The concentrations are given in parts per billion by volume (ppbv). (c) The distribution of aerosol particles in the atmosphere shows evidence of dust and biomass burning in Africa and the Amazon region. Also visible are aerosols from industrial activity in eastern China. The optical depth shown here is a logarithmic measure of the amount of radiation that is absorbed or reflected by atmospheric aerosols. (d) The sulfur dioxide emissions seen here originated with the 30 September 2007 eruption of the Jebel al-Tair volcano in the Red Sea. As the map shows, they were carried over a large portion of Asia and the western Pacific Ocean. Volcanic emissions can affect air travel, reduce air quality, and act as a precursor for sulfate aerosols that can exert a significant influence on climate.

CloudSat's and *CALIPSO*'s active radar and lidar sensors add a vertical dimension to *Aqua* observations by probing the internal structure of cloud and aerosol layers along a narrow strip near the center of the much wider *Aqua* swath. The complementary multi-angle measurements of *PARASOL* in the visible and IR enable climate scientists to infer the size, shape, orientation, and even chemical composition of atmospheric aerosols. Together, the data from the four satellites yield new information about the three-dimensional structure of clouds and aerosols in Earth's atmosphere. Armed with those data, scientists can quantitatively determine how clouds and aerosols influence global energy balance.

The caboose of the A-Train is the *Aura* satellite. Launched in 2004, its primary focus is atmospheric composition.[5] *Aura*'s instruments provide high-resolution maps that show the vertical distributions of greenhouse gases and gases central to ozone depletion. Its observations provide an additional source of aerosol and thin-ice-cloud information that complements similar measurements obtained from the other instruments aboard the train.

A thorough discussion of the A-Train's measurements and how they are applied is beyond the scope of this article, but a complete list of the convoy's sensors and their primary purposes is included in an online supplement. Here we offer examples of observations grouped around two central themes: the global water cycle and atmospheric composition. Figure 2 depicts A-Train measurements of surface evaporation, surface rainfall, and water-vapor and cloud distributions. The data allow scientists to monitor each of those major components of the water cycle and quantify water exchanges between the ocean, atmosphere, and land.

Atmospheric aerosols and trace gases can play an important role in global energy balance. But they also affect public health both regionally in the form of pollution and on larger scales through chemical processes such as the catalyzation of ozone depletion. For those reasons, scientists and policymakers are eager to obtain accurate assessments of atmospheric composition. The instruments aboard the A-Train provide complementary, near-global measurements of aerosols and many atmospheric trace gas species from which climate scientists glean new information about the chemical and physical processes in Earth's atmosphere. Figures 3a and 3b, for example, show a clearly visible ozone hole over Antarctica and distributions of chlorofluorocarbons implicated in ozone destruction. Such observations should become an invaluable resource for monitoring the anticipated ozone recovery in the coming years. Figures 3c and 3d show how aerosol or trace gas distributions offer a unique perspective on such natural events as forest fires or volcanic eruptions.

RAPID CHANGE IN THE ARCTIC

In addition to supporting a diverse population of marine mammals and several human cultures, Arctic sea ice significantly affects the climate system. It reflects solar energy during the summer; it modifies the exchange of heat, gases, and momentum between the atmosphere and the Arctic Ocean; and it affects ocean circulations by modifying the distribution of fresh water. (See the article by Josefino Comiso and Claire Parkinson in PHYSICS TODAY, August 2004, page 38.) As a result, observed changes in the concentration, extent, thickness, and growth and melt rates of Arctic sea ice have important social and climate implications. Climate scientists estimate, for example, that the area covered by Arctic sea ice in September, the month when coverage is at a minimum, has decreased by an average of approximately 60,000 km^2 per year since satellite observations of the region began in 1979.

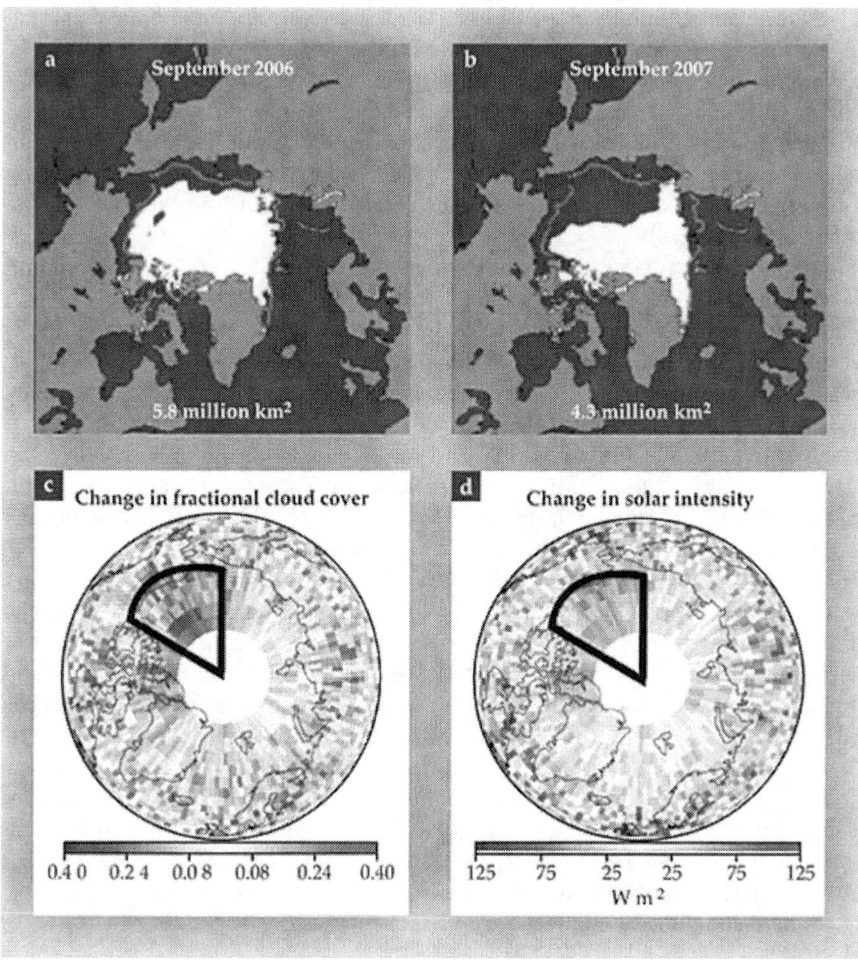

FIGURE 4. Clouds likely were a factor in the dramatic retreat of Arctic sea ice in the summer of 2007. (a) In 2006 the extent of Arctic sea ice as measured by the *Aqua* satellite was significantly less than the 1979–2000 long-term average shown in pink. (b) In 2007 *Aqua* measurements revealed sea ice coverage to be at an all-time low. (Panels a and b courtesy of the National Snow and Ice Data Center.) (c) Average cloud coverage over the Beaufort Sea region (within the black slice) decreased dramatically between summer (June–August) 2006 and the same period in 2007, according to *CloudSat* and *CALIPSO* observations. (d) Associated with the decreased cloud coverage was an increase in solar intensity reaching the surface. (Panels c and d adapted from ref. 7.)

Recent sea ice extents have been especially low. In September 2007, for example, sea ice covered just 4.3 million km^2, the smallest value in recorded history.[6] The observed rate of sea ice retreat in 2007 far exceeded that predicted by climate models, and the discrepancy initially fueled a great deal of concern in the climate community. The startling 2007 ice loss was captured in detail by A-Train sensors. Those observations, especially welcome because in situ measurements covering the Arctic Ocean are difficult to make, provided new insights into the processes that connected atmosphere, ocean, and sea ice and contributed to the Arctic ice melt. Basing their analysis in part on A-Train observations, climate scientists are now in wide agreement that a perfect storm of anomalous weather conditions was responsible for the rapid decline observed in 2007.

Figure 4 shows A-Train observations associated with the 2007 sea ice minimum. In both 2006 and 2007, sea ice coverage was significantly less than the 1979–2000 average, but the observations from 2007 dramatically reveal the sudden melting of a large fraction of the ice that normally blankets the Beaufort Sea. Anomalously high winds in the summer of 2007 contributed to the extreme ice loss by causing a relatively rapid compression of sea ice and its quicker- than-normal transport into warmer waters outside the Arctic.[6] The lower panels of figure 4, however, give evidence that the summertime melting may have been enhanced by another mechanism:[7] Measurements from *CloudSat* and *CALIPSO* indicate that summertime cloud cover in the region decreased by 16% from 2006 to 2007. Irradiance calculations based on those observations suggest that, on average, clearer skies in the summer of 2007 allowed an additional 32 W/m^2 of sunlight to reach the surface.

Back-of-the- envelope calculations suggest that the additional energy delivered in the summer of 2007 could increase surface ice melt by 0.3 m. It could also warm the surrounding ocean's near-surface mixed layer by 2.4 K and thus significantly enhance basal ice melt. Moreover, atmospheric temperature and moisture observations from *Aqua* sensors indicate that the decrease in cloudiness in 2007 was related to increased air temperatures and decreased relative humidity associated with persistent high pressure in the region. In sum, many factors seem to have combined to cause the rapid decline in Arctic sea ice in 2007. A-Train measurements provide evidence that increased solar energy at the surface, associated with reduced cloudiness, was one of the important components contributing to the event.

AEROSOLS: NOT TO BE SNEEZED AT

In addition to their effects on air quality, aerosol particles surely affect Earth's radiation budget, but it's not clear just what that effect is. Depending on their composition, some aerosols scatter solar radiation and tend to cool the planet, whereas others absorb radiation and potentially warm Earth. Moreover, the impact of aerosols strongly depends on the presence or absence of clouds. For instance, a thick dust layer over a dark ocean surface can significantly increase the amount of sunlight reflected back to space, but the same layer above an already bright cloud probably won't have much effect. The combination of active and passive sensors on the A-Train have dramatically improved scientists' understanding of the spatial covariation of clouds and aerosols. We have recently discovered, for example, that absorbing aerosols produced from biomass burning exert a net cooling on the environment when the underlying cloud coverage is less than 40% but lead to warming in more overcast conditions.[8]

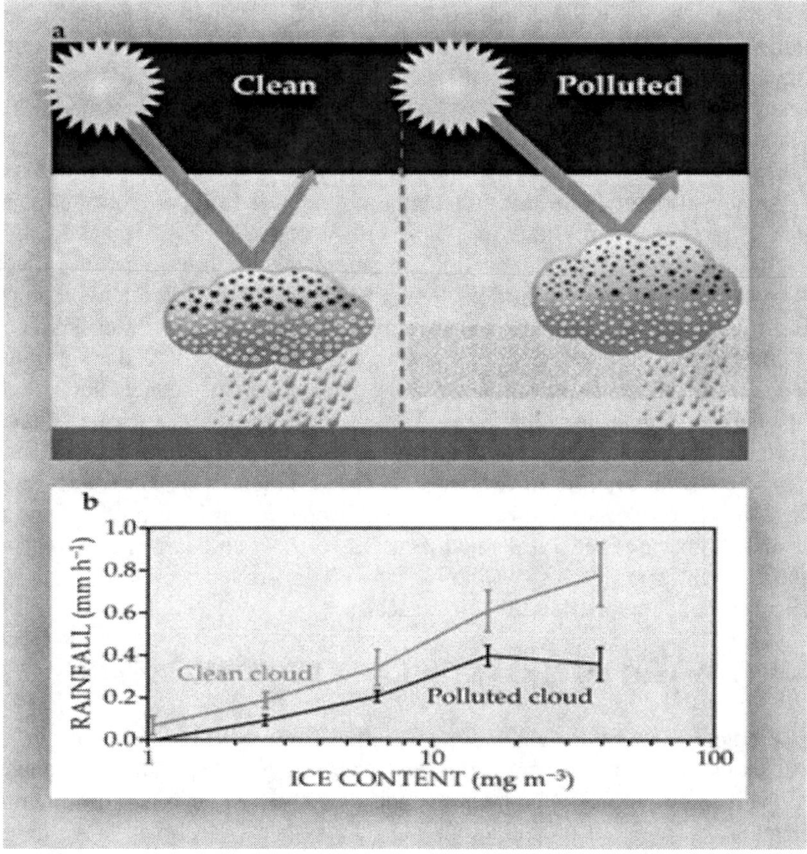

FIGURE 5. Pollution changes cloud properties. (a) Clouds in polluted environments tend to have smaller water drops and ice crystals than those in cleaner environments. As a result, dirty clouds are less likely to generate rainfall and are generally brighter than clean clouds. (b) Data from the *Aura* and *Aqua* satellites, in conjunction with rainfall observations from the *Tropical Rainfall Measuring Mission*, demonstrate that for a given ice content, clouds in polluted environments produce less intense rainfall. (Adapted from ref. 13.)

Aerosols can also substantially modify the characteristics of clouds. Atmospheric physicists have long recognized, for example, that large concentrations of sulfate aerosols might lead to smaller-sized droplets in a cloud; as a result, a cloud with a given amount of water would be brighter in a more polluted environment.[9] That so-called first aerosol indirect effect may be enhanced by the increased concentration of smaller cloud droplets inhibiting precipitation and thus increasing cloud lifetime and cloud cover.[10] Given that low clouds account for about half of the solar energy Earth reflects back to space, the combination of brighter and longer lived clouds could cause a significant cooling that partially offsets the warming from increased greenhouse gas concentrations. Exactly how much cooling is realized has been a topic of considerable debate in the climate community; the sensitivity of clouds to aerosol concentration is a strong function of atmospheric dynamics, local temperature and humidity, and even cloud properties themselves.[11]

To help resolve the debate, the A- Train's diverse instruments are measuring the bulk response of cloud systems to changes in aerosol concentrations; figure 5 shows some of what we have learned. Clouds made from drops of liquid grow deeper, contain smaller droplets, rain less frequently, and appear brighter from above in the presence of large concentrations of small aerosol particles.[12] Moreover, A-Train sensors have furnished groundbreaking measurements of how aerosols affect ice clouds. *Aura* observations of carbon monoxide, a pollutant that often accompanies aerosols from biomass burning, have been combined with *Aqua* measurements of clouds located at the same positions as the CO. Together, they demonstrate that polluted ice clouds generally contain smaller particles than cleaner clouds and are accompanied by weaker precipitation.[13] A knowledge of aerosol–cloud interactions is important for climate prediction, but it is admittedly difficult to prove cause and effect by correlating satellite measurements. The A-Train, though, with its ability to simultaneously measure a wide range of cloud properties in both polluted and clean environments, provides several distinct measures of how clouds respond to aerosols. That the different perspectives are all generally consistent lends credence to A-Train-based analyses of aerosol effects on global scales.

TOWARD IMPROVED CLIMATE FORECASTING

Robust predictions of future climate are essential if the world's policymakers are to develop sound strategies for mitigating and adapting to future climate change. Yet despite the marked progress in climate models over the past 20 years, uncertainties in cloud feedbacks, regional precipitation, and other aspects of climate have improved little since the Intergovernmental Panel on Climate Change's first assessment in 1990. The A-Train carries tools for evaluating how well climate models represent several aspects of present-day energy and water cycles, atmospheric composition and transports, and surface–atmosphere exchanges. Such tests are critical because accurate prediction of climate variability on decadal and longer time scales requires that models be capable of simulating current climate and short-term variations such as the diurnal and annual solar cycles and the year-to-year variations associated with the El Nino Southern Oscillation.

Climate scientists have shown, for example, that models generally fail to accurately predict the ice content of high altitude cirrus clouds in the tropics.[14] Given the warming effect of such clouds on the atmosphere, improperly estimating their ice content is a potentially serious shortcoming. A-Train measurements of cloud temperatures and ice and water vapor content allow modelers to examine specific processes related to cirrus cloud formation in large-scale models. Hopefully, such investigations will lead to significant advances in our ability to represent those clouds, an important component of the climate system. More generally, the A-Train enables climatologists to determine quantitatively the relationships between a wide variety of cloud properties and the surrounding environment on scales of several to hundreds of kilometers.15] Such studies can provide insights that ultimately serve to improve model simulations. A-Train measurements of ozone-depleting trace gases and polar stratospheric clouds also help modelers of stratospheric chemistry make quantitative assessments of polar ozone depletion[16] and evaluate models of polar processes affecting ozone recovery.

THE FUTURE OF CONSTELLATION MISSIONS

The division of instruments among the satellites of the A-Train mitigates the problems inherent in complex multi-instrument payloads without compromising the sensors' ability to make simultaneous measurements. In the near future, at least one new satellite will join the train. Scheduled for launch in November of this year, *Glory* will extend the long-term record of total solar irradiance and will observe natural and anthropogenic aerosols. Japan's *Global Change Observation Mission–Water*, which will carry the successor to one of the microwave radiometers aboard *Aqua*, may join the train in 2012.

Of course, the A-Train cannot be maintained indefinitely. But its contributions to addressing questions about atmospheric composition and the integrated energy and water cycles offer a strong argument for adapting the constellation template to future missions with common themes.

In 2007 a National Research Council committee comprising experts from all areas of the scientific community issued its detailed decadal survey *Earth Science and Applications from Space: National Imperatives for the Next Decade and Beyond*, in response to requests from NASA's Earth science program, the National Oceanic and Atmospheric Administration, and the US Geological Survey to summarize Earth-observing needs in the next 20 years. The National Research Council report is an important road map that outlines and prioritizes 17 new space missions that will become the cornerstone of Earth observation. But the report fails to explicitly outline plans for coordinating future satellites, even though it advocates for several missions with common themes. As NASA and other space agencies plan the future of Earth observation, they should strongly consider adopting the A-Train paradigm, which, we believe, will maximize the scientific impact of their missions.

ACKNOWLEDGMENT

Release time to prepare this article was provided by NASA's Energy and Water Cycle Study program and by Caltech's Jet Propulsion Laboratory, under a contract with NASA. We thank Charles Ichoku, Hal Maring, Mark Schoeberl, and Graeme Stephens for valuable discussions concerning the background and history of the A-Train.

REFERENCES

1. S. Solomon et al., eds., *Climate Change 2007: The Physical Science Basis*, Cambridge U. Press, New York (2007), online at http://www.ipcc.ch/ipccreports/ar4-wg1.htm.
2. R. P. Allan, B. J. Soden, *Science* 321, 1481 (2008); D. E. Waliser et al., *J. Geophys. Res.* 114, D00A21 (2009), doi:10.1029/2008JD010015.
3. G. L. Stephens, *J. Clim.* 18, 237 (2005); S. Bony et al., *J. Clim.* 19, 3445 (2006).
4. G. L. Stephens et al., *Bull. Am. Meteorol. Soc.* 83, 1771 (2002).
5. M. R. Schoeberl et al., *IEEE Trans. Geosci. Remote Sens.* 44, 1066 (2006).
6. S. V. Nghiem et al., *Geophys. Res. Lett.* 34, L19504 (2007), doi:10.1029/2007GL031138.
7. J. E. Kay et al., *Geophys. Res. Lett.* 35, L08503 (2008), doi:10.1029/2008GL033451.
8. D. Chand et al., *Nat. Geosci.* 2, 181 (2009).
9. S. Twomey, *J. Atmos. Sci.* 34, 1149 (1977).
10. B. Albrecht, *Science* 245, 1227 (1989).
11. B. Stevens, G. Feingold, *Nature* 461, 607 (2009).
12. T. S. L'Ecuyer et al., *J. Geophys. Res.* 114, D09211 (2009), doi:10.1029/2008JD011273; M. Lebsock, G. L. Stephens, C. Kummerow, *J. Geophys. Res.* 113, D15205 (2008), doi:10.1029/2008JD009876.
13. J. Jiang et al., *Geophys. Res. Lett.* 35, L14804 (2008), doi:10.1029/2008GL034631.
14. J. Jiang et al., *J. Geophys. Res.* 115, D15103 (2010), doi:10.1029/2009JD013256.
15. H. Su et al., *Geophys. Res. Lett.* 35, L24704 (2008), doi:10.1029/2008GL035888.
16. A. R. Douglass et al., *Geophys. Res. Lett.* 33, L17809 (2006), doi:10.1029/2006GL026492; M. C. Pitts et al., *Atmos. Chem. Phys.*7, 5207 (2007). ■

The *Physics Today* online version of this article includes a list of A-Train sensors and brief descriptions of their primary purposes, www.physicstoday.org.

SESSION C: DECARBONIZING TRANSPORTATION

The Future of Low-Carbon Transportation Fuels

Christopher Yang and Sonia Yeh

Institute of Transportation Studies
University of California, Davis
Davis, CA 95616

Abstract. Petroleum fuel uses make up essentially all of transportation fuel usage today and will continue to dominate transportation fuel usage well into future without any major policy changes. This chapter focuses on low-carbon transportation fuels, specifically, biofuels, electricity and hydrogen, that are emerging options to displace petroleum based fuels. The transition to cleaner, lower carbon fuel sources will need significant technology advancement, and sustained coordination efforts among the vehicle and fuel industry and policymakers/regulators over long period of time in order to overcome market barriers, consumer acceptance, and externalities of imported oil. We discuss the unique infrastructure challenges, and compare resource, technology, economics and transitional issues for each of these fuels. While each fuel type has important technical and implementation challenges to overcome (including vehicle technologies) in order to contribute a large fraction of our total fuel demand, it is important to note that a portfolio approach will give us the best chance of meeting stringent environmental and energy security goals for a sustainable transportation future.

INTRODUCTION

Petroleum fuel uses make up essentially all of transportation fuel usage today and will continue to dominate 95% of transportation fuel usage in 2035 according to the Energy Information Administration's Annual Energy Outlook projection [1]. Biofuels make up the largest increase in the use of alternative fuels, to about 4% of fuel usage in 2035. The same projection also shows a 12% increase in greenhouse gas (GHG) emissions from transportation between 2010 and 2035 due to demand growth. As discussed in earlier chapters, fossil fuel use has many economic and environmental externalities, including our reliance on imported energy source that weakens our energy security, air pollution that impacts health and, of more recent concern, GHG emissions that contribute to changes in climate. While the focus of this chapter is on reducing fossil fuel use and GHG emissions, tackling all of the issues associated with petroleum dependence in transportation requires a coordinated effort involving reducing growth in travel demand, improving vehicle efficiency and a switch to cleaner, lower carbon fuels.

This chapter focuses on low carbon transportation fuels, specifically, biofuels, electricity and hydrogen, which are emerging options to displace petroleum-based fuels. Fuels such as electricity and hydrogen are intimately tied to the vehicle platform, though this chapter primarily focuses on the fuels themselves. This chapter

Physics of Sustainable Energy II: Using Energy Efficiently and Producing it Renewably
AIP Conf. Proc. 1401, 259-270 (2011); doi: 10.1063/1.3653857

also emphasizes fuel use for light-duty vehicles (i.e. passenger cars and trucks), which make up around 55% of energy use in the transportation sector, though fuels used in other modes of transportation will also be briefly discussed. There needs to be considerable technology development involved with widespread use of these fuels, including the development and deployment of vehicle platforms and infrastructure for production, transport and refueling. Each fuel has multiple "pathways" for production and delivery, which will determine their energy and emissions footprint and the benefits associated with their adoption (Figure 1). This chapter discusses the transition challenges to alternative fuels, particularly the infrastructure challenges, and the insights of making a transition to sustainable transportation over the long run.

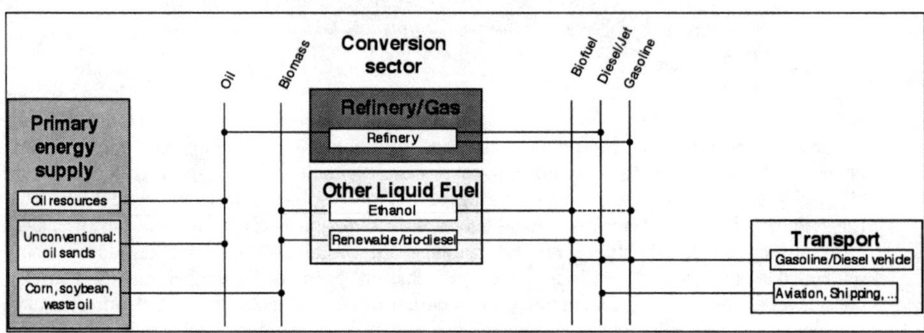

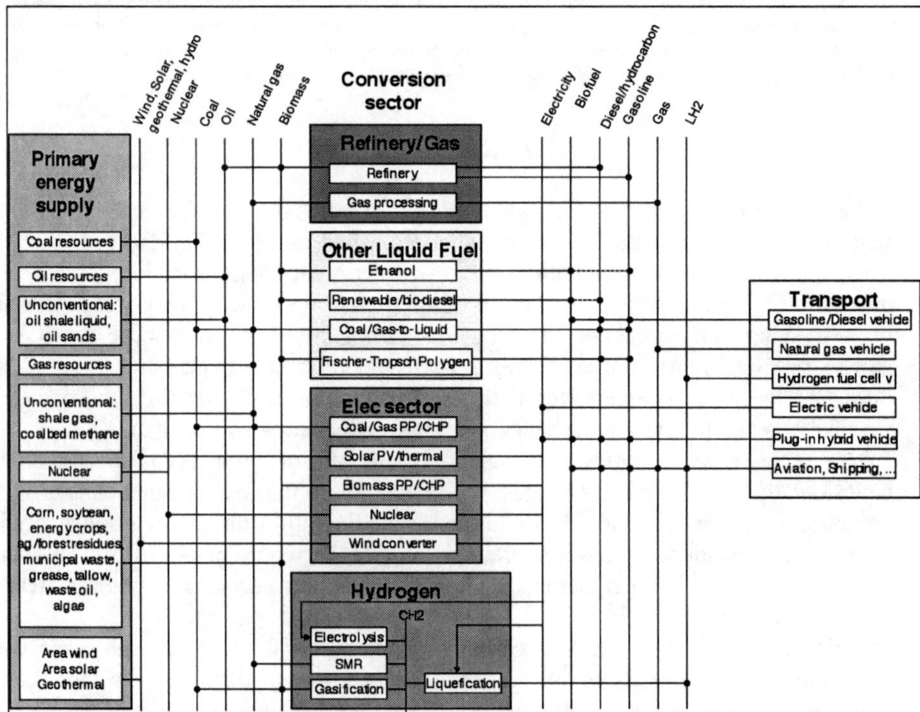

FIGURE 1. Current (top) and potential future (bottom) transportation fuel sources, conversion technologies, fuel types and vehicle technologies.

BIOFUELS AS A TRANSPORTATION FUEL

While biofuels can comprise a range of forms, including liquidskj, solids and gaseous forms, this chapter focuses only on high energy-density liquid fuels used as a substitute for petroleum-based fuels. As shown in Table 1, biofuels can be produced from a wide array of potential biomass feedstocks and technology. Thus, biofuels can have potentially very different fuel properties, energy use, emissions and other impacts (such as environmental impacts in land use and water requirements) throughout their production lifecycle, including cultivation, transport and conversion to biofuels. Biomass feedstocks for liquid fuel production can be categorized into four types: lignocellulosic biomass, sugars/starches, oils and animal fats, and algae. These feedstocks can come from a variety of sources including grain-based crops (such as corn or soy), oilseeds and plants (such as oil palm and sugarcane), agricultural residues, energy crops, forestry resources, industrial and other wastes, and algae. The technology for the conversion of these feedstocks to a liquid fuel can also take several different forms, including biological, chemical and thermochemical processing (Table 1). First generation biofuels, those that are commercially available today, include sugar- and starch- based ethanol (and other alcohol fuels), which is a gasoline substitute and vegetable oils and biodiesel, which are diesel substitutes. Advanced biofuels are derived from pathways currently in development and include alcohol fuels from cellulose, algal-based fuels, and thermochemical conversion of biomass to hydrocarbon that can be converted to a full range of fuels including gasoline, diesel fuel and jet fuel that meet the same specifications as today's petroleum fuels.

TABLE 1. Biofuel feedstock and production pathways. Adapted from Parker et al. [2]

Feedstock category	Feedstock type	Conversion technologies
Starch-based and sugar-based biomass	Corn, sugarcane, sugar beet, sweet sorghum	Bioethanol through hydrolysis and fermentation
Ligocellulosics	Forest biomass, herbaceous energy crops, agricultural and food production residues, municipal solid wastes	Cellulosic ethanol through hydrolysis and fermentation, upgrading of pyrolysis oils to gasoline, Fischer Tropsch diesel
Lipids	Seed oils, yellow grease, animal fats	Fatty acid to methyl esters (FAME), hydro- treatment of fatty acids to hydrocarbons (FAHC)
Algae		Transesterification

Conversion of biomass into a biofuel can take many forms. Commercially available conversion processes used for first generation biofuels include biological fermentation (via yeast) of sugars into ethanol, and chemically catalyzed transesterification of oils/fats and alcohols into biodiesel (and glycerol co-product). More advanced processes are currently being developed and refined, including ethanol production from lignocellulosic biomass (e.g. wood, grass, and straw) requiring conversion of cellulose and hemicellulose into component sugars, thermochemical conversion of lignocellulosic biomass via gasification and a Fischer-Tropsch synthesis process into diesel fuel, and algae biofuels, which require development of cultivation, separation of cells and oils and conversion to useful fuels.

While biomass resources are plentiful, not all is technically, economically and environmentally viable for conversion to transportation fuels. Estimates of US biomass indicate that it could be sufficient to supply somewhere around 80-100 billion gallons of gasoline equivalent of biofuels per year. Depending upon vehicle efficiency and projections of future travel demand, this could be anywhere from approximately 1/3 of transportation fuel demand in 2050 in a business as usual case to nearly all transportation fuel demand in highly efficient and electrified demand future [3]. Limiting the use of specific biomass resources because of sustainability concerns will reduce the availability of these fuels even further. Limitations on sustainable biofuel supply will play an important role in determining the extent to which petroleum fuels can be displaced by biofuels [4].

The sustainability of biofuels is an important question and is dependent on the specific feedstock and conversion pathway to produce the biofuel. While there is no agreed upon definition of sustainability, many different metrics and potential impacts have been proposed and can be considered as contributors to a fuel's sustainability or lack thereof – ecosystem/habitat disruption, deforestation, soil quality impacts, direct and indirect GHG emissions, other air and water pollution, water usage, competition with food crops, and land conversion [4]. These negative impacts are important to quantify because they can mitigate or even exceed the environmental benefits that using biofuels is supposed to provide. The challenge in assessing these impacts is that there is both a wide variety of potential biomass feedstocks and conversion technologies as well as the fact that agriculture is widely variable based upon land quality, soil quality, precipitation patterns, etc, which make generic discussions of feedstock/conversion pathways less useful.

While some transportation modes can be electrified (such as light-duty plug-in electric vehicles, PEVs, or fuel cell vehicles, FCVs), other modes, specifically aircraft, marine shipping, and heavy-duty trucks are most likely to use liquid fuels for the next few decades because of vehicle range and fuel energy density issues. Given that these modes are projected to have significant travel demand growth [1], a low carbon biofuel is perhaps the only option for lowering the GHG intensity in these transportation modes.

ELECTRICITY AS A TRANSPORTATION FUEL

Plug-in electric vehicles are powered, at least in part, by electricity from the electric grid that is stored in an onboard battery. They can be either plug-in hybrid electric vehicles (PHEVs), which can run on electricity or gasoline or battery electric vehicles (BEVs), which run entirely on electricity. PEVs are much more efficient than conventional internal combustion engine vehicles (ICEVs) running on petroleum fuels. In addition, electricity is a decarbonized energy carrier that provides solutions to US petroleum dependence and local air pollution concerns. In addition, electricity can be made from a wide range of domestic resources, including low carbon resources, which can reduce the carbon intensity of fuels. PEVs are beginning to be commercialized in 2011 and make up a tiny fraction of vehicle sales.

While batteries are the key technology for the success of PEVs, the fuel electricity supply and infrastructure side of the equation is also important to understand from a technology and deployment perspective.

PEVs need to be plugged in to "refuel" the onboard batteries, which can range from about 3 kWh for a low-range PHEV to over 25 kWh for a longer range BEV. While current PEVs can be recharged at a conventional 120V outlet (often called level 1 charging), the rate of energy transfer is quite slow (~1-2 kW). To recharge more quickly, it is necessary to use higher voltage and current and a dedicated PEV charger (often called electric vehicle supply equipment or EVSE). Level 2 charging is 240 V and up to 40 amps for up to 9 kW while level 3 charging is being designed to allow for very fast charging (up to 80% of battery capacity in less than 30 minutes). An EVSE will use a standardized plug (e.g. SAE J1772).

Given the low penetration of PEVs, it is not surprising that there are very few PEV chargers deployed. However, there is concern that deployment of home-based charging equipment could be an issue if PEVs are to be widespread and electricity is to be a primary fuel for light-duty transportation. A survey by Axsen and Kurani [5] found that only about 50% of new vehicle buyers have a 120 V outlet within 25 feet of their household vehicle parking space and only 35% within 10 feet. Others have noted that in urban areas such as San Francisco, less than 20% of cars are parked overnight in dedicated off-street parking. Beyond home-based charging, deployment of public infrastructure is likely needed to increase the utility of PEVs and to ease drivers' "range anxiety". Public infrastructure would be useful at the workplace, at retail establishments, along major highways, and other activity centers. Charging times will be much longer than refueling a gasoline tank (30 minutes to several hours), and argue for co-locating charging while drivers are engaged in other activities (e.g. shopping, work). Studies are underway to understand the best locations for public infrastructure to minimize costs while maximizing utility and utilization.

The supply of electricity is an important part of the equation for electrified transportation. Electricity is already produced in large quantities and in the near-term, the amount of electricity that would be demanded from PEVs would be a tiny fraction of total electricity generation [6]. In California for example, charging of one million PEVs (about 4% of total LDVs) would only require about 1% additional electricity generation. Regardless, charging a PEV requires the grid to respond by providing more electricity and the operation of the grid is such that timing of when PEVs charge can be an important factor.

The electricity grid is collection of power plants and transmission and distribution facilities that produces and delivers electricity to end users and is structured to meet continually changing electricity demands by using a number of different power plant types. Some are baseload facilities (often large coal or nuclear plants) that are designed to operate continuously and at low cost, while peaking power plants (often fired with natural gas or oil) are operated only a handful of hours per year when demand is highest and are more costly to operate. The mix of power plants that make up the grid varies significantly from one region to another—based on local demand profiles, resource availability and cost, and energy policy. The timing of conventional electricity demands and of PEV charging demands will impact the types of power plants that are used to meet the additional demands and their associated emissions.

Charging during off-peak hours will tend to flatten the demand profile, reducing the need for additional generating capacity and lowering the average cost of electricity. Charging at peak demand times will increase capacity requirements, while lowering the utilization of existing plants and increasing electricity costs. If charging could be controlled to occur when it was most optimal, PEV demand could respond to grid conditions. Given that cars are parked approximately 95 percent of the time and potentially plugged in for a large fraction of the time they are parked, this is a real possibility. The smart grid, incorporating intelligence and communication between the supply and demand sides of the electricity equation, is needed in order to realize the full benefits of this vehicle charging flexibility. Managing vehicle recharging requires a smart charging system that enables communication between the customer and utilities. Consumers may give the utility or system operator some control over their charging in exchange for lower rates. This type of charging interface can also permit vehicle charging emissions to be appropriately tracked and allocated, which will become increasingly important as states and countries adopt low-carbon fuel standards and impose caps on GHG emissions in different sectors.

The carbon intensity of average US electricity is higher than for gasoline, but this is due to the fact that electricity is an intermediate energy carrier, which has already been converted from a primary energy resource, and can be used with very high efficiency (conversion of stored energy in a battery into mechanical work on the vehicle). Gasoline and diesel, on the other hand, are fuels that have been slightly modified from the original primary energy resource (crude oil) to achieve specific properties suitable for internal combustion engines. These fuels are converted at much lower efficiency to mechanical work on board the vehicle. Thus, while the carbon intensity of electricity (measured in grams of carbon dioxide equivalent emissions per mega joule, gCO2/MJ) is higher than that of gasoline and diesel, the carbon per mile of travel from a PEV can be much lower than that of a conventional or even hybrid vehicle. In addition, the carbon intensity of electricity will gradually be reduced as renewable generation increases and due to other carbon policies.

Of course, the supply of electricity differs in different regions of the country. Some areas, such as the west coast and the Northeast states have lower electricity carbon intensity (due to higher hydropower resources) than other areas, such as the Midwest States where relatively higher portions of electricity are generated from coal. These differences in regional electricity will impact the relative benefits of PEVs vs gasoline vehicles.

While there is the potential for electrification of light-duty vehicles, other transportation sectors are less likely to electrify. Rail, buses and delivery trucks also offer some potential for running on grid electricity. However, after LDVs, the main energy and emissions contributions from transportation come from aviation, heavy-duty long-haul trucking and marine shipping. These sectors present significant challenges to electrification and will likely rely on high-density liquid biofuels (or potentially hydrogen) in order to reduce their fuel carbon intensity.

HYDROGEN AS A TRANSPORTATION FUEL

Hydrogen has been widely discussed as a long-term fuel option to address environmental and energy security goals [7]. FCVs that use hydrogen are significantly more efficient than conventional vehicles, using less energy to produce a mile of vehicle travel. Additionally, the fuel can be made from a wide variety of domestic and low carbon resources, providing solutions to the oil dependency and carbon challenges. While FCVs have not yet been commercialized, several automakers have announced plans to introduce vehicles in the 2015 timeframe.

Like electricity, hydrogen is an energy carrier that is produced from a primary energy resource. Almost any energy resource can be converted into hydrogen, although some pathways are superior to others in terms of cost, environmental impacts, efficiency, and technological maturity. Currently in the US, about 9 million tonnes of H_2 are already produced each year (enough to supply about 30 million FCVs), mainly for industrial or refinery purposes. Natural gas reforming accounts for 95% of current H_2 production in the US and in the near-term, along with coal, should continue to be the least expensive method to produce H_2. In the longer-term, continued use of fossil resources to produce H_2 would necessitate the use of carbon capture and storage (CCS) technologies to minimize the GHG emissions from H_2 production. Additionally, H_2 can be produced from biomass in a production process similar to that from coal (gasification). Electrolytic hydrogen production can also be an important H_2 production technology in the longer-term and offers the potential for zero carbon production from renewables such as solar and wind.

Hydrogen infrastructure includes all of the components associated with producing, delivering and providing H_2 to the vehicle at a refueling station and can generally be categorized into two types: on-site and central production. Onsite production uses existing energy distribution methods for electricity or natural gas to allow for H_2 production at the refueling station (via electrolysis or natural gas steam reforming). Central production of hydrogen would require delivery of hydrogen, via compressed gas trucks, liquefied H_2 trucks or gaseous pipelines, to the refueling station. Over the near- to medium-term, H_2 infrastructure is likely to be comprised primarily of onsite H_2 stations, while over time it is expected to transition to an infrastructure primarily composed of central production and delivery [7], which is lower cost when demand is high enough [8].

One of the key issues regarding the deployment of hydrogen infrastructure is the so-called "chicken-and-egg" problem, which deals with the problem of ensuring that both the H_2 refueling infrastructure and FCVs will have access to the other as they are being deployed. One approach to dealing with this issue of near-term station infrastructure is to coordinate the deployment of vehicles and fuels in targeted locations or "lighthouse" regions. This will ensure that the few stations that are built will have sufficient FCVs that will demand H_2 and vice versa. Additionally, a "cluster strategy" is an even more targeted, coordinated introduction of hydrogen vehicles and refueling infrastructure in a few focused geographic areas such as smaller cities within a larger region [9]. This approach provides acceptable customer convenience (in terms of driving distance to the nearest station) and reliability for FCV customers, while reducing infrastructure costs.

Over the longer-term, if H_2 and FCVs are widely used, the H_2 infrastructure will become a massive energy system that will rival the current oil and gas infrastructure for production, delivery, storage and refueling. This system could consist of a number of large H_2 production facilities and a large distribution network (pipelines and trucks) that supply a widespread refueling station network. Each of these technologies are at a reasonable state of technology development, given the widespread use of H_2 in industrial settings and H_2 station demonstrations. However, these individual components have not been combined into an optimized, reliable energy system. This along with implementation and deployment challenges associated with large-scale infrastructure (i.e. investment, permitting and public acceptance) are the main challenges to H_2 infrastructure, rather than purely technical issues.

Estimated costs for hydrogen fuel at the large scale indicate that H_2 could be cheaper per mile than even advanced gasoline vehicles[7]. However, the challenge is that in the near term, both H_2 fuels and FCVs will be more expensive than conventional vehicles running on gasoline. This provides an important policy challenge to incentivize investments in lower cost, lower carbon outcomes in the face of potentially many years of higher costs.

Like electricity, hydrogen may be most useful in the light-duty sector, but has some applicability in other transportation subsectors. Fuel cell buses and delivery trucks are also potentially viable technologies. However, low energy storage density for H_2 is likely to limit its use as a fuel in long-haul trucks, aircraft and marine applications. Because of these limitations, low-carbon liquid fuel, such as biofuels, are needed to meet transportation energy demands in these sectors.

SUMMARY

Considering the technologies and resources that are available to us, there are several alternative fuel sources that can significantly reduce our reliance on imported oil, improve air quality, and GHG emissions. However, the transition to cleaner, lower carbon fuel sources will need significant technology advancement, and sustained coordination efforts among the vehicle and fuel industry and regulators over long period of time in order to overcome market barriers, consumer acceptance, and unaccounted externalities of imported oil in their fuel price. In addition, policies are also needed to ensure that the environmental performance of these new fuel sources, including GHG emissions and environmental impacts, perform better than fossil fuel and to avoid any unintended consequences that these new fuel sources may present [4, 10].

Unique Fuel Infrastructure Challenges

There are varying degrees of challenges associated with infrastructure design and deployment for each of the three fuel types, biofuel, hydrogen and electricity. These are shown in Table 2.

The infrastructure challenges for biofuels center on biomass feedstock production, collection and transport [2], and in the short term, the delivery of bioethanol and biodiesel to refueling stations and building up the refueling infrastructure for dispensing biofuels that can only be used in flex-fuel vehicles, including E85 (ethanol mixed with gasoline up to 85% by volume) and high-blend of biodiesel such as B80, B90 and B100. Over time, however, when advanced biofuels such as FT fuels and hydrocarbon-based biofuels mature, the infrastructure needs for distributing, delivering and refueling will go away.

Electricity is a widely used energy carrier so fuel electricity will primarily use components of this well-established supply chain. The key infrastructure issue for electricity is the deployment of home and public charging equipment. Another important concern for electricity infrastructure is that distribution systems (i.e. transformers and substations) may need to be upgraded in order to handle the additional demands from PEVs at the circuit level.

H_2 production, transport and refueling stations are the primary infrastructure components that will require continued technology development and significant investment. Since H_2 will use existing primary energy resources, their collection and transport are not major issues. Most of the infrastructure issues are discussed in the hydrogen section above.

TABLE 2. Comparison infrastructure design and deployment challenges for alternative fuel sources, including hydrogen, electricity and biofuels. Grayed out boxes are areas that present special challenges that require more attentions and efforts.

	Central Hydrogen	Electricity	Biofuels
Resource collection extraction	Use existing infrastructure for fossil resources (natural gas, coal)	Existing infrastructure	Wastes require collection, energy crops require dedicated operation, part of larger Ag system
Resource transport	Existing infrastructure	Existing infrastructure	Low energy density limits transport distances
Conversion facility	Large-scale reformers/gasifiers	Existing infrastructure	Biorefinery (including feedstock processing and conversion)
Fuel transport	Trucks or pipelines	Existing infrastructure	Expanded infrastructure needed for conventional biofuels (ethanol and bio-diesel) including rail, trucks and barge as well as inter-modal facilities for the transfer of feedstock or fuel
		distribution may require upgrades	Existing infrastructure for biogasoline, FT fuel, and bio-hydrocarbon fuels
Fuel refueling	New H2 refueling stations	Widespread vehicle chargers	Dedicated refueling stations for conventional biofuels (ethanol and biodiesel)
			Existing infrastructure for biogasoline, FT fuel, and bio-hydrocarbon fuels

Table 3 summarizes the key transitional challenges for the alternative transportation fuels, hydrogen, electricity and biofuels. From the resource perspective, because of the diversity of resources available for both hydrogen and electricity, there should not be any resource limitations associated with these fuels, whereas biomass availability is a key issue surrounding the widespread use of low-carbon biofuels.

From a fuel infrastructure technology perspective, H_2 production, storage and delivery and biofuel production are key areas that need further development, while electricity has no major infrastructure technology needs. From an economic perspective, fuel infrastructure is quite costly, especially when demand for fuel is low and economies of scale are not realized. Large production facilities for hydrogen and biofuels are major economic considerations for investment and technology development. Hydrogen stations are another key area with relatively high near-term costs [11]. While PEV chargers can be expensive on a per vehicle basis, because they are introduced incrementally, they require relatively modest total costs in the near term. Finally, for H_2 and electric vehicles, the rate of vehicle adoption is likely to determine the rate of infrastructure deployment, while the use of biofuels can be less dependent on specific vehicle sales and deployment of fuel infrastructure can be more rapid when advanced biofuel technologies mature.

TABLE 3. Comparison resource, technology, economic and transitional issues for hydrogen, electricity and biofuels.

	Hydrogen	Electricity	Biofuels
Resources	Diversity of resources available for H_2 production	Diversity of resources available for electricity production	Limits on providing enough low-carbon and sustainable biomass
Technologies	Hydrogen production (fossil conversion and electrolysis) and storage are critical technology	No major technology limitations for infrastructure	Biorefineries are critical technology
Economics	High initial costs – large economies of scale associated with stations and central production	Relatively low initial investment costs for home charging compared to other fuels	Biorefineries are primary cost and scale dependent
Transitions	Vehicle adoption will determine the rate of infrastructure deployment, requires significant coordination	Vehicle adoption will determine the rate of infrastructure deployment	Rapid deployment of biofuels in next few decades due to federal policy

CONCLUSIONS AND RECOMMENDATIONS

Alternative fuels, particularly those reviewed in this chapter, are an essential tool for helping to reduce the overall impacts of transportation fuel use, including the dependence on imported oil, air pollution and GHG emissions. While each fuel type has important technical and implementation challenges to overcome (including vehicle technologies) in order to contribute a large fraction of our total fuel demand, it is important to note that a portfolio approach will give us the best chance of meeting stringent environmental and energy security goals for a sustainable transportation future. It will be important to nurture all technologies along because we do not yet know which technologies will provide the most cost effective emissions and petroleum usage reduction while appealing to consumer preferences.

The following are the main recommendations with respect to making the transition to a future of sustainable transportation fuels:

Research is important - Fundamental and applied research in needed by academic communities and stakeholders to improve technologies associated with fuel production, conversion, storage, and utilization as well as scientific understanding of sustainability impacts of these fuels. This research can help to guide R&D as well as investment decisions by government, industry and other stakeholders.

Policies can help level the playing field - Policy incentives are needed to directly incentivize the development and use of low-GHG/sustainable fuels through performance-based standards and market mechanisms. Policies such as GHG emissions standards for automobiles or the low carbon fuel standard (LCFS) [12] are essential for putting the different fuels (and vehicle platforms) into a common framework with which they can be assessed. They allow industry the flexibility to choose different options and approaches to ensure that the targets are met with lower compliance costs.

Sustainability standards should be developed - Effective sustainability policies are needed to prevent impacts on ecologically sensitive areas, air and water pollution, and competition with food resources. Continuous monitoring and assessments of unintended consequences within or beyond the production areas will be essential for the successful transition to a sustainable transportation future.

REFERENCES

1. Energy Information Administration, *Annual Energy Outlook 2011 with Projections to 2035*. EIA, 2011; Available from: http://www.eia.doe.gov/oiaf/aeo/aeoref_tab.html.
2. Parker, N., et al., "*Development of a biorefinery optimized biofuel supply curve for the Western United States.*" Biomass and Bioenergy, 2010, **34**, pp. 1597-1607.
3. McCollum, D. and C. Yang, "*Achieving deep reductions in US transport greenhouse gas emissions: Scenario analysis and policy implications.*" Energy Policy, 2010, **37**(12), pp. 5580 - 5596.
4. National Academy of Sciences, *Liquid Transportation Fuels from Coal and Biomass Technological Status, Costs, and Environmental Impacts*, 2009, NAS, Washington, DC.
5. Axsen, J. and K.S. Kurani, *Anticipating plug-in hybrid vehicle energy impacts in California: Constructing consumer-informed recharge profiles,* Transportation Research Part D, 2010, **15**, pp. 212-219.
6. Lemoine, D.M., D.M. Kammen, and A.E. Farrell, "*An innovation and policy agenda for commercially competitive plug-in hybrid electric vehicles,*" Environmental Research Letters, 2008(1), pp. 014003.
7. National Research Council, *Transitions to Alternative Transportation Technologies--A Focus on Hydrogen*, 2008, NRC, Washington, DC.
8. Johnson, N., C. Yang, and J.M. Ogden, "*A GIS-Based Assessment Of Coal-Based Hydrogen Infrastructure Deployment In The State Of Ohio,*" International Journal of Hydrogen Energy, 2008. **33**(20), pp. 5287-5303.
9. Nicholas, M.A. and J.M. Ogden, *An Analysis of Near-Term Hydrogen Vehicle Rollout Scenarios for Southern California. Research Report UCD-ITS-RR-10-03*, 2010, Institute of Transportation Studies, University of California, Davis.
10. Searchinger, T., et al., "*Use of U.S. Croplands for Biofuels Increases Greenhouse Gases Through Emissions from Land Use Change,*" Science, 2008. **319**(5867), pp. 1238 - 1240.
11. Ogden, J. and C. Yang, *Chapter 15. "Build-up of a hydrogen infrastructure in the US.* in *Perspectives of a hydrogen economy*," eds. M. Ball and M. Wietschel, 2009, Cambridge.
12. Sperling, D. and S. Yeh, "*Low Carbon Fuel Standards,*" Issues in Science and Technology, 2009(2), pp. 57-66.

Technology Status and Expected Greenhouse Gas Emissions of Battery, Plug-In Hybrid, and Fuel Cell Electric Vehicles

Timothy E. Lipman

Transportation Sustainability Research Center
University of California – Berkeley
2150 Allston Way, Suite 280
Berkeley, CA 94704

Abstract. Electric vehicles (EVs) of various types are experiencing a commercial renaissance but of uncertain ultimate success. Many new electric-drive models are being introduced by different automakers with significant technical improvements from earlier models, particularly with regard to further refinement of drivetrain systems and important improvements in battery and fuel cell systems. The various types of hybrid and all-electric vehicles can offer significant greenhouse gas (GHG) reductions when compared to conventional vehicles on a full fuel-cycle basis. In fact, most EVs used under most condition are expected to significantly reduce lifecycle GHG emissions. This paper reviews the current technology status of EVs and compares various estimates of their potential to reduce GHGs on a fuel cycle basis. In general, various studies show that battery powered EVs reduce GHGs by a widely disparate amount depending on the type of powerplant used and the particular region involved, among other factors. Reductions typical of the United States would be on the order of 20-50%, depending on the relative level of coal versus natural gas and renewables in the powerplant feedstock mix. However, much deeper reductions of over 90% are possible for battery EVs running on renewable or nuclear power sources. Plug-in hybrid vehicles running on gasoline can reduce emissions by 20-60%, and fuel cell EV reduce GHGs by 30-50% when running on natural gas-derived hydrogen and up to 95% or more when the hydrogen is made (and potentially compressed) using renewable feedstocks. These are all in comparison to what is usually assumed to be a more advanced gasoline vehicle "baseline" of comparison, with some incremental improvements by 2020 or 2030. Thus, the emissions from all of these EV types are highly variable depending on the details of how the electric fuel or hydrogen is produced.

1. INTRODUCTION

Electric-drive vehicles (EVs) are widely promoted for their environmental and energy efficiency benefits. Since their first "re-introduction" in the 1990s, vehicles running

Physics of Sustainable Energy II: Using Energy Efficiently and Producing it Renewably
AIP Conf. Proc. 1401, 271-298 (2011); doi: 10.1063/1.3653858

on batteries and other types of electrified drivelines have steadily improved in terms of cost and performance to the point where they are now commercially viable. Several major original equipment manufacturers (OEMs) and smaller companies are now making battery-electric, plug-in hybrid, and fuel cell vehicles available for sale or lease, with several additional models expected in the next few years.

Battery-electric vehicles (BEVs) and plug-in hybrid vehicles (PHVs) can together be considered "plug-in electric vehicles" (PEVs) as they derive some or all of their energy from plugging in to electricity grids. Fuel cell vehicles (FCVs) typically are fueled entirely with hydrogen, but could in principle by PEVs as well as they typically contain battery energy storage systems to help complement the operation of the fuel cell system and allow for regenerative braking (where some braking energy can be re-captured through the use of an electrical generator). Finally the term "electric vehicle" or "EV" can be used to describe the set of all of these electrified vehicle types.

PHVs with true "all-electric range" (AER) could allow drivers to make some trips without the engine turning on at all (or at least very little), where the trip can be made almost entirely on the energy stored in the battery. However, some PHVs are not designed for this and instead employ "blended mode" operation, where the design would be for the engine to turn of and on periodically. And in other cases, even for "series type" PHVs with extensive AER, engine operation is to be expected both on longer trips and in other cases where the PHV battery becomes discharged before it can be charged again.

With regard to their emissions performance, unlike conventional gasoline vehicles, whose emissions of greenhouse gases (GHGs) are a combination of "upstream" emissions from fuel production and distribution and "downstream" emissions from vehicle operation, emissions from EVs are more heavily or even entirely (in the case of BEVs and FCVs) upstream in the fuel production and distribution process. In some ways this makes emissions from these vehicles easier to estimate, but there still are many complexities involved. This is particularly true for PHVs that use a combination of grid electrical power and another fuel that is combusted onboard the vehicle.

In the case of battery-electric and fuel cell vehicles, the emissions from the vehicles are entirely dependent on the manner in which the electricity and/or hydrogen are produced, along with the energy-use efficiency of the vehicle (typically expressed in "watt hours per mile/kilometer" for BEVs and "miles/kilometers per kilogram" for hydrogen-powered vehicles). In the case of PHVs, an upstream emission component results from the use of electricity from the wall plug or charger (along with upstream emissions from the production of the vehicle's other fuel), but there can also significant tailpipe emissions depending on travel patterns and the type of plug-in hybrid.

Greenhouse gas emissions from conventional and alternative vehicles have been extensively studied over the past twenty or more years, but are still not fully understood and accounted for. This is particularly so with regard to some of the subtler aspects of vehicle three-way catalyst operation and emissions of trace gases, as well as some potentially important nuances of upstream emissions from virtually all fuels due to the complexity of estimating various aspects of upstream emissions. These include for example difficulty in accurately assessing the true marginal emissions impacts from the increased use of power plants to charge battery EVs and PHVs, and some secondary aspects of upstream emissions that can be important such as the "indirect land use" change effects of the production of biomass for biofuel vehicles. GHG emission reductions from vehicles and fuels represent an important set of strategies for reducing emissions from the transportation sector, among various other options [1].

1.1 Paper Scope and Organization

This paper presents an overview of the current technology status of EVs and their expected potential to reduce emissions of GHGs. The paper is organized as follows. First, the status of EV vehicle technology and automaker commercialization status is briefly reviewed. Second, some background on the issue of GHG emissions from EVs is presented and previous research is reviewed. Third, recent estimates of GHGs from EV fuel cycles are reviewed and compared. Fourth, the potential for EVs to rapidly scale-up to meet the climate challenge is examined. Finally, key uncertainties, areas for further research, and conclusions are discussed.

Also, please note that researchers generally distinguish emissions related to the lifecycle of fuels and energy used to power the vehicle (the "fuelcycle") from emissions related to the lifecycle of the vehicle and the materials it is made from (the vehicle lifecycle). Where emissions are discussed in this paper the focus is mainly but not exclusively on fuelcycle emissions. There has been relatively little work on vehicle lifecycle emissions, but a few noteworthy analysis efforts have been made and it is generally clear that while there are some changes in lifecycle emissions by vehicle type (and the types of materials and manufacturing processes used) overall emissions from vehicles tend to be dominated by the in-use or fuelcycle emissions component. The exception is where, in the case of some EV types, lifecycle emissions can be very low (such as with electricity or hydrogen made from solar, wind, and some biomass resources) and thus the vehicle emissions become proportionally more significant -- but still relatively low in an absolute sense.

2. ELECTRIC-DRIVE VEHICLE TECHNOLOGY STATUS

Re-introduced in the 1990s after a period of success early in the century, BEVs captured a great deal of public attention but were not very successful in the marketplace even with significant government policy support for industry R&D and vehicle sales. The BEVs that were sold or leased by the major OEMs during this period gained a loyal following of drivers, but only totaled approximately 5,000 units produced. Many of these were later recalled and destroyed, much to the dismay of their drivers and famously reported in a popular movie.[1] Hybrid-electric vehicles have since become somewhat popular, with a few million units now sold globally, but EVs still remain a relatively "boutique" product in the global automotive landscape. However, varying projections estimate that there could be from several hundred thousand to a few million EVs on the roads of the U.S. alone by the 2015-2020 timeframe, and the Obama Administration has stated a goal of 1 million PEVs on U.S. roads by 2015.

[1] "Who Killed the Electric Car?" Papercut Films, 2006.

Along with a limited re-release of freeway capable BEVs by most notably Nissan, with its LEAF vehicle, as well as Mitsubishi and several smaller companies, PHVs are now commercially available with the Chevy Volt. Additional models are expected soon with a plug-in Toyota Prius planned for 2012 and several other PHV models rumored as well for near-term commercialization. The most notable differences between these vehicles and the ones that were available in the 1990s include the recent availability of lithium-based batteries, with much higher energy density than available in most previous battery types (such as lead acid and nickel-metal hydride) along with good power characteristics, along with significant advances and cost reductions in other drivetrain components such as electric motors, power inverters, and electrically operated accessory systems.

The Chevy Volt is a mostly series-type PHV, where the gasoline engine is mainly used as a range-extender for a fully electric driveline. Meanwhile vehicles such as the Prius PHV employ a "blended" mode that uses a smaller battery system, a split-power scheme where both the electric motor and gasoline engine can directly power the drive wheels, and less potential for AER operation. The Chevy Volt is capable of driving 30-40 miles using only electricity, while the Toyota Prius PHV is expected to achieve approximately half that amount and only through more gentle operation. Finally, FCVs remain in prototype/pre-commercial status with several hundred vehicles currently being operated, and limited leasing available in a few metropolitan regions. Several OEMs appear to be targeting the 2015-2017 timeframe for wider commercialization of these vehicles, concurrent with plans to slowly expand hydrogen-refueling infrastructure in California and other places.

Figure 1 below presents one vision of how these various EV types might be arrayed across a duty-cycle and driving range spectrum. BEVs are shown to be most appropriate for lighter duty and lower driving ranges, where PHVs (shown as "PHEVs" in the figure) are shown for light/medium-duty cycles and moderate driving ranges and FCVs are more suitable for heavier duty applications and/or longer driving ranges. Finally more conventional combustion vehicles are expected to remain the choice for heavier duty and very long-range applications, but possibly could be replaced in many of the other regions of the spectrum.

In summary, EVs are currently experiencing a renaissance with more types of vehicles available than has historically been the case and significantly improved technology underlying their design. There is, however, much speculation about how successfully the current and planned vehicles will be commercialized "this time around" and significant uncertainty in this related to: 1) consumer response to the vehicles given their perceived performance end economics relative to other types; 2) prevailing gasoline and other fuel prices in the future; and 3) the level and type of government policy and incentive support to be offered. What is clear is that there is ongoing concern about the emissions produced by motor vehicles, and a growing consensus that more need to be done to curb these emissions for climate protection. The expected GHG emissions performance of these EV types is discussed next.

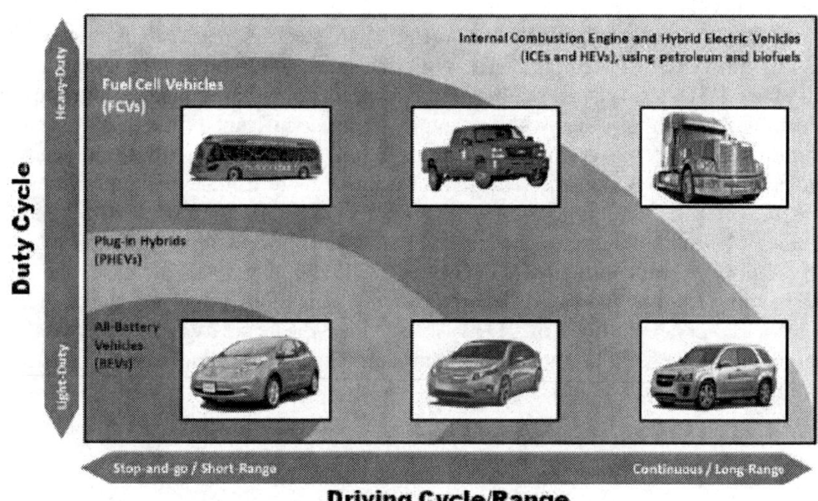

FIGURE 1.: Potential Duty-Cycle and Driving Range Niches for EVs (Source: [2])

3. ELECTRIC-DRIVE VEHICLE EXPECTED GREENHOUSE GAS EMISSIONS PERFORMANCE

Emissions from various types of alternative-fuel vehicles (AFVs) including EVs using electricity and hydrogen as fuels have been reviewed and analyzed over the past few decades. This has occurred along with growing scientific and public recognition of the climate change problem and the role of transportation in contributing to it. This work has revealed some interesting dynamics of GHG formation and emission from vehicle fuel cycles, particularly with regard to various nuances of upstream emissions from fuels and electricity production, secondary effects of fuels production on other aspects of the economy, and emissions of nitrous oxide (N_2O) and other GHGs from combustion engine vehicles [3, 4, 5, 6].

By way of further background it is worth noting that GHGs are a number of different gases and aerosols that have climatic impacts, including three that are commonly analyzed as parts of vehicle fuel and electricity fuel cycles – carbon dioxide (CO_2), methane (CH_4), and N_2O – as well as many refrigerant gases and various other gases and particulates that have climatic impacts (e.g., ozone, carbon monoxide, black soot, volatile organic compounds, etc.). CO_2 has the single largest effect, but various other gases and atmospheric species are significant as well. For example, ozone and aerosols – which are omitted from most analyses of GHG emissions from EVs – have had a greater absolute radiative forcing effect than has nitrous oxide [7].

For EVs of various types that are fueled with electricity and/or hydrogen, the GHGs of most interest are CO_2, CH_4, N_2O, the latest automotive refrigerants (e.g., HFC-134a, HFO-1234yf, etc.), ozone, and secondary particulates from power production. Some other gases with apparently lesser significance (due in part to their relatively weak "global warming potentials") but that also contribute are carbon monoxide and various non-methane hydrocarbons (apart from their contribution to ozone formation).

Table 1 below summarizes the pre-industrial and current levels of atmospheric concentration of four principal GHGs, as well as their total increase and the corresponding level of increased radiative forcing in units of watts per square meter. The increased concentrations of CO_2 have provided the majority of the increases in radiative forcing of these four gases, but the others have also made significant contributions.

TABLE 1. Atmospheric Concentration and Radiative Forcing Increases from Key Greenhouse Gases (1750-2007) (Source: [8])

Gas	Pre-Industrial Level	Current Level	Increase Since 1750	Radiative Forcing (W/m^2)
Carbon dioxide	280 ppm	385ppm	105 ppm	1.66
Methane	700 ppb	1741 ppb	1045 ppb	0.48
Nitrous oxide	270 ppb	321 ppb	51 ppb	0.16
Ozone	25	34 ppb	9 ppb	0.35
CFC-12	0	533 ppt	533 ppt	0.17

3.1. Overview of Previous Research

Research on the GHG emissions from fuel cycles related to EV use dates back to at least the early 1990s, when the introduction of battery EVs by major automakers and a growing concern about climate change spurred interest in comparing the GHG emissions from battery and fuel-cell EVs with the emissions from conventional vehicles. Most studies focused on criteria air pollutants, but some GHGs were occasionally included. Significant research efforts in the 1990s include those by university and government lab research groups [9, 10, 11] and consulting firms [12, 13, 14]. The next decade saw major efforts by automakers [15, 16], industry research organizations [17], and other groups. More recently, major efforts have examined the potential GHG impacts of plug-in hybrid vehicles in a series of efforts that are discussed later in this paper.

This review features the LEM (the Lifecycle Emissions Model) and the GREET (Greenhouse gases, Regulated Emissions, and Energy use in Transportation) model as they are both well developed with long histories, and also are relatively well documented. They therefore are relatively "transparent" tools for analyzing emissions from a wide range of vehicle and fuel combinations. Along with other efforts, these models are described below as being among the key sources of information for comparing emissions of different types of EVs. Other studies have examined more specific vehicle and fuel pathways involving EVs with regard to their GHG emissions, but that still have yielded interesting insights.

4. FORMATION OF GHG EMISSIONS FROM EV FUEL CYCLES

A key feature of GHG emissions from the production of transportation fuels and electricity is that emissions of CO_2 are comparatively easy to estimate: they can be approximated as the carbon content of the fuel multiplied by 3.66 (the ratio of the molecular mass of CO_2 to the molecular mass of carbon), on the assumption that virtually all of the carbon in fuel oxidizes to CO_2.

In contrast, combustion emissions of all the other greenhouse gases are a function of many complex aspects of combustion dynamics (such as temperature, pressure, and air-to-fuel ratio) and of the type of emission control systems used, and hence cannot be derived from one or two basic characteristics of a fuel. Instead, one must use published emission factors for each combination of fuel, end-use technology, combustion conditions, and emission control system. Likewise, non-combustion emissions of greenhouse gases (for example, gas flared at oil fields, or N2O produced and emitted from fertilized soils), cannot be derived from basic fuel properties, and instead must be measured and estimated source-by-source and gas-by-gas. Lipman and Delucchi [18] provide a compendium of many of these emission factors, but we note that some of them have been updated based on more recent data than were available at the time that study was published.

As indicated above, GHG emissions from the lifecycle of fuels for battery EVs and hydrogen fuel cell EVs are entirely in the form of upstream emissions, with no emission from the vehicles themselves (except for water vapor in the case of fuel cell vehicles). As such, the GHG emissions from battery and fuel cell EVs are entirely related to the production of electricity or hydrogen. Emissions from electricity generation processes are generally well known and well studied; this is less true for hydrogen production but in most cases these emissions are well understood as well. Some novel hydrogen production methods, and those that are based on conversion from biofuels, have somewhat complex and certainly not completely understood and established levels of emissions of GHGs.

In contrast, for PHVs, emissions are a complex combination of upstream and in-use emissions. Emissions from these vehicles are more complex than for conventional vehicles or EVs, because these vehicles combine features of internal-combustion engine vehicles with those of EVs. Various vehicle design and operational strategies are available for PHVs, and these can have important emissions implications. For example, PHVs can be designed to be either "charge depleting" or "charge sustaining" and this affects the relative levels of electricity and gasoline used. (See Gonder and Markel [19] and Katrasnik [20] for further discussion of operating strategies for PHVs.)

5. ESTIMATES OF GHG EMISSIONS FROM EV FUEL CYCLES

Various efforts have examined the emissions of GHGs from EV fuel cycles, but looking at different types of vehicles, with one or more fuel feedstock options, and at varying levels of detail. See Appendix A to this paper for a compilation and "typology" of sorts of several of these studies and their scope. Here we briefly describe a few of the major modeling efforts, emphasizing the well-developed LEM and GREET models that are most familiar, and compare their results below.

5.1. LEM – Overview

An extensive effort to assess GHG emissions from motor vehicle fuels and electricity production and use is the LEM project at UC Davis. The LEM uses lifecycle analysis (LCA) to estimate energy use, criteria air-pollutant emissions, and CO2-equivalent greenhouse-gas emissions from a wide range of energy and material lifecycles. It includes lifecycles for passenger transport modes, freight transport modes, electricity, materials, heating and cooling, and more. For transport modes, it represents the lifecycle of fuels, vehicles, materials, and infrastructure. It calculates energy use and lifecycle emissions of all regulated air pollutants plus so-called greenhouse gases. It includes input data for up to 30 countries, for the years 1970 to 2050, and is fully specified for the U.S.

For motor vehicles, the LEM calculates lifecycle emissions for a variety of combinations of end-use fuel (e.g., methanol, hydrogen, electricity, etc.), fuel feedstocks (e.g., petroleum, coal, corn, etc.), and vehicle types (e.g., internal combustion engine vehicle, fuel-cell vehicle, etc.). For light-duty vehicles, the fuel and feedstock combinations included in the LEM are shown in Table 2. The cells with BEVs and FCVs are highlighted (the LEM does not yet include analysis of PHVs).

The LEM estimates emissions of CO_2, CH_4, N_2O, carbon monoxide (CO), total particulate matter (PM), PM less than 10 microns diameter (PM10), PM from dust, hydrogen (H_2), oxides of nitrogen (NOx), chlorofluorocarbons (CFC-12), non-methane organic compounds (NMOCs), (weighted by their ozone-forming potential), hydro-fluorocarbons (HFC-134a), and sulfur dioxide (SO_2) (Table 3). These species are reported individually, and aggregated together weighted by CO_2 equivalency factors (CEFs).

These CEFs are applied in the LEM the same way that global warming potentials (GWPs) are applied in other LCA models, but are conceptually and mathematically different from GWPs. Whereas GWPs are based on simple estimates of years of radiative forcing integrated over a time horizon, the CEFs in the LEM are based on sophisticated estimates of the present value of damages due to climate change. Moreover, whereas all other LCA models apply GWPs to only CH_4 and N_2O, the LEM applies CEFs to all of the pollutants listed above. Thus, the LEM is unique for having original CEFs for a wide range of pollutants.

TABLE 2. Fuel and Feedstock Pathways and Vehicle Types Analyzed in the LEM

Fuel --> ↓ Feedstock	*Gasoline*	*Diesel*	*Methanol*	*Ethanol*	*Methane (CNG, LNG)*	*Propane (LPG)*	*Hydrogen (CH2) (LH2)*	*Electric*
Petroleum	**ICEV, FCV**	**ICEV**				**ICEV**		**BPEV**
Coal	**ICEV**	**ICEV**	**ICEV, FCV**				**FCV**	**BPEV**
Natural gas		**ICEV**	**ICEV, FCV**		**ICEV**	**ICEV**	**ICEV, FCV**	**BPEV**
Wood or grass			**ICEV, FCV**	**ICEV, FCV**	**ICEV**		**FCV**	**BPEV**
Soybeans		**ICEV**						
Corn				**ICEV**				
Solar power							**ICEV, FCV**	**BPEV**
Nuclear power							**ICEV, FCV**	**BPEV**

Note: ICEV = internal-combustion-engine vehicle, FCV = fuel-cell vehicle; BPEV = battery-powered electric vehicle

5.2. LEM – Emission Results for BEVs and FCVs

As noted above, BEVs and FCVs fueled with onboard hydrogen are unique among motor vehicle options in that their emissions are entirely upstream. They are often called "zero emission" vehicles, when in fact this means an absence of tailpipe emissions (which is important from a human health/exposure perspective, to be sure). Most BEV and FCV fuel options do entail significant reductions in GHG and criteria pollutant emissions compared with conventional gasoline vehicles, but this is not always the case (e.g., if coal without carbon capture is the sole feedstock for the electricity for BEV charging).

Table 4 presents the final gram-per-km emission results by vehicle/fuel/feedstock, and percentage changes relative to conventional gasoline vehicles for the U.S. for the years 2010 and 2050. For additional results for Japan, China, and Germany, see Lipman and Delucchi [21].

In the U.S. in the year 2010, BEVs reduce fuel lifecycle GHG emissions by 20% (in the case of coal) to almost 100% (in the case of hydro and other renewable sources of power). If the vehicle lifecycle is included, the reduction is less, in the range of 7% to 70%, because emissions from the battery vehicle lifecycle are larger than emissions from the gasoline internal-combustion engine vehicle (ICEV) lifecycle, on account of the materials in the battery. The emission reduction percentages generally are larger in the year 2050, mainly because of the improved efficiency of vehicles and power plants. Emission reductions in Japan, China, and Germany (again see [21] for details) are similar to those in the U.S., except that in those cases the reduction using coal-power is larger, due the greater efficiency of coal plants in Germany and Japan, and to high SO_2 emissions from coal plants in China (recall that SO_2 has a negative CEF [Table 3]).

TABLE 3. LEM CEFs Versus IPCC GWPs. (Source: IPCC GWPs from the IPCC [30])

Pollutant	LEM CEFs (year 2030)	IPCC 100-yr. GWPs
NMOC-C	3.664	3.664
NMOC-0_3/CH_4	3	not estimated
CH_4	14	23
CO	10	1.6
N_2O	300	296
NO_2	-4	not estimated
SO_2	-50	not estimated
PM (black carbon)	2,770	not estimated
CFC-12	13,000	8,600
HFC-134a	1,400	1,300
PM (organic matter)	-240	not estimated
PM (dust)	-22	not estimated
H_2	42	not estimated
CF_4	41,000	5,700
C_2F_6	92,000	11,900
HF	2000	not estimated

Note: CEF = CO_2-equivalency factor; GWP = global warming potential.

TABLE 4. Gram-per-Kilometer emissions and percentage changes vs. gasoline ICEV (LEM CEFs): U.S. 2010 and 2050 (Source: [21])

Battery EVs - By Type of Power Plant Fuel

Year 2010	Coal	Fuel oil	NG boiler	NG turbine	Nuclear	Biomass	Hydro	Other
Fuel lifecycle (g/km)	266.0	231.9	141.3	143.6	14.6	24.2	10.4	7.7
Fuel lifecycle (% change)	-20.0%	-30.2%	-57.5%	-56.8%	-95.6%	-92.7%	-96.9%	-97.7%
Fuel and vehicle lifecycle* (g/km)	365.9	331.8	241.2	243.5	114.5	124.0	110.3	107.6
Fuel and vehicle lifecycle* (% change)	-6.9%	-15.5%	-38.6%	-38.0%	-70.9%	-68.4%	-71.9%	-72.6%
Year 2050								
Fuel lifecycle (g/km)	227.5	197.2	105.9	107.8	7.8	(-3.2)	5.2	3.0
Fuel lifecycle (% changes)	-18.7%	-29.6%	-62.2%	-61.5%	-97.2%	-101.1%	-98.1%	-98.9%
Fuel and vehicle lifecycle* (g/km)	262.4	232.1	140.8	142.7	42.7	31.7	40.1	37.9
Fuel and vehicle lifecycle* (% change)	-17.1%	-26.7%	-55.5%	-54.9%	-86.5%	-90.0%	-87.3%	-88.0%

Fuel Cell EVs - By Fuel and Feedstock

General fuel -->	Gasoline	Methanol	Methanol	Ethanol	Hydrogen	Hydrogen	Hydrogen	Hydrogen
Fuel specification -->	RFG-Ox10	M100	M100	E100	CH2	CH2	CH2	CH2
Year 2010 Feedstock -->	*Crude oil*	*NG*	*Wood*	*Grass*	*Water*	*NG*	*Wood*	*Coal*
Fuel lifecycle (g/km)	163.9	164.1	47.9	85.4	35.7	135.1	47.8	83.6
Fuel lifecycle (% changes)	-50.7%	-50.7%	-85.6%	-74.3%	-89.3%	-59.4%	-85.6%	-74.8%
Fuel and vehicle lifecycle* (g/km)	223.6	224.1	107.9	145.3	96.6	196.0	108.7	144.6
Fuel and vehicle lifecycle* (% change)	-43.1%	-43.0%	-72.5%	-63.0%	-75.4%	-50.1%	-72.3%	-63.2%
Year 2050								
Fuel lifecycle (g/km)	134.0	122.6	18.3	13.2	27.5	113.3	24.3	61.6
Fuel lifecycle (% changes)	-52.1%	-56.2%	-93.5%	-95.3%	-90.2%	-59.5%	-91.3%	-78.0%
Fuel and vehicle lifecycle* (g/km)	163.7	152.5	48.1	43.0	59.7	145.6	56.6	93.9
Fuel and vehicle lifecycle* (% change)	-48.3%	-51.8%	-84.8%	-86.4%	-81.1%	-54.0%	-82.1%	-70.3%

In the U.S. in 2010, LEM results suggest that FCVs using gasoline or methanol made from natural gas offer roughly 50% reductions in fuel lifecycle GHG emissions. FCVs using methanol or hydrogen made from wood reduce fuel lifecycle GHG emissions by about 85%; FCVs using hydrogen made from water reduce emissions by about 60%, and FCVs using hydrogen made from water (using clean electricity) reduce fuel lifecycle GHG emissions by almost 90%. Again, the reductions are slightly less if the vehicle lifecycle is included, and slightly larger in the year 2050. The patterns in Japan, China, and Germany are essentially the same, because the vehicle technology and the fuel production processes are assumed to be the same as in the U.S.

5.3. GREET Model - Overview

The GREET model has been under development at Argonne National Laboratory (ANL) for about 15 years. The model assesses over 100 fuel production pathways and about 75 different vehicle technology/fuel system types, for hundreds of possible permutations of combinations of vehicles and fuels. It is used by over 10,000 users worldwide and has been adapted for use in various countries around the world [4, 8, 15, 16, 22]. GREET estimates CO_2-equivalent emissions of CO_2, CH_4, and N_2O from the lifecycle of fuels and the lifecycle of vehicles, using the IPCC's GWPs to convert CH_4 and N_2O into CO_2 equivalents.

The latest version of GREET (GREET 1.8c released in 2009) is noteworthy for its much expanded treatment of PHVs along with updated projections of electricity grid mixes in the U.S. based on the latest projections by the Energy Information Administration. This latest version of the model analyzes PHVs running on various fuels along with electricity, not just gasoline and diesel. Additional fuels analyzed for use in PHVs and other vehicle types include corn-based ethanol (E85 – 85% blend with gasoline), biomass-derived ethanol (E85), and hydrogen produced from three different methods: 1) steam methane reforming of natural gas (distributed, small scale); 2) electrolysis of water using grid power (distributed, small scale); and 3) biomass-based hydrogen (larger scale). The analysis also examines different regions of the U.S., and the U.S. on average, for power plant mixes and emission factors for BEV and PHV charging and other electricity demands.

5.4. GREET – GHG Emission Results for EV Types

The GREET model results for BEVs and FCVs are broadly similar to the LEM results discussed above. These results are briefly reviewed here. Recent GREET results for various types of PHVs are shown in Figure 2, below (showing a wide range of potential emissions impacts by PHV fuel and vehicle type), and discussed in Section 5.8 below along with the results of other studies of PHV emissions.

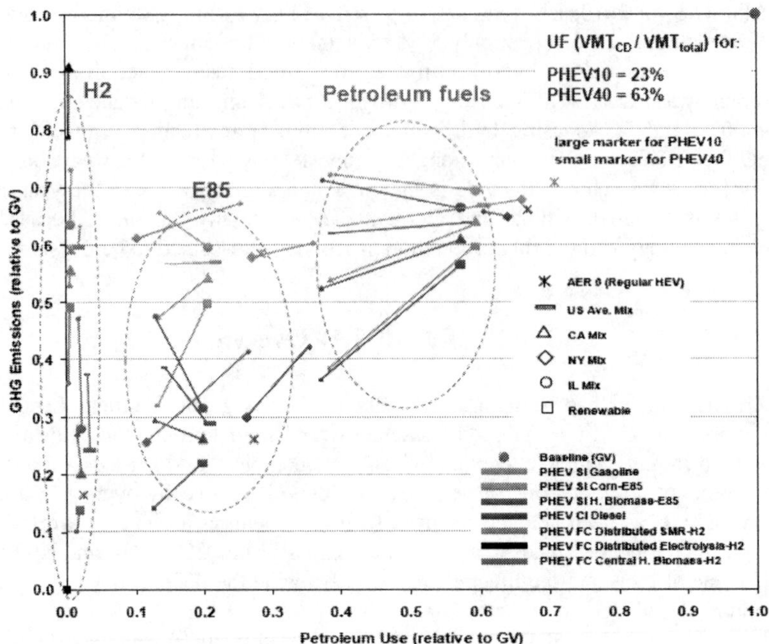

FIGURE 2: Relative GHG emission and petroleum use impacts of PHVs from various fuels Using the GREET Model [23]

GREET results show that emission reductions of about 40% can be expected from BEVs using the average electricity grid mix in the U.S., compared with emissions from conventional vehicles. In comparison, GREET results suggest that BEVs using a California electricity grid mix would produce reductions of about 60%. Meanwhile FCVs using hydrogen derived from natural gas would reduce emissions by just over 50%. FCVs using the average grid mix of U.S. electricity to produce hydrogen through the electrolysis process would result in an *increase* in emissions by about 20%. As shown in LEM as well, BEVs and FCVs using entirely renewable fuels to produce electricity and hydrogen would nearly eliminate GHGs [24].

5.5. Other EV GHG Emission Modeling Efforts

Various other studies of the relative GHG emission benefits of different types of EVs have been done by other university and national laboratory research groups, consulting firms, government agencies, non-government organizations, and industry research groups. Many of these are discussed and compared in the sections that follow.

Key organizations that have been involved in previous efforts include many in the U.S. (some listed below), the Japanese Ministry of Economy, Trade, and Industry,

Japanese research universities including the University of Tokyo, the International Energy Agency, the European Union, Natural Resources Canada, and many other government and research organizations around the world. In the U.S., in addition to the national laboratories and the University of California, key efforts have been led by the Massachusetts Institute of Technology, Carnegie Melon University, Stanford University, the Pacific Northwest Laboratory, the National Renewable Energy Laboratory, General Motors, and the Electric Power Research Institute (EPRI), among others. Once again, several of these efforts are discussed and compared below.

5.6. Comparison of GHG Emissions Estimates for BEVs and FCVs

The study of emissions from battery EVs dates back to the 1990s and even earlier, and several studies were performed on emissions from FCV fuel cycles starting in the late 1990s and through the early 2000s. Early BEV emission studies tended to emphasize criteria pollutants, while later studies have also included CO_2, CH_4 and N_2O. Of course we note that the criteria air pollutants are also GHGs, or go on to help form them through secondary atmospheric chemistry processes (e.g. in the case of ozone and secondary particulates).

With regard to estimates of GHG from BEVs, the emissions of course depend strongly on the mix of powerplants that is in use in a region, and the specific power plant (or alternately average emission factor) that is assumed to be applicable during the assumed period of vehicle charging. Several studies were conducted for California in the 1990s when the introduction of BEVs was being mandated by the state. These included studies by consulting firms conducted for the California Air Resources Board and other efforts such as the "Total Energy Cycle Assessment of Electric and Conventional Vehicles" (or EVTECA) effort in the late 1990s that combined the efforts of three U.S. national laboratories to examine the LCA of BEVs compared with conventional vehicles. These studies generally found significant benefits from BEVs in terms of GHG emission reductions, along with the more mixed results for the criteria pollutants that were the main focus of the studies [25, 26]. However the studies were often limited to CO_2 only as far as the GHGs examined, sometimes along with CH_4 and several air pollutants that were more of concern at the time).

Other studies have been done in the past several years comparing BEVs and FCVs as alternatives to ICEVs, with results based on more "modern" assumptions that are better comparisons to the recent work on emissions from PHVs. One such study by MIT concludes that conventional ICEVs emit about 252 grams of CO_2e per km and that by 2030 this might be reduced to about 156 grams per km. In comparison, 2030 FCVs could emit about 89 grams per km, BEVs could emit 116 grams per km, and a PHV-30 (with a 30 mile/50 km AER) might emit about 86 grams of CO_2e per km. Thus, the study finds that the PHV-30s and FCVs have the largest emission reductions relative to the 2030 ICEV (44% and 42%), followed closely by the BEVs (26%). Hence, all three options (as well as a 2030 advanced conventional hybrid in this analysis) are significantly better than the advanced 2030 ICEV [27].

Another recent such comparison of BEVs and FCVs found that GHG emissions from lithium-ion BEVs were much lower than from either nickel-metal hydride or lead-acid battery based vehicles, ranging from about 235 grams per km for a 100 km range vehicle to about 375 grams per km for a 600 km range vehicle. Meanwhile, FCV emissions are relatively unchanged by driving range, at about 180 grams per km. This assumes the electricity is from the U.S. marginal grid mix and that hydrogen for the FCVs is made from natural gas. Hence this study suggests that FCVs operating on hydrogen from natural gas can have lower GHG emissions than even relatively low-range BEVs in the U.S. [28], a finding that is consistent with most other studies.

A major, ongoing European study makes detailed estimates of lifecycle GHG emissions from alternative-fuel ICEVs, hybrid vehicles, and fuel-cell vehicles [29]. The study uses the Advisor model to simulate vehicle energy use, and a detailed original analysis by LBST (which also worked on the GM/ANL [16] analysis for Europe) to estimate upstream or "well to tank" emissions. The EUCAR study [29] estimated lifecycle emissions for methanol FCVs, using wood, coal, and natural gas as feedstocks, and for compressed-hydrogen vehicles, using wood and natural gas as feedstocks. FCVs using hydrogen made from natural gas had about 55% lower well-to-wheel GHG emissions than a conventional gasoline ICEV, and FCVs using hydrogen made from wood had about 90% lower emissions.

The results of various studies with regard to GHG emissions reductions from BEVs and FCVs, including LEM, GREET, and the studies mentioned above, are compared with several PHV studies in Section 5.8, below. These results should be interpreted with some caution because of the many different assumptions made in the studies, but do provide some sense of the relative emission benefits that EV types can provide.

5.7. Overview of GHG Emissions Estimates for PHVs

PHVs generate GHG emissions from three distinct sources: the lifecycle of fuels used in the ICE, the lifecycle of electricity used to power the electric drivetrain, and the lifecycle of the vehicle and its materials. A number of studies, reviewed below, have estimated GHG-emission reductions from PHVs relative to conventional ICEVs considering the lifecycle of fuels and the lifecycle of electricity generation. Because energy use and emissions for the vehicle lifecycle are an order of magnitude smaller than energy use and emissions for the fuel and electricity lifecycle [30, 31, 32, 33], and because there are relatively few studies of emissions from the PHV vehicle lifecycle, once again we do not consider the vehicle lifecycle in much detail here, but point interested readers to the efforts referenced above for more details.

With regard to GHG emissions from the lifecycle of petroleum fuels used in ICEs for PHVs, these depend mainly of the fuel use of the engine and the energy inputs and emission factors for the production of crude oil and finished petroleum products. A number of studies estimate the fuel use of ICEs in PHVs; for example, Table 2 of Bradly and Frank [34] reports the gasoline consumption reduction (%) and the gasoline consumption of the conventional ICEV (l/100-km) for a variety of simulated and tested PHVs. The PHVs were found to reduce gasoline consumption by 50% to 90%. To estimate the GHG emissions from the petroleum-fuel lifecycle, many studies use the ANL GREET model, discussed elsewhere in this paper.

GHG emissions from the use of electricity by PHVs depend mainly on the energy use of the electric drivetrain, the efficiency of electricity generation, and the mix of fuels used to generate electricity. The energy use of the electric drivetrain in a PHV is a function of the size and technical characteristics of the electric components (battery, motor, and controller), the vehicle driving and charging patterns, and the control strategy that determines when the vehicle is powered by the battery and when it is powered by the ICE. The studies reviewed here and tabulated in Table 5 consider two basic control strategies, discussed earlier in the paper: "all-electric" or "blended." In the all-electric strategy, the PHV is designed to have a significant all-electric range (AER) over a specified drive cycle, and the vehicle runs on the battery until the battery's state of charge (SOC) drops to some threshold value (e.g., 40%), at which point the engine turns on and the vehicle operates as a charge-sustaining hybrid (e.g., the Prius) until the next recharging.

In a PHV with a large AER, the electric drivetrain is sized to have enough power to be able to satisfy all power demands over the drive cycle without any power input from the engine. By contrast, in the blended strategy, the electric drivetrain and the engine work together to supply the power over the drivecycle. The blended strategy can be either "engine dominant," in which case the battery is used to keep the engine running at its most efficient torque/rpm points, or "electric dominant," in which case the engine turns on only when the power demand exceeds the capacity of the electricity drivetrain [19]. (See Katrasnik [20] for a comprehensive formal framework for analyzing energy flows in hybrid electric vehicle systems.)

The efficiency of electricity generation can be estimated straightforwardly on the basis of data and projections in national energy information systems, such as those maintained by the Energy Information Administration (for the U.S.) (www.eia.doe.gov/fuelelectric.html) or the International Energy Agency (for the world) (www.iea.org). Lifecycle models, such as GREET and the LEM, also have comprehensive estimates of GHG emissions from the lifecycle of electricity generation for individual types of fuels.

However, it is not straightforward to estimate the mix of fuels used to generate the electricity that actually will be used to charge batteries in PHVs. The "marginal" generation fuel mix depends on the interaction of supply-side factors, such as cost, availability, and reliability, with anticipated hourly demand patterns, and can vary widely from region to region [35]. This supply-demand interaction can be represented formally with models that attempt to replicate how utilities actually dispatch electricity to meet demand. A few studies, reviewed below, have used dispatch models to estimate the mix of fuels used to generate electricity for charging PHVs. However, as dispatch models generally are not readily available, most researchers either have assumed that the actual marginal mix of fuels is the year-round average mix, or else have reported results for different fuel-mix scenarios.

5.8. Comparison of GHG Emissions Reductions From PHVs

Table 5 presents the results of several key PHV emission studies. The studies summarized in the table indicate that PHVs have 20% to 60% lower GHG emissions than their counterparts ICEV counterparts, with the lower-end reductions corresponding mainly to relatively low-carbon fuel mixes for electricity generation.

To put the grid GHG emission factors of Table 5 into perspective, the LEM estimates that in the U.S. in the year 2020, lifecycle emissions from coal-fired plants are 1030 grams of CO_2e per kWh-generated, and lifecycle emissions from gas-fired plants are 520 grams of CO2e per kWh-generated, using IPCC GWPs. Studies using dispatch modeling of the electricity grid indicate a narrower range of reductions, 30% to 50%. By comparison, studies tabulated by Bradley and Frank [34] indicate slightly greater reductions, about 40% to 60%.

TABLE 5: PROJECTED PHEV GREENHOUSE GAS EMISSIONS IMPACTS

Report	Emissions Estimation[(a)]	CD Range (km)	Control Strategy	Year	Grid GHGs (gCO2e/kWh)[(b)]	PHEV GHGs (gCO2e/km)[(b)]	ICEV GHGs (gCO2e/km)[(b)]	% Reduction (vs. ICEV)
EPRI 2001 [17]	Average	32.2[(c)]	AE	2010	427	144	257	44%
		96	AE	2010	427	112	257	57%
Samaras and Meisterling [36]	Scenario	30	AE	NR	200	126	257	51%
					670	183	269	32%
					950	217	276	21%
		90	AE	NR	200	96	257	63%
					670	183	269	32%
					950	235	276	15%
Kromer and Heywood [27]	Average	48	Blended	2030	769	86.2	156	45%
		96	Blended	2030	769	89.8	156	43%
Silva et al. [38]	Average	~57[(d)]	AE	n.s.[(e)]	543 (U.S.)	~110[(d)]	n.s.	n.e.
					387 (Europe)	~105[(d)]	n.s.	n.e.
					428 (Japan)	~108[(d)]	n.s.	n.e.
Jaramillo et al. [39]	Scenario	60	AE	n.s.[(e)]	883 (coal)	~125 to 220[(d)]	~230[(d)]	~4% to ~46%
PNNL [40]	Simplified dispatch	53	n.s.[(f)]	2002	94% NG/6% C	n.s.	n.s.	40%
					1% NG/99% C	n.s.	n.s.	-1%
					US average	n.s.	n.s.	27%
Stephan and Sullivan [37]	Scenario	63	n.s.[(f)]	current /long term	598 (current NG)	184/119[(g)]	432	57%/72%[(g)]
					954 (current coal)	274/192[(g)]	432	37%/56%[(g)]
					608 (US average)	177	432	59%
Parks et al. [41]	Dispatch	32	Blended	2004	454	154	251	39%
ANL [23]	Dispatch/	32	Blended	2020	US average	146	233	37%

	Scenario				California	140	233	40%
					Illinois	162	233	30%
					Renewable	115	233	51%
EPRI and NRDC [42]	Dispatch/ Scenario	16	AE	2050	97	140	233	40%
					199	143		39%
					412	147		37%
		32.2	AE	2050	97	103	233	56%
					199	109		53%
					412	119		49%

Notes:

CD = charge depleting; GHG = greenhouse gas; CO2e = CO_2 equivalent; ICEV = internal combustion engine vehicle; PHEV = plug-in hybrid electric vehicle; AE = all electric (meaning the vehicle operates solely on the battery until a certain state of charge is reached); blended = vehicle is designed to use both the engine and battery over the drivecycle; n.s. = not specified; n.e. = not estimated; NG = natural gas; C = coal.

(a) "Average" means annual-average emissions from the entire national electric grid. "Scenario" means the study considered different fuel-mix and hence emission scenarios for the electric grid. "Dispatch" means the study estimated marginal fuel mixes and emissions for PHEV charging based on a dispatch model.

(b) GHG emissions and CO_2 equivalency are estimated as follows:

EPRI 2001, Parks et al., Silva et al., and Stephan and Sullivan: CO_2 only.

ANL: 2007 IPCC [43] GWPs for CH_4 and N_2O.

Samaras and Meisterling, EPRI/NRDC, and Jaramillo et al.: 2001 IPCC [30] 100-year GWPs for CH_4 and N_2O.

Kromer and Heywood and PNNL: IPCC 1995 GWPs for CH_4 and N_2O.

Note that Samaras and Meisterling and Jaramillo et al. do not explicitly state which GHGs they include in their CO2e measure; however, they refer to CO2e estimates from the GREET model, which considers CH_4 and N_2O. Similarly, PNNL does not state which CO2e measure they use, but they do state that they use GREET version 1.6, with year 2001 documentation, so we assume that the 1995 IPCC GWPs apply.

(c) Using the "unlimited" case, which allows the maximum amount of electric miles.

(d) We estimated these from a graph: Figures 2 and 4 of Silva et al., and Figure 4 of Jaramillo et al.

(e) Year of analysis not specified, but appears to be roughly current.

(f) PNNL and Stephan and Sullivan estimate emissions from electric operation only; they do not estimate emissions from the ICE in a PHEV.

(g) The number before the slash is the result for "current technology" electricity generation, and the number after the slash is the result for "new technology" electricity generation, in the long term. The new technology is more efficient than the current technology

Some of the results of Table 5 merit further explanation. For example, Kromer and Heywood [27] report a higher grid GHG intensity than several other cases, but lower emissions per kilometer than Samaras and Meisterling [36] and Graham/EPRI [17]. The high grid GHG intensity comes from DOE-EIA projections, and the lower emissions per kilometer is likely due to the assumed improvement in efficiency and emissions in the 2030 ICEV. The relatively large reductions estimated by Stephan and Sullivan [37] are due to several factors: 1) they start with a relatively high-emitting gasoline vehicle; 2) they consider electric operation of the PHV only; 3) they assume relatively efficient power plants in the long term; and 4) they consider only CO_2 emissions.

In sum, PHVs promise significant reductions in GHG emissions in most regions and conditions. This is especially the case in the longer term, when the electricity grid is likely to be cleaner, and vehicles are likely to have higher battery storage capacities.

5.9. Comparison of GHG Emissions Reductions From EV Types

By way of an overall comparison of the emission reductions estimated for the various types of EVs, see Figures 3 and 4 below. As shown in Figure 3, BEVs have the potential to reduce well-to-wheels GHG emissions by about 55-60% using either natural gas power plants or the California grid mix (which is heavily dependent on natural gas). Using coal-based power, BEVs may reduce emissions by about 20% or slightly increase them (i.e., model results vary somewhat), and using the U.S. grid mix (which is about half coal-based) emission reductions on the order of 25-40% appear possible. For FCVs using hydrogen produced from natural gas steam reformation, GHG emissions can be reduced by 30-55% according to the various studies. Once again, when entirely or nearly-entirely powered by completely renewable fuels such as wind, solar, and hydropower, GHG emissions from both BEVs and FCVs can be almost entirely eliminated.

As shown in Figure 4, emission reductions possible from PHVs are somewhat more modest than for some BEV and FCV configurations. For a PHV type considered in several studies that has a 30-mile/50-km electric range, GHG emission reductions compared with a conventional vehicle are estimated to be in the range of 30-60% using the U.S. grid mix. For the California electricity mix, a range of 40-55% has been estimated. Also, one estimate shown in the figure estimates a 50% reduction potential with PHV30s running on renewables based electricity. We note that for PHVs in particular, these relative emission reduction results vary by assumed driving patterns and distances as well as underlying emission factors for electricity and gasoline used. This leads to further sources of potential variation amongst the studies, along with other variables such as the assumed driveline efficiencies, upstream emission factors, and the type and size of the vehicle itself.

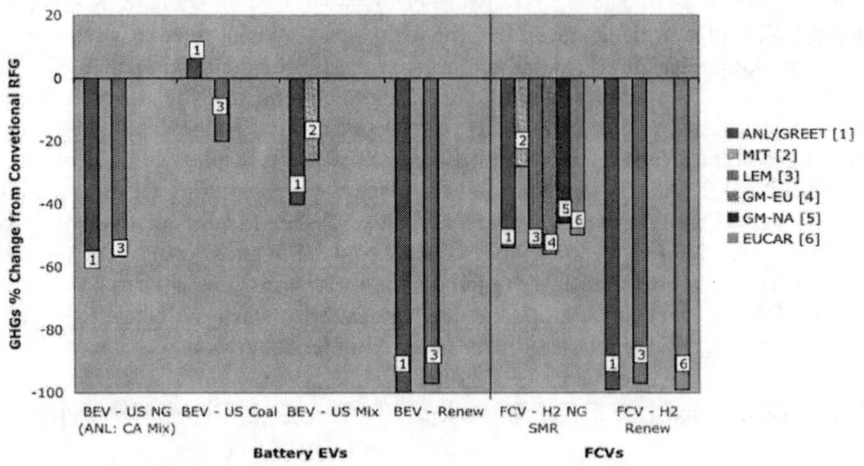

Figure 3: Comparison of Well-to-Wheels GHG Emission Reduction Estimates From BEVs and FCVs (*Sources: 1) ANL/GREET [22, 24]; 2) MIT [27]; 3) LEM [Table 4]; 4) GM-EU [16]; 5) GM-NA [15]; 6) EUCAR [29]*)

Notes: BEV = battery electric vehicle; CA = California; FCV = fuel cell vehicle; H2 = hydrogen; NG = natural gas; renew = renewable fuel; SMR = steam methane reforming; US = United States

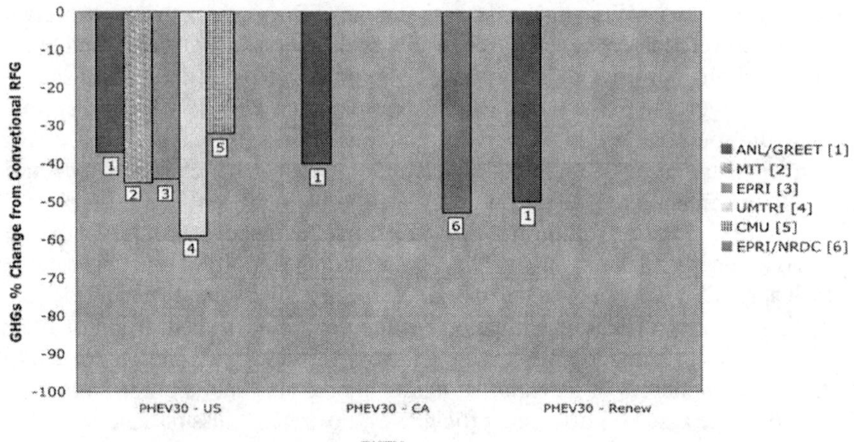

FIGURE 4. Comparison of Well-to-Wheels GHG Emission Reduction Estimates From PHVs (*Sources: 1) ANL/GREET [22, 44]; 2) MIT [27]; 3) EPRI [17]; 4) EPRI/NRDC [42]; 5) UMTRI [37]; 6) CMU [36]*)

Notes: CA = California; PHV30 = plug in hybrid vehicle with 30 mi/50 km electric range; renew = renewable fuel; US = United States

6. MAGNITUDE OF POSSIBLE GHG REDUCTIONS: SCALING UP THE EV INDUSTRY

As shown above, EVs can offer significant GHG benefits when compared on a one-to-one basis with conventional vehicles. However, a major issue with a rapid scale-up of EVs is the availability of advanced electric vehicle battery packs in the numbers needed for a major commercial launch of vehicles by several automakers at once. The need to scale-up battery production in the cell sizes and configurations needed for different types of EVs is accompanied by several other needs to support the introduction of EVs into consumer households. These include: 1) improving the procedures for installing recharging facilities for EVs at household and other sites; 2) better understanding of the utility grid impacts of significant numbers of grid-connected vehicles; 3) better understanding the consumer and utility economics of EV ownership (and/or leasing of car or battery); and 4) better education for consumers and tools to assist them to determine if their driving habits would be good "fits" for the characteristics of the different types of EVs. These and other related issues are being explored by the University of California and other groups as new EVs are being introduced into the market [45, 46]

Additional issues related to vehicle scale-up include provision of hydrogen for FCVs, currently an expensive proposition for low volumes of dispensed fuel, development and dissemination of appropriate safety procedures for first responders in dealing with accidents with vehicles with high voltage electrical systems and/or hydrogen fuel storage, and additional education and outreach programs for mechanics and fleet managers [47]. These measures will be needed to help EVs become more established and acceptable to consumers in various market segments.

6.1. Scaling Up the EV Industry – How Fast Can it Be Done?

A recent analysis examined the potential of various options to scale up to become a "Gigaton solution" by 2020, meaning that it could account for a gigaton of CO_2 reduction on a global annual basis by 2020 [48]. The study found that achieving "Gigaton Scale" with a strategy based largely on a massive introduction of grid-connected EVs would require about 1,000 times as many batteries in the near term as are expected to be available (i.e., tens of millions globally rather than tens of thousands), growing to needs for hundreds of millions of battery packs by 2020. This implies a massive investment in battery production capacity at a time when battery designs are still being improved and perfected to the point where commercially acceptable PHVs and BEVs can be produced. This suggests that achieving Gigaton

Scale with EVs is not possible by 2020. However, much larger gains are possible by 2030 and especially 2050, given the relatively slow dynamics of motor vehicle fleet stock turnover [49].

The EPRI/NRDC study noted above includes scenario estimates of future GHG reductions from vehicles fleets in the U.S., and finds that reductions of up to about 500 megatons per year are possible by 2050, depending on the level of PHV fleet penetration and the CO_2 intensity of the electricity sector.

It is important to note, however, that more generally PHVs and other EVs are technologies that can scale fairly rapidly. Typical automotive volumes run to several hundred thousand units per year for individual popular models (e.g. the combined U.S. and Japanese sales of the Toyota Prius are around 275,000 to 300,000 per year), and there is the potential to incorporate electric drive technology into many vehicle models. The rate of scaling is mainly limited by the growth of supplier networks and supply chains, and by the dynamics of introducing new vehicles with 15-year lives into regional motor vehicle fleets, along with economic and market response constraints on the demand side.

7. KEY UNCERTAINTIES AND AREAS FOR FURTHER RESEARCH

The analysis of lifecycle CO2e GHG emissions from advanced electric vehicles involves many uncertainties. Some of these are relatively clear (e.g., what exactly are the driveline efficiencies of various types of alternative fuel vehicles, what efficiencies are involved in key upstream fuel production processes, and so on) and many others are subtler but potentially of significance. These include secondary impacts of fuel cycles, such as the "indirect land use change" impacts of biofuels, where production of biofuels implies cultivation of land that in some cases can displace its use for other purposes, and how emissions from power plants and other combustion sources actually result in exposures and potential harm to humans and the environment.

Exploring these uncertainties in much detail is beyond the scope of this paper, but is discussed in some of the studies referenced here. However below we briefly mention a few key sources of remaining uncertainty in LCA of EV fuel cycles. We also note that the GREET model in particular now includes the ability to include estimates of the levels of uncertainty in key input variables, and incorporates this capability through a graphical user interface (GUI) version of the model interface that runs in a PC Windows environment. This can be useful but of course still can benefit from additional efforts to characterize and attempt to narrow the remaining uncertainties themselves.

7.1 Key Uncertainties in LCA Analysis of GHGs from EV Fuel Cycles

Because GHGs are produced in myriad ways from EV fuel cycles, including both upstream and vehicle-based emissions (in the case of PHVs and HEVs), and because EV technologies are still evolving, there are considerable uncertainties involved in present-day or prospective analysis of their impacts. Over the course of the past twenty years many of these uncertainties have been narrowed – for example the manufacturing cost and performance of electric vehicle motors and motor controllers has become better established – but many still remain.

Some of the key remaining uncertainties include:

- uncertainties in the emission rates of high GWP value gases (e.g., N_2O, CH_4, refrigerants, etc.) that are emitted in lower quantities than CO_2 from vehicle fuel cycles, but that can still be significant;
- secondary impacts such as indirect land use change, macroeconomics, etc.;
- climate impacts of emissions of typically overlooked but potentially important pollutants such as oxides of sulfur, ozone precursors, and particulate matter;
- rate of future vehicle and fueling system performance improvements;
- potential "wild cards" in future fuels production processes, such as the successful introduction of carbon capture and sequestration; and
- breakthroughs in electricity, advanced biofuel, or hydrogen production.

As time goes on, we can expect more to be learned about these key areas, and for the remaining uncertainties to be narrowed. At the same time, new fuel cycles based on evolving technology are likely to become available but with potentially significant uncertainties until more is learned about them in turn (e.g., diesel type fuels from algae, new types of PHVs running on various fuels, other new types of synthetic Fischer-Tropsch process and bio-based fuels, etc.). The significant amount of research currently underway is encouraging, but given the pressing nature of the energy and climate challenges facing many nations, one could argue that more attention should be paid to this critical area.

8. CONCLUSIONS

EVs of various types are experiencing a commercial renaissance but of uncertain ultimate success. Many new EVs are being introduced by different automakers with significant technical improvements from earlier models, particularly with regard to further refinement of drivetrain systems and important improvements in battery and fuel cell systems.

The various types of hybrid and all-electric EVs can offer significant GHG reductions when compared to conventional vehicles on a full fuel-cycle basis. In fact, most EVs used under most condition are expected to significantly reduce lifecycle CO2e GHG emissions. Under certain conditions, EVs can even have very low to zero emissions of GHGs when based on renewable fuels. However at present this is more expensive than other options that offer significant reductions at lower costs based on the use of more conventional fuels. It is important to note that when coal is heavily used to produce electricity or hydrogen (e.g. with the U.S. grid mix that is about half composed of coal), GHG emissions tend to increase significantly compared with conventional fuel alternatives. Unless carbon-capture and sequestration becomes a reality, using coal-based fuels even in conjunction with electric drive systems offers little or no benefit.

In general, various studies show that BEVs reduce GHGs by a widely disparate amount depending on the type of powerplant used and the particular region involved, among other factors. Reductions typical of the U.S. would be on the order of 20-50%, depending on the relative level of coal versus natural gas and renewables in the powerplant feedstock mix. However, much deeper reductions of over 90% are possible for BEVs running on renewable or nuclear power sources. PHVs running on gasoline reduce emissions by 20-60%, and fuel cell EV reduce GHGs by 30-50% when running on natural gas-derived hydrogen and up to 95% or more when the hydrogen is made (and potentially compressed) using renewable feedstocks. These are all in comparison to what is usually assumed to be a more advanced gasoline vehicle "baseline" of comparison, with some incremental improvements by 2020 or 2030. It is important to note once again, however, that emissions from all of these EV types are highly variable depending on the details of how the electric fuel or hydrogen is produced. This is true despite the fact that GHG and air pollutant emissions are typically mostly or entirely "upstream" rather than from the vehicle's tailpipe. This makes these emissions in principle easier to control but also may mean that they are far removed from where the fuel is actually used in the vehicle.

Overall, EVs offer the potential for significant and even dramatic reductions in GHGs from transportation fuel cycles. Pursuing further development of this promising set of more efficient technologies would thus seem to be of paramount importance, given the rapidly spiraling growth in motor vehicle ownership and use around the globe and the declining natural resource base remaining to support it.

ACKNOWLEDGEMENTS

The author would like to thank Dr. Mark Delucchi of UC Davis for assistance with a previous paper assessing likely levels of electric vehicle greenhouse emissions upon which this paper is partly based. The author would also like to thank David Hafemeister, emeritus Professor of Physics at Cal Poly University, for his assistance with the March 5-6 2011 American Physical Society "Second Conference on the Physics of Sustainable Energy" event at UC Berkeley, which prompted the development of this paper.

REFERENCES

1. Shaheen, SA, Lipman, TE, *Internat'l Association of Traffic and Safety Sciences* 31 (2007) 6-20.
2. U.S. Department of Energy, *The Department of Energy Hydrogen and Fuel Cells Program Plan* (Draft) (2010).
3. Delucchi, MA, *A Lifecycle Emissions Model (LEM): Lifecycle Emissions From Transportation Fuels, Motor Vehicles, Transportation Modes, Electricity Use, Heating And Cooking Fuels, And Materials*, UCD-ITS-RR-03-17 (2003).
4. Wang, M, *Journal of Power Sources* 112 (2002) 307-321.
5. Weiss, RF, Craig, SE, *Geophys Research Letters* 3 (1976) 751-753.
6. Prigent M and de Soete, GD, SAE Tech Paper Series #8904922 (1989).
7. Forster, PM, Ramaswamy, V, Artaxo, P, Berntsen, T, Betts, R, Fahey, DW, Haywood, J, Lean, J, Lowe, DC, Myhre, G, Nganga, J, Prinn, R, Raga, G, Van Dorland, R (2007) In: Solomon, S, Qin, D, Manning, M, Chen, Z, Marquis, M, Avery, KB, Tignor, M, and Miller, HL (Eds) *Climate Change 2007: The Physical Science Basis*, Cambridge University Press.
8. Oak Ridge National Laboratory, Carbon Dioxide Information Analysis Center, http://cdiac.ornl.gov (2009).
9. DeLuchi, MA, *Emissions of Greenhouse Gases from the Use of Transportation Fuels and Electricity*, ANL/ESD/TM-22, Argonne National Laboratory (1991).
10. Wang, MQ, *GREET 1.0 – Transportation Fuel Cycles Model: Methodology and Use*, ANL/ESD-33, Argonne National Laboratory (1996).
11. Delucchi, MA, Emissions of Non-CO2 Greenhouse Gases From the Production and Use of Transportation Fuels and Electricity, UCD-ITS-RR-97-05 (1997).
12. Acurex Environmental Corporation, *Evaluation of Fuel-Cycle Emissions on a Reactivity Basis: Volume 1 - Main Report*, FR-96-114 (1996).
13. Unnasch, S, Greenhouse Gas Analysis for Fuel Cell Vehicles, 2006 Fuel Cell Seminar (2006).
14. Thomas, CE, James, BD, Lomax, F, *Journal of Power Sources* 25 (2000) 551-567.
15. General Motors and Argonne National Lab, *Well-to-Wheel Energy Use and Greenhouse Gas Emissions of Advanced Fuel/Vehicle Systems* (2001).
16. General Motors et al., *Well-to-Wheel Analysis of Energy use and Greenhouse Gas Emissions of Advanced Fuel/Vehicle Systems – A European Study*, (2002).
17. Graham R, *Comparing the Benefits and Impacts of Hybrid Electric Vehicle Options,* Electric Power Research Institute Report 1000349 (2001).
18. Lipman, TE and Delucchi, MA, *Climatic Change* 53 (2002) 477-516.
19. Gonder, J and Markel, T, SAE Technical Paper Series, #2007-01-0290, Society of Automotive Engineers (2007).
20. Katrasnik, T, *Energy Conversion and Management* 50: 1924-1938 (2009).
21. Lipman, TE and Delucchi, MA, "Expected Greenhouse Gas Emission Reductions by Battery, Fuel Cell, and Plug-In Hybrid Electric Vehicles" In G. Pistoia (Ed.) *Battery, Hybrid, and Fuel Cell Vehicles*, Elsevier Press, ISBN: 978-0-444-53565-8 (2010).
22. U.S. Environmental Protection Agency, EPA420-F-05-001 (2005).
23. Elgowainy, A, Burnham, A, Wang, M, Molburg, J, Rousseau, R, *Well-to-Wheels Energy Use and Greenhouse Gas Emissions Analysis of Plug-In Hybrid Electric Vehicles*, ANL/ESD/09-2, Argonne National Laboratory (2009).

24. Wang, M, "Well to Wheels Energy Use Greenhouse Gas Emissions and Criteria Pollutant Emissions – Hybrid Electric and Fuel Cell Vehicles," *SAE Future Transp. Tech. Conference* (2003)
25. Wang, M, Wu, Y, Elgowainy, A, *GREET1.7 Fuel-Cycle Model for Transportation Fuels and Vehicle Technologies*, Argonne National Laboratory (2007).
26. Rau, NS, Adelman, ST, Kline, DM, *EVTECA - Utility Analysis Volume 1: Utility Dispatch and Emissions Simulations*, TP-462-7899, National Renewable Energy Laboratory (1996).
27. Kromer, MA, and Heywood, JB, *Electric Powertrains: Opportunities and Challenges in the U.S. Light-Duty Vehicle Fleet*, MIT Pub. LFEE 2007-03 RP (2007).U.S. Environmental Protection Agency, Compilation of Air Pollutant Emission Factors (AP-42) (2009).
28. Thomas, CE, *International Journal of Hydrogen Energy* 34 (2009) 6005-6020.
29. EUCAR and ECJRC, *Well-To-Wheels Analysis of Future Automotive Fuels and Powertrains in the European Context* (2007).
30. Keoleian, GA, Lewis, GM, Coulon, RB, Camobreco, VJ, Teulon, HP, SAE Technical Paper Series, #982169, Society of Automotive Engineers (1998).
31. Sullivan, JL, Williams, RL, Yester, S, Cobas-Flores, E, Chubbs, ST, Hentges, SG, Pomper, SD, "Life Cycle Inventory of a Generic U.S. Family Sedan: Overview of Results USCAR AMP Project," SAE Technical Paper Series #982160 (1998).
32. Delucchi, MA, *World Resources Review* 13 (1) 25-51 (2001).
33. Burnham, A, Wang, M, Wu, Y, *Development and Applications of GREET 2.7 – The Transportation Vehicle-Cycle Model*, ANL/ESD/06-5, Argonne National Laboratory (2006).
34. Bradly, TH, and Frank, AA, *Renewable and Sustainable Energy Reviews* 13: 115-128 (2009).
35. Edison Electric Institute (EEI), "Diversity Map," http://www.eei.org, (2009).
36. Samaras, C. and Meisterling, K, *Environmental Science & Technology* 42: 3170-3176 (2008).
37. Stephan, CH, Sullivan, J, *Environmental Science and Technology* 42: 1185-1190 (2008).
38. Silva, C, Ross, M, Farias, T, *Energy Conversion and Management* 50 (2009) 1635-1643.
39. Jaramillo, P, Samaras, C, Wakeley, H, Meisterling, K, *Energy Policy* 37 (2009) 2689-2695.Bohm, MC, Herzog, HJ, Parsons, JE, Sekar, RC, *Int'l Journal of Greenhouse Gas Control* 1 (2007) 113–120.
40. Kintner-Meyer, M, Schneider, K, Pratt, R, *Impact Assessment of Plug-in Hybrid Vehicles on Electric Utilities and Regional U. S. Power Grids*, Pacific Northwest National Laboratory (2006).
41. Parks, K, Denholm, P, Markal, T, *Costs and Emissions Associated with Plug-In Hybrid Electric Vehicle Charging in the Xcel Energy Colorado Service Territory*, NREL/TP-640-41410, National Renewable Energy Laboratory (2007).Dones, R, Heck T, Hirschberg S, In *Encyclopedia of Energy* (2004) Vol 3. 77-95.
42. Electric Policy Research Institute and National Resources Defense Council (2007) *Environmental Assessment of Plug-In Hybrid Electric Vehicles, Volume 1: Nationwide Greenhouse Gas Emissions*, Rpt. No. 1015325.
43. Intergovernmental Panel on Climate Change, *Climate Change 2007: The Physical Science Basis*, Contribution of Working Group I to the Fourth Assessment Report of the IPCC, ed. by S. Solomon, D. Qin, et al., Cambridge University Press (2007).
44. Behrentz, E, Ling, R, Rieger, P, Winer, AM, *Atmospheric Environment* 38 (26) (2004) 4291-4303.
45. Martin, E, Shaheen, SA, Lipman, TE, Lidicker, J, *International Journal of Hydrogen Energy* 34 (20): 8670-8680 (2009).
46. Kurani, KS, Turrentine, TS, Sperling, D, *Transportation Research Part D* 1 (2): 13 –150 (1996).
47. California Environmental Protection Agency, *Hydrogen Blueprint Plan: Volume 1* (2005).
48. Paul, S and Tompkins, C, (Eds.) *Gigaton Throwdown: Redefining What's Possible for Clean Energy by 2020* (2009).
49. Lipman, T, Plug-In Hybrid Electric Vehicles, In: Paul, S and Tompkins, C, (Eds.) *Gigaton Throwdown: Redefining What's Possible for Clean Energy by 2020* (2009).

SESSION D: ENHANCED EFFICIENCY OF BUILDINGS

Exploring the Limits of Energy Efficiency in Office Buildings

David E. Claridge and Oleksandr Tanskyi

Energy Systems Laboratory and
Department of Mechanical Engineering
Texas A&M University
402 Harvey Mitchell Parkway South
College Station, TX 77843-3581

Abstract. This paper explores the limiting energy efficiency for the energy uses in a particular office building. This limit might be viewed as the "Carnot efficiency" for the entire building. It assumes that all energy-consuming services in the building are provided at the minimum energy value that does not violate physical law. Rigorous physical limits such as the Carnot efficiency are used where applicable; in other cases, a plausible approximation has been adopted. Based on the assumptions made, it would be possible to provide all of the energy-based services currently provided in the building using approximately 0.2 kWh/(ft^2-yr). This limiting value is less than 1% of the energy used by a typical office building in the United States. Examination of expected advances in individual technologies suggests that it may be possible to construct a building that uses approximately 0.8 kWh/(ft^2-yr) within the next decade.

First and second law efficiency limits are very familiar for the heating and cooling devices used in buildings. However, the authors have not seen them used to describe the overall energy efficiency of buildings. This is doubtless due to the fact that different buildings have different energy-using devices and require a range of services that require energy input. The energy efficiency of buildings is sometimes treated by considering buildings that are 15% or 30% "above code". For example, ASHRAE has published a series of six design guides that "are designed to provide recommendations for achieving 30% energy savings over the minimum code requirements of ANSI/ASHRAE/IESNA Standard 90.1-1999." [1]

Just as it is not possible to build an air conditioner that operates at the Carnot COP, it will not be possible to build a building that operates at the thermodynamic limit of efficiency. But the assumption of adiabatic surfaces, reversible thermodynamics, and Carnot devices are idealizations that are very useful in the study thermodynamics; they are also valuable in helping determine the potential for improvements in heating and cooling devices. Likewise, a similar examination of an entire building should be instructive in developing an understanding of the minimum energy that might be used by a building that provides a particular set of energy-using services. Comparison of this minimum with the current performance of even the most efficient buildings can provide a framework for considering future targets for energy consumption of buildings.

Physics of Sustainable Energy II: Using Energy Efficiently and Producing it Renewably
AIP Conf. Proc. 1401, 301-312 (2011); doi: 10.1063/1.3653859

The limit to the energy efficiency of office buildings is explored in this paper by considering a very specific building. The Energy Systems Laboratory (ESL) and the Texas Center for Applied Technology recently occupied the Valley Park II Office Building (VP II) in College Station, Texas. VP II is a leased 25,774 square foot building located just off the Texas A&M University campus. It is a simple rectangular single-story building approximately 100-feet wide and 250 feet long as shown in Figure 1.

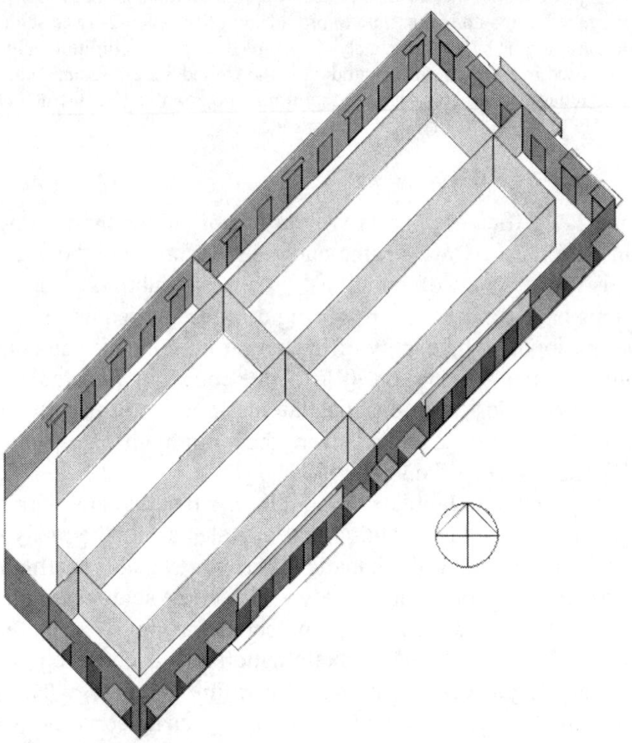

FIGURE 1. Front view and schematic floor plan of the Valley Park II office building.

The VP II office building requires most of the energy-using services found in a typical office building. It must be cooled and heated to provide comfort for the occupants. Ventilation air is required to maintain the indoor air quality needed for healthy working conditions. Lighting is required to enable the occupants to work and move about the building. There are numerous computers and printers throughout the building to meet the needs of the engineers, programmers and administrative staff in the building. There are two large copiers in the building and water fountains that provide cooled water. There is a lunch room where the staff can heat lunch in a microwave oven or make a cup of coffee and hot water for hand washing in the restrooms.

MINIMUM ENERGY REQUIRED TO PROVIDE VP II BUILDING ENERGY SERVICES

The energy efficiency limit for this building is explored by assuming that relevant comfort, lighting and air quality codes are met and that other services are provided at the level currently used as shown in Table 1. Building conditions will be maintained at 73 °F with a maximum relative humidity of 50% which falls approximately in the middle of the ASHRAE comfort chart [2]. Outside air will be provided at a rate of 5 cfm per occupant plus 0.06 cfm/ft^2 of building area in accordance with ASHRAE Standard 62-2010 [3]. The Illuminating Engineering Society of North America (IESNA) recommends lighting levels of 20-50 fc for offices, depending on the age of the occupants [4]. The occupants broadly range in age from approximately 20 to approaching 70 years of age, so 35 fc will be used, assuming that the needs of each person are met.

There are not standards for the use of computers, monitors, copiers, printers, water coolers, cooking, and hot water, so these uses will be based on the best information available regarding current usage. There is slightly more than one computer per person throughout the building, but this will be rounded down to one and since most occupants have two monitors, this will be rounded up to two 23-inch monitors for each person. There are two large copiers in the building, but the number is relatively unimportant for this exploration. The copier count for the past year indicates that approximately 2,000 copies were made for each student and staff member in the ESL offices, the copiers will be assumed to make 2,000 copies/(person-year). There is approximately one printer per person and lacking hard data, it is assumed that printers make the same number of copies as the copiers, or 2,000 pages/(person- year) as well. There is no hard data on water-cooling, so It is conservatively assumed that one quart of water is cooled from 70°F to 50°F for each person daily. For cooking it is assumed that all food/drink heating is equivalent to one cup of water/(person-day) heated from 70°F to 212°F. Hot water in the restrooms is assumed to be ½ gallon/(person-day) heated from 70°F to 105 °F. The first people typically arrive for work about 7:30 am and the building is generally empty by 7:00 pm on weekdays. There is a small amount of weekend occupancy. It is assumed the building is occupied from 7 am to 7 pm, five days per week and unoccupied on the weekends, for 60 occupied hours per week as a close approximation to the current building use.

TABLE 1. VP II Building Energy-Using Services.

Service	Level Provided	Notes
Comfort	73°F/50% RH	ASHRAE Standard 55
Ventilation	5 cfm/person + 0.06 cfm/ft^2	ASHRAE Standard 62-2010
Lighting	35 fc when occupied	IESNA Standard
Computers	One/person	Typical
Computer Monitors	Two 23-in/person	Typical
Copiers	Two	2000 pages/(person-year)
Printers	One/person	2000 pages/year
Water Coolers	1 qt/(person-day)	Cooled from 70°F to 50°F
Cooking	1 cup water/(person-day)	Heated from 70°F to 212°F
Hot Water	½ gal/(person-day)	Heated from 70°F to 105°F

The energy required to provide heating and cooling of a building strongly depends on the heat introduced into the building by the energy-using devices in the building, so the minimum energy required to provide comfort and ventilation will be discussed after the subsequent items listed in Table 1.

Lighting. It is assumed that the building will use the minimum lighting energy if it is equipped with ideal light systems that distribute the light flux equally all over the office zone area from an ideal light source that generates lighting energy only in the visible spectrum.

According to The Illuminating Engineering Society of North America [5], illuminance requirements for an office area are 20-50 foot-candles (200-500 lux) largely depending on the age of the occupants. As noted above, 35 foot-candles is assumed as the average illuminance requirement in the building. There is no straightforward way to recalculate the illuminance requirements in lux ($\mathrm{lm/m^2}$) (photometric unit) corresponding to irradiance (radiometric unit) based on physical power. In radiometric units all wavelengths are weighted equally, while photometric units take into account the fact that the human eye's visual system is more sensitive to some wavelengths than others. To balance this difference, every wavelength is given a different weight according to the luminosity function [6]. Based on the luminous efficacy of radiation and assumption that the light source is an ideal "white" light source from a black body with temperature 5800 K and truncated to 400–700 nm, an ideal lighting system efficiency limit is 251 lm/W [3].

This value is three times higher than the 80 lm/W sunlight efficiency (for Class G star with surface emission temperature 5800 K) or two times higher than the 124 lm/W spiral tube fluorescent lamp with electronic ballast [7,8].

This luminous efficacy value is estimated according to the equation [9]

$$F(\lambda_1,\lambda_2)=683\int_{\lambda_1}^{\lambda_2}\Phi_\lambda(\lambda)\cdot V(\lambda)\cdot d\lambda\,. \tag{1}$$

where: 683 – constant from the definition of the candela;

$\lambda_1 = 400nm$; $\lambda_2 = 700nm$ - borders of visible spectrum range;

$\Phi_\lambda(\lambda)$ - spectral emissive power (Planck distribution), in the band between λ_1 and λ_2;

$V(\lambda)$ - photopic luminous function.

Based on a luminous emittance of 35 foot-candles (350 lux) energy provided by the lighting system to the office space is 1.4 W/m^2 or 0.13 W/ft^2. Energy that is consumed by the lighting system will be transferred to the conditioned area and become part of internal heat gain. It is further assumed that the lighting control system will turn lights off when the occupants are not present such that the average light will be on 6 hours per day during the occupied period so the average lighting power is 1,676 W if no daylighting is used. If daylighting is used, it will simply be assumed to meet half the lighting requirements, reducing the average occupied lighting power to 838 W. During unoccupied hours, average lighting power of 0.01 W/ft^2 or 250 W is assumed. It will be assumed that daylighting is used in the building.

Computers. There are physical limits to the power required for computation but considering the computation power of a simple iPhone, it is assumed that this is adequate computation power for the average ESL worker and that is only 2.5W for 1 GHz processor. Most engineers will require more computational power than that on occasion, but it is a plausible value for the average power that is needed during the 30 hours per week that the average worker is assumed to be physically present in his/her office. For the 128 VP II occupants, that corresponds to an average computer power of 160 W during the 12 hours daily that the building is occupied.

Computer Monitors. For computer monitors, it is assumed that the physical limit is the light power emitted by the monitors. Assuming a monitor area of 0.14 m^2 for a monitor with 23 in diagonal, a light flux of 250 cd/m^2 [10] and an ideal "white" light source with an energy efficiency of 251 lm/W, the energy consumption of an ideal monitor is

$$P_{monitor} = 250\frac{cd}{m^2}\cdot\frac{4\pi\cdot lm}{cd}\cdot\frac{W}{251lm}\cdot 0.14\text{m}^2 = 1.752W. \quad (2)$$

With two monitors per person that are assumed to be on for 6 hours per day, the 256 monitors then consume 224 W average in the building during the 12 occupied hours.

Printers and Copiers. The physical limit on the energy required for printers is not obvious, so the authors simply took an old ink jet printer and measured the power that it consumed while it printed one page. It required 0.07 W-hours to print one page. This is sufficiently small that if it is assumed that the standby power can be eliminated in an ideal printer, the printer power is no longer a major item in the building. Printing the assumed two thousand pages per person per year then requires an average of 7 W to power printers when the building is occupied. The 2000 copies per person per year are assumed to require another 7 W average during occupied hours for the copiers in the building.

Heated and Cooled Water and Cooking. If drinking water is cooled from 70°F to 50°F, a Carnot refrigerator will have a COP of 28.3. This will require 4.3 W on average when the building is occupied to cool one quart of water/(person-day). Heating water from 70 °F to 212 °F, an ideal Carnot heat pump will have a COP of only 4.66, so 53 W is required on average when the building is occupied to heat one cup of water or food equivalent per person each day. For heating water in the restrooms, it is assumed that ½ gallon/(person-day) is heated from 70°F to 105°F using a Carnot heat pump with a COP of 15.65. This requires an average of 31 W when the building is occupied.

FIGURE 2. This IPad4 consumes only 2.5 W of electricity.

Heating and Cooling Requirements. To determine the requirements for cooling and heating of the building, the following sources of heat gains and losses must be considered:

- Internal Gains
- Occupants
- Solar
- Ventilation
- Envelope

Internal Gains

The internal gains are due to the electricity used in the building from the lights, computers, monitors, printers, copiers, water cooling, cooking and water heating that have just been discussed.

TABLE 2. Average occupied and unoccupied - hour internal gains.

Gain Source	Occupied-Hour (W)	Unoccupied-Hour (W)
Lights	838	240
Computers	160	0
Monitors	224	0
Printers/Copiers	14	0
Water Coolers	4.3	0
Food Heating	53	0
Restroom Hot Water	31	0
Total	1,325	240

Table 2 shows that the internal gains total 1325 W on average while the building is occupied and 240 W when the building is unoccupied.

Occupant, Solar, Ventilation and Envelope Gains/Losses

The occupants are assumed to be moderately active office workers. According to the American Society of Heating Refrigeration and Air Conditioning Engineers, the typical heat gain for these workers is 250 Btu/hr (~73 W) per person of sensible gain and 200 Btu/hr (~59 W) per person of latent gain [2]. It is assumed that combining the travel schedule of numerous staff members in the building and the time in excess of 40 hours per week worked by many that the average staff member is in the building for 40 hours per week. This results in average sensible occupant gains of 6,252 W on average and 5,002 W of latent gain during occupied hours.

The theoretical limit for solar gains is zero. However, it is assumed that 838 W of solar gain are used on average for day-lighting during occupied hours.

ASHRAE Standard 62-2010 [3] requires 2,190 cubic feet per minute of ventilation air when the building is occupied. The outdoor air used for ventilation needs to have the temperature increased or decreased and in many cases, the humidity lowered to meet the interior comfort requirements. However, in keeping with looking for the theoretical minimum amount of energy required to achieve this condition, there is no law of nature that makes perfect heat exchangers or perfect moisture exchangers impossible. Hence, if it is assumed that such devices are used, and that there is no air leakage into or out of the building, then the only energy requirement to provide ventilation at the assumed indoor conditions will be the power required to operate the ventilation fans. It would be possible in principle to use devices driven by the indoor/outdoor temperature difference to provide the ventilation air. It will be assumed that a perfect fan is used that raises the pressure of the ventilation air by 5 Pa. This requires an average of 5.1 W of fan power to provide the 2,190 cfm of ventilation air needed when the building is occupied.

An adiabatic envelope is possible in theory, so the envelope is assumed to be adiabatic and envelope gains and losses are then zero. This results in the building heat gains shown in Table 3 for our ideal office building.

TABLE 3. Heat gains for the ideal VP II Building.

Gains	Occupied (W)	Unoccupied (W)	Annual Total (kWh_{th})
Internal	1,325	240	5,357
Occupant – Sensible	6,252	0	18,757
Occupant – Latent	5,002	0	15,006
Solar	838	0	2,515
Ventilation Load	5.1	0	15
Envelope Load	0	0	0
Totals	13,413	240	41,650

Building Cooling and Heating Energy

For cooling and heating, it is assumed that outside air is used to provide cooling when the conditions permit using a perfect heat exchanger. A Carnot chiller operating between the wet bulb temperature and the room temperature is used for cooling when economizer cooling is not possible and any heat needed is provided with a Carnot heat pump. Table 3 shows that the total heat that must be removed from the building is 41,650 kWh of thermal energy over the course of a year. The use of a perfect water-side economizer that is able to remove heat at the wet bulb temperature, and with Houston, Texas weather will remove 32,787 kWh of this cooling load. Then the Carnot chiller will have to remove only 8,864 kWh of heat. With an average COP of 222, it requires only 40 kWh of electricity.

The heating load in this building is zero. The envelope has no conduction heat loss, so the only remaining source of a heating load would be from the ventilation air, and since this load is met with a perfect heat exchanger, there is no heating load, so a heat pump is not needed and heating electricity for this building is zero.

These energy flows are shown in Figure 3. The figure shows that while there are still substantial heating and cooling loads of 24,900 kWh_{th} heating and total cooling loads of 41,635 + 15 + 39,270 = 80,920 kWh_{th}, the heating load is entirely handled by the air-to-air heat exchanger and the almost 90% of the cooling load (all but 8,864 kWh_{th}) is handled by the combination of the air-to-air enthalpy recovery device and the water-side economizer. And the Carnot chiller operating between the room temperature and the ambient wet-bulb temperatures requires only 40 kWh. This shows it is theoretically possible for this 25,774 ft^2 office building to provide all of the services described using only 5,357 kWh inside the building, 15 kWh of fan power, and 40 kWh of chiller power, for a total of 5,462 kWh!

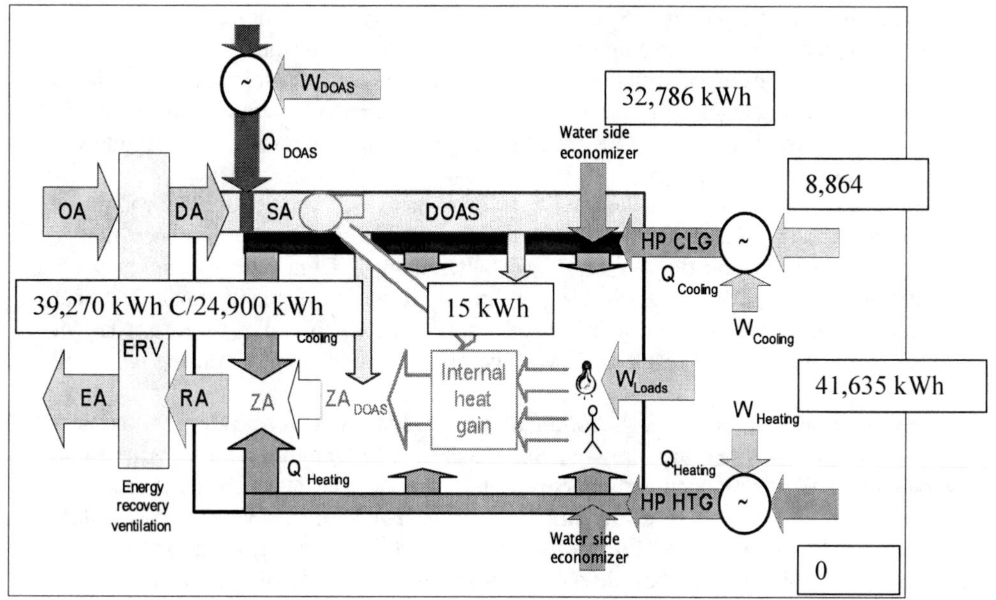

FIGURE 3. Energy flows for the "ideal" building case considered in the text.

Comparison with U.S. Office Buildings

U.S. office buildings use 24kWh/(ft^2-yr) of energy on site [11]. A very good new office building uses something like 3.5 kWh/(ft^2-yr) of energy on site [12]. Yet, what we have said here is that it is theoretically possible to power this building with 5,462 kWh, or 0.21 kWh/(ft^2-yr). There are approximately 40 ft^2/capita of office space in the United States using approximately 2.9×10^{11} kWh of site energy (thermal plus electric). If the world had 40 ft^2 of office space per capita that used only 0.21 kWh/(ft^2-yr), world-wide office consumption for 6.9 billion people would be 5.8×10^{10} kWh for office space, or only 20% of the energy used today in U.S. offices. The 200 ft^2 of gross space (office plus hallways, rest rooms, conference rooms, etc.) occupied by the average occupant of the VP II Building could generate as much energy as it uses each year (or be Net Zero) with an array of photovoltaic cells the size of two ordinary sheets of copy paper!

What Might Be Possible in the Next 10 Years?

Let's consider the state-of-the-art in the individual technologies today and make a few projections.

Lighting. If there is proto-type fluorescent lighting reportedly yielding between 120 and 125 lumens/W and the authors have seen reports of laboratory LED lighting achieving 200 lumens per watt, but have been unable to verify these reports or learn the spectral distribution. If we assume that it will be possible to achieve a complete lighting system that yields 125 lumens per watt level, including fixtures, etc., within

10 years, then the target lighting consumption for the office building would be just 1,676 W when occupied on average and 480 W when it was vacant.

Computers. The assumption made for computers was made based on today's technology. The target will be the same and implicitly assume that increased capability is provided at the 2.5 W/computer consumption level. The target will remain at 160 W for the building.

Monitors. The authors do not have a firm basis for projecting the room for improvement here. Based on the fact that lighting already has an efficiency that is nearly half the limit and that chillers have better than half the Carnot efficiency for production of chilled water at design conditions, it seems plausible to project that monitors should be able to achieve one third the theoretical limit as Target consumption. This would result in an average of 672 W for monitors during occupied hours.

Printers and Copiers. The "theoretical limit" assumptions for printers and copiers were based on levels that are already achieved except for standby losses, so the same average of 14 W during occupied hours is assumed as the Target consumption.

Water Coolers, Food Heating and Restroom Hot Water. While large chillers get within a factor of two of Carnot efficiency, it will be assumed that small water coolers get within a factor of three of Carnot efficiency, resulting in an average consumption for water cooling of 13 W. The same Target will be used for food heating and restroom hot water heating efficiency, resulting in an average consumption of 159 W for food heating and 93 W for hot water.

TABLE 4. "Target" electric use for internal equipment.

Gain Source	Theory (W)	Targets (W)
Lights	838/240	1,676/480
Computers	160	160
Monitors	224	672
Printers/Copiers	14	14
Water Coolers	4.3	13
Food Heating	53	159
Restroom Hot Water	31	93
Total	1,325/240	2,787/480

The total Target consumption for these functions is now slightly more than two times the theoretical minimum at 2,787 W, or 2.1 times the theoretical minimum.

Ventilation Energy Use Target

2,190 cfm of ventilation air is required by ASHRAE Standard 62 when the building is occupied. At least two membrane-based technologies are being developed that show promise of very effective enthalpy recovery [13].The Target ventilation load for the VP II building is based on use of 90% effective enthalpy recovery devices with insignificant air leakage in the building. This would result in 3,927 kWh_{th} of cooling load and 2,490 kWh_{th} of heating load from the ventilation air. It is assumed that fan pressurization of 0.25 inWG (62 Pa) is used compared with the 5 Pa assumed for the ideal case with a fan efficiency of 80% and with a motor efficiency of 95%. That results in a requirement for 135 watts of fan power when the building is occupied.

Target Envelope Characteristics

For the target envelope characteristic, advanced window technologies that been demonstrated will be assumed with a window U-value of 0.1 Btu/(hr-ft^2-°F) (1.42 W/m^2-K). U-values for the walls and roof of 0.033 Btu/(hr-ft^2-°F) and 0.02 Btu/(hr-ft^2-°F), respectively, will be assumed that correspond to about 6-inches (15 cm) of polystyrene insulation for walls and 10-inches (25 cm) for the roof.

Heating and Cooling Equipment

For cooling and heating, the water-side economizer will still be used when conditions permit, but now the water-side economizer will meet the cooling loads only when the wet bulb temperature is 58°F or below. An average cooling COP of 10 and an average COP of 5 for heating are assumed. These assumptions result in a total cooling load of 59,921 kWh of which the economizer only meets 5,984 kWh, so the chiller provide 49,936 kWh. The average chiller COP of 10 results in 4,994 kWh of electricity use for cooling vs the 40 kWh in the theoretical limit. The heating load that must be met by the heat pump is now 19,345 kWh, requiring 3,869 kWh of electricity.

Target Energy Use

The energy use in the target building is shown in Table 5. The total is now 20,517 kWh or 0.8 kWh/(ft^2-yr). This is almost four times the theoretical limit, but is still 30 times smaller than a typical U.S. office building, and less than one-fourth that of a very good building today.

TABLE 5. Energy use in "Target" building.

Use	Target Consumption (kWh/year)
Internal Uses	11,249
Fans	406
Cooling	4,994
Heating	3,869
Total	20,517

CONCLUSIONS

This exploration of the limits of energy efficiency for the energy uses in a particular office building finds that the minimum energy use to provide all of the energy-based services currently provided in the building would require approximately 0.2 kWh/(ft^2-yr) which is less than 1% of the energy used by typical office buildings in the United States. This would permit the building to generate as much energy as it consumes on an annual basis if it had a photovoltaic array corresponding to approximately 1.5 ft^2 of cells for each occupant in the building. It may be possible to construct a building that uses approximately 0.8 kWh/(ft^2-yr) within the next decade.

REFERENCES

1. ASHRAE, "Advanced Energy Design Guide CD," American Society of Heating, Refrigerating and Air-conditioning Engineers, Atlanta, GA, 2010.
2. ASHRAE. *ASHRAE Handbook: Fundamentals.* Atlanta, GA: American Society of Heating, Refrigerating, and Air-Conditioning Engineers, Inc., 2009.
3. ASHRAE, *ANSI/ASHRAE/IESNA Standard 62.1-2010: Ventilation for Acceptable Indoor Air Quality.* Atlanta, GA: American Society of Heating, Refrigerating, and Air-Conditioning Engineers, Inc., 2010.
4. IESNA, *The IESNA Lighting Handbook: Reference and Application.* Ninth Edition. New York: Illuminating Engineering Society of North America, 2000.
5. K. Chen, *Energy Management in Illuminating Systems.* New York: CRC Press, 1999, 158 pp.
6. W.R. McCluney, *Introduction to Radiometry and Photometry.* Boston: Artech House, 1994, 404pp.
7. Panasonic, Panasonic Spiral Fluorescent Ceiling Lamp EcoNavi Used for Residential Lighting. (In Japanese) http://panasonic.co.jp/corp/news/official.data/data.dir/jn100609-1/jn100609-1.html. Accessed: 08/16/2010.
8. EIA, *Annual Energy Review.* Energy Information Administration, U.S. Department of Energy, http://www.eia.doe.gov/emeu/aer/pdf/aer.pdf. Accessed: 5/10/2010.
9. A.V. Arecchi A.V., T. Messadi, and R.J. Koshel. *Field Guide to Illumination.* SPIE Field Guides. Volume FG II. Bellingham, WA: SPIE Press, 2007, 139 pp.
10. Samsung, "23" Touch of Color HDTV Enabled LCD Monitor," http://www.samsung.com/us/computer/monitors/LS23ELNKF/ZA-features. Accessed: 08/20/2010.
11. EIA, "Annual Energy Outlook 2010," DOE/EIA-0383(2010), U.S. Energy Information Agency, 2010..
12. S. Plesser and M.N. Fisch, "The New House of the Region of Hannover: Energy Efficiency in a Public Private Partnership," Proceedings 2007 International Conference for Enhanced Building Operation, San Francisco, 2007, http://repository.tamu.edu/bitstream/handle/1969.1/6241/ESL-IC-07-11-40.pdf?sequence=1.
13. ARPA-E Program, Presentations by Dais Analytics and Advance Materials at U.S. DOE ARPA-E BEETIT Kick-Off Meeting, December 6-7, 2010, Arlington, VA.

Energy Simulation Tools for Buildings: An Overview

Philip Haves

Environmental Energy Technologies Division
Lawrence Berkeley National Laboratory
Berkeley, CA 94720

Abstract. Energy simulation is used in building design and has the potential to be used to support building operation. It is also used in product development and in scenario analysis to inform policy development. The wide variety of phenomena in buildings that need to be modeled in order to simulate building performance are discussed briefly, along with the variety of numerical methods that are used. There is a need to improve the computational speed of building simulation, particularly for use in optimization and parametric studies. Examples of the use of simulation in support of both building design and building operation are presented. The needs for further development of both tools and practices in order to improve the accuracy, robustness and effectiveness of simulation modeling are summarized.

INTRODUCTION

Energy use in buildings is responsible for ~40% of the carbon dioxide emissions in the developed world and ~33% globally [1]. In addition, the US spends ~$1 trillion/yr on the construction, renovation and operation of buildings. In order to reduce this level of consumption, the US Department of Energy (DOE) and its precursors, together with industry and universities, have been working since 1970 to develop accurate and reliable computer simulations that can enhance energy-efficient designs before construction and enhance operation of the buildings once they are completed. In this chapter, we will discuss a range of topics that relate to energy simulation for buildings:

- History of building simulation
- Simulation program architecture
- Modeling and numerical methods
- Numerical methods and computational challenges
- Validation

Physics of Sustainable Energy II: Using Energy Efficiently and Producing it Renewably
AIP Conf. Proc. 1401, 313-327 (2011); doi: 10.1063/1.3653860

- Example applications:
    - San Francisco Federal Building: natural ventilation system design
    - Naval Station Great Lakes: real-time simulation

Uses of simulation include analysis of the expected energy performance of design alternatives, demonstrating conformance with buildings energy codes, such as California's Title 24 [2], energy performance standards, such as ASHRAE Standard 90.1 [3] and 'green' building rating systems, such as LEED [4]. In each case, simulation is used to compare the design to a reference design. The prediction of the actual energy consumption of the building is more challenging because of uncertainties in how it will be constructed, occupied and operated. Simulation is starting to be used to monitor and analyze actual building performance and to detect and diagnose faults by comparing expected and measured performance. In addition, simulation is used to inform the development of new products for buildings, such as high performance glazings and energy-efficient air conditioning equipment, by defining cost and performance targets. It is also used in the development of energy policy, where it is used for scenario analyses involving models of different prototypical buildings that together are representative of the building stock.

A BRIEF HISTORY OF BUILDING SIMULATION

Significant simulation programs for buildings were not developed until after the oil embargo of 1973-74, which raised national awareness of U.S. energy shortfalls and the associated potential for economic disruption. The first programs were limited to a single thermal zone and calculated room temperature and humidity and heating and cooling loads [5, 6]. In the late 1970's and early 1980's, these modeling capabilities were extended to the treatment of multiple thermal zones and the capacity and efficiency of heating, ventilating and air conditioning (HVAC) equipment under different operating conditions [7, 8, 9, 10]. There was a largely parallel evolution of modular programs for system simulation, first for active solar systems [11] and then for HVAC control systems [12]. By the 1990's, PC software technology had evolved to the point where graphical user interfaces could be developed for simulation programs [13, 14], making them easier and quicker to use. The mid 1990's saw the start of the development of DOE's current flagship simulation program, EnergyPlus [15], beginning with the merging of two existing government-sponsored programs [7, 8], which had complementary strengths and weaknesses. The mid 1990's also saw the development of a building simulation program in China [16] and the start of the development of the Modelica system simulation language and associated tools [17], which first found applications in the aeronautical and automotive engineering fields but is now starting to be used for in building systems.

SIMULATION PROGRAM ARCHITECTURE

Simulating the energy performance of buildings involves the following steps, as illustrated in Figure 1:

- Collect and input information on the building design; both the fabric (walls, windows *etc.*) and the systems, e.g. HVAC
- Identify appropriate weather data for the location(s) of interest – either 'typical' weather data, selected using predefined statistical criteria, or data for a particular period of interest, e.g. to enable comparison of simulation results with measured performance
- Determine the schedules that define the operation of the building, e.g. thermostat settings *vs.* time of day, week and year.
- Set up a heat flow network model of the building for solution using finite difference methods or for pre-calculating transfer functions
- As a preliminary step in the simulation phase, explicitly or implicitly initialize the state variables, such as the temperatures of elements of the building fabric that have significant thermal capacity. This is typically done by repeating the first day of the simulation until a quasi-static periodic solution is obtained.
- As another preliminary step, run simulations of selected 'design days' to calculate the output capacity of the HVAC components required to meet the control set-points in order to achieve thermal comfort during the occupied periods.
- Run the initialized simulation for the desired period – often a year consisting of typical months for the desired location.
- Examine the time series outputs for expected behavior under different operating conditions to provide some qualitative and semi-quantitative confirmation that the simulation model has been set up correctly.
- Examine the predefined reports that summarize performance metrics of interest, e.g. energy consumption, thermal comfort, peak electricity demand, carbon dioxide emissions and water consumption.

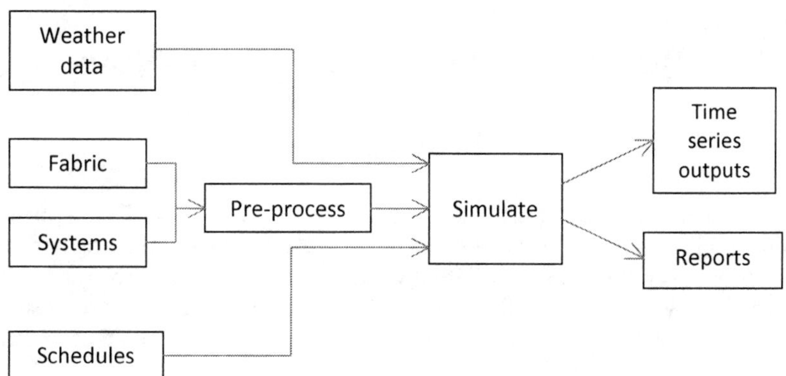

FIGURE 1. Block diagram representing the high-level flow of information in building simulation.

DATA MODEL

As shown in Figure 2, a multi-zone building simulation program has a hierarchical internal data model that defines how the description of the building is stored and accessed during the simulation. The relationships and inheritance between the objects are pre-defined but the number of instances of each class of object depends on the design of the particular building being simulated. For example, a building includes walls and windows (Building Elements) that have geometrical form and, typically, multiple material layers, e.g. siding + insulation + sheetrock (Material Assemblies). Similar data models are found in other computer tools used in building design, e.g. cost estimating tools and Computer-Aided Design (CAD) tools. Efficient and robust exchange of building description data between such tools requires either a common data model [18] or a well-defined, consistent mapping between the data models in the different tools.

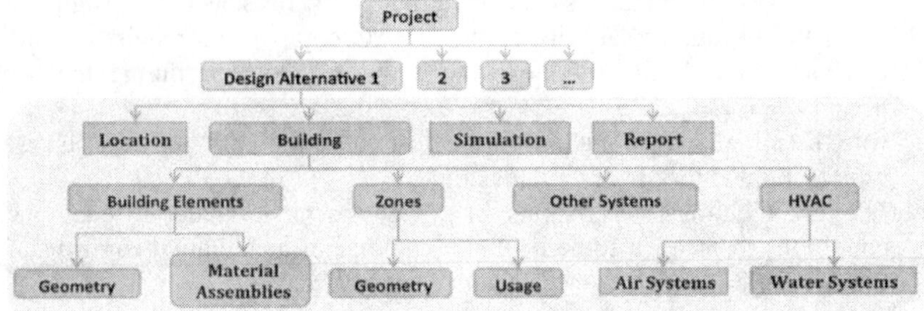

FIGURE 2. Block diagram showing a partial generic data model for building simulation. The shaded area represents one instance of a building. A user interface for a simulation program will typically manage multiple instances, representing different design alternatives.

MODELING AND NUMERICAL METHODS

Modeling

Building energy simulation programs model a wide variety of physical phenomena. The time scales vary from decades (effect of climate change on building performance), year (seasonal variations), days (diurnal variations), minutes (control system response, occupant behavior, short-term weather fluctuations), to effectively instantaneous (optical performance). Phenomena in the building envelope and occupied spaces include:

- Heat and mass transfer – radiation, convection, conduction, absorption
- Air flow – mechanical and natural ventilation
- Optical behavior of glazing systems and enclosures
- Solar radiation:
    - Solar position, shading, atmospheric turbidity

- Sky radiation: sky dome, reflections from ground and other buildings
- Pollutant transport and fate – gas, particles
- Indoor environmental quality – air pollution, odor, health, glare, noise, lighting
- Occupant behavior – operation of thermostats, windows, blinds
- Occupant comfort – thermal stress, draft risk, adaptive behavior

Phenomena in HVAC and other systems include:

- Heat and mass transfer – heating and cooling systems:
  - refrigeration cycles – vapor compression, absorption
  - heat exchangers – sensible heat transfer, condensation, evaporation
- Fluid flow networks – turbulent and transitional flow:
  - duct and piping systems
  - natural ventilation and infiltration
- Control systems:
  - single input single output cascaded control - local loop and supervisory control levels
  - optimization-based control, including model predictive control

Different levels of detail are used in:

- Scenario modeling – sets of representative buildings as proxies for the whole building stock
- Whole building modeling:
  - dynamic heating and cooling load calculations, including thermal storage in the building fabric
  - quasi-static models of HVAC systems and components – energy storage neglected (except for thermal storage tanks)
  - simple zoning and generic systems in the early stages of design
  - room-level zoning, specific equipment in the later stages of design
- Room airflow:
  - inter-zonal networks: each zone homogeneous in terms of temperature, humidity, contaminant concentrations
  - Computational Fluid Mechanics (CFD): velocity, temperature and contaminant fields in single spaces or groups of adjacent spaces.
- HVAC:
  - annual simulation: predefined flow rates, quasi-static component models based on manufacturers' data,
  - control system design, e.g. demand response: dynamic component models, explicit modeling of local loops.

Numerical Methods

Building simulation programs employ a variety of numerical methods. A common approximation is to assume that the surface of each element (wall, window, floor *etc.*) is isothermal. Cases where this approximation is unacceptable are dealt with by subdividing the surface. Different phenomena, and the associated numerical methods, include:

- Thermal diffusion in the building fabric (walls, slabs *etc.*) – typically 1-D; 2-D or 3-D for window frames, foundations, ground, modeled with either:
    - transfer functions – fast but cannot represent non-linearities (e.g. phase change materials)
    - finite difference – explicit or implicit
- Long wave radiation exchange:
    - dependence on temperature may be modeled exactly or, more often, is linearized: $(T_i^4 - T_j^4) \approx 4T_{ave}^3(T_i - T_j)$ – temperature differences are typically small relative to the absolute temperatures
    - use view factors between surfaces (more accurate) or exchange between each surface and a fictitious area-weighted mean radiant temperature.
- Surface convection is also often linearized:

  $$q_i = h_c(\Theta, v, L, \Delta T)\,(T_i - T_{air})$$

  where h_c, the convective heat flux at surface i, is a function of the tilt of the wall, Θ, the velocity of the room air away from the boundary layer, v, the length scale of the surface, L, and the temperature difference $\Delta T = (T_i - T_{air})$. Since h_c typically depends on a fractional power of ΔT, the previous value of ΔT can be used, enabling the linearization.
- Room temperatures and heating and cooling loads: simultaneously solve heat balances on each surface and on the room air:
    - 7(+) simultaneous equations per room: room air + 6(+) surfaces (floor, ceiling and at least four walls)
    - iterate or linearize to obtain solution directly
    - solve for air and surface temperatures if the room is floating or unconditioned
    - solve for heating/cooling loads and surface temperatures if the room air temperature is controlled
- Air flow:
    - inter-zonal networks - two port non-linear elements connected at nodes that implicitly enforce mass balances
    - CFD: Navier Stokes equations + turbulence models:
        - large eddy models for airflow around buildings
        - simpler turbulence models for simple flows in interior spaces,
- Daylight distribution in rooms:
    - simple method: use daylight factors (ratio of interior to exterior illuminance), pre-calculated for simple geometries
    - detailed method: ray tracing – forward (source to scene) or backward (scene to source)
- HVAC system simulation:
    - component models: non-linear differential and algebraic equations
    - solver matches inputs and outputs → well-posed problem
    - problem reduction methods → small set of iteration variables

There are a number of computational challenges for simulation programs:

- Slow execution is a problem for:
  - CFD
  - ray tracing
  - parametric studies
  - optimization
- Both low speed and poor robustness are problems for system simulation, which typically involves large sets of non-linear differential and algebraic equations
- Parallel computing can use either multi-core processors, graphical processing units (GPU's) or supercomputers:
  - some problems are 'embarrassingly parallel': e.g. parametric studies, some optimization methods
  - multi-threading can help with: ray tracing, CFD, radiant exchange … (hand crafting of multi-threaded code is often required)
- Visualization of building performance – fast, interactive display of large data sets

VALIDATION

There are two aspects of energy simulation program validation:

1. Validation of model algorithms under controlled conditions
2. Validation of the ability of simulation programs to predict actual energy performance in the context of conventional design, construction and operation.

There are four complementary methods of model validation:

1. **Analytical tests:** very simple test cases where an analytical solution can be determined and compared with the simulation results, e.g. 1-D heat diffusion in a uniform material with a sinusoidal excitation.
2. **Code comparison:** comparison of the code of various programs with the aim of identifying algorithm differences by inspection.
3. **Comparison of results:** different programs are used to model and simulate the same building under the same conditions and the results from each program are compared with each other.
4. **Empirical validation:** one or more programs are used to model and simulate a particular building that has been well instrumented and carefully monitored. The simulated performance and the measured performance are then compared.

Empirical validation is time-consuming and expensive and requires skilled, experienced personnel and a high level of attention to detail. As a result, there are relatively few high quality empirical validation data sets. Most of the existing empirical validation data sets are for simple buildings; these data sets are useful in various ways but do not address the range of interactions between systems that occur in complex commercial buildings.

A significant amount of effort has been applied to the *comparison of results* approach. Figure 3 shows a sample comparison of the results from eleven different simulation programs. Similar differences are seen between the different programs for the test cases shown and similar trends are found in other comparisons [19].

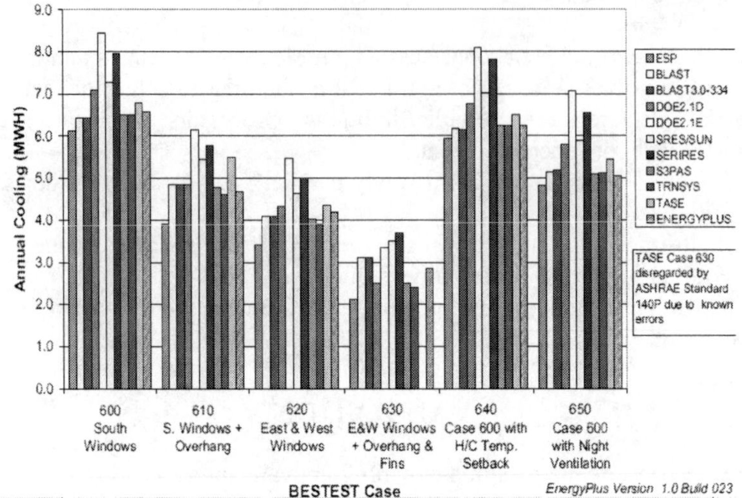

FIGURE 3. Cross validation of various simulation codes for six building configurations

The validation of the ability of simulation programs to predict actual energy performance in the context of conventional design, construction and operation is a more open-ended process and has become a source of controversy in the building energy efficiency community. Figure 4 is a plot, taken from [20], that shows how the results of energy simulations performed in support of submissions for a LEED [4] rating compare with the corresponding utility bills for the second year of operation. A critique of this study is presented in [21].

A number of factors can give rise to differences between design simulation results and measured performance, such as those shown in Figure 4:

- Design simulations do not model real conditions but use the standard inputs required by building codes and/or rating schemes, including:
  - occupancy schedules
  - plug loads
  - weather
- Real buildings often do not perform as expected by their designers:
  - faulty construction
  - malfunctioning equipment
  - incorrectly configured control systems
  - inappropriate operating procedures

- Models always contain idealizations, simplifications and approximations, and may contain errors:
  - approximations and errors in model algorithms introduced by the developers
  - user simplifications of the building design to limit the complexity of the simulation
  - user mistakes in generating the input data for the simulation from the (simplified) building design

There is a need for standard protocols for use in simulating building performance for different applications. There is a related need to standardize the characterization of input uncertainties, their propagation in simulation and their representation and interpretation in simulation results. Some of these uncertainties can be characterized by continuous distribution functions, e.g. variations in thermophysical properties of building materials due to variations in composition. Other, more complex variations may be better represented by different scenarios, e.g. the effect of different levels of maintenance standards on HVAC system efficiency.

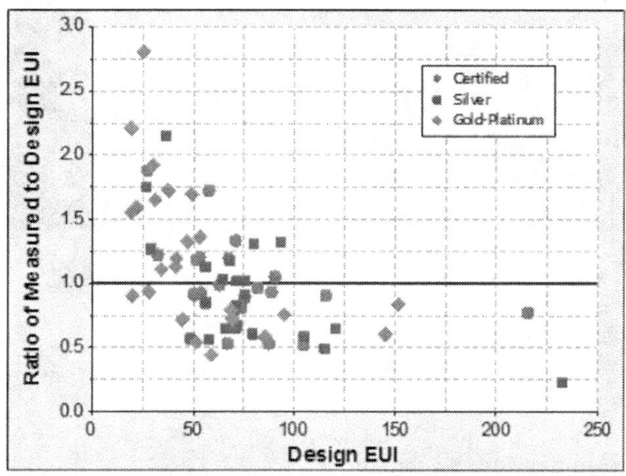

FIGURE 4. Ratio of **measured to design energy use intensity (EUI) for a sample of new, LEED-certified commercial buildings.**

EXAMPLE APPLICATIONS

Design of the Natural Ventilation System for the San Francisco Federal Building

The new San Francisco Federal Office Building, shown in Figure 5, includes an eighteen-storey tower in which the open plan office areas are naturally ventilated. The natural ventilation is driven by the prevailing wind from the west north-west. Summertime temperatures in San Francisco mostly range between 55 °F and 70°F; however, for a few days of a typical year, the maximum temperature rises to ~90°F. One question for the design team was: 'how much exposed thermal capacity ("thermal mass") is required to maintain a comfortable interior environment during these hot spells?' Other design questions related to the natural ventilation system included:

- Is buoyancy needed to supplement wind?
- If so, are external chimneys needed to supplement internal buoyancy?

FIGURE 5. Computer-generated image of the new San Francisco Federal Office Building. The main façade faces south-east.

A coupled thermal and airflow simulation [22] was used to predict the performance of different design options:

- wind-driven cross-flow ventilation
- internal stack with no wind
- internal plus external stack with no wind
- internal stack with wind
- internal and external stack with wind

In each case, cooling is performed by ventilating the building during the day when the outside temperature is low enough. When the daytime outside temperature is too high, the windows are kept closed during the day and opened at night to allow the cooler night air to cool the exposed thermal mass, which then absorbs heat the next day.

Table 1 shows the number of hours when the listed base temperature was exceeded during the occupied hours during the cooling season for the different design options. The main conclusions are:

1. Wind-driven night ventilation produces reasonable comfort conditions during the day for all but a few days of a typical year.
2. Internal stack-driven night ventilation resulting from low-level openings on the NW and high-level openings on the SE is less effective than wind-driven ventilation, resulting in internal temperatures on hotter days that are ~1°F higher than for the wind-driven case.
3. A combination of wind-driven and internal stack-driven ventilation produces a modest improvement in performance compared to the wind only case. The contribution of internal stack may be more significant if/where there is significant reduction is wind pressure due to shielding by adjacent buildings.
4. Addition of external 'chimneys,' formed by adding a second glazing layer to the south-east façade and partitioning it horizontally and vertically, does not improve the performance of the combination of wind-driven and internal stack-driven ventilation, and may be slightly counter-productive, due to the increased flow resistance caused by the chimney. In the absence of wind, addition of external chimneys helps the internal stack somewhat.

TABLE 1. Occupied hours above various base temperatures

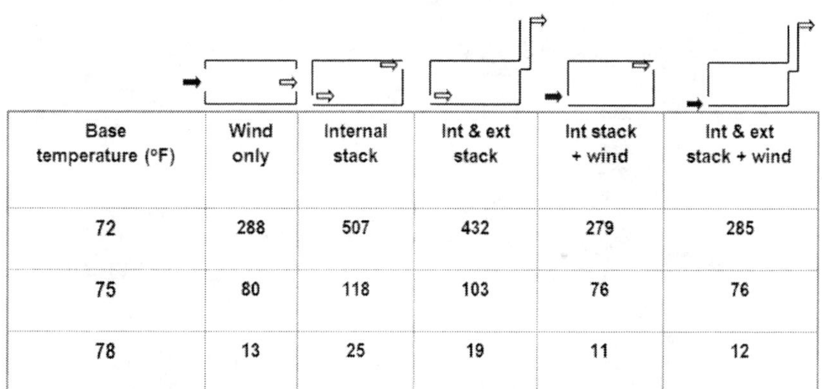

Base temperature (°F)	Wind only	Internal stack	Int & ext stack	Int stack + wind	Int & ext stack + wind
72	288	507	432	279	285
75	80	118	103	76	76
78	13	25	19	11	12

The simulations helped the design process by:

- Giving designers and client confidence that natural ventilation will work
- Informing the selection of a system ('wind only') that combines good performance with the lowest first cost.

Further details of the design, the design process and the simulation analysis are given in [23, 24].

Whole Building Performance Monitoring and Fault Detection

Simulation can also be used to support the operation of buildings by providing a real-time baseline for measured performance. A proof-of-concept whole building performance monitoring tool has been implemented in two buildings at Naval Station

Great Lakes, near Chicago. EnergyPlus simulations of the buildings are connected to the building control systems using the Building Controls Virtual Test Bed [25]. A weather station that measures outdoor air dry bulb and relative humidity, wind speed and direction, and direct and diffuse solar irradiance is connected to the control system and the measurements used as the time-dependent boundary conditions for the simulation. Dedicated electrical power sub-meters were installed to measure the lighting power, plug load power and chiller power in order to compare the simulation results with the actual performance of the main building sub-systems. Significant differences between predicted and measured performance indicate possible equipment faults or operating problems.

Figure 6 shows part of the display for the building operator for a situation in which there are significant differences between the simulated and measured electric power consumption, shown in the upper plot. The lower plot shows the cooling plant end-use consumption, showing that the differences between the simulated and measured electric power are due to unexpected operation of the cooling plant, which was traced to a control problem in the building. Further details are presented in [26, 27].

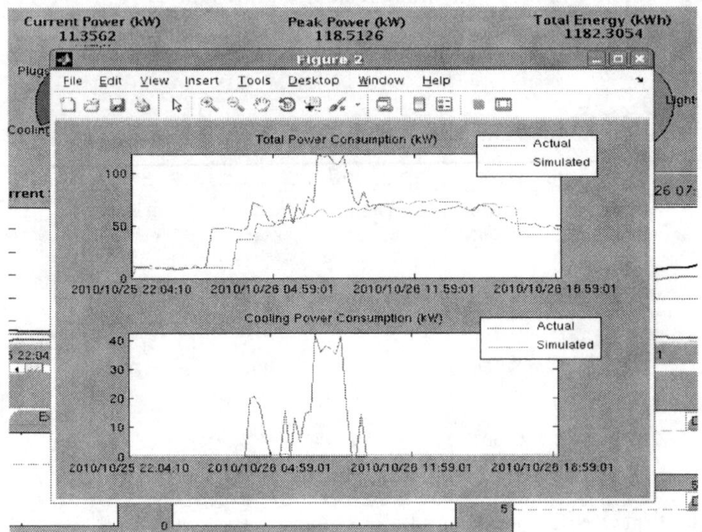

FIGURE 6. The screen-shot of the interactive display of simulated and measured performance for the building operator.

FUTURE DEVELOPMENTS

There are a number of areas where further development of simulation methods and tools is required. These include:

- Usability:
    - better user interfaces to support the workflow used by designers

- software interoperability – seamless data exchange between different types of software, e.g. CAD, cost estimation and energy simulation
- Computational efficiency: improve both single threaded and multi-threaded performance
- Software architecture: separation of models, numerical solvers and user interfaces into separate modules
- Stochastic modeling: develop models of occupant behavior, weather variability and other non-deterministic phenomena
- Model validation:
  - empirical validation – testing in laboratories and real buildings
  - inter-model comparisons
  - characterization of modeling uncertainties
- Process validation – develop standard practices, QA
  - development of standard practices and protocols for common simulation applications
  - extend simulation programs to generate output uncertainties based on specified input data uncertainties and model uncertainties
- Education and training:
  - develop and deliver curriculum on the generic principles of simulation,
  - develop training methods and materials for specific tools.

SUMMARY AND FURTHER INFORMATION

Energy simulation is used in building design and has the potential to be used to support building operation. It is also used in product development and in scenario analysis to inform policy development. There is a wide variety of phenomena in buildings that need to be modeled in order to simulate building performance and a variety of numerical methods that are used in simulation programs. There is a need to improve the computational speed of building simulation, particularly for use in optimization and parametric studies. Examples of the use of simulation in support of both building design and building operation have been presented. There is a need for further development of both tools and practices in order to improve the accuracy, robustness and effectiveness of simulation modeling.

A comprehensive review of the state-of-the art in building simulation is presented in [28]. One book that can serve as a textbook is [29]. The International Building Performance Simulation Association (IBPSA) has held bi-annual conferences on building simulation since 1985 [30] and the papers are downloadable at no charge. IBPSA-USA has started a Building Energy Modeling Book of Knowledge (BEMBOOK) wiki [31], the intention being to delineate a cohesive body of knowledge for building energy simulation.

ACKNOWLEDGEMENTS

This work was supported by the Assistant Secretary for Energy Efficiency and Renewable Energy, Office of Building Technology, State and Community Programs of the U.S. Department of Energy under Contract No. DE-AC02-05CH11231 and by the U.S. General Services Administration and the ESTCP program of the U.S. Department of Defense.

REFERENCES

1. IPCC. *Emissions Scenarios*. Nebojsa Nakicenovic and Rob Swart (Eds.). Cambridge University Press, 2000.

2. California's Energy Efficiency Standards for Residential and Nonresidential Buildings http://www.energy.ca.gov/title24/ 2011.

3. ***ASHRAE Standard 90.1 -- Energy Standard for Buildings Except Low-Rise Residential Buildings,* American Society of Heating, Refrigerating and Air-Conditioning Engineers, Atlanta, GA. 2010.**

4. US Green Building Council. *Leadership in Energy and Environmental Design.* http://www.usgbc.org/LEED 2011.

5. **T. Kusuda, *NBSLD, the Computer Program for Heating and Cooling Loads in Buildings. Building Science Series 69, National Bureau of Standards, Washington, DC, 1976.***

6. G. A. Bennett et al., *CAL-ERDA Program Manual, Los Alamos Scientific Laboratory,* Los Alamos, NM. 1977

7. F.W. Buhl, R.B. Curtis, S.D. Gates, J.J. Hirsch, M. Lokmanhekim, S.P. Jaeger, A.H. Rosenfeld, F.C. Winkelmann, B.D. Hunn, M.A. Roschke, H.D. Ross, G.S. Leighton. DOE-2: A New State-of-the-Art Computer Program for Energy Utilization Analysis of Buildings, *Proc. Second International CIB Symposium on Energy Conservation in the Built Environment*, Copenhagen, Denmark, May 1979.

8. BLAST Support Office.. *BLAST 3.0 UsersManual*.: BLAST Support Office, Department of Mechanical and Industrial Engineering, University of Illinois. Urbana-Champaign, IL. 1992.

9. J.A. Clarke and D. McLean. *ESP - a Building and Plant Energy Simulation System, Energy Simulation Research Unit,* University of Strathclyde, Strathclyde, 1988.

10. L. Palmiter, T. Wheeling and R. Judkoff, *SERIRES: Solar Energy Research Institute—Residential Energy Simulator*, SERI (now NREL) Report. 1982.

11. Solar Energy Laboratory, *TRNSYS Version 15 User Manual and Documentation*, Solar Energy Laboratory, Mechanical Engineering Department, University of Wisconsin, Madison, WI, 2000.

12. G.E. Kelly, C. Park, D.R. Clark and W.B. May, Jr. , HVACSIM+, a dynamic building/HVAC control system simulation program. In: *Proc. Workshop on HVAC Controls Modeling and Simulation.* Georgia Institute of Technology. (February 2 – 3), 1984.

13. *eQUEST Version 3.55, February 2005.* www.doe2.com/equest

14. Visual DOE user manual, Architectural Energy Corporation. http://www.archenergy.com/products/visualdoe/ 2004.

15. D. B. Crawley, L. K. Lawrie, F. C. Winkelmann, W. F. Buhl, Y. J. Huang, C. O. Pedersen, R. K. Strand, R. J. Liesen, D. E. Fisher, M. J. Witte and J. Glazer. *EnergyPlus: creating a new-generation building energy simulation program.* Energy and Buildings, **33**(4), pp 319-331, April, 2001.

16. Designer's Simulation Toolkits (DeST), Tsinghua University, Beijing, China, 2005. www.dest.com.cn (*Chinese version only*)

17. P. Fritzson, Principles of Object-Oriented Modeling and Simulation with Modelica 2.1, Wiley. 2003.

18. BuildingSMART. *Industry Foundation Classes*. http://buildingsmart-tech.org/ 2011

19. M. J. Witte, R.J. Henninger, J. Glazer, and D.B. Crawley. Testing and validation of a new building energy simulation program. In *Proceedings of Building Simulation 2001*. Rio de Janiero, Brazil: IBPSA. 2001

20. M. Frankel and C. Turner. How Accurate is Energy Modeling in the Market? *Proc. 2008 ACEEE Summer Study on Energy Efficiency in Buildings*, Asilomar, CA. 2008.

21. J.H. Scofield. Do LEED-certified buildings save energy? Not really. *Energy and Buildings*, **41**(12): 1386. 2009.

22. Y.J. Huang, F.C. Winkelmann, W.F. Buhl, C.O. Pedersen, D.E. Fisher, R.J. Liesen, R. Taylor, R.K. Strand, D.B. Crawley, L.K. Lawrie, Linking the COMIS multi-zone airflow model with the EnergyPlus building energy simulation program, in: *Proceedings of Building Simulation'99*, IBPSA, Vol. II, Kyoto, Japan, pp.1065–1070, September 1999.

23. E. McConahey, P. Haves and T. Christ. The Integration of Engineering and Architecture: a Perspective on Natural Ventilation for the new San Francisco Federal Building, *Proc. 2002 ACEEE Summer Study on Energy Efficiency in Buildings*, Asilomar, CA, August, 2002

24. P. Haves, G. Carrilho da Graça and P.F. Linden. Use of Simulation in the Design of a Large Naturally Ventilated Commercial Office Building, *Proc. Building Simulation '03*, Eindhoven, Netherlands, IBPSA, August 2003.

25. BCVTB Documentation. http://simulationresearch.lbl.gov/bcvtb 2011.

26. Z. O'Neill, M. Shashanka, X. Pang, P. Bhattacharya, T. Bailey, and P. Haves. Real time Model-based Energy Diagnostics in buildings. Submitted to *Proceedings of Building Simulation 2011*. Sydney, Australia. IBPSA. November, 2011.

27. X. Pang, P. Bhattacharya, Z. O'Neill, P. Haves, M. Wetter and T. Bailey, Real-time Whole Building Energy Simulation Using EnergyPlus and the Building Controls Virtual Test Bed, Submitted to *Proceedings of Building Simulation 2011*, Sydney, Australia, IBPSA. November, 2011.

28. *Building Performance Simulation for Design and Operation*, edited by J. L. M. Hensen and R. Lamberts, Spon Press, 2011.

29. *Energy Simulation in Building Design (2nd edition),* J.A. Clarke, Elsevier, 2001.

30. International Building Performance Simulation Association conference papers, http://www.ibpsa.org/m_papers.asp

31. *Building Energy Modeling Book of Knowledge,* IBPSA-USA http://bembook.ibpsa.us/

Smart Buildings and Demand Response

Sila Kiliccote, Mary Ann Piette, and Girish Ghatikar

Lawrence Berkeley National Laboratory
1 Cyclotron Rd. MS90-3111, Berkeley CA 94720

Abstract. Advances in communications and control technology, the strengthening of the Internet, and the growing appreciation of the urgency to reduce demand side energy use are motivating the development of improvements in both energy efficiency and demand response (DR) systems in buildings. This paper provides a framework linking continuous energy management and continuous communications for automated demand response (Auto-DR) in various times scales. We provide a set of concepts for monitoring and controls linked to standards and procedures such as Open Automation Demand Response Communication Standards (OpenADR). Basic building energy science and control issues in this approach begin with key building components, systems, end-uses and whole building energy performance metrics. The paper presents a framework about when energy is used, levels of services by energy using systems, granularity of control, and speed of telemetry. DR, when defined as a discrete event, requires a different set of building service levels than daily operations. We provide examples of lessons from DR case studies and links to energy efficiency.

INTRODUCTION

The objective of this paper is to explore a conceptual framework and a set of definitions that link building energy efficiency, control system features, and daily operations to electric grid management and demand response in smart buildings. Demand response (DR) can be defined as mechanism to manage the electric demand from customers in response to supply conditions, such as through prices or reliability signals. Such concepts and definitions are needed as the building industry and the electric utility industry become more integrated in supply demand side operations. It is critical for the energy industry to more strongly link demand-side performance objectives with electricity supply-side concepts.

One motivation for this framework is to facilitate understanding of automation of DR in demand side systems. The examples in this paper draw from research on commercial buildings, though the concepts are relevant to industrial facilities and residential buildings. This framework also emphasizes existing buildings but the ideas are applicable to new buildings and may help guide concepts to move DR into building codes and standards.

A key theme of this work is to understand not just how much energy a building uses, but when it uses energy and how quickly it can modify energy demand. This is not a new concept, but as more sophisticated controls are installed in buildings, the opportunities to better link demand and supply side systems are improving. Previous papers have discussed definitions of energy efficiency, daily peak load management,

Physics of Sustainable Energy II: Using Energy Efficiently and Producing it Renewably
AIP Conf. Proc. 1401, 328-338 (2011); doi: 10.1063/1.3653861

and DR [1 & 2]. This paper discusses the different speeds of DR, automation basics, and related control system features and telemetry requirements.

One objective of this DR research is to evaluate building electric load management concepts and faster scale dynamic DR using open automation systems. Such systems have been developed by the California Energy Commission's Public Interest Energy Research Program (PIER). The PIER Demand Response Research Center (DRRC) has led this effort and developed and deployed systems throughout California and the Northwest in a technology infrastructure known as OpenADR [3]. The intention of the signaling infrastructure is to allow building and industrial control systems to be pre-programmed, enabling a DR event to be fully automated with no human in the loop. The standard is a flexible infrastructure design to facilitate common information exchange between utility or Independent Systems Operator (ISO), and end-use customer. The concept of an open standard is intended to allow anyone to implement the signaling systems, providing the automation server or the automation clients. These standardized communication systems are being designed to be compatible with existing open building automation and control networking protocols to facilitate integration of utility/ISO information systems and customer electrical loads [4].

The next section of this paper outlines the six key elements of the conceptual framework for traditional energy management and emerging demand responsiveness. This is followed by a section that discusses levels of building services in relation to the six key elements. This section also discusses control systems and the speed of telemetry. Next we present an example of how this framework can be applied to advanced lighting controls and we reference the New York Times Building in New York as an example of an as-built advanced multi-functional lighting control system. We conclude with a brief summary and key research issues associated with the framework.

DEMAND-SIDE MANAGEMENT: LINKING ENERGY EFFICIENCY AND DEMAND RESPONSE IN SMART BUILDINGS

We provide a brief description of six energy and demand management concepts. The first three concepts we classify as "traditional" energy management. The second three concepts are "emerging"demand responsiveness. Following each of the six concepts is a comment on the role of automation and timescales. These six sections are:

Traditional Energy Management:

1. Continuous Energy Management
2. Monthly peak demand management
3. Daily time-of-use energy management

Emerging Demand Responsiveness:

1. Day-ahead Demand response
2. Day-of Demand Response
3. Ancillary Services Demand Response

Traditional Energy Management

Continuous Energy Efficiency

Energy efficiency can be defined as providing some given level of building services, such as cooling or lighting, while minimizing energy use. A strategy or technology that provides the same amount of service with less energy is a more efficient technique. A good example is to compare the lumens per watt of a fluorescent versus incandescent light. At the whole building level a more efficient building is one that provides HVAC, lighting, and miscellaneous plug load services using less energy for the same services than a comparison building. To actually achieve high levels of energy efficiency in a complex commercial building requires energy efficient components combined with well-commissioned controls and good operational practices.

The key point about energy efficiency is that building control strategies and operations should be optimized with energy use minimized every hour of the year for the given "service" the building is providing at any moment. Our success in reducing energy use in commercial buildings is strongly linked to our improved ability to measure the services the buildings systems provide while ensuring that energy waste is reduced as much as possible. We need to

We have not, however, described more advanced strategies such as thermal storage or pre-cooling that allow for variations in charging and discharging of thermal systems. To optimize building performance we will want to consider what we are trying to minimize. Optimal control strategies to minimize energy costs may differ from strategies to minimize total energy use or CO_2 emissions (as CO_2/kWh may vary between the day and night). Ideally one can achieve both low energy use and low energy costs!

Automation: The automation of continuous energy management is provided by energy management and control systems (EMCS).

Timescale: Thousands of hours per year.

Monthly Peak Electric Demand Management

The majority of large commercial buildings in the US pay peak electric demand charges. These charges often represent about one-third of the monthly electricity costs, yet they are not as well understood or as well managed as total (monthly or annual) electricity use. Peak electric demand charges typically have a time period they are associated with, such as the afternoon from noon to 6 pm. Some tariff designs have peak demand charges that apply to the monthly peak during on, partial or mid-peak, and off peak periods. Others have demand ratchets that may result in a peak demand that occurs in one month to set charges for 12 months. The key issue here is it is not how much energy is used, but when the most demand for electricity occurs. Efforts to reduce these charges require understanding rates, building controls, weather sensitivity and occupancy patterns.

Automation: Historically many energy management systems have offered demand-limiting features to reduce the peak demand by "limiting" electricity use when demand is high. While these are in limited use, they are available in many EMCS platforms and they require integrating whole- building electric use data with the EMCS.

Timescale: A few hours per month

Daily Time-of-Use Management

Similar to the presence of peak electric demand charges, most large commercial buildings have time-of-use (TOU) charges where electricity during the day time hours is more expensive than nighttime use. TOU energy management techniques involve careful consideration of scheduling equipment to reduce use of expensive electricity if possible.

Automation: Most EMCS provide scheduling of HVAC and lighting systems including programming of demand shifting strategies. As mentioned below most buildings do not use thermal storage so they do not "charge" energy systems during off peak periods. Some facilities do, however, modify energy use patterns to reduce expensive on-peak energy.

Timescale: Key periods of the day

The above three basic concepts are applicable to most commercial buildings with TOU and peak demand charges.

Emerging Demand Response Management

As we move toward a future in which the electric grid has greater communication with demand-side systems, it is useful to define and explore the time-scales of energy management and DR.

Day-ahead ("Slow" DR)

Day-ahead DR involves informing a demand-side customer the day before a DR event that the DR is pending the following day. In the case of manual DR this notification allows the facility manager to prepare a facility to participate in DR for the given schedule. Day-ahead real- time pricing can be an example of Day-ahead DR. Some RTP designs issue 24 electricity prices for each hour of the following day. This allows facility managers to schedule their loads and manage their electricity costs.

Automation: Most DR in US commercial buildings is manually initiated. However efforts to develop and deploy open DR automation standards have shown that most buildings with EMCS are good candidates for DR automation. Day-ahead signals allow the EMCS to schedule next-day DR events and are sometimes used to automate pre-cooling [6]. The DR program evaluations in California showed that about 15% of the time the person responsible for the manual response did not act [6].

Timescale: 50-100 hrs/yr (though day-ahead hourly real- time prices can be continuous, high price events are fewer hours per year.)

Day-of DR

Day-of DR can be defined as DR events that occur during the day when the event is called. These DR events typically have a scheduled time and duration. Day-of DR may also be an hour-ahead or 15-minute ahead real time price. A facility manager has less notice to prepare to participate in such events.

Automation: Similar to Day-Ahead DR, Day-of DR is often initiated manually. The more "real time" the DR, the more compelling is the need to automate DR because the notification for a person in the loop is more problematic with faster time scales of DR. Pre-cooling may not be possible in "Day-of" DR events.

Timescale – 30-60 hrs/yr (though hour-ahead real-time prices can be continuous, high price events are fewer hours per year.)

Fast DR

A third class of DR is ancillary services. There are several classes of ancillary services such as load following systems, spinning and non-spinning reserves, and regulation capability [7]. Fast DR can be thought of DR that is available quickly and the DR may not last long but it can be harvested quickly. The DR event may only be five minutes in duration. There are several recent research projects that have explored such "fast" DR [7].

Automation: Fast DR requires automation because people often cannot "jump" to action when notified of a fast DR event. These fast DR events may not last long. The electric loads are often restored within five to ten minutes of when they were curtailed [8]. The existing Internet-based DR automation systems are being considered for their speed and applicability to this class of DR.

Timescale: 5-10 hrs/yr

SERVICE LEVELS, CONTROLS AND TELEMETRY

There are three key features of demand-side systems to consider as commercial buildings begin to participate in all six of the electricity value chains listed above. These are, Levels of Service, Granularity of Controls, and Speed of Telemetry.

Levels of Service

There is a tremendous opportunity to better link DR and energy efficiency by improving understanding of the levels of service provided by existing buildings and building end- use systems. Take the example of an office building, which is designed to provide ventilation to support good indoor air quality, indoor climate control, lighting, and other services such as hot water, office equipment plug loads, and vertical transport (elevators). Good energy management practices assume that there is not much energy wasted. The building is heated, ventilated, lit, and cooled at optimal levels to provide comfort, but energy waste is minimized.

Given this as the baseline, to participate in DR requires that the service level that is provided in normal operations is minimized. Common examples are to change temperature set points or reduce lighting levels. Better measurement and monitoring of actual temperatures and lighting level distributions will improve our ability to change servicelevels since we want to ensure "optimal energy efficiency" as the starting point for DR.

Granularity of Advanced Controls

Similar to the desired ability to "measure" levels of services provided in a building is the desire to "control" the level of service. To participate in DR events we do not want to simply "turn off" a service, rather we'd like to "reduce" the service. This ability to improve control can provide features important for continuous energy management, monthly peak demand management, and daily TOU control. Further examples are provided below.

Speed of Telemetry and Response

This final category of infrastructure moves us from manual DR to fully automated systems. Research and automated DR programs in California have shown that existing Internet systems are fast enough to provide a signaling infrastructure for Day-ahead and Day-of DR [8]. Research is beginning to explore the capabilities of such systems for fast DR.

Table 1 below summarizes the key concepts explored in this framework:

TABLE 1. Summary of demand-side systems features electricity value chains

Concept	Automation	Timescale	Level of Service	Speed
Continuous Energy Management	Provided by EMCS	1000 hrs/year	Optimize each hour	Slow
Daily TOU Energy Management	Provided by EMCS	Select time of the day	Optimize for TOU	Slow
Monthly Peak Demand Management	Provided by EMCS	Few hrs/month	Minimize demand charges	Slow
Day-ahead DR	Can be automated	50-100 hrs/year	Temporarily Reduced	Medium
Day-of DR	Can be automated	30-60 hrs/year	Temporarily Reduced	Medium/Fast
Ancillary Services	Can be automated	5-10 hrs/year	Temporarily Reduced	Fast

SMART BUILDINGS ARE SELF-AWARE AND GRID-AWARE

Smart buildings are equipped with integrated end-use systems to deliver maximum level of service for each kilowatt hour consumed. They have sensors and monitoring equipment that collect data and are used in fault detection and commissioning. They are responsive to the occupants' needs and their environment. They are self-aware and should be grid-aware. The integration of technologies to link energy efficiency and OpenADR must meet the requirements of the electricity value chains and key features of demand-side systems, namely levels of service, granularity of controls, and speed of telemetry. These technology requirements vary based on the type and use of energy management. For example, the EMCS and technologies used for continuous and TOU energy management and peak demand management can be well integrated and interoperate with the needs of OpenADR. Subsequently, the same OpenADR system infrastructure could be integrated and enhanced to meet the requirements of ancillary services. This essentially means that the underlying technology should be designed to meet the context-setting framework of varied demand-side requirements. The figure below (Figure 1) shows linkages between the electricity value chains and their key features those are necessary for a robust technology framework.

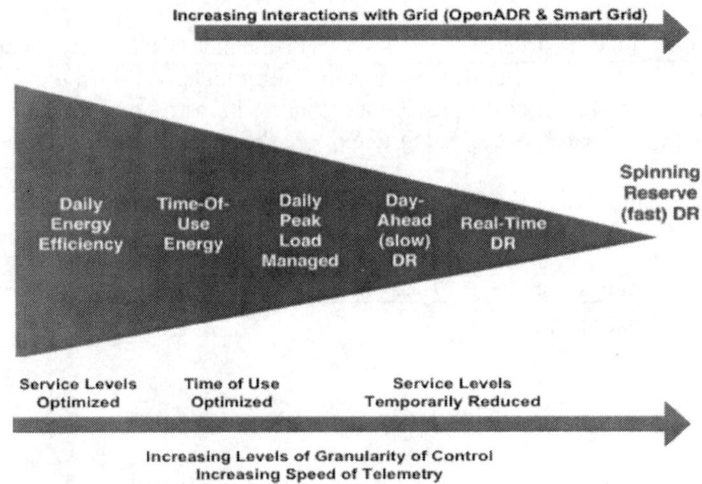

FIGURE 1. Service levels, controls and telemetry in grid-aware smart buildings.

The left side of the figure above (Figure 1) shows that most hours of the year we are concerned with continuous energy efficiency. Each hour energy use can be optimized relative to the energy services begin delivered. As we move to the right, few hours of the year are included and we begin to reduce building service levels in DR periods.

From left to right, there is increasing need for granularity of controls and speed of telemetry. Our ability to provide fine grain controls into end-use building systems improves both energy management and demand responsiveness. Further examples are provided below using dimming lighting and DR capabilities. As we move to the left toward faster DR systems, increasing speeds of telemetry are needed to initiate the DR. While this paper does not go into the details of all of the functional requirements of such systems, we acknowledge that the end-use controls within the building become a key component of the end-to-end system for DR.

The use of Internet-based signals and information technology (IT) with a Service Oriented Architecture (SOA) using web services and well- designed IT systems for DR can meet the demand-side systems' needs in relation to the electricity value chain. SOA, which uses eXtensible Markup Language (XML), a widely accepted standard for communication, and an Internet-based platform, can facilitate communications interoperability and ease of sharing structured data among complex systems. Such interoperability needs are in use by the Building Automation and Controls Network (BACnet) protocol in form of BACnet web services (BWS) [9]. Thus, the OpenADR standards that deliver both price and reliability signals, are an important step toward integration and automation of DR. The context-setting framework defined by GridWise to meet technical, informational, and organizational requirements for interoperability within DR systems is well studied and developed for OpenADR and is being commercialized throughout California. While OpenADR primarily facilitates technical and informational needs among DR systems (both human to machine and machine to machine), the information model also considers facility or end-user's needs when signals and data pertaining to DR events are sent and the facility determines the optimal DR strategy based on that information. OpenADR is also being evaluated for ancillary services in new research efforts on Fast DR.

ADVANCED LIGHTING SYSTEM EXAMPLES

Today's dimming lighting systems are perhaps the best example of an advanced emerging technology that provides daily continuous energy minimization with excellent DR capability. By drawing less when there is abundance of daylight or reducing electricity from the grid when electricity costs are highest, dimming ballasts are an enabling technology that allows building lighting loads to become more elastic. Concerns for electricity disruptions and power outages have stimulated the industry to re- examine and re-design dimming controls to implement DR and energy efficiency measures. Advances in lighting technologies coupled with the pervasiveness of the Internet and wireless technologies have led to new opportunities to realize significant energy saving and reliable demand reduction using intelligent controls [10].

Many manufacturers now produce electronic lighting control equipment that are wirelessly accessible and can control dimmable or multilevel lighting systems while complying with existing and emerging communications protocols. These controllers are well-suited to retrofit applications where it may be less cost-effective to add wiring to communicate with downstream lights. The lighting industry has also developed new technology with improved performance of dimming lighting systems. The system efficacy of today's dimming ballasts compare well with non-dimming ballasts, where historically there was an energy penalty for dimming.

As a result, from an energy efficiency perspective, dimming ballasts can provide seamless integration of indoor lighting and daylighting delivering continuous low energy use with optimized lighting levels. From a DR strategies perspective, dimmable ballasts can be utilized for demand limiting and demand shedding. Often times, even when dimming strategies are detectable, they can still be acceptable by the occupants [11]. In the newly built New York Times building, the installation of individually addressable dimming ballasts provides highly flexible lighting systems which can minimize energy use for lighting when there is adequate daylight. Advances in lighting control algorithms also facilitated demand shedding of lighting loads to allow good participation in regional DR programs [3].

The process to develop an automated DR strategy based on which lighting control features and layout one has in their building is as follows: A building operator can use either a manual or automated approach. If central control of lighting is available, the next step is to evaluate the "granularity" of the lighting control which is determined through a set of yes/no questions. Advanced lighting controls and increased levels of granularity allow us to define explicit steps in building lighting that can potentially be exercised during DR events.

Research is also beginning to explore the possible role of dimmable lighting for regulation products in the ancillary services market. Regulation products are generation units that are on-line, and synchronized with the ISO so that the energy generated can be increased or decreased instantly through automatic generation control (AGC) to balance the grid in real-time. While there are many technical challenges this research will address, the main objective is to explore whether the reserve markets may be better served if the ISO can obtain small load reductions from many distributed loads, rather than megawatts of power from a few generators.

DISCUSSION AND RESEARCH NEEDS

As we begin to explore the functional requirements for linking buildings to the electric grid we must ensure that we understand the fundamental concepts to support optimal and continuously monitored energy efficiency. Many of the technologies required for DR can benefit energy efficiency and advances in controls and service level monitoring will provide greater flexibility in energy management. As energy markets become more complex and there is a growing urgency for greater levels of energy efficiency, facility managers will need to explore better control of demand-side systems.

Facility engineers will need tools and systems to understand their existing systems and how it can participate in these new DR markets. Many energy markets will see dynamic prices and DR programs that provide economics incentives for facilities that can modify their end-use loads.

As we enhance our experience and understanding with the dynamic energy management concepts described above, our next technical challenge will be to quantify the performance metrics associated with each of the domains. For example, whole-building energy benchmarking is widely practiced and well understood process. Whole-building peak demand benchmarking is not! Electric load factors that compare average energy use and peak demand help characterize how "peaky" a building load shape is. Such load factors could be developed for different times of the day. Beyond the whole-building benchmarks are the opportunities to move into end-use benchmarks. Lighting system benchmarks are likely to be more straightforward than HVAC because of the lack of climate sensitivity.

SUMMARY

This paper has described a framework for characterizing energy use and the timescales of energy management for both energy efficiency and DR. This work builds on our experience using a standard set of Internet signals to trigger DR events in buildings. The development of advanced controls for energy management has also helped improve the ability of commercial building loads to be good DR resources. Further work is needed to develop tools and methods to help building owners and facility managers evaluate investments in advanced controls for both energy efficiency and DR.

ACKNOWLEDGEMENTS

This work was sponsored by the Demand Response Research Center (http://drrc.lbl.gov) which is funded by the California Energy Commission (Energy Commission), Public Interest Energy Research (PIER) Program, under Work for Others Contract No.150-99-003, Am #1 and by the U.S. Department of Energy under Contract No. DE- AC02-05CH11231.

An earlier version of this paper appeared at Grid Interop Forum, Atlanta 2008.

REFERENCES

1. Kiliccote S., M.A. Piette, and D. Hansen. *Advanced Controls and Communications for Demand Response and Energy Efficiency in Commercial Buildings*. Proceedings of Second Carnegie Mellon Conference in Electric Power Systems, Pittsburgh, PA. LBNL Report 59337. January 2006.
2. Kiliccote S., M.A. Piette, D.S. Watson, and G. Hughes. *Dynamic Controls for Demand Response in a New Commercial Building in New York.* Proceedings, 2006 ACEEE Summer Study on Energy Efficiency in Buildings. LBNL-60615. August 2006.
3. Piette, M.A., S. Kiliccote and G. Ghatikar. *Design and Implementation of an Open, Interoperable Automated Demand Response Infrastructure*. Presented at the Grid Interop Forum, Albuquerque, NM. November 2007. LBNL- 63665.
4. Holmberg, D.G., S.T. Bushby, and J.F. Butler. *BACnet® for Utilities and Metering.* American Society of Heating, Refrigerating, and Air-Conditioning Engineers, Inc. (ASHRAE) Journal, Vol. 50, No. 4, PP. 22-30, April 2008.
5. Xu P. and L. Zeagrus. *Demand Shifting with Thermal Mass in Light and Heavy Mass Commercial Buildings.* LBNL-61172. 2006.
6. Quantum Consulting Inc. and Summit Blue Consulting, LLC. 2004. *Working Group 2 Demand Response Program Evaluation – Program Year 2004 Final Report.* Prepared for Working Group 2 Measurement and Evaluation Committee. Berkeley CA and Boulder CO, December 21.
7. Eto, J., J. Nelson-Hoffman, C. Torres, S. Hirth, B. Yinger, J. Kueck, B. Kirby, C. Bernier, R. Wright, A. Barat, D. Watson. *Demand Response Spinning Reserve Demonstration.* LBNL-62761. May 2007.

8. Piette, M.A., G. Ghatikar, S. Kiliccote, E. Koch, D. Hennage, and P. Palensky. *Open Automated Demand Response Communication Standards: Public Review Draft 2008-R1*. LBNL number forthcoming. May 2008.
9. ANSI/ASHRAE 135-2001. *BACnet: A Data Communication Protocol for Building Automation and Control Networks.* American Society of Heating, Refrigerating, and Air-Conditioning Engineers (ASHRAE). June 2001; REPLACED by ANSI/ASHRAE 135-2004.
10. Rubinstein F. and S. Kiliccote. *Demand Responsive Lighting: A Scoping Study*. DRRC Report to California Energy Commission (CEC). LBNL-62226. May 2006.
11. Newsham G. *Detection and Acceptance of Demand Responsive Lighting in Offices with and without Daylight.* Leukos, Vol. 4, No. 3, pp 139-156. January 2008.

Appliance Standards and Advanced Technologies

Louis-Benoit Desroches

Energy Efficiency Standards Group
Energy Analysis Department
Lawrence Berkeley National Laboratory
1 Cyclotron Road, Berkeley, CA 94720
http://efficiency.lbl.gov/

Abstract. Energy efficiency has long been considered one of the most effective and least costly means of reducing national energy demand. The U.S. Department of Energy runs the appliances and commercial equipment standards program, which sets federal mandatory minimum efficiency levels for many residential appliances, commercial equipment, and lighting products. The Department uses an engineering-economic analysis approach to determine appropriate standard levels that are technologically feasible and economically justified (*i.e.*, a net positive economic benefit to consumers and the nation as a whole). The program has been very successful and has significantly reduced national energy consumption. Efficiency is also a renewable resource, with many new, even more efficient technologies continuously replacing older ones. There are many promising advanced technologies on the horizon today that could dramatically reduce appliance and commercial equipment energy use even further.

INTRODUCTION

The United States consumes about 100 quads (quadrillion Btu) per year of energy from a mixture of coal, natural gas, gasoline, nuclear, and other sources.[1] As shown in Fig. 1, approximately 42% of this energy is consumed by the residential and commercial sectors (*i.e.*, buildings), with the reminder split evenly between the transportation (29%) and industrial (30%) sectors. Thus the energy needed for buildings and appliances is close to one-half of the total source energy used by the United States. Of the 42% consumed by buildings, 30% is from electrical energy (split equally between commercial and residential). The remaining 11% is used by these sectors for non-electric energy, mostly natural gas. The breakdown of energy consumption in residential and commercial buildings is shown in Figs. 2 (residential) and 3 (commercial), where primary energy use is disaggregated into the main end-uses. U.S. residences consumed approximately 22 quads in 2010 whereas commercial buildings consumed approximately 18.3 quads. The predominant end-uses are space conditioning, water heating, lighting, refrigeration, and other miscellaneous electric loads.

Physics of Sustainable Energy II: Using Energy Efficiently and Producing it Renewably
AIP Conf. Proc. 1401, 339-352 (2011); doi: 10.1063/1.3653862

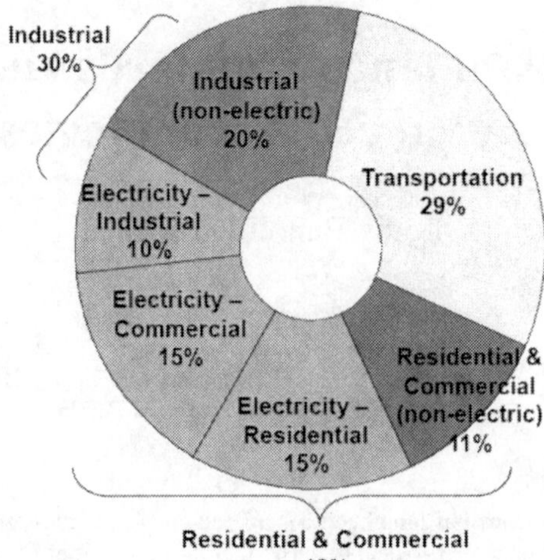

FIGURE 1. Primary energy demand in transportation, residential, commercial, and industrial sectors, as a percentage of total U.S. demand. Energy demand is also separated by electric and non-electric end-uses.[1]

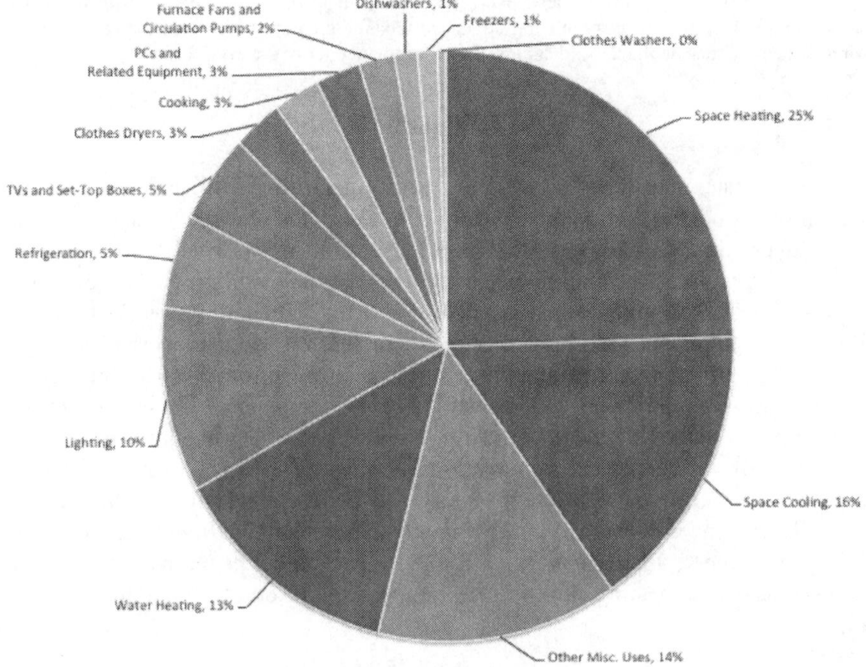

FIGURE 2. Primary energy demand in the residential sector for the U.S. in 2010. Demand is separated by major end-uses.[1]

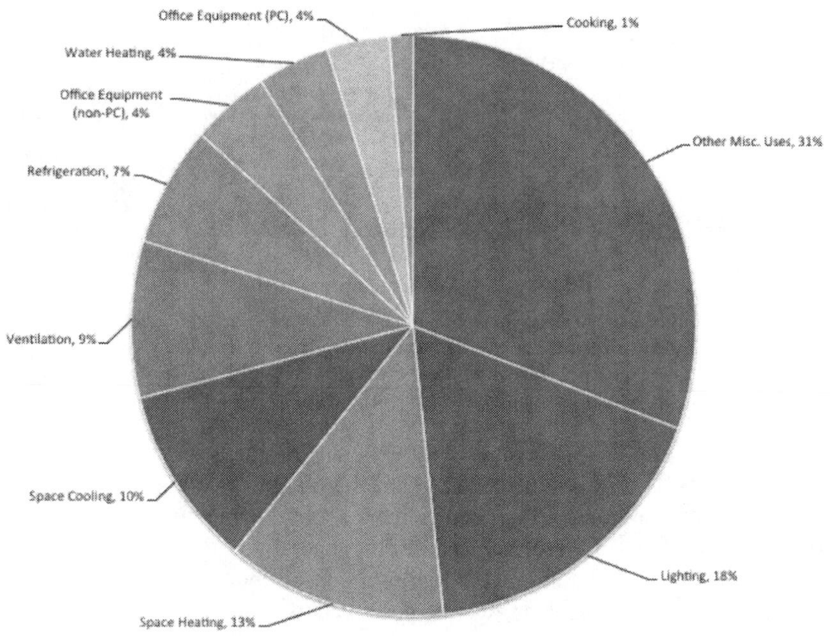

FIGURE 3. Primary energy demand in the commercial sector for the U.S. in 2010. Demand is separated by major end-uses.[1]

With increasing constraints on energy supplies and concerns over climate change impacts, the interest in sound energy policy has risen dramatically. Several reports have been produced in recent years highlighting the importance of energy efficiency in policy and the large savings potential it represents.[2,3,4] The point often emphasized is the cost-effectiveness of energy efficiency measures. For most appliances, the increased initial capital costs are quickly offset by reduced operating costs, such that payback periods are generally only a few years for many efficiency improvements. Thus, over the course of an appliance's full lifetime, the efficiency measures are often cost-negative (*i.e.*, cost of conserved energy is less than average electricity or natural gas retail prices), as shown in Fig. 4.

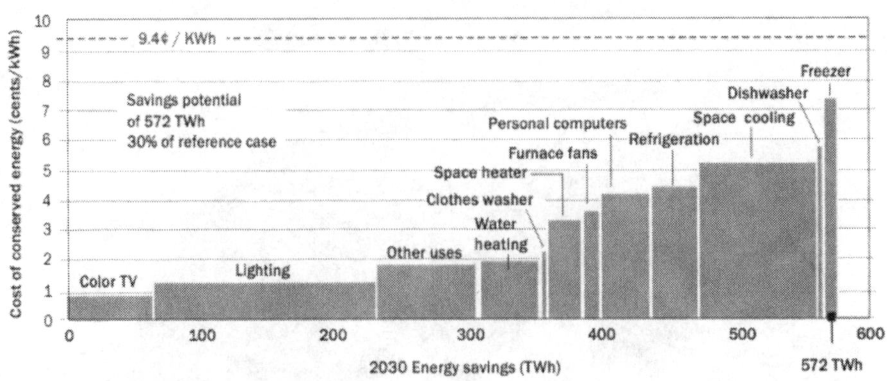

FIGURE 4. Cost of conserved energy vs. potential energy savings for a variety of appliance efficiency measures in the residential sector. The average retail price of electricity in the U.S. is 9.4 cents/kWh, as shown by the dashed red line. Thus all of these efficiency improvements are cost-effective. Figure reproduced from original source.[3]

There are a number of reasons why efficient products have not penetrated the market as easily as they should have, given the obvious economic benefits. There is often inadequate communication to consumers about the operating costs of appliances, both at the moment of purchase and during operation. Even if they have proper information, consumers often have a high implicit discount factor, well above market rates, placing greater weight on present costs versus future savings.[5] There continues to be a persistent principal-agent problem, where those who select the equipment (*e.g.*, landlords, developers, contractors) differ from those who pay the operating costs (*e.g.*, tenants, future occupants). These market failures hinder the adoption of more efficient technologies, requiring a regulatory approach to transform the market.

This chapter examines the role of mandatory federal energy efficiency standards for residential appliances and commercial equipment (including lighting) in reducing national energy use and promoting more efficient technologies. Appliance standards are often referred to as a market push program. In addition, this chapter will highlight some promising new technologies on the horizon with the potential to significantly reduce energy consumption in appliances and commercial equipment. This chapter will not discuss voluntary standards programs such as ENERGY STAR or tax incentives, which pull the high-efficiency end of the market, nor will it discuss mandatory labeling standards such as the Federal Trade Commission's EnergyGuide label.

U.S. APPLIANCE STANDARDS PROGRAM

History

In response to the oil crisis of 1973-1974, California became the first jurisdiction in the nation (including the federal government) to prescribe mandatory standards for appliances. Shortly thereafter Congress passed the 1975 Energy Policy and

Conservation Act that established a federal energy conservation program for major household appliances, and directed the U.S. Department of Energy (DOE) to develop appropriate efficiency targets. Unfortunately, little progress was made and no federal standards were established for many years. In response, several other states began adopting their own appliance efficiency standards programs, similar to California's. Appliance manufacturers became concerned about the potential for a heterogeneous set of state standards, and negotiated for a uniform set of federal appliance standards that would preempt state standards. The negotiations resulted in the 1987 National Appliance and Energy Conservation Act (NAECA). NAECA established the first set of federal standards for refrigerators, refrigerator-freezers, and freezers; room air conditioners; fluorescent lamp ballasts; incandescent reflector lamps; clothes dryers and clothes washers; dishwashers; kitchen ranges and ovens; pool heaters; television sets (withdrawn in 1995); and water heaters. Although Congress set the initial federal energy efficiency standards, it also established schedules for DOE to periodically review and update these standards. The 1992 Energy Policy Act (EPAct-1992), the 2005 Energy Policy Act (EPAct-2005), and the 2007 Energy Independence and Security Act (EISA) have all expanded the list of products covered under the federal appliance standards program, which now includes a variety of water-consuming products and some industrial equipment as well. The current program[6] affects over 80% of residential primary energy consumption and over 60% of commercial primary energy consumption.

Standard-Setting Process

When developing a minimum efficiency standard, the Secretary of Energy must weigh the benefits and burdens of a potential standard when selecting the level of stringency. The standards should achieve maximum efficiency that is technologically feasible and economically justified, and to have significant energy savings (though none of these terms have precise meaning in the law). Developing standards is intended to be an open process that involves all stakeholders as active participants, fostering consensus in a back and forth process. DOE relies heavily on input from manufacturers, consumers, and experts from industry and non-governmental organizations. The earliest standards were a reasonable consensus agreement between industry and non-governmental organizations representing consumer and societal interests.

As part of the analysis to determine the appropriate standard level, DOE takes the appliances apart and studies each component individually and collectively. A cost-efficiency curve is developed, which characterizes the cost increment for a corresponding efficiency increment. This is known as an engineering analysis. The cost-efficiency curve is then fed into a larger economic analysis, which includes a consumer life-cycle cost analysis, a national impacts analysis, a manufacturing impacts analysis, a utility impacts analysis, and an environmental assessment. Ultimately, the overall analysis looks at the trade-offs between an increased purchased price for a more efficient appliance versus a decreased operating expense. The Secretary can then make an informed decision that addresses the need for the nation to conserve energy and takes into account consumer impacts, manufacturer impacts, any potential lessening of product utility, any potential lessening of competition, and other relevant factors. The adopted standard must be technology neutral (*i.e.*, more than one design option can potentially achieve the efficiency level), must not require proprietary technology only possessed by a subset of manufacturers, must be practical to produce at scale, must not have significant impact on utility to consumers, and must not impact safety.

Since DOE uses an engineering-economic approach, it has the ability to prescribe standards that go beyond current models in the market, as long as the standard is feasible and shown to be economically justified. This is an extremely valuable approach to regulatory analysis. Some international jurisdictions rely on statistical approaches that eliminate a certain percentage from the bottom of the market. Other jurisdictions use the top-performing models as a basis for the next generation of standards. Neither of these approaches, however, can result in standards that are more stringent than the most efficient models on the market. Fig. 5 highlights this effect with respect to refrigerators. The 1990 standard was prescribed by NAECA. DOE published a Final Rule in 1990, based on engineering data from 1989, with the first year of compliance set for 1993. DOE's approach produced a more stringent, yet economically justified, standard compared to either a statistical or top-performer approach.

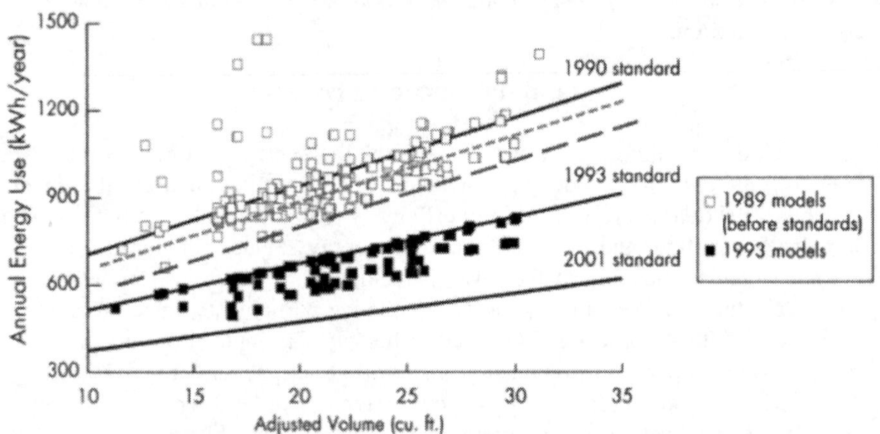

FIGURE 5. Illustration of DOE's economic-engineering approach to setting efficiency standards, which can often exceed the best available models on the market. The 1993 standard was based on 1989 engineering data. DOE's method produced a more stringent standard than either a statistical approach (red short-dashed line) or a top-performer approach (blue long-dashed line). A statistical approach eliminates a certain percentage of models from bottom of the market, whereas a top-performer approach sets an efficiency standard equal to the best units on the market.

The analysis process generally takes approximately 3 years before DOE issues a Final Rule. The first year of compliance is usually set at 3-5 years after publication of a Final Rule, to give manufacturers sufficient time to adjust their designs and production lines. The update frequency of appliance standards is approximately once every 6-10 years, depending on the product category.

Prior to 2009, the average publication pace was roughly 1-2 Final Rules per year, but in 2009 DOE published 6 Final Rules:

- 14 products with standards prescribed by EISA;
- Ranges and ovens;
- General service fluorescent lamps and infrared lamps;
- Commercial package boilers and very large commercial package air-conditioners and heat pumps;
- Refrigerators beverage vending machines;
- Commercial clothes washers.

In 2010 DOE published 4 Final Rules, with 1 Final Rule scheduled for completion in 2010 but currently pending:

- Residential water heaters;
- Direct heating equipment;
- Pool heaters;
- Small electric motors;
- Refrigerators and freezers (pending).

In 2011, DOE is scheduled to complete 10 Final Rules, in addition to the refrigerators and freezers Final Rule:

- Residential clothes dryers (completed);
- Room air conditioners (completed);
- Residential furnaces;
- Residential central air conditioners and heat pumps;
- Fluorescent lamp ballasts;
- Battery chargers;
- External power supplies (Class A);
- Ellipsoidal reflector (ER), bulged reflector (BR), and small diameter incandescent reflector lamps;
- Microwave ovens;
- Residential clothes washers.

Finally, DOE is currently conducting ongoing analysis for the following products:

- Furnace fans;
- Distribution transformers;
- Metal halide (MH) lamp fixtures;
- High intensity discharge (HID) lamps;
- Medium electric motors;
- Residential dishwashers, and more.

Energy and Economic Impacts from Standards

The cumulative effect of decades of successive state and federal standards has dramatically reduced energy consumption in residential appliances and commercial equipment. Fig. 6 illustrates this impact for residential furnaces, central air conditioners, and refrigerators. The case of refrigerators is particularly striking; since 1972, the average annual energy use of a new refrigerator has decreased by an impressive 70%, and will decrease even further once the latest refrigerator Final Rule comes into effect. In addition, the decline in refrigerator energy use occurred while average capacity was increasing, premium features became more prevalent, and retail price was decreasing.

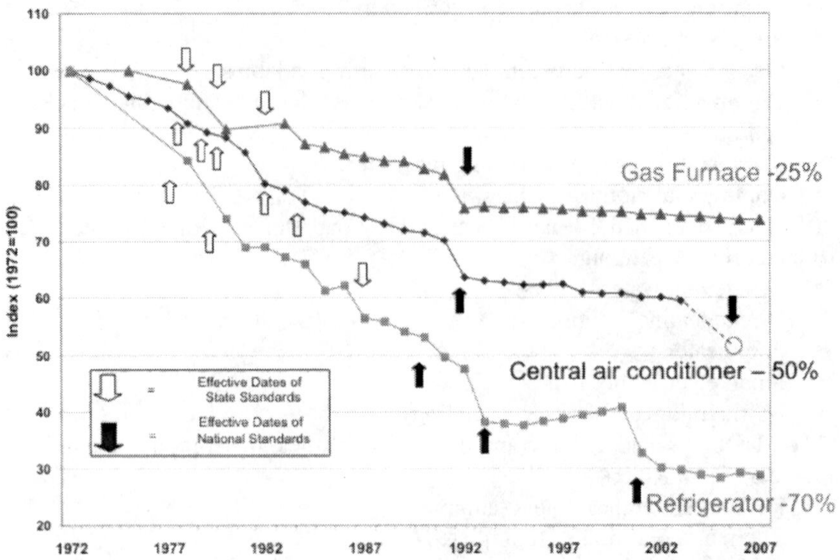

FIGURE 6. Illustration of the impact of state and federal efficiency standards on annual energy usage for residential gas furnaces (brown triangles), central air conditioners (blue diamonds), and refrigerators (pink squares). Annual energy usage is indexed to 100 in the year 1972. As an example, modern refrigerators use 70% less energy than early 1970's models, due to successive state and federal standards.

When DOE publishes a Final Rule, it includes the estimated energy savings of the new standard over a 30-year period, as well as the national economic benefit to consumers in the form of a net present value (NPV). When summing up future costs/savings, those dollar amounts must be converted to a present value by taking into account financing rates and opportunity costs of future cash flows. This is achieved by progressively discounting future dollar values in the total sum. For national impacts, DOE publishes NPVs using both a 3% per year discount rate and a 7% per year discount rate. DOE selects standard levels that maintain a positive NPV – in other words, averaged over the nation, consumers save money over the full lifetime of the appliance compared to pre-standard appliances. As an example, the combined water heaters, direct heating equipment, and pool heaters Final Rule (published jointly) is estimated to save the nation 2.81 quads of energy, 164 million metric tons (MMT) of CO_2, and save consumers $2.0 billion (using a 7% discount rate) over 30 years. Similarly, the small electric motors Final Rule will save an estimated 2.2 quads of energy, 112 MMT of CO_2, and save consumers $5.3 billion (using a 7% discount rate) over 30 years.

The impacts of standards on individual consumers will of course vary, depending on regional energy prices, regional weather, household size, individual usage behavior, individual appliance lifetime, specific consumer discount rate, and other factors. This distribution of impacts is included in the life-cycle analysis, which utilizes Monte Carlo simulations to capture the full range of potential impacts. As an example, the clothes washer standard of 2004 reduced energy use by 22%, but the individual consumer impacts exhibited significant variation (Fig. 7). The economic impacts varied between $808 in net savings to $126 in net costs per household. The maximum household savings of $808 reduced the average life-cycle cost by 50%. Ninety percent of households had a net savings, and only ten percent had a net cost, with a mean impact of $103 in net savings (a reduction in life-cycle costs of 6.3%). More generally, the overall ratio of consumer benefits to consumer costs for all appliance standards passed to date is estimated to be approximately 2.7:1.[7,8]

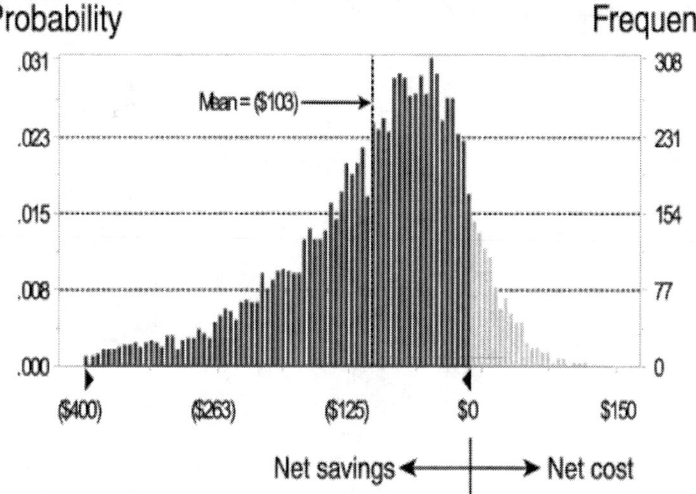

FIGURE 7. Illustration of the potential distribution of consumer impacts. Shown here is the probability distribution of consumer life-cycle impacts resulting from DOE's 2004 efficiency standard for residential clothes washers (a 22% reduction in energy use). Consumer impacts can vary significantly depending on many household factors, such as usage patterns, local climate, and local electricity rates. For this clothes washer standard, 90% of consumers had a net savings (blue bars) and 10% had net costs (yellow bars), and the overall mean impact was $103 in life-cycle savings.

From 1978-2000, DOE spent $7.3 billion on energy efficiency technology research and development, of which less than 5% was used to run the appliance standards program (from 1979-2010 inclusive, DOE has spent $0.3 billion on the program). In 2001, the National Academy of Sciences and the National Research Council estimated the economic benefits of this energy efficiency research to date.[9] They assigned $30 billion in savings due to the development of efficient technologies such as compact fluorescent lamps (CFLs) and advanced window coatings. In contrast, $48 billion in savings were attributable to mandatory efficiency standards for 9 residential products, illustrating the enormous benefit and programmatic cost-effectiveness of a national appliance, lighting, and commercial equipment standards program. The ratio of national economic benefits to program costs continues to exceed, on average, 100:1. In total, those standards were estimated to reduced annual residential primary energy consumption by 9% and annual carbon emissions by 132 MMT of CO_2 by 2025.

More recently, the American Council for an Energy-Efficient Economy (ACEEE) studied the impacts of energy efficiency standards on total U.S. electricity consumption and projected those impacts out to 2030.[10] Existing standards are estimated to save 273 TWh in 2010 (7%), and projected to save 498 TWh in 2020

(11.5%) and 563 TWh in 2030 (12%), as shown in Fig. 8. Upcoming standards (scheduled for publication between 2009 and 2013) are projected to increase electricity savings by 100 TWh in 2020 and 177 TWh in 2030, as shown in Fig. 9. For comparison, a typical 500 MW coal plant delivers about 3 TWh of electricity per year. The savings in 2010 alone are equivalent to shutting down nearly 100 typical coal plants. Despite all the savings achieved to date, there are still plenty of substantial opportunities left from updating current standards and expanding coverage to include previously unregulated products.[11]

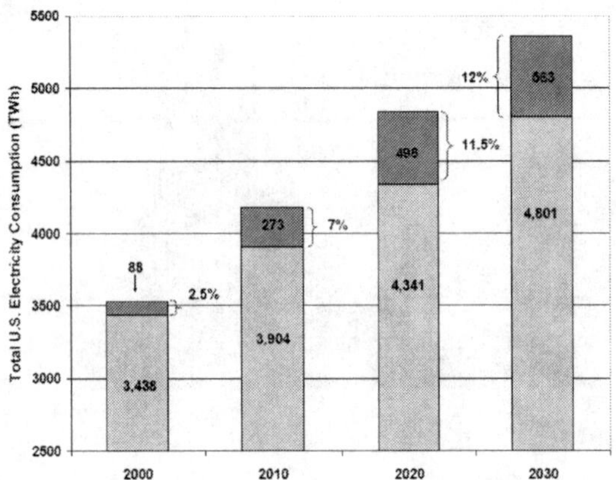

FIGURE 8. Impacts on total U.S. electricity consumption resulting from appliance and commercial equipment standards promulgated to date. Shown are total U.S. electricity consumption (green) and avoided electricity consumption due to standards (orange). In 2010, the appliance standards program is estimated to have reduced total national electricity consumption by 7%, and is projected to reduce consumption by 12% in 2030. Figure reproduced from original source.[10]

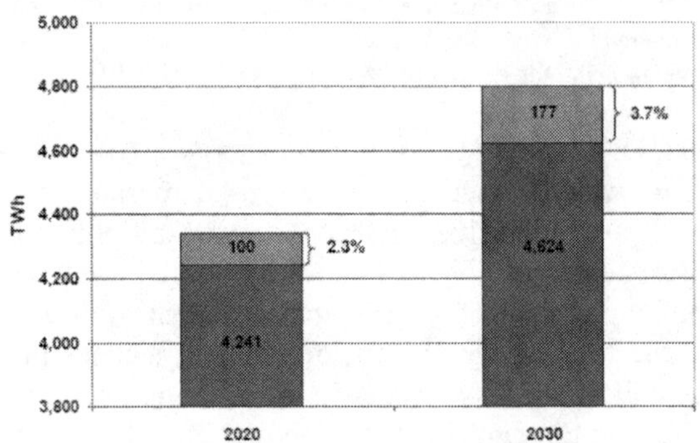

FIGURE 9. Impacts on total U.S. electricity consumption resulting from new appliance and commercial equipment standards scheduled for publication between 2009 and 2013. Shown are estimated total U.S. electricity consumption (blue) and avoided electricity consumption due to new standards (green). By 2030, the new standards are projected to reduce consumption by almost 4% in 2030. Including current standards already in effect (*i.e.*, Fig. 8), the appliance standards program is projected to reduce national electricity consumption by nearly 16% in 2030. Figure reproduced from original source.[10]

FUTURE APPLIANCE TECHNOLOGIES

There are several technologies in development that promise to substantially reduce the energy consumption of residential appliances and commercial equipment. Some are already commercially available in niche markets, while others are still in the prototype phase. This section outlines a handful of such technologies that show significant energy-savings potential, and may eventually become part of a future energy efficiency standard analysis. These technologies, as well as many others, are explored in more detail in a recent report.[12]

In the case of refrigerators, there are a multitude of individual components that could be used to improve energy efficiency. Such components include vacuum insulated panels (sides and door), linear compressors capable of operating at variable speeds (responding to variations in use and outdoor temperature), improved condensing coil designs to improve heat transfer (possibly eliminating the need for cooling fans), adaptive anti-sweat heaters and defrost, and potentially using 2 smaller compressors for the fridge and freezer compartments instead of a single larger compressor. Even the most efficient refrigerators sold today do not include all such options, so there is plenty of room for improvement. Vacuum insulated panels and linear compressors in particular are very effective, and are likely to be more prevalent after the pending refrigerator standard comes into effect. In the longer term, several institutions are working on developing magnetic refrigeration for residential use. Magnetic refrigeration uses the magnetocaloric effect to drive a refrigeration cycle, and is already used for scientific and industrial purposes when refrigeration temperatures near absolute zero are needed. The main energy use is in cycling a magnetic field on and off, which is less demanding than cycling a compressor.

Other household appliances could benefit from more radical changes. Conventional clothes dryers have barely evolved in the last decades, and the only significant improvement has been the introduction of heat/moisture sensors that can automatically terminate a drying cycle. In Europe, heat pump clothes dryers are quickly gaining an appreciable market share, and use roughly half the energy of a conventional dryer.[13] Furthermore, heat pump clothes dryers are ductless systems, and can thus be easily placed anywhere in the home (perfect for retrofits and apartments). More radically, some recent studies illustrate the savings potential of integrating many appliances into one. An integrated air-source heat pump and water heater, for example, can deliver the heat pump's waste heat (when in cooling mode) to the water heater, improving the efficiency of both appliances.[14]

Heating, ventilation, and air conditioning (HVAC) systems are a dominant energy end-use in both the residential and commercial sectors. In the residential sector, using a variable-speed compressor in a central air conditioning system, along with a few other improvements, can reduce energy usage by nearly half (since losses are nonlinear with air flow speeds and temperature gradients). In the commercial sector, there are significant savings opportunities in using indirect evaporative cooling systems, variable refrigerant flow systems, and adsorption cooling systems. Evaporative cooling is an ancient technology, but increases the relative humidity of air in the process. Indirect evaporative cooling uses the same latent heat cooling principle, but does not increase the humidity of air. Variable refrigerant flow systems are popular in other parts of the world (especially Asia), and use pipes of refrigerant instead of ducts of air to distribute cooling throughout a commercial building. Duct losses are often very significant, so this approach may substantially improve a building's energy performance. Adsorption cooling (distinct from absorption cooling) is relatively rare, but is very well suited to applications with waste heat by-products or significant solar heating potential. Adsorption cooling systems are less toxic, less corrosive, and easier to maintain than absorption cooling systems.

Lighting systems have a wide range of potential energy-saving technologies. Many use control systems to lower the light output when the lighting service isn't needed (*e.g.*, an unused office hallway at night, an empty parking garage). A reduction in brightness of 50% doesn't appreciably impact lighting for safety reasons, but substantially lowers the power draw. Other technologies coming onto the market include high-efficiency plasma lamps (in place of high-intensity discharge lamps) and light-emitting diode (LED) lamps. DOE launched a competition called the L Prize to produce a highly efficient replacement for the standard 60W incandescent bulb, without compromising color quality.[15] The maximum prize is $10 million. The first submission was a LED bulb that used less than 10W for the equivalent lumen output of a 60W incandescent bulb, while maintaining a color rendering index of 90 or greater (incandescent bulbs have a CRI of 100). Another promising lighting technology just emerging on the market uses quantum dots to filter light. Quantum dots are nearly lossless absorbers and emitters of light, and can be fined tuned to emit specific wavelength distributions of light. As a result, they can be used with bluish-white LED bulbs (highly efficient) to convert the unappealing color to a more pleasing color similar to incandescent light. Quantum dots may ultimately enable efficient LED bulbs to be readily accepted by consumers.

The most efficient televisions currently on the market utilize LED backlights that can be individually turned down when the picture content is dark (or dark in that region of the screen). This not only reduces the power draw of the television, it also improves the contrast ratio, resulting in a more vivid image. Quantum dots could also be used in future displays such as televisions, reducing the optical losses incurred as light travels through the optical stack in a liquid crystal display (LCD). Typical LCD TVs only manage to transmit about 5% of the light generated by the backlight unit to the viewer, so there is much room for improvement. Organic LEDs (OLEDs) promise to revolutionize the display market in the near future and significantly reduce energy usage. OLEDs are an emissive technology, as opposed to LCDs, which are a transmissive technology requiring a backlight. Individual OLEDs can be manufactured at sufficiently small sizes to become individual pixels in a television display, something impossible with ordinary LEDs. OLEDs are already used in some small displays such as cell phones, though quality issues remain when manufacturing OLED displays at larger sizes.

Electronic products, such as computers, printers, modems, routers, and the newest generation of televisions, are likely to require nearly constant Internet or network access for full functionality. This has the potential to induce enormous energy usage, since products will be constantly left on and awake. To address this, the IEEE has developed a new industry standard for Energy Efficient Ethernet (IEEE P802.3az), which allows network ports to power down when unused and then periodically waken for brief periods in order to check network traffic. Another technique known as network proxying transfers the network-aware functionalities of a device (such as a computer or printer) to the network card. This allows the device to go to sleep, and the network card can wake the device when needed. Since the network card uses far less energy than the device as a whole, the savings potential is large without compromising utility. Some computers are already being sold with this capability today.

Finally, another generic design strategy applicable to many appliances and equipment is to replace a single, large component with several smaller ones coupled with a control system. An example was discussed above with respect to refrigerators and compressors. Another example is distribution transformers. Transformers have a very nonlinear efficiency curve with respect to load, and tend to operate most efficiently when operating at a large percentage of their full rated load. Distribution systems, however, are designed to meet the highest possible load, not the average or typical load. Thus, most distribution transformers are tested (for minimum efficiency standards) at 35% of full rated load, and are rarely subject to loads greater than 20% of full rated load in field usage. Thus, replacing a single large transformer with two or three smaller transformers, coupled with a control system, allows a single smaller transformer to operate at a higher percentage of full load most of the time, and a secondary (and perhaps tertiary) transformer to energize when needed during periods of high demand. This strategy, while specific to transformers in this example, can be used for many other end-uses, including HVAC, motors, pumping systems, etc.

CONCLUSIONS

Over the past thirty years, the U.S. Department of Energy's appliance standards program has been tremendously successful in producing significant energy savings for individual consumers and the nation as a whole. The standards are technologically feasible and economically justified, and consumers have benefited enormously from standards to date, in the form of reduced operating costs. Additionally, the aggregate national economic benefits vastly exceed the programmatic costs, by over 100:1, making the federal appliance standards program a highly efficient, cost-effective usage of taxpayer funds.

Affordable energy efficiency is a renewable resource. New innovations, new materials, and new design approaches continuously improve the efficiency of residential appliances, commercial equipment, and lighting. The appliance standards program is a vital component in this innovation cycle. Additional savings potential continues to be identified, and there exist many interesting advanced technologies on the horizon that can deliver significant energy savings in the future.

ACKNOWLEDGMENTS

I would like to thank Jim McMahon and Greg Rosenquist for their help in providing some of the background material. I also wish to thank David Hafemeister for assistance on an early draft of this chapter. The Energy Efficiency Standards group at Lawrence Berkeley National Laboratory is funded by the U.S. Department of Energy's

Office of Energy Efficiency and Renewable Energy, Building Technologies Program under Contract No. DE-AC02-05CH11231.

REFERENCES

1. U.S. Energy Information Administration's Annual Energy Outlook. http://www.eia.doe.gov/forecasts/aeo/
2. J. Creyts et al., "Reducing U.S. Greenhouse Gas Emissions: How Much at What Cost?", McKinsey & Company, 2007.
3. B. Richter et al., "Energy Future: Think Efficiency", American Physical Society, 2008.
4. H.C. Granade et al., "Unlocking Energy Efficiency in the U.S. Economy", McKinsey & Company, 2009.
5. A. Sanstad and J. McMahon, "Aspects of Consumers' and Firms' Energy Decision-Making: A Review and Recommendations for the National Energy Modeling System (NEMS)", Lawrence Berkeley National Laboratory, 2008.
6. Appliances and Commercial Equipment Standards Program. http://www1.eere.energy.gov/buildings/appliance_standards/
7. S. Meyers et al., *Energy*, **28**, 755-767 (2003).
8. S. Meyers, J. McMahon, and B. Atkinson, "Realized and Projected Impacts of U.S. Energy Efficiency Standards for Residential and Commercial Appliances", Report No. LBNL-63017, Lawrence Berkeley National Laboratory, 2008.
9. R.W. Fri et al., "Energy Research at DOE: Was It Worth It?", National Academy Press, 2001.
10. M. Neubauer et al., "Ka-BOOM! The Power of Appliance Standards", Report No. ASAP-7/ACEEE-A091, Appliance Standards Awareness Project and American Council for an Energy Efficient-Economy, 2009.
11. G. Rosenquist et al., *Energy Policy*, **34**, 3257-3267 (2006).
12. L.-B. Desroches and K. Garbesi, "Max Tech and Beyond", Lawrence Berkeley National Laboratory, 2011. http://efficiency.lbl.gov/bibliography/max_tech_and_beyond
13. J. Nipkow and E. Bush, "Promotion of Energy-Efficient Heat Pump Dryers", 5th International Conference on Energy Efficiency in Domestic Appliances and Lighting, 2009.
14. V. Baxter et al., "Development of a Small Integrated Heat Pump for Net-Zero Energy Homes", 9th International IEA Heat Pump Conference, 2008.
15. L-Prize. http://www.lightingprize.org/

Energy on the Home Front

Thomas W. Murphy, Jr.

University of California, San Diego, 9500 Gilman Drive–0424, La Jolla, CA 92093-0424

Abstract. This article explores a variety of ways to measure, adjust, and augment home energy usage. Particular examples of using electricity and gas utility meters, power/energy meters for individual devices, whole-home energy monitoring, infrared cameras, and thermal measurements are discussed—leading to a factor-of-four reduction in home energy use in the case discussed. The net efficiency performance of a stand-alone photovoltaic system is also presented. Ideas for reducing one's energy/carbon footprint both within the home and in the larger community are quantitatively evaluated.

INTRODUCTION

It is clear that we cannot continue to rely on cheap fossil fuel energy for the indefinite future. The coming decades will see increased efforts to replace conventional energy sources with renewables—both as an effort to control carbon dioxide emissions and to mitigate economic disruptions caused by resource depletion. Many of the alternatives lack the reliability and convenience of our fossil mainstays. There do not appear to be any one-size-fits-all silver-bullet solutions to the energy challenge. Reliance on future technological breakthroughs (fusion, the ideal battery, a superconducting smart-grid, etc.) feels more precarious as time marches forward. After all, were we not all supposed to have jet packs by now?

Among the strategies discussed for meeting future challenges, one option with perhaps the largest potential impact is seldom discussed: choosing to use less. As concerned citizens of this world, we make choices all the time about how we might have less impact on world resources. We might recycle household waste, try to buy goods with less packaging, take our own bags to the grocery store, join a community-supported-agriculture cooperative, or any number of other behavioral changes. We might also (or instead) focus our attention on using less energy at home. This is the challenge I took up in 2007—made possible by a willing wife—and we have cut our domestic consumption of energy by about *a factor of four* in the process. If many of us were willing to change behavior on this scale, the enormity of our energy challenge would melt to a more obviously manageable beast. And even if few people adopt such behaviors, I am reassured to know how little it takes to meet my own needs, and that I am personally much less of a drain on the system.

In this article, I aim to provide the reader with tools, knowledge, and examples for accomplishing large reductions in energy usage. As a caveat, not all of the approaches I describe for our San Diego home will work for all people in all locations. On the other hand, we started from a comparatively slight energy profile and were able to trim this substantially. The theme of this article is more about *reduction* than *efficiency*.

Physics of Sustainable Energy II: Using Energy Efficiently and Producing it Renewably
AIP Conf. Proc. 1401, 353-366 (2011); doi: 10.1063/1.3653863

The latter is undoubtedly a useful tool in our arsenal, but gains in efficiency can sometimes lead to more profligate usage of energy. This phenomenon is known as Jevons' Paradox, or the Khazzoom-Brookes postulate.

TAKING STOCK

Real change must start with measurement. Without quantifying the scale of energy use, one has no baseline for improvement, and no context for where improvements can be most effective. A variety of inputs have guided my own exploration, as described below.

Utility Bills

Utility bills are a good place to start. The resolution is (typically) monthly, but good enough to establish a baseline. Because energy use and type tends to be seasonal, start by making a plot of electricity, gas, and any other inputs, such as heating oil. It is tremendously useful to use common units for the various inputs (I have tried without success to get my utility company to do this). Using the common unit of kWh, we list the conversions in Table 1.

TABLE 1. Home Energy Units.

Energy Source	Billing Unit	kWh
Electricity	1 kWh	1.0
Natural Gas	1 Therm	29.3
Gasoline	1 U.S. gallon	36.6
Heating Oil	1 U.S. gallon	40.6

For the purpose of assessing environmental impact, it is important to understand the local portfolio of energy sources used for electricity production, as this impacts the conversion efficiency and carbon footprint of the resource. For instance, a natural gas electricity plant---common in California—may deliver 0.35 kWh of electricity to my home for every 1 kWh of energy derived from natural gas.

The average American household (of which there are about 115 million) has a daily use of 30 kWh of electricity, and 35 kWh (1.2 Therms) of natural gas [1]. Our starting point was substantially lower. In the 12-month period preceding my energy "awakening," our household of two used 3828 kWh of electricity (10.5 kWh/day) and 9000 kWh (307 Therms; 24.5 kWh/day) of natural gas. Given that the source of our electricity is dominated by natural gas turbines, the factor of three conversion/delivery efficiency puts the two sources roughly at parity in terms of energy use and CO_2 production.

Reading the Meters

Utility bills provide the bird's-eye view of a household's consumption habits. Some utilities also allow online monitoring of "smart" meter activity at hourly resolution—although often delayed by a day or more. Even given this welcome capability, neither form easily translates into action items for demand reduction. This is where metering can help. Digital meters are replacing analog meters in some parts of the country, but it is helpful to be able to read both types. The primary function of the meter is to measure cumulative usage for billing purposes. Thus the "odometer" aspect is the meter's main function. But for home assessment, the rate of usage is far more important—whether for electricity or natural gas. The odometer is often too crudely displayed to be of much use in determining instantaneous rate. For instance, the new "smart" electricity meter on our house only displays electricity usage in increments of 1 kWh—which is not very useful when daily consumption is knocked down to about 2 kWh per day. The biggest sin of our smart meter is the absence of an instantaneous power reading. Internally, it measures household current and multiplies (temporally) by voltage to get instantaneous power. But this information is not made available to the curious physicist.

Ignoring the odometer function of the meter, which is straightforward to figure out—even for the tricky alternating-direction analog gauges, I will describe how to extract useful rate information.

Analog Electricity Meter: Almost all analog electricity meters have a spinning disk whose average rate is proportional to power usage. A constant, usually called Kh, is typically printed on the faceplate of the meter, and is often 7.2 or some multiple thereof. This relates to the number of watt-hours (Wh) per turn of the meter. So if the spin period is T seconds, the power in Watts is $3600 \times \mathrm{Kh}/T$. Be aware that the disk rate can vary by as much as a factor of two during one revolution even at constant input power. For accurate results, an integral number of rotations should be observed.

Digital Electricity Meter: The rotating disk is no longer available, and my digital meter refuses to provide a measure of instantaneous power. But it does have a "simulated" disk function, consisting of little blocks that appear and disappear in a way that mimics the left-to-right motion of the analog disk. After hours on the phone with my utility provider (during which time I explained the difference between kW and kWh several times), I learned that each time a block appears or disappears, one Wh of energy has been used. My meter cycles through six states of its disk symbols, so that one cycle represents 6 Wh. If this takes time, T, the power in Watts is $3600 \times 6/T$.

Natural Gas Meter: My analog gas meter has an odometer with a resolution of one-hundred cubic feet (hcf), which is 1.02 Therms. But it also has two dials for which full revolutions measure 0.5 cf and 2.0 cf. The dials have ten tick marks around the periphery, so that one may achieve 0.01 cf of resolution with a modest amount of interpolation. But since the gap between the fine scale and odometer dials is so large, meaningful monitoring must be frequent enough to ensure no dial wraps went unnoticed.

By studying the rate of the electricity meter, one can learn how much power particular devices consume, what a typical "on" state of a house is, and—very importantly—what the "off" state base load is for the house. By studying the natural gas meter, one can understand the constant rate of usage from pilot lights, and how much energy goes into a shower, for instance.

The Biggest Reduction

The single-biggest energy reduction in my household derived from monitoring the natural gas meter. I watched the 2 cf dial make a steady 0.72 revolutions per hour during a period when there were no demands for gas. Pilot lights were solely responsible for this consumption, amounting to 10 kWh/day, or almost half of our total natural gas usage. I turned off the furnace pilot light to learn that it was 70% of the problem. In fact, during summer months, the furnace pilot light used more gas than we used for all of our hot water! Needless to say, I left the pilot light off, as spring had arrived and the heating season was over. The big savings came the next year, when we planned to hold off on re-lighting the pilot light until it became unbearably cold in the house. That day never came. We tolerated temperatures that occasionally dipped to 13°C (55°F), but in San Diego, this is the worst we saw.

Admittedly, not everyone can tolerate these temperatures—and it took a few years before it seemed normal to us. But besides pointing out that humans did not evolve simultaneously with heating, ventilation, and air conditioning (HVAC), a key observation is that I don't care how warm my bookcase is; *I* want to be warm. And there are plenty of ways to maintain thermal comfort in a cool environment. Down slippers, blankets on the couch, and a 60 W (max) dual-control mattress pad go a long way. Heat the body, not the house.

The Kill-A-Watt®

A very handy device for taking stock of electricity demands is a unit known as the Kill-A-Watt®, which can be inserted between an outlet and the plug of a device to monitor instantaneous power at 1 W resolution (up to about 1800 W), accumulated energy at 0.01 kWh resolution, and elapsed time.[1] This is extremely useful for characterizing refrigerators, washing machines, computer and entertainment devices, etc. I have kept one on our efficient refrigerator for the last year, finding an average power consumption of 37 W. This device was crucial for deciding what devices I wanted to place on the solar circuit (described below), and has also been responsible for our unplugging or getting rid of devices that used far more energy than their function warranted. Because of the Kill-A-Watt, I know how much power each setting of the electric mattress pad consumes, what my laptop uses when asleep, charging, etc., and how much energy our always-on wireless internet access demands.

Knowledge is power. And knowledge *of* power is even better. For the investment, the Kill-A-Watt is bound to be worthwhile. To set the scale, 1 W of power over a year consumes 8.8 kWh of electricity. Depending on local cost of electricity, this might typically translate into about one dollar per year. If just 20 W of constant power drain is eliminated, the unit pays for itself in the first year. But I hesitate to make strictly financial judgments, as these often miss larger points.

[1] This, and other commercial devices named in the text, does not represent an endorsement over competitive models. I simply report successful use of certain products, without having explored other—possibly superior—models.

My New Friend, TED®

Most friends cannot be bought. TED®, The Energy Detective, is an exception. TED is a whole-home electricity monitoring device, fitted into the main circuit-breaker box and providing 1 W resolution. My wife and I arrived at the house of our friends for Thanksgiving, bringing the just-purchased TED as an uninvited guest. While some of us (kids included) had great fun minimizing power output of the house, our hostess—expending substantial energy on meal preparation—was less enamored of the service interruptions. TED soon found refuge in our house, and despite my wife's initial attitude that the last thing we need is *another* energy monitoring device, within a few days she admitted that TED is "pretty cool."

TED provides an optional LCD unit that displays—among other options—instantaneous power. Finally I have access to the single-most important information pertaining to home usage without having to go outside and deal with the indirect (and not instant) information provided by the electricity meter. Now it is straightforward to measure appliances that the Kill-A-Watt could not (due to high power or integrated wiring). Logging features allow captures of activities at one-second resolution for a few hours, and several days at a time at one-minute resolution. It also tracks minimum, maximum, and cumulative usage across days, among other measures. Configuring a home's internet router appropriately allows remote access to TED's instantaneous and stored data.

By the time TED arrived at our house, we had already trimmed our usage substantially. But even so, having a number constantly on display motivates further change. Within days, we trimmed our utility base load ("off-state" power) from 52 W to 36 W—mostly by recognizing that our HVAC system demanded 13 W to sit idle, even when the system is turned off. Figure 1 shows an example data capture from TED at one-second resolution, in this case highlighting the secret life of the washing machine.

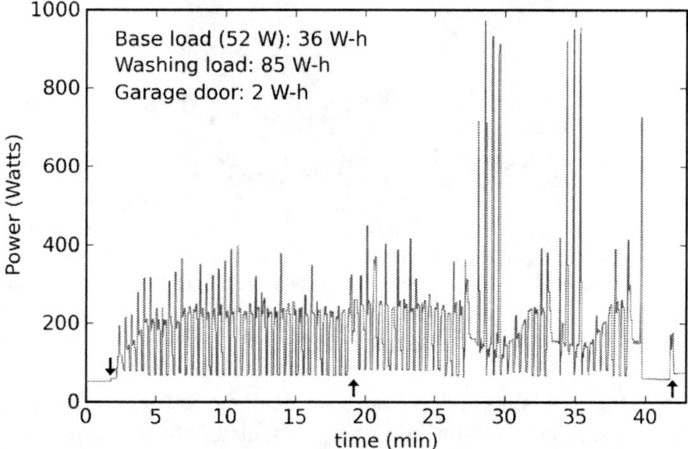

FIGURE 1. The energy cycle of a front-load washing machine, from TED, at one-second resolution. The down arrow is when the washing machine was plugged in (7 W phantom load), and the up arrows indicate times when the garage door was opened, then closed, each followed by a period when the

garage door light remained on. Otherwise all activity is due to the washing machine's rotation-rest cycles and final spin cycles. Interestingly, 75% of the 6000 J energy associated with opening or closing the garage door is due to the compact fluorescent light bulb, operating at 15 W.

Infrared Camera

Motivated by the fascinating article by Woolaway on infrared camera technologies in the first publication of the APS Sustainable Energy Conference in 2008 [2], I obtained a 120×120 format thermal infrared camera. While in my case, no single use could justify the expenditure, the range of discovered applications has made it a valued asset. In addition to learning about home energy issues, the camera is useful as an educational tool, to assist thermal engineering associated with my research, troubleshooting electronic component failure, and in general developing a solid intuition about the radiative world.

Within my home, the infrared camera has revealed a number of energy problems. It has exposed gaps in insulation and identified walls without insulation. It has helped to identify electronic devices that are consuming power unexpectedly. It has pinpointed the origin of drafts around windows and doors. It is also responsive to evaporative cooling that accompanies water leaks. The camera is therefore a great tool for evaluating a house's thermal state. But its expense seldom justifies its purchase for such an application in a single house. Figure 2 shows an example image from the infrared camera, in which an active dishwasher behind thin wood paneling reveals the path of the plumbing.

FIGURE 2. Example infrared view of a dishwasher in action. The dishwasher is situated within a counter-top "peninsula," so that this view is of the back side, covered by thin wood paneling. Note the hot water supply line, the warmed counter-top, the laptop on the counter at right, and that one of the two stools has recently been sat on. The temperature range in this image runs from about 22°C to 35°C.

ThermoChrons®

For assessing the thermal state of a house, it is hard to beat a time series of temperatures at key locations. But outfitting a home with a data acquisition system can be impractical. The iButton® Thermochron®, made by Maxim, makes the job easy. These are devices about the size of a stack of four nickels (17.4 mm diameter; 5.9 mm thick), containing an internal battery and no wires or ports. Communication is via the 1-Wire protocol, the device clicking into a USB-connected snap-ring socket for programming and readout. "Missions" can be set up to acquire temperatures on a sampling cadence from one second to hours, and can store 4096 records at 0.06°C resolution. An optional start-of-mission delay can also be set, and the device can either wrap or stop logging when the memory is full.

The most attractive aspect of the thermochron is that it can go anywhere it will fit, with no wires to complicate matters. They have metal exteriors, and therefore low emissivity. To couple effectively to the surrounding air, I hang them from string and use aluminum tape to provide a low-emissivity attachment. The hanging thermochrons provide yet another conversation-starter in our little house of data. Figure 3 provides an example of what can be learned from such information. For me, this will serve as a useful baseline if I make changes to insulation, roof albedo, ventilation, etc.

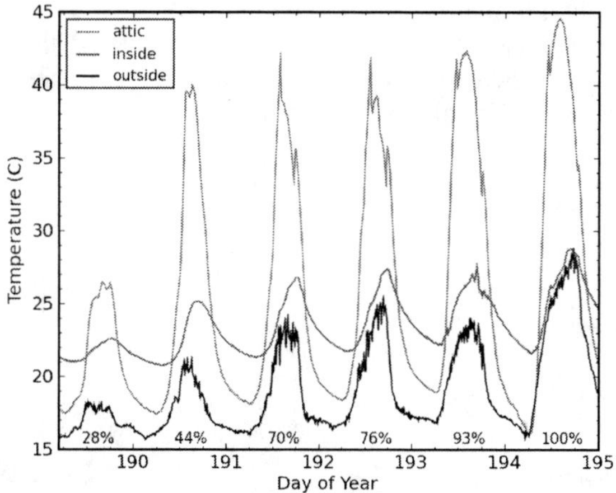

FIGURE 3. Part of a Thermochron campaign in July 2010, sampled at ten-minute intervals. The inside temperature (smoothest, intermediate curve) is not air conditioned, although an attic fan comes on when the attic temperature exceeds 42°C (then off at 33°C). The first day was completely overcast, and each day thereafter was progressively sunnier, until the last perfectly sunny day (photovoltaic yields for each day appear at bottom). Even days with no direct sun impose a thermal load on the roof. The attic fan appears to have only a few-degree impact. Most nights (and mornings) were accompanied by a marine layer, but the record indicates a clearing of the marine layer midway through the last night, as the roof radiated to space and turned sharply cooler before a clear sunrise.

DOMESTIC SOLAR ENERGY

As a physicist teaching courses on energy technologies, I felt woefully uneducated on the practical ins and outs of domestic energy production, and solar energy in particular. Because I was renting at the time, installing a grid-tie system—like almost all photovoltaic (PV) installations where utility power is available—was not a viable option. I embarked instead on the rewarding journey of building my own stand-alone system from commercial parts. I started very small, building two independent systems, each using one solar panel: a 64 W multi-junction thin-film panel, and a 130 W poly-crystalline silicon panel. The system is described in detail in the July 2008 issue of *Physics Today* [3].

The PV system has marched through a series of evolutionary steps since the publication of its description in 2008. The 64 W panel was set aside (now used to power a pump in a rainwater catchment system) and the system consolidated to a single one using a growing number of 130 W panels, a larger, smarter inverter, and more batteries. In the process, we moved to a different house and I wired a few dedicated breaker-protected AC circuits within the house for PV power distribution. The system now consists of 8 130 W panels; four 12 V, 150 Ah golf-cart batteries arranged in a 2×2 24 V configuration; a 3500 W inverter with the ability to switch in utility power when it senses low battery voltage; and internet-accessible monitoring.

The stand-alone PV system runs the refrigerator, entertainment console, cable modem and wireless router, two laptop computers, printer, and the attic fan. In all, we pulled 52% of our electricity in the last year from the PV system. The attic fan is a particularly good match, in that it tends to kick on right around the time the batteries reach full charge and begin refusing available energy. Plus, the attic fan is only needed on sunny days, when the batteries are likely to reach the full-charge state. Figure 4 shows a day in the life of the PV system.

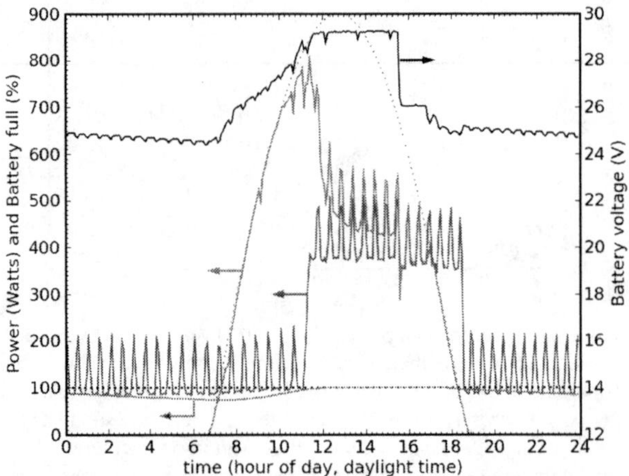

FIGURE 4. One day of PV generation in early May 2011, at five-minute resolution. The arc-shaped curve is the solar input (the dotted line is a section of a cosine curve for comparison). The load curve is dominated by the cycling of the refrigerator, and the attic fan in the afternoon. The battery (at top) achieved its "absorb" state voltage just before noon, after which the charge controller accepted a diminishing amount of the available sunlight. If not for the attic fan, far less solar energy would have been utilized. The curve at bottom represents the state of charge of the battery.

Efficiency Report

Over the last year, the batteries in my PV system were powering their share of the house 92% of the time, switching to utility input during long periods of rain and clouds. The duty cycle could be substantially increased with more batteries. In the last year, the PV system received an average of 4.1 kWh/day and distributed 2.5 kWh/day to appliances within the house. The resultant 62% efficiency breaks down as shown in Table 2.

TABLE 2. PV System Efficiency.

Source	% of Input Energy	Comments
MPPT Charge Controller	4.8	In series with input from panel
Net Battery Loss	7.8	Not the same as battery energy efficiency
System Power	4.7	9 W for PV devices, monitors, communications
Inverter Base-Load	9.6	20 W to keep inverter on (92% of time, last year)
Inverter Inefficiency	11.2	88.5% efficient at converting DC to AC

The battery system constitutes a net 7.8% drain on the system, but since the battery is only a drain during the day when it charges, the net drain under-represents the battery energy efficiency. In my case, 53% of the energy delivered to appliances originated from the battery, while the remaining 47% came directly from the solar input while the battery charged. Given that roughly half the time is daytime, this balance makes sense. The actual energy efficiency of the battery bank was 84%—meaning that 84% of the energy delivered into the battery re-emerged in a useful form. This is different from charge efficiency, measured as amp-hours in versus out, which is 93% in my system. The difference is due to the difference in voltage during charge versus discharge states. The batteries characterized here are in their first year of operation, so that these numbers are at the high end of their lifetime performance.

An additional loss is incurred relative to a grid-tie system because potential solar input is turned aside when the batteries are fully charged. Compared to a friend's house with an excellently-exposed grid-tie system in the same neighborhood, my system soaked up 76% of the available 5.2 equivalent full-sun hours averaged over the past year. A properly designed stand-alone system needs some margin to deal with less-than-perfect weather, meaning that untapped solar potential on clear afternoons is inevitable. Devices like attic fans can partially fill this void. Given the ~90% inverter efficiency on a grid-tie system, and putting the numbers together, my stand-alone system delivers to appliances about 53% of the available energy that a grid-tie system would do.

The lesson is that a stand-alone system incurs a substantial efficiency hit compared to a grid-tie system. Moreover, the cost of periodically replacing batteries roughly offsets the cost-avoidance of the utility electricity. But I do not regret in the least the experience and independence my system has given me. I now know much more about the practical side of PV, and can understand better the challenges our nation will face if relying on large inputs of intermittent power, requiring storage. As I write this, the food in my mother's refrigerator/freezer is spoiling due to tornado/storm-induced power outages in Tennessee. I have peace of mind that this would not happen to me, and I may be very popular with neighbors if I offer to store some of their food in a crisis.

A Side Benefit of Solar

At least as important as the direct energy input from the PV array is the behavioral conditioning it has fostered. Building my own system—especially a stand-alone system—makes my energy very personal. I have worked to catch and store that energy, and I'm not about to waste it frivolously. I would never think of leaving the television or lights on when the batteries are losing charge. Standing in front of an open refrigerator door is a new kind of sin. The magic part is that this mentality carries over to parts of the house that are not on the PV distribution—and then beyond the boundaries of home. I consider this level of awareness to be a perk. Rather than take electricity for granted, I have come to value it on a more personal level.

REDUCTION TACTICS

Thus far, I have discussed ways to monitor energy consumption and identify power hogs, describing also a stand-alone photovoltaic system to offset utility electricity. A few examples of reduction have been offered along the way, but here I present a comprehensive list of the changes we have made to reduce our energy consumption. The net effect is that today we have a daily use of 5.0, 2.3, and 2.5 kWh in natural gas, utility electricity, and PV electricity, respectively, averaged over the past year.[2]

Compared to our baseline performance in 2007–2008, we cut our gas usage by a factor of five, and our utility electricity usage by a factor of 4.5. Thus our domestic energy/carbon footprint (ignoring personal travel) was reduced by roughly a factor of five—counting the photovoltaic contribution as carbon-free. Actual electricity usage in the house—supplemented by PV—was reduced by a factor of 2.2, so that the total household energy went from 35 kWh/day to 9.8 kWh/day. Figure 5 illustrates our electricity and natural gas history through this transition.

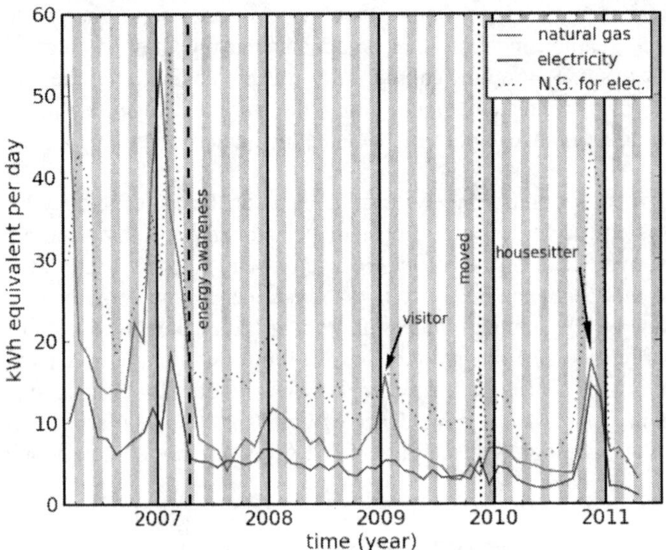

[2] An adjustment to the natural gas and utility usage was required to account for the presence of a housesitter during the past year, who, in a three-month period used 1043 kWh of electricity and 1250 kWh of gas. For these months, values were taken from the previous year.

FIGURE 5. Usage of utility electricity and natural gas in my household across five years. The dashed vertical line indicates the moment at which I began paying attention to energy use. The dotted vertical line represents a move from a condominium to a house. The lower solid curve is electricity usage, and the dotted curve is the same thing multiplied by 3 to represent natural gas needed at the power plant to generate the electricity used. Note the difference in natural gas usage (upper solid curve) between the summers of 2006 and 2007, due to the furnace pilot light! A visitor during January 2009 prompted us to resume heating the household temporarily, and a housesitter in the fall of 2010 produced another departure. Given the changes in awareness, household, and occupants during this time span, it becomes clear that the reduction is tied to *people* and *behaviors*, not the house itself

What specific actions did we take to achieve this goal? Table 3 lists the various contributions, along with an estimate of the impact to average daily energy consumption.

TABLE 3. Household Energy Reduction Efforts.

Action	Source	Daily Saving (kWh)
Set Thermostat to 13°C (55°F)	Gas	15
Take less frequent showers	Gas	3
Take shorter showers, water off, mostly	Gas	2
Replace all incandescent bulbs with CFL	Elec.	1.5
Line dry clothes	Elec.	1
Be vigilant about turning unused devices off	Elec.	1
Swap refrigerator (75 W avg. → 37 W avg.)	Elec.	1
Disconnect large phantom loads	Elec.	0.5

Exporting Impact

Shaving demand at home is a direct and obvious way to reduce one's energy/carbon footprint. But consider that the total American energy diet consumes 3 TW continuously, amounting to 10 kW per person, on average. This totals 240 kWh/day per person and is far in excess of average household amounts. Assuming an average of three persons per household (300 million people in 115 households), and that every kWh of electricity delivered consumed 3 kWh in energy resources, the average American only expends about 42 kWh per day in domestic electricity and natural gas—only 17% of the total. In this context, working hard to shave domestic consumption may seem ineffective. But consider that:

- it is the sector over which we have the most direct control;
- the mentality of demand reduction at home propagates to external choices/behaviors;
- choices we make as consumers of domestic goods have an external impact.

So what other household decisions can have a big impact? Table 4 lists some possibilities (all of which are employed to some degree by me and my wife).

The numbers in Table 4 should not be taken as definitive, but rather suggestive. For this audience, it may be beneficial to illustrate the techniques employed to develop numbers for the table, so that readers can refine or apply similar quantitative analyses to suit personalized situations.

TABLE 4. Example Estimated External Reductions

Action	Estimated Daily Saving, averaged over year
Commute via public transportation	25 kWh per gallon of gas used in one round-trip
Decide against some air travel (even for work)	20 kWh per 10,000 miles avoided annually
Eat less meat	12 kWh vegetarian, 18 kWh vegan
Consolidate errands	5 kWh per gallon of gas avoided weekly
Refrain from buying luxury/unnecessary goods	2.5 kWh per $1000 avoided annually
Join a Community-Supported-Agriculture unit	2 kWh
Keep office lights off during the day	1 kWh

For automotive travel, we simply use the fact that each gallon of gas contains 36.6 kWh, and scale this to the frequency of the travel event. Conveniently, each yearly gallon used/saved corresponds to 0.1 kWh each day. Therefore, a commute occurring 250 days per year saves 25 kWh per day for each gallon that would be used for the round-trip commute. For air travel, a typical passenger jet gets 50 miles per gallon per passenger. Each 10,000 miles therefore consumes 200 gallons per passenger, so 20 kWh per day averaged over the year.

The impact of dietary choices on energy is a fascinating and rich topic. A 2002 study [4] concluded that agriculture accounts for 17% of fossil fuel use in the U.S. Since 85% of our individual 240 kWh/day net energy use derives from fossil fuels, we each expend 35 kWh/day on food. A 2100 kcal daily diet corresponds to 2.44 kWh/day, but overproduction and waste actually results in a per capita domestic food production (after subtracting exports) of about 3800 kcal daily, or 4.4 kWh/day. Immediately we see a factor of ten disparity in the amount of fossil fuel energy used compared to the energy content of the food that is produced/consumed. A study by Eshel and Martin in 2006 [5] evaluated the relative energy efficiencies of different dietary choices. The average American diet—consisting of 28% animal products (15% from meat, 13% from dairy and eggs)—consumes about twice as much fossil fuel energy to produce as does a vegetarian diet getting 15% of its calories from dairy and eggs. A vegan diet is better still (though perhaps not as easy), using about 20% of the fossil fuel energy that the average American diet uses in production. The energy efficiency of grains tends to be about ten times that of milk, chicken, and eggs, and about 40 times that of red meat and tuna [6]. But most of the energy is still expended on transport, refrigeration, and preparation. Modest assumptions lead me to guess that a vegetarian and vegan diet use 80% and 65% of these resources, respectively. The net effect is that a vegetarian diet in America uses about two-thirds the energy of the average diet, while a vegan diet may cut investment in food energy in half. Joining a CSA, and getting 500 kcal/day in vegetables might save something like 2000 kcal of fossil fuel energy per day, or about 2 kWh.

The estimated energy impact of consumer goods is very crude, and based on the estimate that 10% of an item's cost went into energy (mining, manufacture, distribution, etc.). This is in rough parity with our overall energy expenditure as a fraction of gross domestic product. Assuming an energy cost of $0.1/kWh, $1000 of purchased consumer goods in a year embodies about 1000 kWh, averaging to a little over 2.5 kWh per day. Buying used items is greatly facilitated by online tools, and is one of our primary tactics to relieve demand on the manufacture of new products.

Office lights are estimated to run at about 160 W (four 40 W fluorescent tubes), 8 hours per day for 250 days in the year.

Based on this table and the behavioral changes that I have adopted, I estimate a personal daily savings of approximately 60 kWh outside the home. This is about twice the impact that my wife and I have made on the household energy front.

THE NET EFFECT

Having reduced my household energy demand from 35 kWh/day to 7 kWh/day (not including the stand-alone PV contribution of 2.5 kWh/day), my personal saving is approximately 15 kWh/day on domestic energy—since these savings are shared by two people. By adopting a portfolio of the behavioral changes listed in Table 4, I estimate a personal reduction of 60 kWh/day. The combined effect is almost a third of the average American energy allocation of 240 kWh/day. It is rare to see efficiency gains of similar magnitude. To the extent that I was already a below-average energy consumer, the fractional gain may be even higher than stated, in my case.

Choices to reduce consumption can be difficult, and are indeed heretical to the growth-oriented narrative of our society. But the latter aspect is the key point. Only in the face of concern that growth is nearing physical limits does it make sense to modify our behaviors. To the extent that the limits are real, reduction may be our sharpest available tool. Voluntary reductions are far more palatable than mandatory reductions based on curtailed resource availability and associated unaffordability, and can also represent a fun challenge.

This article is intended to illustrate some examples for how reduction in energy demand might be accomplished, but should not be taken as a template to be copied. Each individual and situation will be different. A program of reduction is far more attractive if it is devised personally, and not handed down from some "authority."

ACKNOWLEDGMENTS

I thank my wife for being a *mostly* willing partner in this energy reduction experiment. I'll take the thermochrons down soon—I promise. I also am grateful to Brian Pierini, whose nearby grid-tie photovoltaic system provides me with very useful comparison data.

REFERENCES

1. http://www.eia.doe.gov/tools/faqs/, accessed May 2011.
2. J. T. Woolaway,, "Infrared technology trends and implications to home and building energy use efficiency," in *Physics of Sustainable Energy*, edited by D. Hafemeister et al., AIP Conference Proceedings **1044**, American Institute of Physics, Melville, NY, pp. 217–231, (2008).
3. T. W. Murphy, Jr., "Home photovoltaic systems for physicists," *Physics Today*, **61**, issue 7, pp. 42–47, (July 2008).

4. L. Horrigan,, R. S. Lawrence, and P. Walker, "How sustainable agriculture can address the environmental and human health harms of industrial agriculture," *Environ. Health Persp.*, **110**, 445–456., (2002).
5. G. Eshel, and P. A. Martin, "Diet, Energy, and Global Warming," *Earth Interactions*, **10**, 9, 2006.
6. Pimentel, D., and Pimentel, M., *Food, Energy, and Society*, University Press of Colorado, 77–84, (1996).

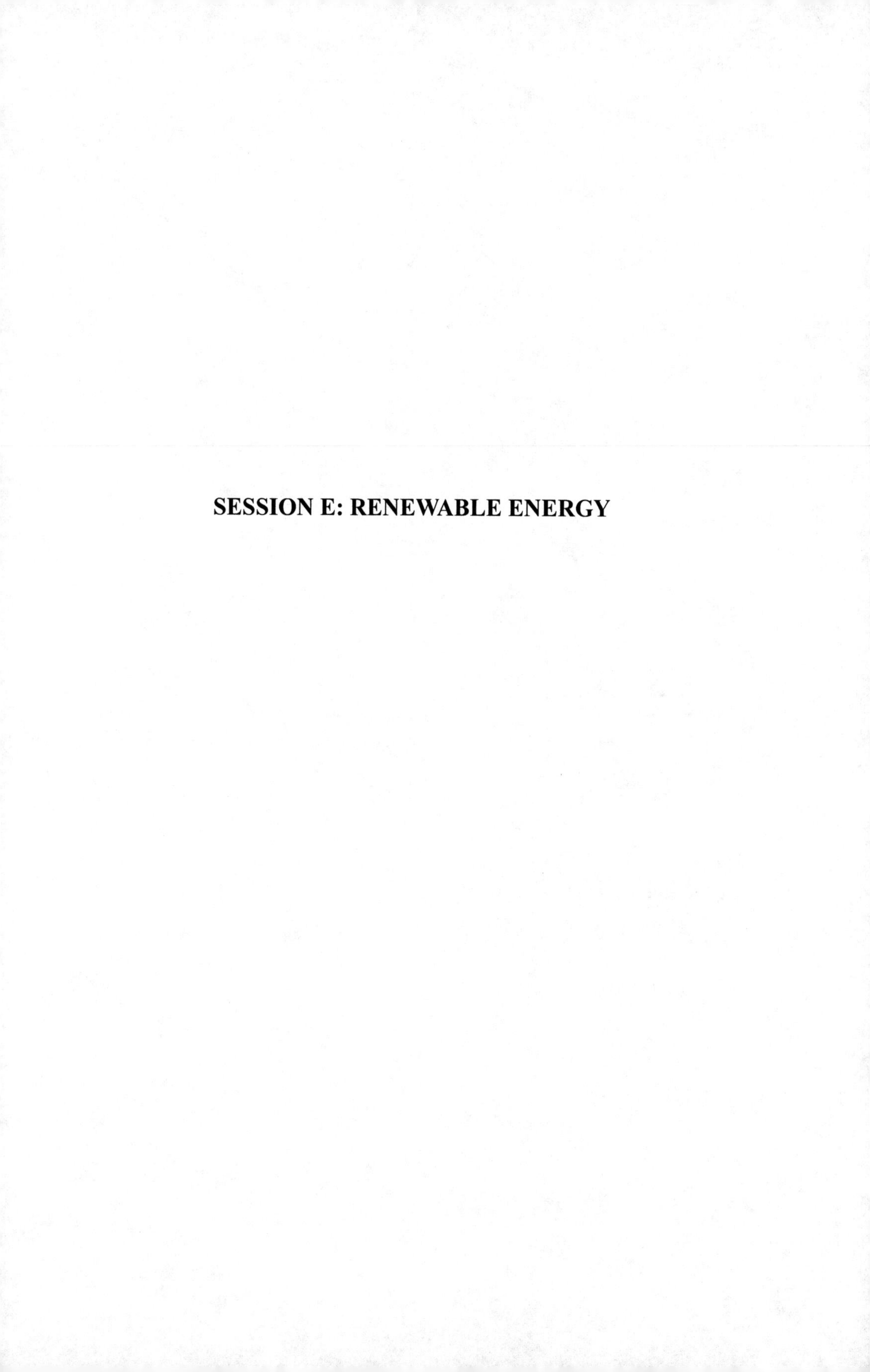

SESSION E: RENEWABLE ENERGY

Renewable Electricity in the United States: The National Research Council Study and Recent Trends

K. John Holmes[a] and Lawrence T. Papay[b]

[a] *Associate Director*
Board on Energy and Environmental Systems
National Research Council
500 5th Street, NW
Washington, DC

[b] *CEO, PQR, LLC and Sector Vice President for Integrated Solutions (Ret.)*
Science Applications International Corporation
La Jolla, CA

Abstract. The National Research Council issued *Electricity from Renewables: Status, Prospects, and Impediments* in 2009 as part of the America's Energy Future Study. The panel that authored this report, the Panel on Electricity from Renewable Sources, worked from 2007 to 2009 gathering information and analysis on the cost, performance and impacts of renewable electricity resources and technologies in the United States. The panel considered the magnitude and distribution of the resource base, the status of renewable electricity technologies, the economics of these technologies, their environmental footprint, and the issues related to scaling up renewables deployment. In its consideration of the future potential for renewable electricity, the panel emphasizes policy, technology, and capital equally because greatly scaling up renewable electricity encounters significant issues that go beyond resource availability or technical capabilities. Here we provide a summary of this report and discuss several recent trends that impact renewable electricity.

INTRODUCTION

During the rapid rise and fluctuations of energy prices that occurred in 2007, the National Academies of Sciences and Engineering began the America's Energy Future (AEF) study. Designed to be a foundational piece of the cost, performance, and impacts for energy supply and demand technologies, the AEF study took place in the context of rapidly changing and generally increasing energy costs, international insecurity, and ultimately a major recession. From the initiation of this study (2007) to the present (2011), there has been continuing concerns about short-term economic

Physics of Sustainable Energy II: Using Energy Efficiently and Producing it Renewably
AIP Conf. Proc. 1401, 369-386 (2011); doi: 10.1063/1.3653864

prosperity and long-term competitiveness, with energy at the core of many policy and scientific deliberations. Here we will review the results of one component of the AEF study, the report of the Panel on Renewable Electricity (Renewables panel), and note some trends that have occurred since this report was issued. Results from the full AEF study were published in 4 reports in 2009 [1-4].

AEF RENEWABLES PANEL'S STUDY APPROACH

The Renewables panel's mandate was to examine the technical potential for electric power generation from wind, solar photovoltaic, geothermal, solar thermal, hydroelectric, and other renewable sources. The panel evaluated renewable electricity technologies based on estimated times to initial commercial deployment and their costs, performance, and impacts. The primary focus of the study was on the quantitative characterization of technologies with deployment times from the present to approximately 25 years from the present, the renewable technologies that show the most opportunities and promise for commercial development within the short to mid-term that potentially could have substantial impacts on the US electricity system. In keeping with the overall focus of the AEF Study, the panel was to assess the state of development of technologies, not develop at policy recommendations.

Fundamental to the AEF study was separating the assessment of energy supply and demand technologies into three time periods: an initial period out to year 2020 that considered the costs, performance, and impacts of present renewable electricity technologies; a second period from 2020 to 2035 that considered the cost and performance characteristics of current technologies and the barriers, costs, performance, and research and development challenges for nascent technologies; and the third period that looked at the long-term barriers and research and development challenges, especially basic research needs, for renewables technologies beyond 2035. The first two time frames were the primary focus of the AEF study and the Renewables panel.

As opposed to organizing its report around individual renewable resources (i.e., having separate chapters on solar, wind, geothermal, biomass, and hydropower), the Renewables panel organized its report around the consideration of the resource bases, technologies, economics, impacts, and deployment characteristics for renewables. This organization emphasizes the degree to which renewable resources share common considerations when it comes to estimating the resource base, technologies, impacts and deployment issues. It also can help understand the benefits of integrating a portfolio of renewables into the electricity system. By necessity, much of the discussion is directed at attributes of each renewable source, such as technologies, state of readiness, and costs for individual types of solar, wind, or geothermal sources. Further, as the Renewables panel considered the issues related to deployment of new, non-hydropower renewables to a level where they make a signification impact on total electricity generation, the focus turns primarily to solar and wind technologies. This could have caused the report's basic storyline to read like a puzzle, with each renewable having its own characteristics in terms of its resource base, technology readiness, economics, and impacts. For example, solar electricity has the largest resource base and some well-developed technologies for tapping the resource base, but those technologies are still relatively expensive and evolving compared to other electricity sources. Wind power has a fairly large resource base, though not as large as solar, and well-developed technologies that are relatively cost competitive with other electricity sources. However, the organization of the report allowed the panel to discuss the opportunities that the full portfolio of renewables provides as well as the major deployment and integration issues.

SUMMARY OF RENEWABLES PANEL RESULTS

Contribution to Electricity Generation

As of 2009, renewable electricity generation made up about 10.5 percent of all electricity (5). The majority of this is from traditional hydropower (~ 7 percent) and wind (~2 percent). However, growth rates for renewables are high, with wind power generation growing at a compounded annual growth rate of almost 30% from 2000-2009 and installations of solar PV growing at similar rates, though solar's growth comes from a smaller base. Biomass (wood and waste), geothermal, and hydropower also contribute to renewable electricity generation, though these sources are not a large component of renewable electricity's current growth.

Resource/Technology Assessment

The panel concluded that the renewable resource base is large compared to electricity demands and many renewables technologies are sufficiently developed so that this growth will continue to be significant over the first time period. In particular, there is a great deal of wind and solar resources and lesser amounts of geothermal, biomass, and hydropower. The size of the resource base makes the key resource barriers related to not the quality of the base, but how to integrate the spatial and temporal variability of the resources and how to resolve the relationship of the resource bases to electricity demand and transmission. In terms of technologies, there are clearly a wide array of renewable electricity technologies that are sufficiently developed and are being deployed over the first time period, including wind turbines, solar photovoltaics, concentrating solar power, geothermal technologies tapping hydrothermal resources, and traditional hydroelectric technologies. The costs and performance of these already-developed technologies will be driven by incremental improvements in the technologies, learning curve cost reductions, and cost reductions associated with manufacturing economies of scale. This is not to imply that the first period technologies are at the same point in development or will not require technological innovations. Table 1 shows the areas of potential wind power technology improvements, which indicates the focus on technology developments are in improving the performance of individual components within the wind turbine system. There also is the trend towards larger turbines. Figure 1 shows the general push in solar technologies to improve the overall efficiency of solar cells as well as the competition among flat plate, crystalline, and concentrating PV cell technologies. Though solar, wind, and other first period technologies are all unique in terms of their

TABLE 1. Areas of Potential Wind Power Technology Improvements [6]			
		Performance and Cost Increments (Best/Expected/Least Percentages)	
Technical Area	Potential Advances	Annual Energy Production	Turbine Capital Cost
Advanced Tower Concepts	• Taller towers in difficult locations • New materials and/or processes • Advanced structures/foundations • Self-erecting, initial, or for service	+11/+11/+11	+8/+12/+20
Advanced (Enlarged) Rotors	• Advanced materials • Improved structural-aero design • Active Controls • Passive controls • Higher tip speed/lower acoustics	+35/+25/+10	-6/-3/+3
Reduced Energy Losses and Improved Availability	• Reduced blade soiling losses • Damage-tolerate sensors • Robust control systems • Prognostic maintenance	+7/+5/0	0/0/0
Drivetrains (Gearboxes and Generators ad Power Electronics)	• Fewer gear stages or direct-drive • Medium/low speed generators • Distributed gearbox topologies • Permanent-magnet generators • Medium-voltage equipment • Advanced gear tooth profiles • New circuit topologies • New semiconductor devices • New materials (gallium arsende [GaAs], SiC)	+8/+4/0	-11/-6/+1
Manufacturing and Learning Curve	• Sustained, incremental design and process improvements • Large-scale manufacturing • Reduced design loads	0/0/0	-27/-13/-3
Totals		+61/+45/+21	-36/-10/+21

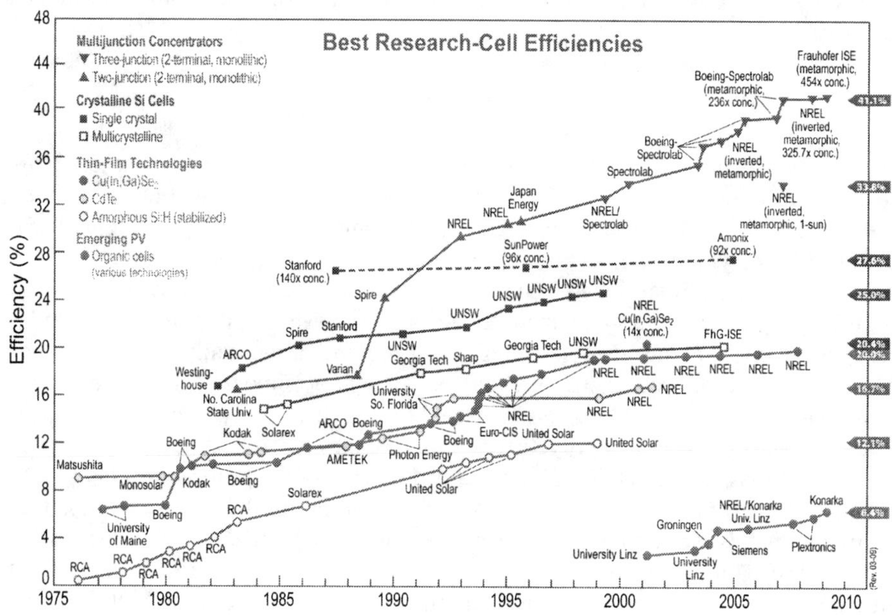

FIGURE 1. Historical progress of solar cell efficiencies (L. Kazmerski, National Renewable Energy Laboratory [NREL]).

cost, deployment, and integration characteristics, they are clearly much further along the development and deployment pathway compared to what the panel considered to be second period technologies. It should be noted that, although biomass may be appealing in terms of its technology status and cost, the future of biomass has been impacted greatly by mandates within the recent Energy Independence and Security Act of 2007, which greatly expand mandates for biofuels use. Because of the issues associated with biofuels development, the biomass resource base and its use in liquid transportation fuels was covered in the AEF study by the Panel on Alternative Liquid Fuels [3].

Technologies that will not be commercially deployed on a large scale until the second time period, if ever, include hot dry rocks or enhanced geothermal systems that use the heat stored in deep rocks and marine hydrokinetic technologies that use ocean tide and waves. These technologies will require fundamental research and technological innovations before they would be commercial sources of renewable electricity. For example, there are different approaches and designs for capturing wave energy that will need to be developed, deployed, and demonstrated before wave energy might become a major component of the technologies that a utility or other entity might consider as part of increasing its generating capacity along the coast. Finally, there are the more speculative technologies that might not even be fully conceived of at this point, such as high-altitude kites or PV installations in space, that might not even begin to see deployment before the middle of the century but could emerge with the development of new technologies, materials, and other fundamental advances.

Penetration

One element of the AEF study committee and its associated panels was to examine "aggressive but achievable" deployment rates for the future penetrations of energy technologies [1-4]. In general, the AEF study committee defined an aggressive but achievable deployment rate scenario as one that is more accelerated than the base case deployment rates defined by the Energy Information Agency (EIA)'s reference case (AEO2009), but less dramatic than an all-out or crash effort that could result in disruptive economic and lifestyle changes and require substantially new technologies [2]. The EIA produces annual projections of energy supply, demand, and prices conditions and includes a "business as usual" reference case [7]. This reference case, updated annually by the Energy Information Agency (EIA), assumes that existing laws, regulations, and practices are maintained (currently the project goes out to 2030) and provides a "business as usual" baseline scenario for the future of renewable electricity that is used in a wide array of policy and technical settings. In the AEO2009 scenario, renewables are projected to generate 14 percent of U.S. electricity, with 8 percent of that total from non-hydropower renewables. For the first time frame out to 2020, the Renewables panel concluded that there are clearly no current technological constraints for wind, solar photovoltaics and concentrating solar power, conventional geothermal, and biomass technologies to accelerate deployment, with the primary challenges being the cost-competitiveness of the existing technologies relative to most other sources of electricity (with no costs assigned to carbon emissions or other currently unpriced externalities), the lack of sufficient transmission capacity to move distant resources to demand centers, and the lack of sustained policies [1]. In the panel's opinion, an aggressive but achievable deployment level would be to have non-hydropower renewables contributing 10-percent or more of total electricity generation with trends towards continued growth. Combined with hydropower, this would mean renewable electricity would approach 20-percent of total U.S. electricity generation by 2020. In the second time frame out to 2035, the panel concluded that an aggressive but achievable deployment scenario would have continued and even further accelerated deployment resulting in non-hydroelectric renewables providing collectively 20-percent or more of domestic electricity generation.

Continued development and deployment of renewable electricity technologies, especially during the third time period beyond 2035, is expected to provide lower costs and potentially increase the percentage of total electricity generation that comes from renewables. However, increasing penetration of renewables, especially those with time-varying resource bases, poses issues for integrating those resources into the electricity system In considering issues associated with integrating renewables, the panel focused on two different penetration "endpoints", 20 percent or less and 50 percent or more. Integration of up to approximately 20-percent of non-hydropower renewables into the electricity system would require improvements to the transmission and distribution grid, additional transmission capacity and fast-responding generation, but no electricity storage. Clearly, this 20 percent is dependent on the generation mix and other factors within the transmission and distribution systems. Utilities that have substantial amounts of hydropower or natural gas-fired generation or well-integrated into the wholesale electricity market being able to absorb more variable renewable electricity compared to utilities that rely on nuclear and coal-fired generation. At the other extreme, achieving a predominant (50-percent or more) level of renewable electricity penetration will require new scientific advances including electricity storage and dramatic changes in the way that electricity is generated, transmitted and used. This includes some combination of intelligent, two-way electric grids; scalable and cost-effective methods for large scale and distributed storage (either direct electricity energy storage or generation of chemical fuels), widespread implementation of rapidly dispatchable fossil electricity technologies, use of synergies between two or more renewable resources and/or greatly improved technologies for cost-effective long-distance electricity transmission will be required [1]. The panel used these two endpoints to illustrate the varying levels of complexity associated with different penetration levels.

Deployment Factors

Since there are sufficient resources and technologies to allow renewables to greatly expand their contribution, the actual level of deployment will depend on other factors, including economics and issues related to integrating renewables into the electricity system. In general, the costs of electricity from renewables are greater than that for electricity from existing fossil and nuclear sources. Table 2 shows an estimate of the levelized cost of energy (LCOE) for various sources. The LCOE takes into account capital, fuel, operations and maintenance, and transmission costs as well as the capacity factor [1]. In addition, the spatial and temporal variability of renewables must be integrated into the overall electricity system. Figure 2 shows how one source, wind power, various in its spatial distribution and Figure 3 shows how that variability in the wind resources impacts the costs over different classes of resources. There have also been and continues to be much interests in electricity system costs for integration the variable characteristics of wind at different penetration levels. This includes both the costs of transmission and costs to integrate the variable nature of renewables into the electricity system. A recent report looked at 40 different studies of the costs of transmission for wind power projects that cover a broad geographic area and found that the costs ranged from $0 to $.079 per kWh, with the majority below $.025 per kWh, and a median of $.015 per kWh [9]. The costs to integrate renewable electricity include the need for other generation to help track load, provide voltage support and capacity reserves. Generally the costs rise with the levels of penetration and service areas that rely on fast responding natural gas or hydropower having lower integration costs than these that had a base load of coal and nuclear sources. The panel concluded that, at least for the best studied resource of wind, when the wind generation would be about $.080 per kWh, the impact of grid integration costs appeared to be less than 15 percent where wind produced 20 percent or less of total electricity generation [1].

There are a myriad of other factors related to capital, labor, materials, and policy that will influence the penetration of renewables into the electricity system. These include: constraints on capacity for larger-scale manufacturing and installation;

integration of variable resources into the existing electricity infrastructure and market; business risk and cost issue; unpredictability of and inconsistency in regulatory policies; and time requirements for building the necessary technical, business, and human infrastructures [1]. Issues such as the impacts of materials and labor shortages on costs, navigating the investment valley of death for deploying new technologies, changes in federal policies, and the individual local and state policies for siting and deploying these resources pose significant challenges.

Scenarios

One approach to understanding the impacts of an accelerated penetration of renewables, including the accelerated but achievable penetration level discussed above, is through the use of scenarios. In particular, there are a variety of scenarios that attempt to define the technological, economic, environmental, and implementation characteristics of a high renewable electricity future. Such scenarios have been developed for wind, solar, and geothermal [6, 10-18]. The most comprehensive in terms of considering the widest array of factors is the 20 percent wind study, a scenario developed by the American Wind Energy Association and DOE's National Renewable Energy Laboratory (NREL) that considered 20 percent of electricity generation would come from wind power by 2030 [6]. The results of this scenario demonstrate the potential costs and opportunities associated with such a future. It would require 300 GW of wind power generation with an installation of a total of 100,000 wind turbines and requiring 140,000 jobs in the manufacturing, construction, and operations of these turbines. Achieving this level of wind power generation would reduce CO_2 emissions by 800 million tons per year in 2030 and result in a direct increased cost for the total electricity sector of $43 billion dollars over the no-new-wind case and estimated transmission costs of $23-$100 billion [6, 19-20]. The high penetration solar scenarios offer a similar assessment of the additional costs and opportunities associated with greatly scaling up electricity production from solar PV [11-13, 17].

Combining multiple renewable technologies could also be used to meet the goal of providing 20 percent of total electricity generation by 2035 from new renewable electricity generation. Deploying an array of renewables might ease electricity system integration, particularly for wind generation. Balancing wind with solar, which do not normally peak together, and geothermal and biopower, which add base-load power, could mitigate the temporal variability in generation and take advantage of the geographical variability in the resource base for renewable electricity. For example, the variation in wind and solar can interact synergistically in locations where solar energy peaks during daylight hours, while wind energy peaks during late-night hours [21]. Other meteorological conditions that cause synergies between solar and wind power occur in regions where low atmospheric pressure fronts create more winds and clouds, and high atmospheric pressure conditions create sunny, stagnant conditions. Table 3 provides one such pathway to reach 20 percent of electricity generation by 2035, using AEO2009 as the basis for estimating the total electricity generation in 2035. This level of generation is achieved by assuming wind power capacity additions

TABLE 2. Levelized Cost of Energy (in $2007 per kWh) for New Plants Coming Online in 2012 from AEO2009 by Technology (Numbers for total LCOE from AEO2008 in parentheses) SOURCE: [1, 7-8].

Technology	Capacity Factor	Levelized Cost
Pulverized Coal	85%	.0871 (.0581)
Conventional Gas Combined Cycle	87%	.0807 (.0727)
Conventional Combustion Turbine	30%	.1317 (.1215)
Concentrating Solar Power	31%	.2508 (.1661)
Wind	36%	.0911 (.0849)
Offshore Wind	33%	.2095 (.1649)
Solar Photovoltaic	22%	.3622 (.3081)
Geothermal	90%	.1033 (.0668)
Biopower	83%	.1066 (.084)

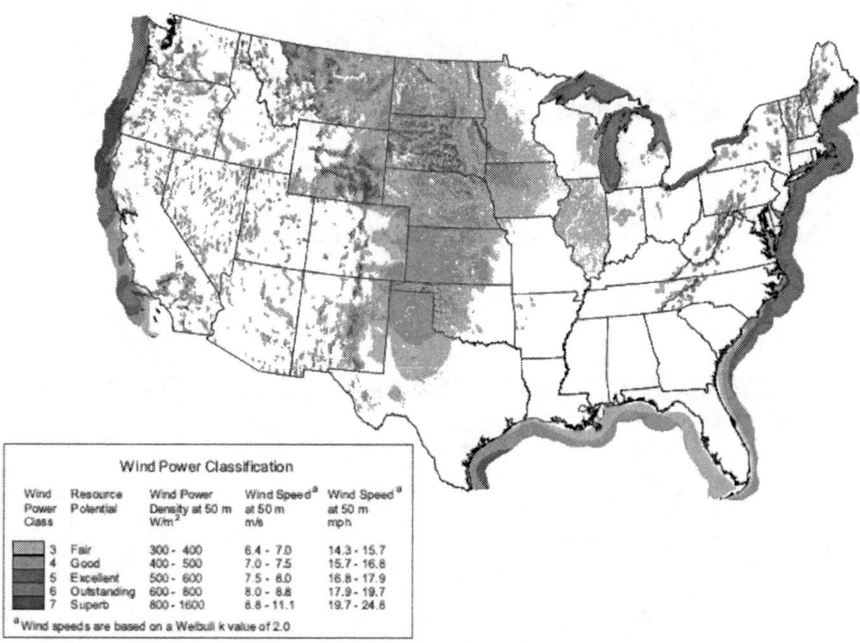

FIGURE 2. US map of wind power classes and 50 meter wind power resources. (DOE United States - Wind Resource Map: http://www.windpoweringamerica.gov/pdfs/wind_maps/us_windmap.pdf)

grow to 9.5 GW per year, a slight increase over the 8.4 GW installed in 2008, and then remain constant at this level out to 2035. This also shows solar generation capacity growing to 50 GW by 2035, a much smaller gain than those projected in the high market penetration solar scenarios [11-13, 15-18]. It assumes that an additional 15 GW would come from geothermal heat by 2035, which is slightly more that the Western Governors' Association's estimated potential resource base of 13 GW in the western United States [22]. It also assumes that an additional 15 GW would come from biomass. There is no attempt to portray the mix of renewables in Table 3 as an optimal mix, but as one mix that might reduce electricity system integration issues while being in line with the natural resource base and the human and financial resources needed to manufacture, deploy and maintain this generation.

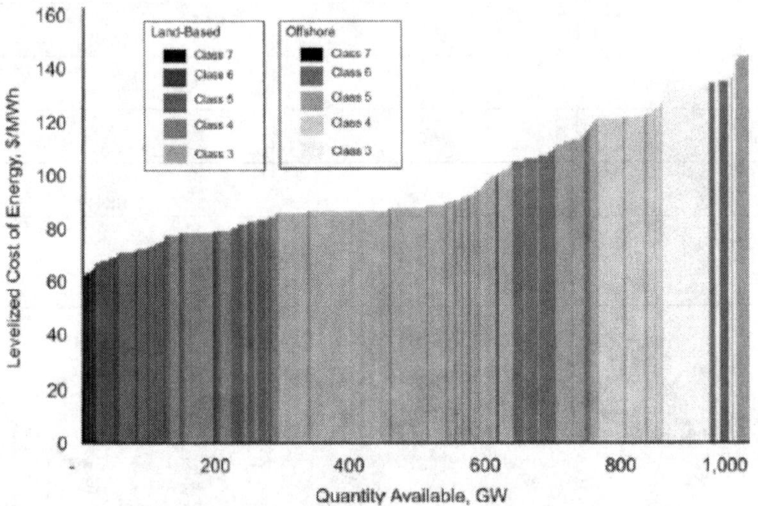

FIGURE 3. 2010 costs of wind power with the federal Production Tax Credit, $1,600/MW-mile transmission, and no integration costs for various wind classes [6].

TABLE 3. Capacity and Generation from Multiple Renewable Resources Sufficient to Meet 20 Percent of Estimated U.S. Electricity Demand in 2035

	Generating Capacity (GW) in 2035	Capacity Factor	Electricity Generation (GWh) in 2020	Electricity Generation (GWh) in 2035
Wind	252	0.35	349,524	786,429
Solar	50	0.15	29,200	65,700
Biomass	15	0.90	52,560	118,260
Geothermal	15	0.90	52,560	118,260

RECENT TRENDS

There have been numerous changes in the electricity sector and overall economy that have impacted the deployment of renewable electricity since 2009. Some of the changes have been factors directly relevant to renewables, such as costs or policies for renewables. However, many of the changes having significant impacts on renewables deployment are external, including the reduction in available capital and investments in due to the economic slowdown or improvements economics and resource base of other sources of electricity. Here we highlight some of these issues, though this list is neither comprehensive nor constant.

Renewable Trends

Factors that are directly related to renewable electricity include insufficient regional transmission capacity for wind power and the declining costs for PV modules. Transmission capacity is critical for moving wind power from some of the high resource, low population areas shown in Figure 2 to demand centers. Figure 4 shows the increasing impacts of transmission shortages on wind power production in Texas. An estimated 17 percent of potential wind energy generation within Texas was curtailed according to the Electric Reliability Council of Texas compared to 8 percent in 2008 and 1 percent in 2007 [23]. The Bonneville Power Administration also will reduce its region's wind power output during extreme high Columbia River flows [24]. This is done to prevent sending excess waters over dam spillways that has negative impacts on protected fish populations [24]. The state of Texas has announced a plan to invest almost $5 billion in additional transmission to bring wind power from the western part of the state to population centers, but the construction of these lines is not expected to begin until 2012. The difficulty in siting new transmission has also brought about a possible revisualization of the necessary transmission infrastructure required for bringing large amounts of wind power into the electricity sector. Central to many high wind power scenarios is the concept of a coast-to-coast "Electricity Superhighway" radiating from the Midwestern wind states (from Kansas north to the Dakotas and Wyoming) to the large cities in the Midwest and population centers along the coasts. However, the difficulty in financing and siting such transmission through multiple states has spurred interests in designs that reduce the size of the Electricity Superhighway and incorporate more offshore wind power where generation can be delivered directly into urban areas and transmission nodes along the coast without the need to site transmission through multiple states. Figure 5 shows such a concept. The Atlantic Wind Connection project being financed by Google and others fits into this scheme by providing a transmission "backbone" for moving 6,000 MW of off-shore wind into the electricity sector along the mid-Atlantic.

During the panel's study, constraints including limited supplies of raw materials, manufacturing capacity, and skilled installers as well as increasing international demands caused an increase in costs for new wind and PV generation systems. As shown in Figure 6, the upward trend for PV prices reversed itself by 2009. This is especially true for the PV modules, which is the assembly of interconnected photovoltaic cells, and may not be true for the other system components. It is unclear how current economic conditions and changes in manufacturing and market conditions will affect wind power costs. Though installed wind power project costs continued to rise in 2009, prices for wind power turbines may have leveled off [23].

Other Electricity Sector and Economic Trends

Two trends that have occurred in the electricity sector and wider economy that are having impacts on the deployment of additional renewable electricity capacity are the expansion of natural gas supplies and general slowing of the economy. Both of these have played roles in the slowing of the wind power generation in particular. The expansion of natural gas supplies due to innovations in extraction of gas from shale has had a major impact on wholesale electricity prices [23]. As shown in Figure 6, this impacts the competitiveness of wind power in the wholesale electricity market. The impacts of an expansion of natural gas supplies on renewables is complex since renewable electricity both competes with natural gas as well as relies on the ability of natural gas generation plants to respond quickly when renewable resources are not available. The economic downturn also has hurt investments in the clear energy sector. Though private investments in this sector have increased greatly over the past decade, there has been a marked slowdown since peaking in 2008. Figure 7 shows this trend for wind power. A recent report indicates that clean energy investment in the US increased in 2010 over 2009 values, but this increase may be due to public stimulus spending [28].

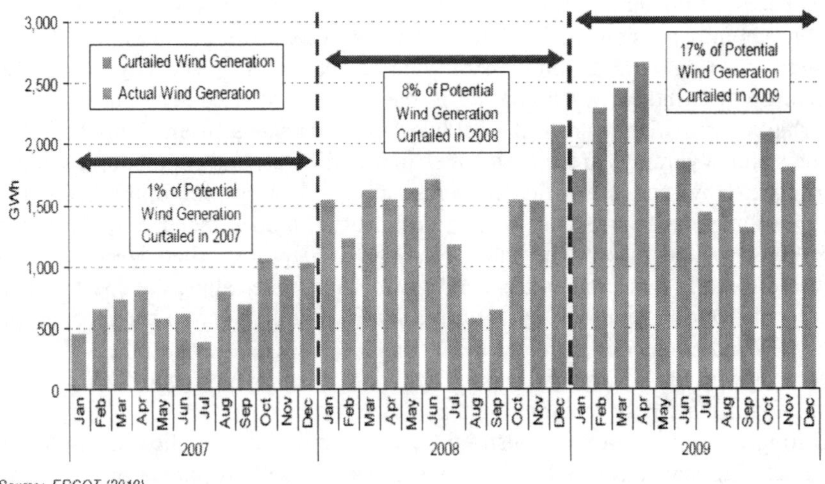

FIGURE 4. Wind power curtailments within Texas [23].

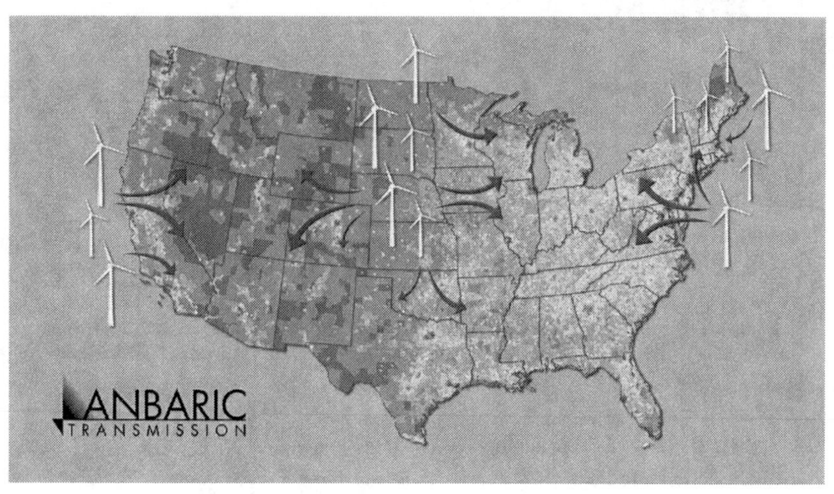

FIGURE 5. Concept of a regional Electric Superhighway [25].

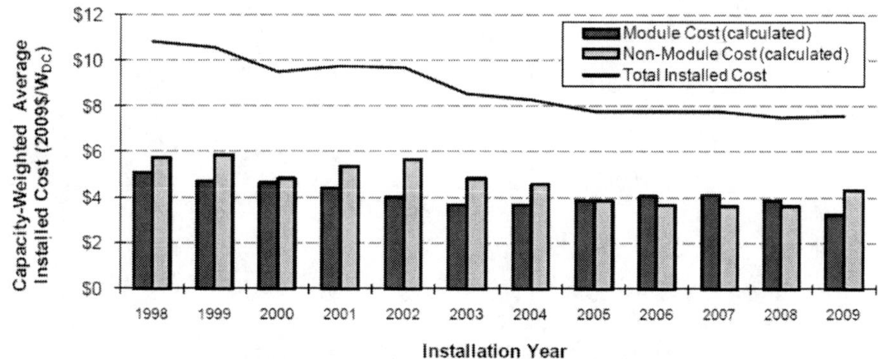

FIGURE 6. Costs for solar modules, other system components, and total PV system costs over time [26].

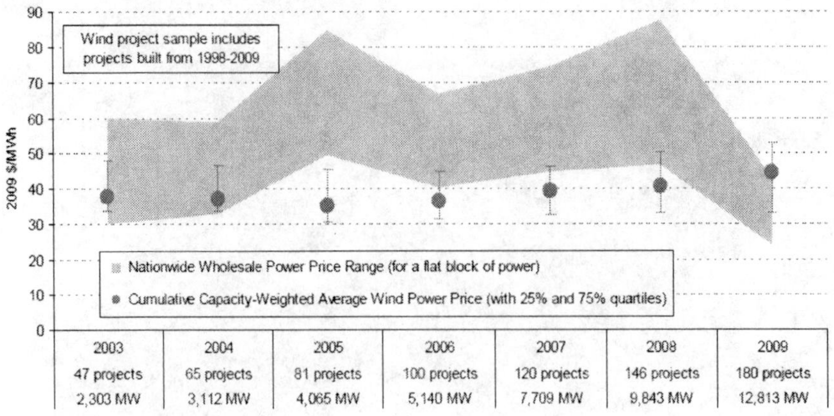

FIGURE 7. Price of wholesale power versus wind power price [24].

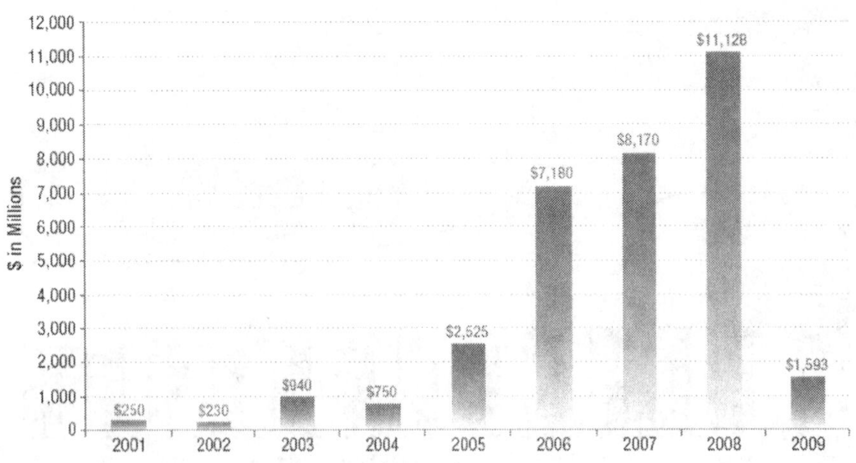

FIGURE 8. US wind energy project financing [27].

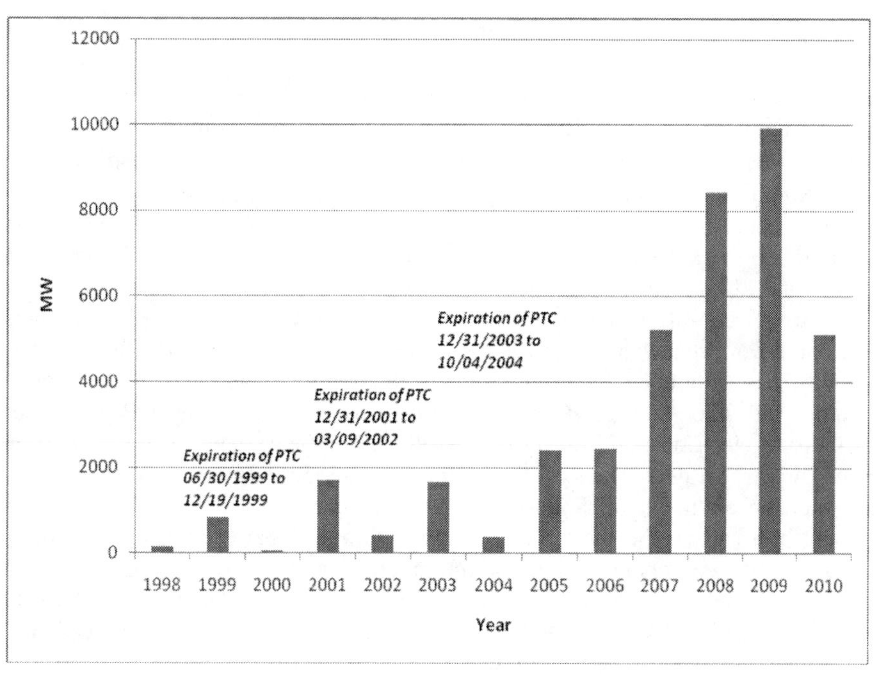

FIGURE 9. Annual installations of new wind power capacity.

Impacts of Recent Trends

Renewable electricity, especially wind and solar power, experienced large growth in new capacity during the panel's activities, even through the beginnings of the economic downturn in 2009. However, after 3 years of large increases in annual installations of wind power, wind power capacity additions fell by almost half to 5.4 GW in 2010 from almost 10 GW in 2009 (Figure 9). This slowdown has continued in the first quarter of 2011 with only 1.1 GW installed, although there was 5.6 GW under construction [29]. PV has not experienced this decline in growth, with a total of almost 900 MW being installed in 2010, up over 400 MW from 2009, despite the economic slowdown [30]. Going forward, there are a complex set of variables that will determine the trends for future growth, some of which will depend on future policies. In particular, Figure 9 shows the impacts of past expirations of the federal Production Tax Credit (PTC) on wind power capacity additions. The PTC currently provides a tax credit of 2.2 cents/kWh for the first 10 years of service of a privately-owned or investor-owned renewable electricity project and is currently scheduled to expire for wind power at the end of 2012.

CONCLUSIONS

Given the resource base and the state of technologies, the Renewables panel concluded that electricity from renewables represent a significant opportunity—with attendant challenges—to provide low CO_2-emitting electricity generation from resources available domestically and generate new economic opportunities for the United States [1]. The panel concluded that an aggressive by achievable deployment of non-hydropower renewables could contribute 10 percent of total US electricity generation by 2020 and 20 percent or more of total electricity generation by 2035. Combined with hydropower, this means that almost 30 percent of total electricity production could come from renewables. Since the panel released its report in 2009, renewable electricity has continued to grow, though this growth has been uneven due to economic and other factors. Going forward, renewable electricity will continue to contribute to electricity generation in the US. Its ultimate level of penetration will depend on an array of factors, including its economic competitiveness, system integration, access to capital, and local, state, and federal policies. In addition, growth in renewables capacity in the US will occur within a global economy where the US and other nations will play roles in the international supply and demand for materials and technologies. Indeed, in recognizing the importance of renewables internationally, the NRC and NAE recently cooperated with the Chinese Academy of Sciences and Chinese Academy of Engineering to produce a report looking at the opportunities and challenges for renewables in both countries [31]. The next decade will prove critical for understanding the potential for renewable electricity to make a significant contribution to the electricity sector in the US. And this future will play out in a dynamic national and international setting, impacted by a wide array of factors that sometimes directly and sometimes only indirectly relate to renewable resources and technologies.

ACKNOWLEDGEMENTS

The authors acknowledge the expertise and dedicated efforts of the NRC's Panel on Electricity from Renewable Sources, upon whose report this paper is taken. The members of this panel were: Lawrence Papay (chair), Allen Bard, Rakesh Agrawal, William Chameides, Jane Davidson, Mike David, Kelly Fletcher, Charles Gay, Charles Goodman, Sossina Haile, Nathan Lewis, Karen Palmer, Jeffery Peterson, Karl Rabago, Carl Weinberg, and Kurt Yeager. The authors also would like to acknowledge the many individuals that made presentations to the NRC panel and reviewers that provided comments on the panel report, which are listed in the panel's report [1].

REFERENCES

1. NAS-NAE-NRC (National Academy of Sciences-National Academy of Engineering-National Research Council). 2009. Electricity from Renewable Resources: Status, Prospects, and Impediments. National Academies Press, Washington, DC.
2. NAS-NAE-NRC. 2009. America's Energy Future: Technology and Transfer. National Academies Press, Washington, DC.
3. NAS-NAE-NRC. 2009. Alternative Liquid Transportation Fuels: Technological Status, Costs, and Environmental Impacts. National Academies Press, Washington, DC.
4. NAS-NAE-NRC. 2009. Real Prospects for Energy Efficiency in the United States. National Academies Press, Washington, DC.
5. EIA (Energy Information Administration). 2008. Annual Energy Review 2007. Washington, DC.
6. DOE (Department of Energy). 2008. 20-percent Wind Energy by 2030: Increasing Wind Energy's Contribution to US Electricity Supply. Washington DC.
7. EIA (Energy Information Administration). 2008c. Annual Energy Outlook 2008 with Projections to 2030, U.S. Department of Energy DOE/EIA-0383(2008). Washington, DC.
8. EIA. 2008d. Annual Energy Outlook 2009 Early Release, U.S. Department of Energy DOE/EIA-0383(2009). Washington, DC.
9. Mills, A. and R. Wiser. 2009. The Cost of Transmission for Wind Energy: A Review of Transmission Planning Studies, Lawrence Berkley National Laboratories. Berkley, CA.
10. Smith, J.C. and B. Parsons. 2007. What does 20% look like? IEEE Power and Energy Magazine 5: 22-33.
11. DOE). 2007. Solar America Initiative – A Plan for the Integrated Research, Development, and Market Transformation of Solar Energy Technologies. Washington, DC.
12. SEIA (Solar Energy Industries Association). 2004. Our Solar Power Future—The US Photovoltaic Industry Roadmap Through 2030 and Beyond. Washington, DC.
13. Pernick, R., and C. Wilder. 2008. Utility Solar Assessment (U.S.A) Study Reaching Ten% by 2025. Clean Edge, Inc.
14. MIT (Massachusetts Institute of Technology). 2006. The Future of Geothermal Energy: Impact of Enhanced Geothermal Systems (EGS) on the United States in the 21st Century. Cambridge, Mass.
15. Zweibel, K., J. Mason, and V. Fthenakis. 2007. A Solar Grand Plan, Scientific American, January Issue.
16. Stoddard, L., J. Abiecunas,. and R. O'Connell. 2006. Economic, Energy, and Environmental Benefits of Concentrating Solar Power in California. National Renewable Energy Laboratory. Golden, CO.
17. Feltrin, A., and A. Freundlich. 2008. Material Considerations for Terawatt Level Deployment of Photovoltaics. Renewable Energy 33: 180-185.
18. Grover, S. 2007. Energy, Economic, and Environmental Benefits of the Solar America Initiative. DOE National Renewable Energy Laboratory, Subcontractor Report NREL/SR-640-41998. Washington, DC.
19. AEP (American Electric Power). 2007. Interstate Transmission Vision for Wind Integration, AEP White Paper. Columbus, OH.
20. JCSP (Joint Coordinated System Plan). 2009. Joint Coordinated System Plan 2008.
21. CEC (California Energy Commission). 2007. Intermittency Analysis Project : Final Report.
22. WGA (Western Governor's Association). 2006. Clean and Diversified Energy Initiative: Geothermal Task Force Report. Washington, DC
23. DOE. 2010. Annual Report on US Wind Power Installation, Cost, and Performance Trends. DOE Energy Efficiency and Renewable Energy. Washington, DC
24. ClimateWire, Bonneville Power to wind generators -- shut down, and you get free power, February 25, 2011.
25. Krapels, E. and J. Edwards. Integrating 200,000MWs of Renewable Energy into the US Power Grid: A Practical Proposal. Anbaric Transmission, LLC.

26. LBNL (Lawrence Berkeley National Laboratory). 2010. Tracking the Sun III: The Installed Cost of Photovoltaics in the United States from 1998-2009
27. DOE. 2009. 2009 Renewable Energy Data Book. DOE Energy Efficiency and Renewable Energy. Washington, DC
28. Pew Charitable Trust. 2011. Who's Winning the Clean Energy Race – 2010 Edition. Washington, DC.
29. AWEA (American Wind Energy Association). 2011. US Wind Industry First Quarter 2011 Market Report. Washington, DC.
30. SEIA (Solar Energy Industries Association). 2011. U.S. Solar Market Insight: 2010 Year in Review.
31. NAP (National Academies Press). 2010. The Power of Renewables: Opportunities and Challenges or China and the United States. Electricity from Renewable Resources: Status, Prospects, and Impediments. National Academies Press, Washington, DC.

Integrating Renewable Electricity on the Grid*

George Crabtree[a], Jim Misewich[b], Ron Ambrosio, Kathryn Clay, Paul DeMartini, Revis James, Mark Lauby, Vivek Mohta, John Moura, Peter Sauer, Francis Slakey, Jodi Lieberman and Humayun Tai

[a]*Materials Science Division*
Argonne National Laboratory
Argonne, Illinois
and
Departments of Physics, Electrical and Mechanical Engineering
University of Illinois at Chicago

[b]*Associate Laboratory Director for Basic Energy Sciences*
Brookhaven National Laboratory
Upton, New York

Abstract. The demand for carbon-free electricity is driving a growing movement of adding renewable energy to the grid. Renewable Portfolio Standards mandated by states and under consideration by the federal government envision a penetration of 20-30% renewable energy in the grid by 2020 or 2030. The renewable energy potential of wind and solar far exceeds these targets, suggesting that renewable energy ultimately could grow well beyond these initial goals.

The grid faces two new and fundamental technological challenges in accommodating renewables: location and variability. Renewable resources are concentrated at mid-continent far from population centers, requiring additional long distance, high-capacity transmission to match supply with demand. The variability of renewables due to the characteristics of weather is high, up to 70% for daytime solar due to passing clouds and 100% for wind on calm days, much larger than the relatively predictable uncertainty in load that the grid now accommodates by dispatching conventional resources in response to demand.

Solutions to the challenges of remote location and variability of generation are needed. The options for DC transmission lines, favored over AC lines for transmission of more than a few hundred miles, need to be examined. Conventional high voltage DC transmission lines are a mature technology that can solve regional transmission needs covering one- or two-state areas. Conventional high voltage DC has drawbacks, however, of high loss, technically challenging and expensive conversion between AC and DC, and the requirement of a single point of origin and termination. Superconducting DC transmission lines lose little or no energy, produce no heat, and carry higher power density than conventional lines. They operate at moderate voltage, allowing many "on-ramps" and "off-ramps" in a single network and reduce the technical and cost challenges of AC to DC conversion. A network of superconducting DC cables overlaying the existing patchwork of conventional transmission lines would create an interstate highway system for electricity that moves large amounts of renewable electric power efficiently over long distances from source to load. Research and development is needed to identify the technical

* This chapter contains the Conclusions (as Abstract), Executive Summary and sections on Energy Storage and Long-distance Transmission from the 2010 APS Panel on Public Affairs report, *Integrating Renewable Resources on the Grid,* http://www.aps.org/policy/reports/popa-reports/upload/integratingelec.pdf. See also a short qualitative summary of the report in G.W. Crabtree and Jim Misewich, *Is the Grid Ready for Renewables?* APS News 19(11) Dec 2010, http://www.aps.org/publications/apsnews/201012/backpage.cfm.

Physics of Sustainable Energy II: Using Energy Efficiently and Producing it Renewably
AIP Conf. Proc. 1401, 387-405 (2011); doi: 10.1063/1.3653865

challenges associated with DC superconducting transmission and how it can be most effectively deployed.

The challenge of variability can be met (i) by switching conventional generation capacity in or out in response to sophisticated forecasts of weather and power generation, (ii) by large scale energy storage in heat, pumped hydroelectric, compressed air or stationary batteries designed for the grid, or (iii) by national balancing of regional generation deficits and excesses using long distance transmission. Each of these solutions to variability has merit and each requires significant research and development to understand its capacity, performance, cost and effectiveness. The challenge of variability is likely to be met by a combination of these three solutions; the interactions among them and the appropriate mix needs to be explored.

The long distances from renewable sources to demand centers span many of the grid's physical, ownership and regulatory boundaries. This introduces a new feature to grid structure and operation: national and regional coordination. The grid is historically a patchwork of local generation resources and load centers that has been built, operated and regulated to meet local needs. Although it is capable of sharing power across moderate distances, the arrangements for doing so are cumbersome and inefficient. The advent of renewable electricity with its enormous potential and inherent regional and national character presents an opportunity to examine the local structure of the grid and establish coordinating principles that will not only enable effective renewable integration but also simplify and codify the grid's increasingly regional and national character.

EXECUTIVE SUMMARY

The United States has ample renewable energy resources. Land-based wind, the most readily available for development, totals more than 8000 GW of potential capacity. The capacity of concentrating solar power is nearly 7,000 GW in seven southwestern states. The generation potential of photovoltaics is limited only by the land area devoted to it, 15–40 MW/km^2 in the United States. To illustrate energy capacity vs. projected demand, the US generated electric power at an average rate of approximately 450 GW in 2009, with peaks over 1000 GW during the summer months. By 2035, electricity demand is projected to rise 30%.

To date, 30 states plus the District of Columbia have established Renewable Portfolio Standards (RPS) requiring a minimum share of electrical generation to be produced by renewable sources. In addition to state policies, federal policymakers have put forward proposals to establish a national RPS, making the need for technological developments more urgent.

However, developing renewable resources presents a new set of technological challenges not previously faced by the grid: the location of renewable resources far from population centers, and the variability of renewable generation. Although small penetrations of renewable generation on the grid can be smoothly integrated, accommodating more than approximately 30% electricity generation from these renewable sources will require new approaches to extending and operating the grid.

The variability of renewable resources due to characteristic weather fluctuations introduces uncertainty in generation output on the scale of seconds, hours and days. These uncertainties affect up to 70% of daytime solar capacity due to passing clouds and 100% of wind capacity on calm days for individual generation assets (Figure 1). Although aggregation over large areas mitigates the variability of individual assets, there remain uncertainties in renewable generation that are greater than the relatively predictable variation in demand that the grid deals with regularly.

Greater uncertainty and variability can be dealt with by switching in fast-acting conventional reserves as needed on the basis of weather forecasts on a minute-by-minute and hourly basis, by installing large scale storage on the grid or by long distance transmission of renewable electricity providing access to larger pools of resources in order to balance regional and local excesses or deficits. At present, renewable variability is handled almost exclusively by ramping conventional reserves up or down on the basis of forecasts. However, as renewable penetration grows, storage and transmission will likely become more cost effective and necessary.

Forecasting

The high variability of renewable generation, up to 100% of capacity, makes forecasting critical for maintaining the reliability of the grid. Improving the accuracy and the confidence level of forecasts is critical to the goal of reducing the conventional reserve capacity, and will result in substantial savings in capital and operating costs.

The variability of renewable energy is easily accommodated when demand and renewable supply are matched—both rising and falling together. However when demand and renewable supply move in opposite directions, the cost of accommodation can rise significantly. For example, if the wind blows strongly overnight when demand is low (as is often the case), the renewable generation can be used only if conventional base-load generation such as coal or nuclear is curtailed, an expensive and inefficient option that may cause significant reliability issues. Alternatively, on calm days when there is no wind power, the late-afternoon peak demand must be met entirely by conventional generation resources, requiring reserves that effectively duplicate the idle renewable capacity. Reducing the cost of dealing with these two cases is a major challenge facing renewable integration.

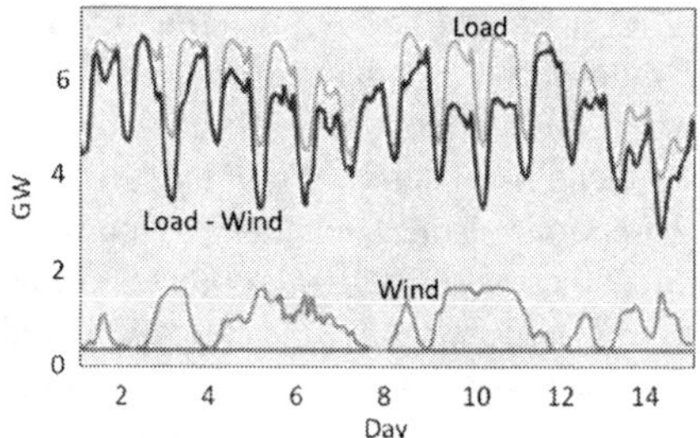

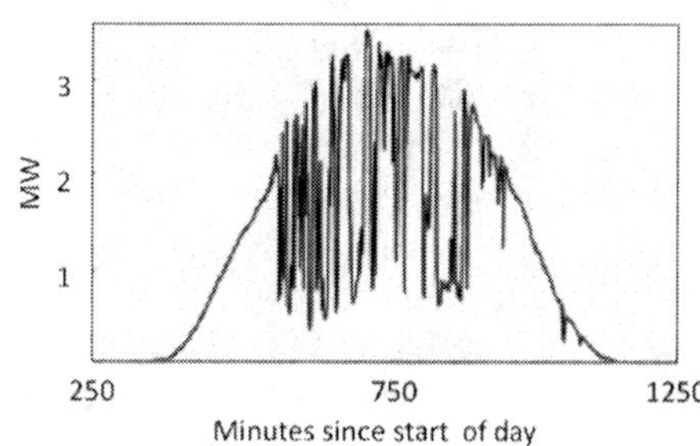

FIGURE 1. Variability of wind generation over a 14 day period, with variation of 100% on calm days, for a 1.5 GW wind plant in a 10 GW capacity system (left). Variability of solar photovoltaic generation due to passing clouds in 3.5 MW capacity system (right).

Recommendations on Forecasting

The National Oceanic and Atmospheric Administration (NOAA), the National Weather Service (NWS), the National Center for Atmospheric Research (NCAR) and private vendors should:

• Improve the accuracy of weather and wind forecasts, in spatial and temporal resolution and on time scales from hours to days. In addition to accuracy, the confidence level of the forecasts must be improved to allow system operators to reduce reserve requirements and contingency measures to lower and more economical levels.

Forecast providers, wind plant operators, and regulatory agencies should:
• Agree on and develop uniform standards for preparing and delivering wind and power generation forecasts.

Wind plant operators and regulatory agencies should:
• Develop and codify operating procedures to respond to power generation forecasts. Develop, standardize and codify the criteria for contingencies, the response to up- and down-ramps in generation, and the response to large weather disturbances. Develop response other than maintaining conventional reserve, including electricity storage and transmission to distant load centers.

Energy Storage

As renewable generation grows it will ultimately overwhelm the ability of conventional resources to compensate renewable variability, and require the capture of electricity generated by wind, solar and other renewables for later use. Transmission level energy storage options include pumped hydroelectric, compressed air electric storage, and flywheels. Distribution level options include: conventional batteries, electrochemical flow batteries, and superconducting magnetic energy storage (SMES). Batteries and SMES also might be integrated with individual or small clusters of wind turbines and solar panels in generation farms to mitigate fluctuations and power quality issues. Although grid storage requires high capacity and a large footprint, it also allows a stationary location and housing in a controlled environment, very different from the conditions for portable or automotive storage. These differing requirements open a wide variety of still-unexplored storage technologies to the grid.

Currently, energy storage for grid applications lacks a sufficient regulatory history. Energy storage on a utility-scale basis is very uncommon and, except for pumped hydroelectric storage, is relegated to pilot projects or site-specific projects. Some states such as New York categorize storage as "generation," and hence forbid transmission utilities from owning it. Utilities are therefore uncertain how regulators will treat investment in energy storage technologies, how costs will be recovered, or whether energy storage technologies will be allowed in a particular regulatory environment.

Recommendations on Energy Storage

The Department of Energy (DOE) should:
• Develop an overall strategy for energy storage in grid level applications that provides guidance to regulators to recognize the value that energy storage brings to both transmission and generation services to the grid;

• Conduct a review of the technological potential for a range of battery chemistries, including those it supported during the 1980s and 1990s, with a view toward possible applications to grid energy storage; and
• Increase its R&D in basic electrochemistry to identify the materials and electrochemical mechanisms that have the highest potential for use in grid level energy storage devices.

Long Distance Transmission

Renewable sources are typically distributed over large areas in the upper central and southwestern US, including the Dakotas, Iowa, Minnesota, Montana, Arizona and New Mexico, far from demand centers east of the Mississippi and on the West Coast. New large area collection strategies and new long distance transmission capability are required to deliver large amounts of power a thousand miles or more across the country. This long distance transmission challenge is exacerbated by a historically low investment in transmission: from 1979–1999 electricity demand grew by 60% while transmission investments fell by more than 50%.. In denser population areas there are community concerns around new right-of-way for above ground transmission towers. The "not in my backyard" arguments are costly to overcome and can delay or stop above-ground transmission construction.

While high voltage DC is the preferred transmission mode for long distances, the drawbacks of single terminal origin and termination, costly AC-DC-AC conversion, and the decade or more typically needed for approval for long lines create problems for renewable electricity transmission. Superconductivity provides a new alternative to conventional high voltage DC transmission. Superconducting DC lines operate at zero resistance, eliminating electrical losses even for long transmission distances, and operate at lower voltages, simplifying AC-DC conversion and enabling wide-area collection strategies. Superconducting DC transmission lines carrying 10 GW of power 1600 km can be integrated into the Eastern and Western grids in the US while maintaining transient and short-term voltage stability.

Recommendations on Long Distance Transmission

DOE should:
• Extend or replace the Office of Electricity program on High Temperature Superconductivity for a period of 10 years, with focus on DC superconducting cables for long distance transmission of renewable electricity from source to market; and
• Accelerate R&D on wide band gap power electronics for controlling power flow on the grid, including alternating to direct current conversion options and development of semiconductor-based circuit breakers operating at 200 kV and 50 kA with microsecond response time.

Business Case

Utility renewable energy investments are typically assessed from regulatory, project finance, and technical perspectives. The regulatory assessment focuses on ensuring utility compliance with renewable portfolio standards (RPS) and that costs are kept within prudent limits. The project finance view looks at the merits of the investment within discrete boundaries of the funding and cash flows exclusive to the project under review. The technical assessment evaluates the engineering and operational risks of the project and specific technologies involved.

While these conventional views are important for investors, utilities, regulators and ratepayers, they do not fully capture the set of benefits that a renewable energy investment can deliver beyond the boundaries of a given project, such as the physical benefits of transmission and storage and the organizational benefit of developing an integrated approach to the grid. Inclusion of these additional benefits in an expanded business case will enhance the profile of the renewables investment, and more importantly, begin to recognize the value of synergies among storage, transmission and renewable generation on the grid.

Recommendations on Business Case

The Federal Energy Regulatory Commission (FERC) and the North American Electric Reliability Corporation (NERC) should:

• Develop an integrated business case that captures the full value of renewable generation and electricity storage in the context of transmission and distribution; and

• Adopt a uniform integrated business case as their official evaluation and regulatory structure, in concert with the state Public Utility Commissions (PUCs).

ENERGY STORAGE (full report section)

As renewable energy penetration grows, the increasing mismatch between the variation of renewable energy resources and electricity demand makes it necessary to capture electricity generated by wind, solar and other renewable energy generation for later use. Storage can help smooth short-time fluctuations in generation inherent in wind or solar energy as well as time-shift renewable generation resources from low-demand periods to high-demand periods.

The Case for Grid-Level Energy Storage

Grid level or stationary utility energy storage includes a range of technologies with the ability to store electricity and dispatch it as needed.[1. 2] Energy storage can enhance the reliability and resilience of the grid through short-term storage for peak-shaving and power quality uses and longer-term storage for load-leveling and load-shifting applications. As larger amounts of intermittent renewable energy sources such as wind and solar energy enter the market, grid energy storage becomes a means of compensating for generation fluctuations of these sources on timescales ranging from seconds to hours.

Large-scale energy storage on the electric grid is not a new concept. The current grid uses pumped hydro and to a lesser extent, compressed air energy storage (CAES) for these purposes. These options could be expanded, but are limited to geographically appropriate sites. They have the advantage of fast response; a few minutes or less for pumped hydro and about 10 minutes for CAES.

Batteries offer another means of grid-level energy storage by converting electricity to chemical energy during times when electrical supply exceeds demand. Unlike pumped hydro and CAES, battery storage is feasible for any geographical location.

Thermal storage using molten salts or other media is effective for concentrating solar power plants like the solar energy generating systems in the U.S. Mojave desert, and the Andasol plants near Granada, Spain.[3] Thermal storage stabilizes fluctuations due to passing clouds and allows electricity to be produced after the hours of peak sunshine.

Flywheels are being effectively used in California and New York for frequency regulation, which will become more important with increased integration of variable power sources. The international fusion community uses flywheels to store 2-4 GJ (~ 1 MWh) and deliver power at several hundreds of megawatts, accumulating many thousands of charge/discharge cycles over their 20-year lifetime.[4] Superconducting magnetic energy storage (SMES) with a capacity of a few MJ is used for regulating power quality. Much higher power and energy SMES—that can deliver 100 MW of power for seconds to minutes—has been developed for fusion applications. The opportunities for lower cost and higher energy storage capacity are related to the cost and maximum magnetic field strength of superconducting wire. Synergies between DC superconducting transmission and SMES, which also operates at DC, offer cost and technology savings opportunities.

Increased interest in large scale storage led ARPA-E to issue a broad call for proposals for utility scale energy storage including each of the categories described above.[5, 6] Funded projects include SMES, flywheels, compressed air and batteries.[7]

The use of energy storage for utility applications can be divided into three categories: (1) for base load bulk power management, (2) for grid support in the form of distributed or load leveling storage, or (3) for power quality and peak power storage, including uninterruptable power supply applications. Within each of these broad categories, different timescales from seconds to hours apply. The purpose of the storage and the timescale of response determine which energy storage technologies are best suited for a given application. Figure 2 depicts a number of energy storage options, including several different battery chemistries.

Currently, the most pervasive use of large-scale chemical energy storage is for power quality in the form of uninterruptible power supplies (UPS). UPS is used to protect expensive electrical assets such as computer data centers and critical infrastructure. Such systems do not require high energy content since most power outages are less than a minute in length. Lead acid and metal hydride batteries are the mainstays of this industry.

For renewable generation, storage can help manage the transmission capacity for wind energy resources. By adding energy storage, wind plants located in remote areas can store energy during peak production periods and release it during peak demand periods. Storing the generated electricity rather than using it in real time lowers the need for transmission lines and also allows retailers to maximize profits by selling power during peak usage periods, which do not usually correspond with peak wind output periods. While these applications will become increasingly important as renewable energy is more widely deployed, less expensive and higher capacity storage must be developed to improve their economic appeal.

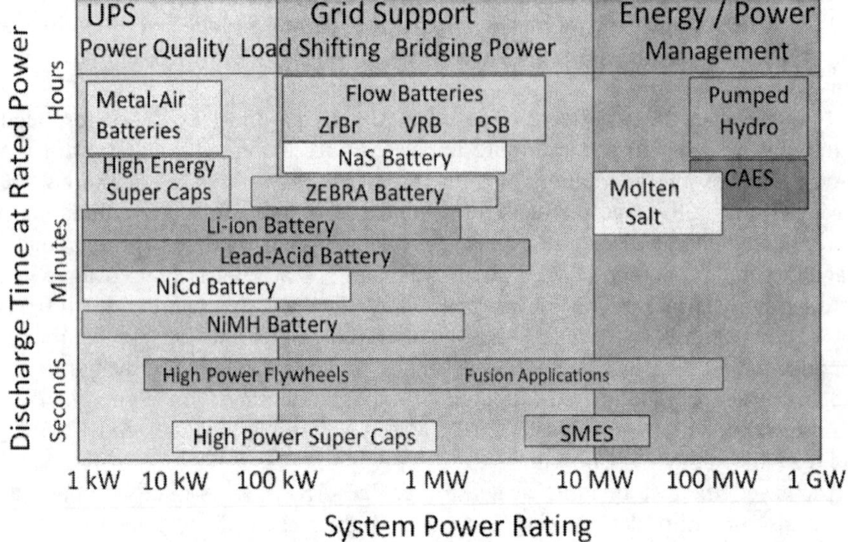

FIGURE 2. The energy storage options sorted by power rating and discharge rate. Sources: Adapted from EPRI Report 1020676, Electrical Energy Storage Technology Options (2010).

The Physical Scale of Grid Energy Storage

The availability of wind and solar energy sources can vary significantly, sometimes in a matter of seconds and at other times over hours or days. The different time frames impose different energy storage requirements: (1) relatively low capacity but fast response for changes that occur within seconds or over a period of a few hours and (2) high capacity but slower response for changes that extend over one or more days. We term the first storage need a "power application" and the second an "energy application." Although storage requirements extend continuously across the time spectrum, and many storage technologies span the two applications, this simplifying classification provides a useful sense of the physical scale of the storage challenge.

In the accompanying table, we illustrate the power application storage need for a 70% reduction in solar photovoltaic (PV) electricity generation or 20% reduction in wind generation, assuming each occurs over a one-hour period. We illustrate the energy application storage need for accommodating 12 hours of solar production and 24 hours of wind production. The table shows the physical sizes of various kinds of storage units required for a 100 megawatt solar installation—the generating capacity

Power Applications

Storage Technology	100 MW Solar PV or CSP 70 MWh Storage Capacity	750 MW Wind 150 MWh Storage Capacity
Lead-acid battery	1170 m^3	2500 m^3
Lithium-ion battery	194 m^3	417 m^3
Sodium-sulfur battery	269 m^3	558 m^3
Flow battery	2340 m^3	5000 m^3
Molten salt thermal	5300 m^3	Not Applicable

Energy Applications

Storage Technology	100 MW Solar PV or CSP 1200 MWh Storage Capacity	750 MW Wind 18000 MWh Storage Capacity
Flow battery	40000 m^3	600000 m^3
CAES	385000 m^3	$5.77 \times 10^6 m^3$
Pumped hydro	$2.14 \times 10^6 m^3$ (500 m head)	$32.1 \times 10^6 m^3$ (500 m head)
Molten Salt thermal	90900 m^3	Not Applicable

TABLE 1. Volume of energy storage systems required for low capacity fast response power applications and high capacity slower response energy applications.[1]

of typical large photovoltaic and moderate concentrating solar power (CSP) plants—and for a 750 megawatt wind farm—the capacity of typical large wind installations. Note that a molten salt thermal storage unit is appropriate only for a CSP plant.

Battery Energy Storage Technologies

Interest in electric vehicles is driving a great deal of investment in energy storage R&D for mobile applications. There is the potential that technological developments for the mobile application will yield benefits for stationary, grid-scale application as well. However, the electric vehicle application is considerably more demanding than the grid-energy storage application. The requirement to store high energy or power per

---

[1] References and Technical Information
Battery energy densities: Electricity Storage Association (www.electricitystorage.org/ESA/technologies/)
CAES energy density: McIntosh, AL Installation—Mathew Wald, New York Times Sept 29, 1991(query.nytimes.com/gst/fullpage.html?res=9D0CEEDE103DF93AA1575AC0A967958260&sec=&spon=&pagewanted=print); Roy Daniel, "Power Storage, Batteries and Beyond," 2009 CERA Week, (www.ihscera.com/aspx/cda/filedisplay/filedisplay.ashx?PK=35893)
Pumped hydro energy density: Assumes 500 m elevation difference between upper and lower reservoirs
Molten salt thermal energy density: Andasol 1 Installation, Spain – David Biello, "How to Use Solar Energy at Night," Scientific American, February 2009 (www.scientificamerican.com/article.cfm?id=how-to-use-solar-energy-at-night);
Solar Millennium (www.solarmillennium.de/Technologie/Referenzprojekte/Andasol/Die_Andasol_Kraftwerke_entstehen_,lang2,109,155.html)

unit weight is less rigorous in stationary applications than in mobile applications. Moreover, vehicle applications require the technology to be highly impervious to a wide range of temperature and humidity variations, as well as to extreme vibration environments. The utility application allows storage systems to be housed in a controlled ambient environment, making the battery design challenges less demanding.

Because utility storage requirements are less stringent than those for transportation, battery technologies developed under the DOE's vehicle technology program in past years—but later discontinued because of their unsuitability for vehicle applications—may, once again, be feasible alternatives for stationary applications associated with the grid. In the 1980s and early 1990s, the DOE maintained a diverse portfolio of battery chemistry technologies for research support under its vehicle technologies program. During the Clinton administration as part of the Partnership for the New Generation of Vehicles (PNGV) program DOE focused on two battery chemistries: nickel metal hydride and lithium ion. However, in light of the coming need for battery storage to accommodate greater integration of renewable energy resources on the grid, it may be useful to revisit the discontinued battery chemistries to assess whether or not any of them are suitable candidates for today's utility applications.

Battery Materials for Energy Storage

Currently, lead acid and sodium sulfur systems have the most extensive track record for large-scale energy storage.

Lead Acid. In the 1980s, lead acid batteries for utility peak shaving were tested, but the economics at that time did not support further deployment. However, continued incremental improvements in lead acid technology and increased energy costs are making use of lead acid more economical. Recent innovations in lead acid technology demonstrated three to four times the energy density with improved lifetimes over conventional lead acid batteries. One promising technology is the combination of ultra capacitors and lead acid batteries into integrated energy storage devices sometimes referred to as "ultra batteries."

Sodium Sulfur. Sodium sulfur batteries use molten sodium and sulfur separated by a ceramic electrolyte. This battery chemistry requires an operating temperature of about 300°C to maintain the active materials in a molten state. These batteries have a high energy density, a high efficiency, and a projected long cycle life. Of emerging battery technologies suitable for utility applications, sodium sulfur batteries are the most technologically mature, and are deployed on a limited scale in Japan and in the United States. A Japanese firm, NGK Insulators, is responsible for most of the development and commercialization of sodium sulfur for utility applications.

With additional research and demonstration, other battery technologies also could prove useful for large-scale energy storage.

Flow Batteries. A flow battery is a rechargeable battery that converts chemical energy to electricity by reaction of two electrolytes flowing past a proton-exchange membrane, illustrated in Figure 3.The principle is similar to a fuel cell except that the reaction is reversible and the electrolytes are reused instead of being released to the atmosphere. Additional electrolyte is stored in external tanks and pumped through the cell to charge or discharge the battery. The energy storage capacity is limited only by the size of the tanks, making scale-up relatively easy, with cost-per-unit of energy storage generally lower than for non-flow batteries, which improves the attractiveness for larger sizes. Flow batteries offer potentially higher efficiencies and longer life than lead acid batteries. Flow batteries such as vanadium and zinc bromide (ZnBr) show great promise. Flow batteries have good efficiencies (over 75%) and long lifetimes (over 10,000 charge discharge cycles) and are scalable because battery size is determined by the size of the electrolyte holding tank.

Vanadium Redox Flow batteries are a relatively new technology. Energy is stored chemically in different ionic states of vanadium in a dilute sulfuric acid electrolyte. The electrolyte is pumped from separate storage tanks into flow cells. Vanadium flow batteries of 800 kW to 1.5 MW are being successfully demonstrated outside of the United States in applications such as UPS for semiconductor manufacturing, island grid capacity firming and grid peak shaving applications.

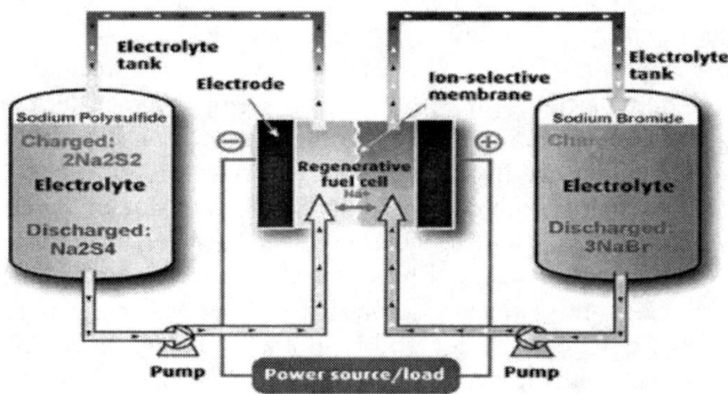

FIGURE 3. The principle of the flow battery where energy is stored in liquid electrolytes and recovered as electricity.

Zinc bromide flow batteries are regenerative fuel cells based on a reaction between zinc and bromide. An aqueous solution of zinc bromide is circulated through two compartments within the cell from separate reservoirs. While zinc bromide batteries use electrodes as substrates for the electrochemical reaction, the electrodes themselves do not take part in the reaction; therefore, there is no electrode degradation with repeated cycling. Several zinc bromide systems in the 200 to 500 kW range have been demonstrated for peak shaving and island grid applications.

Further development of liquid metal batteries, polysulfide bromide cells and metal air batteries could also prove useful. Liquid metal batteries are another class of batteries that potentially could provide up to 10 times the energy storage capacity of current batteries. Like the sodium sulfur battery, liquid metal batteries are a high temperature stationary technology. Polysulfide bromide (PSB) cells are flow batteries, based on regenerative fuel cell technology, that react two salt solution electrolytes, sodium bromide and sodium polysulfide. Metal air batteries have the potential to deliver high energy densities at low cost, if challenges with recharging can be overcome.

Barriers and Recommendations

Energy storage for grid applications lacks a sufficient regulatory history. Utility scale energy storage is very uncommon and, except for pumped hydroelectric storage, is only being used in pilot projects or site-specific projects. Utilities are therefore uncertain how regulators will treat investment in energy storage technologies, how costs will be recovered, or whether energy storage technologies will be allowed in a particular regulatory environment.

Energy storage applications can provide functions related to generation, load and transmission, further confusing the question of regulatory treatment of investments in grid level energy storage. For example, a utility can use bulk energy storage to store electricity generated during a low-cost period, such as late at night, for later use in a time of high-cost generation, such as during peak daytime use. From a regulator's perspective, storage may look like load when it is being charged and like generation when it is discharged. At the same time, storage can reduce transmission congestion, provide voltage support at a time of peak use, and provide other ancillary services that support transmission functions.[8] The ability of energy storage technology to fill multiple roles as load, generation and supporting transmission has created confusion and uncertainty about how energy storage should be regulated.

Moreover, the current system does not fully credit the value of storage across the entire utility value chain. Storage contributes to generation, transmission, and distribution, which have been viewed historically as independent components of the grid system. Because of this structural separation, and because there are relatively few storage precedents, cost recovery for grid-level energy storage investments remains largely undefined. Without clear rules governing cost recovery, utilities tend to under-invest in energy storage. It is comparatively easier for utilities to invest in conventional approaches to grid stability, such as natural gas spinning reserves, for which established precedents for cost recovery are more likely to be included in the utility's rate base.

Vehicle-to-Grid Considerations

If plug-in electric hybrid vehicles (PHEVs) succeed in achieving significant market growth in the coming decades, the potential will exist to use the on-board energy storage of these vehicles as distributed energy storage that would be available to the larger grid while the vehicles are plugged in, or recharging. PHEVs could bring the capability of discharging back to the grid to improve grid utilization, level demand, and improve reliability.

However, one challenge to such an application will be determining how PHEV usage will interact with high levels of renewable energy generation capacity, especially wind and solar power. Both solar and wind power vary diurnally. If the PHEV charging load matches peak renewable energy production—such as wind power generation in areas where the wind blows more consistently overnight—then the PHEV load and renewable source will be well matched temporally. If the PHEV charging does not match daily renewable energy generation cycles well, then the mismatch is problematic. Smart grid technologies that enable time-of-use pricing could encourage consumers to match their vehicle charging with times of higher renewable generating capacity.

Recommendations on Energy Storage

The Department of Energy (DOE) should:

- Develop an overall strategy for energy storage in grid level applications that provides guidance to regulators to recognize the value that energy storage brings to both transmission and generation services to the grid;
- Conduct a review of the technological potential for a range of battery chemistries, including those it supported during the 1980s and 1990s, with a view toward possible applications to grid energy storage; and
- Increase its R&D in basic electrochemistry to identify the materials and electrochemical mechanisms that have the highest potential for use in grid level energy storage devices.

LONG DISTANCE TRANSMISSION (full report section)

The advent of solar and wind renewable energy generation brings new challenges for the collection and long distance transmission of renewable energy, and for distribution of renewable electricity in power-congested urban areas. Renewable sources are typically distributed over large areas in the central and southwestern United States, far from demand centers east of the Mississippi and on the West Coast (see Figure 4). This means new large area collection strategies and new long distance transmission capability are required to deliver large amounts of renewable power a thousand miles or more across the country. Like the US road system before interstate highways, the power grid is designed to serve local and regional customers with local and regional generation and delivery infrastructure. To adequately address our national energy needs in the renewable energy era, the grid must change its character, from a locally-designed, built and maintained system to one that is regionally and nationally-integrated.

Delivery of increased renewables-based power to urban areas also presents new challenges. Today, 82% of the US population lives in urban or suburban settings[9] where power use is high and demand for increased energy is, and will continue to be, strongest. Renewable electricity from remote sources helps to meet this demand without increasing carbon dioxide emissions. However, the additional power currently must be distributed over infrastructure designed and installed to meet much smaller needs. Congestion on existing lines inhibits growth, and as urban areas expand and

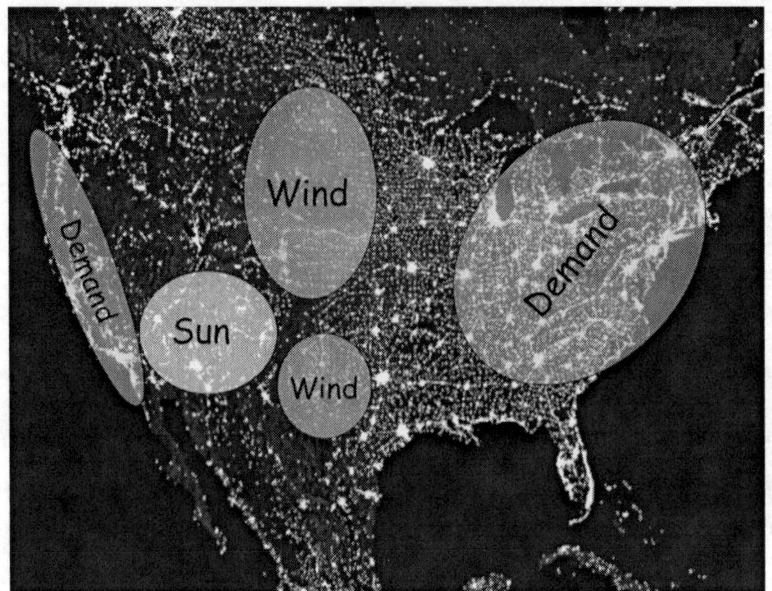

FIGURE 4. The long-distance separation between renewable generation and electricity demand.

merge, the area over which power distribution needs to be coordinated grows. The urban setting makes installation of new lines to meet demand growth expensive and challenging because of the difficulty in securing new "right of way" permits. This delays the installation of new distribution lines up to 10 years and loads the existing lines well beyond their design limits.

Long Distance Transmission Options

Until recently, long distance delivery of electricity over several hundreds of miles remained a specialized area of technology with a fairly small demand and footprint. Most cities are served by nearby fossil coal or gas generation plants, requiring transmission over short distances. An exception is hydroelectric generation in Canada and the northwest US, which produces large amounts of power far from demand centers and justifies long distance transmission. For distances greater than a few hundred miles, direct current (DC) transmission is favored over alternating current (AC) for its lower electrical losses and lower cost. The challenge for DC transmission is the conversion technology from AC sources to DC transmission and back to AC for use. The first commercial high voltage DC transmission lines in 1954 used mercury arc converters for AC-DC conversion, replaced by semiconductor thyristors[10] in 1972, and by insulated gate bipolar transistors (IGBTs) in the 1980s. Although technical progress is reducing the cost of semiconductor power electronics, the cost and technical challenges of AC-DC conversion are still a major barrier for increasing DC transmission.

The mandated growth of wind and solar generation through Renewable Portfolio

Standards (RPS) to 20% or 30% of electricity supply by 2020 or 2030 dramatically changes the landscape of long distance transmission. Such large fractions of renewable power often are not found within 100 miles of urban load centers, and community concern about visual esthetics creates barriers to installation of the large scale wind or solar plants needed to supply such population centers. Rooftop photovoltaics can alleviate some of the need for long-distance transmission, but often at a higher cost than wind or concentrating solar power, and with smaller but significant esthetic concerns. Renewable portfolio standards and the development of large-scale wind and solar resources require a significant investment in raising the capacity and efficiency of long-distance electricity transmission. This long distance transmission challenge is exacerbated by the historically low investment in transmission in the U.S. From 1979–1999 electricity demand grew by 60% while transmission investments fell by more than 50%.[11]

Direct Current Transmission Options

The looming investments in long distance electricity transmission justify a close look at the technology choices available to meet the need. Raising voltage and lowering current reduces losses when transmitting high power over long distances. For example, the largest high voltage direct current transmission project, the Xiangjiaba line terminating in Shanghai, China, operates at 800 kV and delivers 6 GW of power over 2000 km.[12] Such high voltages strain the capability of semiconductor power electronics to interconvert between AC and DC, driving up the cost and limiting the penetration of conventional DC technology. The losses in such a long DC transmission line can be as high as 10%.[13] While high voltage DC is the preferred transmission mode for long distances, there are drawbacks to implementing it for renewable electricity transmission. It requires a single point of origin and termination, precluding wide area DC collection and end user distribution schemes. In addition, the high voltage requires expensive and technically challenging conversion by semiconductor power electronics between AC and DC, and it requires unsightly towers and substantial right of way that can take a decade or more to gain approval in all the relevant—but uncoordinated—regulation zones. Despite these drawbacks, conventional high voltage DC transmission is a mature technology that can be implemented to meet renewable electricity transmission needs over moderate distances. Additional high voltage DC transmission within one- or two-state regions is needed to link regional renewable electricity sources to population centers.

Underground superconducting DC transmission lines are an emerging option that offers a potential route to a national renewable electricity transmission system.[14-17] Superconducting DC lines operate at zero resistance, eliminating electrical losses even for long-distance transmission. Because they eliminate loss and produce no heat, superconductors carry much more current and power than conventional conductors (see Figure 5). Without losses to minimize, there is no need to raise voltage and lower current to extreme levels. Operation at 200 kV–400 kV enables multi-terminal "entrance and exit ramps" that collect power from several wind or solar plants and deliver it to several cities as it makes its way east or west. Recent feasibility studies by

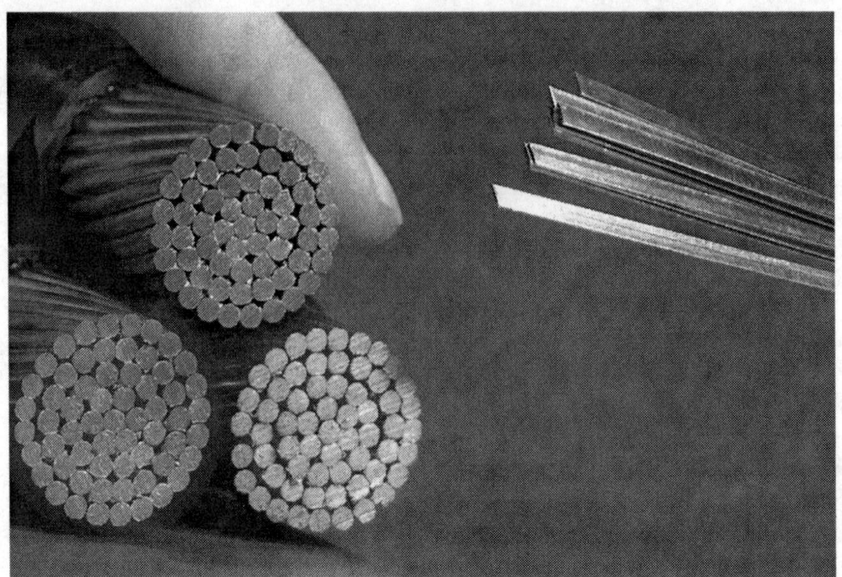

FIGURE 5. The superconducting wires on the right carry the same current as the conventional copper wires on the left. but in much smaller cross-sectional area. Image courtesy of American Superconductor Corporation.

EPRI show that superconducting DC transmission lines carrying 10 GW of power 1600 km can be integrated into the grid, while maintaining transient and short term voltage stability.[18]

While superconducting DC cables have no electrical losses, they require refrigeration to maintain them at superconducting temperatures, often the boiling point of liquid nitrogen, 77 K. Technology development of refrigeration systems and dielectrics for electrical insulation that operate effectively at these temperatures are needed to lower the cost of long-distance superconducting transmission. Superconducting DC transmission couples naturally with superconducting magnetic energy storage (SMES), also a DC system, where electrical energy is stored in superconducting magnets with low loss, deep discharge capability and fast response time. The potential synergies of DC superconducting transmission and SMES are promising and remain to be evaluated. Laboratory demonstration of DC superconducting cable has been carried out at Chubu University in Japan. [19] A proposal for a DC superconducting electricity "pipeline" is shown in Figure 6.

Long distance transmission offers a partial solution to the variability challenge of renewable energy. Balancing generation with load typically takes place within a local or regional balancing area with sufficient dispatchable conventional resources to meet load fluctuations. Increasing the size of balancing areas to aggregate over many wind or solar plants substantially decreases variability, reducing the need for conventional reserves and lowering cost.[20-23]

The complexity of balancing over large areas with many generation and load resources eventually limits the size of the balancing area. Even in this case, however,

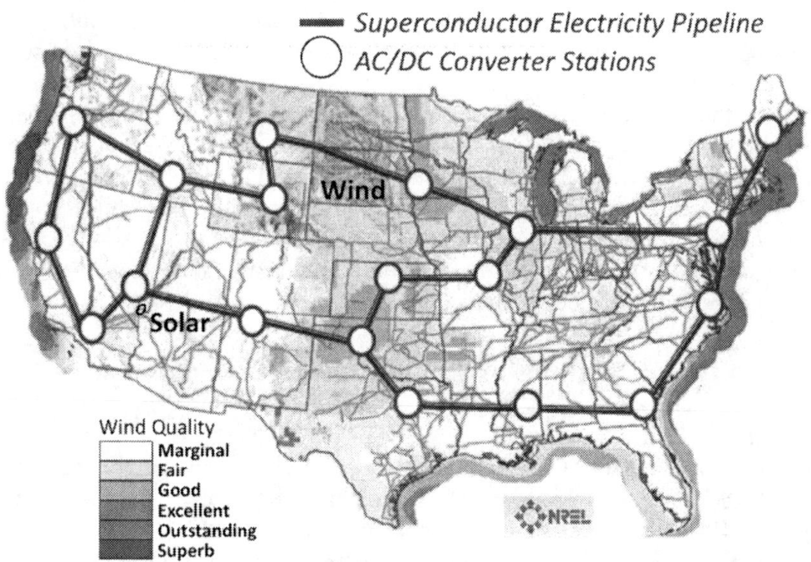

FIGURE 6. A system of DC superconducting transmission lines for carrying renewable electricity from remote sources to population centers. Image courtesy of American Superconductor Corporation.

long distance transmission plays a role. Generation excesses and deficits across the country can be anticipated by forecasting and matched over long distances to balance the system. An excess of wind power in the upper central US might be balanced by transmission to a power deficit in the East. Under these conditions specific excesses and deficits are identified and balanced much like conventional generation is switched in or out to balance load at the local level at present. With adequate forecasting, such specific opportunities can be identified, arranged in advance and executed dynamically as the situation develops.[20, 23] This new level of distant generation balancing requires additional high-capacity long distance transmission that is operator controllable by power electronics, allowing excess generation in one area to be directed to specific targets of deficit far away.

Urban Power Distribution

Urban distribution capacity remains a significant challenge to the user side of the grid. Congestion of power lines in cities and suburbs and the high cost and long permitting times needed to build new lines could all hold back increasing the use of renewable electricity. However, the use of superconducting AC cables that carry five times the power of conventional copper cables in the same cross-sectional area could solve this problem. Three demonstration projects in the US have used superconducting AC cables to deliver electricity in the grid, proving that this approach is technically sound.

For example, the Long Island Power Authority has relied on a superconducting underground AC cable to deliver 574 MW of power since 2008. Replacing key conventional cables in urban grids with superconducting counterparts would provide sufficient capacity for decades of growth without the need for new rights-of-way or infrastructure.[24]

Although the performance of superconducting cables far exceeds that of conventional cables, the cost is still too high to achieve widespread penetration. Research and development are needed to bring this technology to the commercial tipping point. Despite the promise of superconductivity for renewable electricity transmission and for urban power distribution, the DOE Office of Electricity Delivery and Energy Reliability's program for research into high temperature superconductivity for electric applications will be eliminated in 2012.

Recommendations on Long-distance Transmission

DOE should:

- Extend or replace the DOE/OE program on High Temperature Superconductivity for a period of 10 years, with focus on DC superconducting cables for long distance transmission of renewable electricity from source to market; and
- Accelerate R&D on wide band gap semiconductor power electronics for controlling power flow on the grid, including alternating to direct current conversion options and development of semiconductor-based circuit breakers operating at 200 kV and 50 kA with microsecond response time.

Acknowledgement

This work was supported by the Center for Emergent Superconductivity, an Energy Frontier Research Center funded by the US Department of Energy, Office of Science, Office of Basic Energy Sciences under Award Number DE-AC02-98CH1088 (GWC). JM is an employee of Brookhaven Science Associates, LLC under Contract No. DE-AC02-98CH10886 with the U.S. Department of Energy (DOE).

REFERENCES

1. Steven E. Koonin, Testimony before the Senate Energy and Natural Resources Committee, Dec 10, 2009,http://energy.senate.gov/public/index.cfm?FuseAction=Hearings.Hearing&Hearing_ID=df36718-e767-4437-7703-c8dfb3ea58cd, accessed August 29, 2010.
2. B. Roberts, *Capturing Grid Power,* IEEE Power and Energy Magazine, 32, July/August (2009), www.electricitystorage.org/images/uploads/docs/captureGrid.pdf, accessed August 29, 2010.
3. R. Gabbrielli and C. Zamparelli, *Optimal Design of a Molten Salt Thermal Storage Tank for Parabolic Trough Solar Power Plants,* Journal of Solar Energy Engineering 131, 041001 (2009).
4. Joint European Torus (JET) Flywheels, http://www.jet.efda.org/focus-on/jets-flywheels/flywheel-generators/, accessed July 25, 2010.
5. *Grid-Scale Rampable Intermittent Dispatchable Storage (GRIDS),* https://arpa-e-foa.energy.gov/FoaDetailsView.aspx?foaId=85e239bb-8908-4d2c-ab10-dd02d85e7d78, accessed July 25 2010.
6. Mark Johnson, David Danielson and Imre Gyuk, *Grid Scale Energy Storage,* ARPA-E Pre- Summit Workshop, March 1, 2010, http://arpa-e.energy.gov/LinkClick.aspx?fileticket=k-

81lTzfv34%3D&tabid=259, accessed July 25, 2010.
7. http://arpa-e.energy.gov/ProgramsProjects/GRIDS.aspx
8. Bonneville Power Administration, *Balancing Act: BPA grid responds to huge influx of wind power,* DOE/BP-3966, November 2008.
9. CIA World Factbook, https://www.cia.gov/library/publications/the-worldfactbook/fields/2212.html, accessed April 2010.
10. A thyristor is a four-layer semiconductor that is often used for handling large amounts of power.
11. EIA Annual Energy Review 2009, p231; Edison Electric Institute, quoted in http://www.wiresgroup.com/resources/historyofinvestment.html, accessed July 4, 2011.
12. Sonal Patel, *The Age of the 800 kV HVDC,* Global Energy Network Institute, Feb, 2010, http://www.geni.org/globalenergy/library/technical-articles/transmission/powermag.com/the-age-of-the-800-kv-hvdc/index.shtml, accessed July 25, 2010.
13. Peter Hartley, Rice University, *HVDC Transmission: part of the Energy Solution?,* http://cohesion.rice.edu/CentersAndInst/CNST/emplibrary/Hartley%2004May03%20NanoTehConf.ppt, accessed July 25, 2010.
14. Workshop on Superconducting DC Electricity Transmission, January 21-22, 2010, Houston, TX, http://events.energetics.com/DCCableWorkshop2010/, accessed May19 2010.
15. *Program on Technology Innovation: a Superconducting DC Cable,* EPRI Final Report 1020458, S. Eckroad, Principal Investigator, December 2009
16. *Program on Technology Innovation: Study on the Integration of High Temperature Superconducting DC Cables within the Eastern and Western North American Power Grids,* EPRI Report 1020330, T. Overbye P. Ribeiro T. Baldwin, Principal Investigators, November 2009.
17. *Program on Technology Innovation: Transient Response of a Superconducting DC Long Length Cable System Using Voltage Source Converters,* EPRI Report 1020339, S. Nilsson and A. Daneshpooy, Principal Investigators, 2009.
18. *Program on Technology Innovation: a Superconducting DC Cable,* EPRI Final Report 1020458, S. Eckroad, Principal Investigator, December 2009
19. Makoto Hamabe, Tomohiro Fujii, Isamu Yamamoto, Atsushi Sasaki, Yuji Nasu, SatarouYama Yamaguchi, Akira Ninomiya, Tsutomu Hoshino, Yasuhide Ishiguro, and Kuniaki Kawamura, *Recent Progress of Experiment on DC Superconducting Power Transmission Line in Chubu University,* IEEE Transactions on Applications of Superconductivity 19, 1778 (2009).
20. C.L. Archer and M.Z. Jacobson, *Spatial and temporal distributions of U.S. winds and wind power at 80 m derived from measurements,* Journal of Geophysical Research 108, 4289 (2003).
21. J.F. DeCarolis and D.W. Keith, *The economics of large-scale wind power in a carbon constrained world,* Energy Policy 34, 395 (2006).
22. C.L. Archer and M.Z. Jacobson, *Supplying Baseload Power and Reducing Transmission Requirements by Interconnecting Wind Farms,* Journal of Applied Meteorology and Climatology 46, 1701 (2007).
23. Willett Kempton, Felipe M. Pimenta, Dana E. Veron, and Brian A. Colle, *Electric power from offshore wind via synoptic-scale interconnection,* Proceedings of the National Academy of Sciences 107, 7240 (2010).
24. Another promising route for increasing the capacity of existing infrastructure is high performance nano-composite dielectric insulation, which increases rated current carrying ability and occupies less space.

Development of Non-Tracking Solar Thermal Technology

Roland Winston, Bruce Johnston and Kevin Balkowski

UC Advanced Solar Technologies Institute
University of California at Merced
Merced, CA 95343

Abstract. The aims of this research is to develop high temperature solar thermal collectors that do not require complex solar tracking devices to maintain optimal performance. The collector technology developed through these efforts uses non-imaging optics and is referred to as an external compound parabolic concentrator. It is able to operate with a solar thermal efficiency of approximately 50% at a temperature of 200°C and can be readily manufactured at a cost between $15 and $18 per square foot.

INTRODUCTION

Solar thermal technologies have not received the attention that photovoltaic system development has in recent years. Despite tepid interest among many researchers in the United States, the technology has the potential for broad applications including electric power generation, production of process heat and the heating and cooling of medium and large buildings and building complexes. Until recently, high temperature solar thermal system architecture required the capability of tracking the sun's path throughout the day. Tracking systems have high installation and maintenance costs; therefore, the development of a high temperature collector that operates efficiently without a tracking system should have many potential commercial applications. At UC Merced we decided to embark on a journey to develop this type of technology.

The research team included participants from the University of California at Merced, SolFocus Inc. and the United Technologies Research Center. With primary funding from the California Energy Commission the team conducted a number of experiments with the specific aims of developing high temperature solar thermal collectors that are easily manufactured and do not require complex solar tracking devices to maintain optimal performance. The collector technology developed through these efforts is referred to as an external compound parabolic concentrator (XCPC). (See Figure 1.). It is able to operate with a solar thermal efficiency of approximately 50% at a temperature of 200°C. Furthermore, he XCPC can be readily manufactured at a cost between $15 and $18 per square foot.

Physics of Sustainable Energy II: Using Energy Efficiently and Producing it Renewably
AIP Conf. Proc. 1401, 406-412 (2011); doi: 10.1063/1.3653866

FIGURE 1. An Early XCPC Prototype at UC Merced.

EXTERNAL COMPOUND PARABOLIC CONCENTRATOR

The system that the team developed is the External Compound Parabolic Concentrator (XCPC) referred to above. Glass vacuum tubes form the housing in which light energy is converted to heat and transferred to a liquid medium. This vacuum tube technology is well established, and China provides the majority of the tubes used in solar thermal systems that rely on them. Several different tube designs were evaluated.

Two basic tube designs were of particular interest. The first is a standard tube-in-a-tube design. These employ a configuration whereby the fluid heat transfer medium enters the vacuum tube and flows down the length of a smaller tube then returns to the manifold via a slightly larger diameter tube that is concentric with the inlet flow tube. (See Figure 2.) A selective coating surrounds the return channel to facilitate absorption and heat transfer to the fluid. This design is commercially available and has a reasonable degree of reliability.

A second design is referred to here as a U-tube configuration. It was developed by the research team and produced slightly better performance that the tube-in-a-tube design. The inlet and the return segments of the fluid carrying tube a both in contact with sleeve bearing the selective coating. (See Figure 3) As suggested by its name the tube is shaped like the letter "U" with the inlet and outlet at the same end and requiring only one manifold assembly.

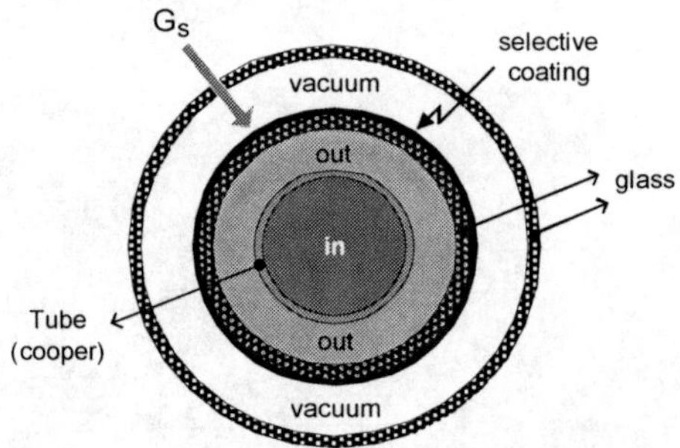

FIGURE 2. Example of a Tube-In-A-Tube Vacuum Tube Configuration.

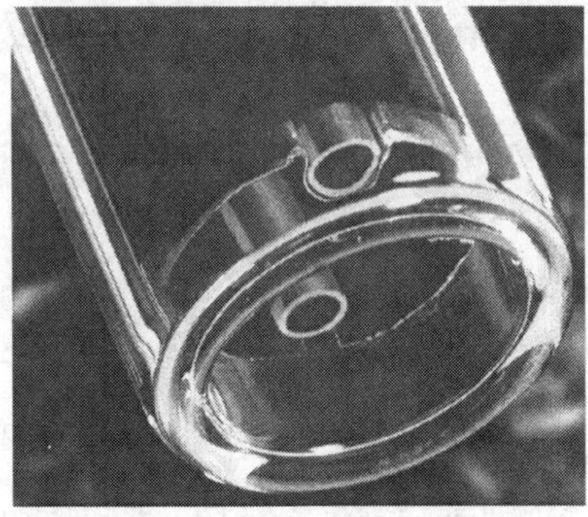

FIGURE 3. Example of a U-Tube Vacuum Tube Configuration.

XCPC TEST RESULTS

Experiments with collector positioning were conducted to determine if any one orientation was favored. Two orientations, North-South (NS) and East-West (EW), were considered. The NS orientation can be described as positioning the collector assembly so that the long dimension is perpendicular to the path of the sun at solar noon. In essence the tubes are oriented North to South as the name implies. The EW orientation places them 90 degrees to the NS orientation.

The EW orientation produced slightly better results at higher temperatures; however, the NS configuration offers greater ease in maintenance with only a small sacrifice in performance. (See Figure 4.)

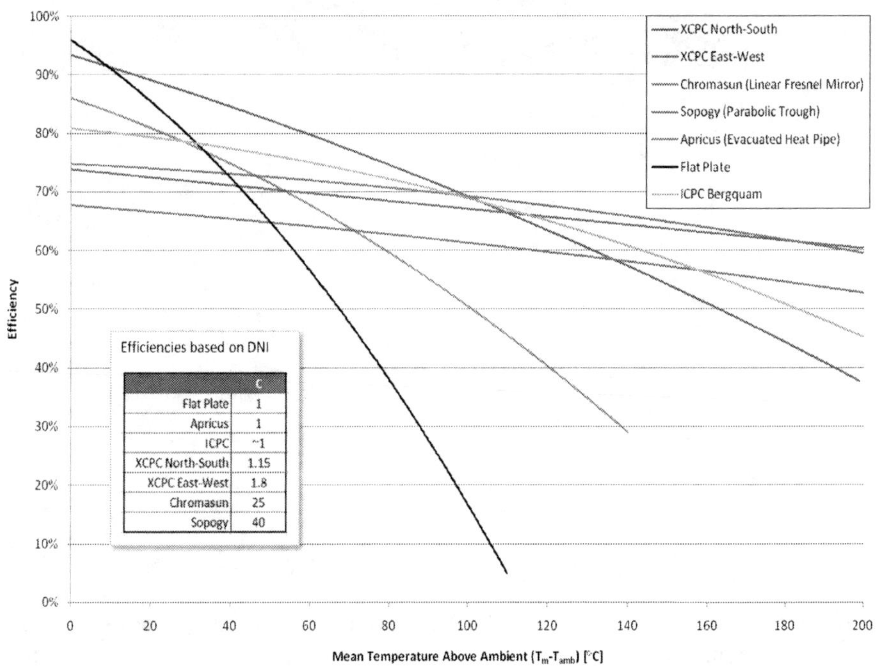

FIGURE 4. Collector Efficiency.

The XCPC relies on non-imaging optics (NIO) and is designed with a 60-degree acceptance angle. This permits uniform heat conversion and transfer to occur for the majority of the day. The contour of the reflector was modeled using LightTools™ software and NIO principles. Typical thermal efficiency as a function of azimuth angle can be seen below. (See Figure 5.) The uniformity in performance across the 120 degree range is accompanied by a very sharp increase and decrease at the extremes of the range. In essence, the system is either on or off.

The XCPC also demonstrated the capability of operating consistently at 200°C. Output at this temperature is sufficient for a wide variety of practical applications; however, of particular interest is the capacity to provide adequate heat to a double effect absorption chiller. Operating such a chiller opens the door for very efficient, lower cost cooling applications.

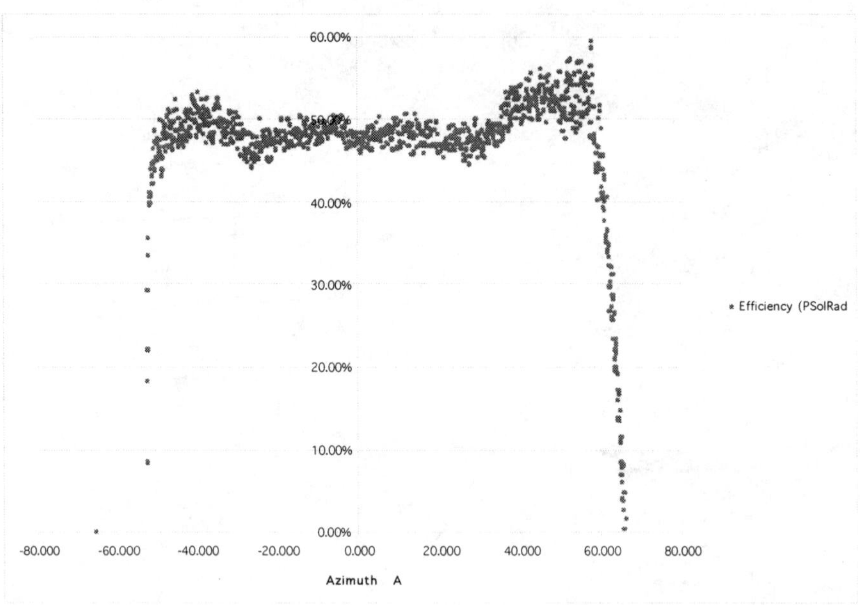

FIGURE 5. Example of Thermal Efficiency.

An efficient non-tracking solar thermal system offers several immediate benefits over a tracking system. First, the initial cost of purchasing a tracking system can be prohibitive. Second, a tracking system requires an electric power source that adds to the day-to-day operating cost. Third, tracking systems have many moving parts that ultimately leads to higher maintenance costs. Part wear and breakage can potentially compromise tracking accuracy and reduce system performance. Finally, the tracker assembly is considerably heavier than a stationary system.

During the course of this project, a total of seven different XCPC configurations were tested, and an initial XCPC prototype was created and tested at UC Merced. After improving the reflector technology and incorporating a new evacuated thermal absorber design, an improved prototype was constructed and tested. A system using East-West collectors (the collectors are oriented horizontally) with U-Tubes and Reflectech-coated reflectors performed the best in our tests with roughly 47% efficiency at 200°C. After further improvements and adjustments, a 10kW prototype using North-South collectors (the collectors are oriented vertically) was manufactured by SolFocus and tested at the NASA/Ames facility. (See Figure 6.)

Although a high temperature XCPC was developed and performed well, improvements in the troughs, tube architecture and selective coatings would be extremely valuable and would improve the performance and reliability of the system.

With respect to the troughs the angular tolerance could be improved from 0.5° to 2.0°. The use of Reflectech and Alanod for the reflective surface is cumbersome in that it is difficult to apply and unlikely to function reliably over extended periods of time; therefore, the development of a new surface materials/treatments is needed.

Finally, the cost of assembly could be reduced by better tooling and focusing on mass production.

FIGURE 6. The 10kW SolFocus Test Loop at NASA/Ames.

The vacuum tubes could be improved by developing innovative flow paths that facilitate better heat exchange. This would add a level of complexity to the manufacturing process; however, significant improvements in performance would outweigh the cost increase. Vacuum seals are also problematic and are generally the point of failure in high temperature systems. The metal to glass seal could be greatly improved to extend the life of the individual tubes. Finally, the selective coatings used to absorb light energy could be greatly enhanced. The key element to coating improvement lies in the property of emissivity. Target values of less than 0.07 at 400°C are needed to achieve a desirable result. Additionally low reflectance ($\rho \approx 0$) at wavelengths less than 2 micron and high reflectance ($\rho \approx 1$) at wavelengths greater than 2 microns is highly desirable. These conditions are believed to be achievable and would measurably improve performance.

A non-tracking high temperature solar thermal system could be used in many applications. Researchers at UC Merced are exploring the possibility of employing this technology to generate process heat that can be used in the agricultural sector, and efforts are underway to utilize the technology in conjunction with double effect absorption chillers to provide structural cooling/air conditioning.

The UC Merced Solar Cooling Demonstration Project is the first of its kind in the USA. It has been designed by students studying within the UC Advanced Solar Technologies Institute and consists of a 23.5-KW solar thermal collector array working in conjunction with a 6.5-ton double effect absorption chiller to cool an office building at the UC Merced Castle Research Facility. The system can operate in 3 different modes. It can be driven entirely by natural gas, entirely by solar thermal collectors or driven as a hybrid using both.

A double effect absorption chiller can be used when the heating fluid temperature is sufficiently higher than the solution temperature in the high-pressure generator (HPG) of the chiller. The solution temperature in the HPG is typically around or below 130°C. Therefore the XCPC heat source providing temperatures in the 150°C – 250°C range should work well with this system. A chiller can be driven by oil or steam.

When the chiller is driven by oil, the oil can reach high temperature of ~250°C while being maintained at atmospheric pressure. The working principle of an oil driven double effect absorption chiller is similar to that of a hot water driven single effect chiller. The oil is heated in the solar collector and then enters an oil storage tank. Hot oil leaving the storage tank enters the high-pressure generator where the solution is heated to produce refrigerant (water) steam. The oil then returns to the hot oil tank at a lower temperature. The water steam enters the low-pressure generator (LPG) where additional refrigerant is produced from the solution in the LPG.

With higher COP, a double effect absorption chiller requires much less heat than a single effect chiller to produce the same amount of cooling load. Even though the solar collector efficiency drops nominally from 0.90 to 0.80 and the oil has smaller specific heat than water, only a small oil flow rate is needed and the storage tank size is smaller as well. The insulation around the tank, however, should be more effective and more expensive than the low temperature tank that can be used in single effect systems.

The oil tank storage and control logic are similar to that mentioned for the solar thermal driven single effect chillers. The reliability considerations are critical because of the possible leakage of oil in the solar collector. In addition, the oil pump and piping, etc., all require good sealing to prevent oil leakage. The oil-based system in general however could be easier to seal than high-pressure steam. Another possible issue with oil is the corrosion of the tubes that are used to transport the oil to the chiller that means that material compatibility issues must be addressed in selection of the oil.

Due to the less common usage of oil as the heating fluid for absorption chillers, customized design is needed for the high stage generator. The development cost associated with designing this feature and the expected relatively small sales volume implies that the cost of the oil driven chiller will be somewhat higher than non-oil driven absorption chiller.

As of this writing, the Solar Cooling Demonstration Project has been installed, and researchers are ready to begin collecting data.

Potential and Innovations in Rooftop Photovoltaics

Ben Bierman

Solyndra
47488 Kato Road
Fremont, California 94539

Abstract. Photovoltaic technology has reached a point where its cost and capability make it one of a handful of carbon-free sources of electrical energy that could meet a meaningful fraction of US energy demand. In this paper we will first compare Photovoltaics with several other carbon free energy technologies, then look at the economics of Solyndra's rooftop photovoltaic solution as an example of the current state of the art, as well as the market dynamics that have resulted in dramatically faster adoption in Germany vs. the United States.

INTRODUCTION: US ENERGY DEMAND AND RENEWABLE ALTERNATIVES

The US Energy Information Administration estimates US average energy consumption of all forms at 3.3 terawatts. Of this, nearly 3 terawatts come from combustion of fossil fuels, including Coal, Oil and Natural Gas. This generates enormous quantities of CO_2 and other greenhouse gases.

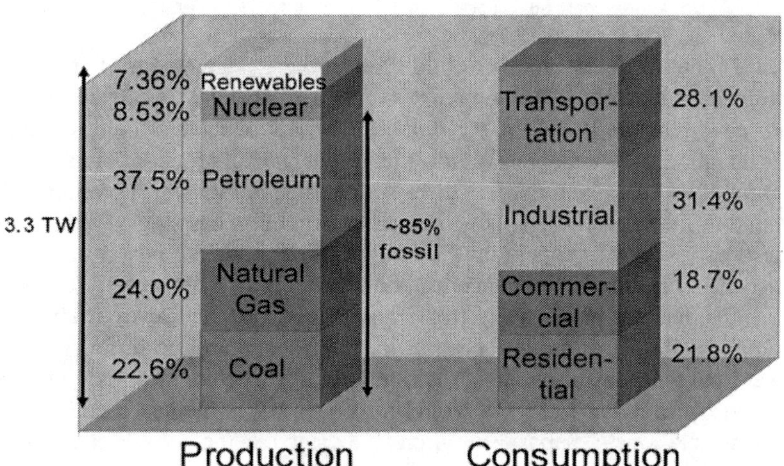

FIGURE 1. US Energy production and consumption, source US DOE: http://www.eia.doe.gov/emeu/aer/contents.html

Physics of Sustainable Energy II: Using Energy Efficiently and Producing it Renewably
AIP Conf. Proc. 1401, 413-435 (2011); doi: 10.1063/1.3653867

A number of carbon-free technologies were evaluated by the David Ginger lab at the University of Washington for their potential to replace 1 terawatt of energy. One option is to raise crops as a source of biomass for fuels. Using typical land productivity and plant energy content, it was determined that approximately 470 million acres would be required to produce 1 terawatt of energy. This is equivalent to approximately one quarter of the land in the United States and also represents the total land currently under cultivation. To pursue such a course would require the United States to refocus agriculture away from food production and towards energy production. This is not a practical solution given the food needs of the US and the rest of the world. That said, it is likely that biomass will be an important contributor to reducing US carbon output, as a number of new technologies can process many forms of organic waste into energy.

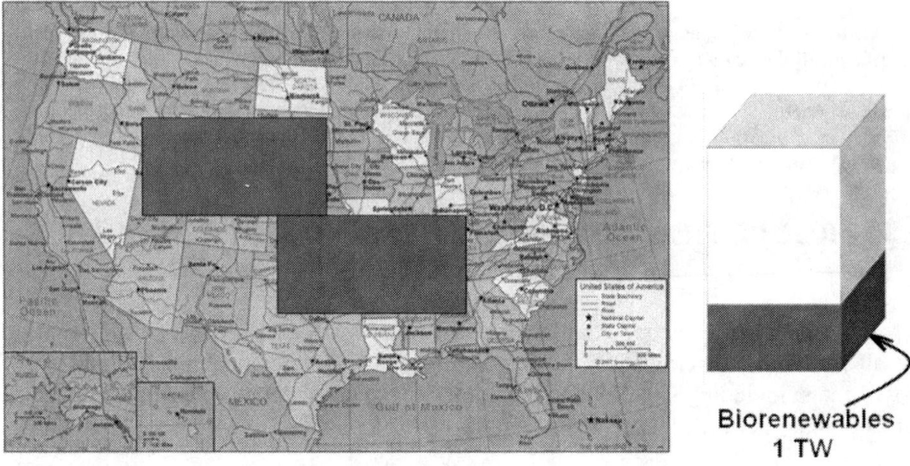

FIGURE 2. US Approximate land area required to produce 1 terawatt of energy from biomass.

Nuclear Fission provides another option for producing a terawatt of energy. With most commercial reactors sized at approximately 1 gigawatt of electrical output, one thousand new reactors would be required. These reactors would consume an insignificant amount of land, and wouldn't require technical breakthroughs. However, construction of nuclear reactors often faces strong local resistance. There are only 104 reactors in the United States, and new reactors construction has slowed dramatically in recent decades. Recent events at the Fukishima Daiichi nuclear power plant in Japan have further increased safety concerns about where to build reactors. As these pressures force nuclear plants away from coastal locations, and away from cities and water supplies, the construction costs will likely increase both for the reactors themselves, and the transmission and distribution infrastructure required to deliver the electricity to consumers. The US would also need to implement a comprehensive, long term nuclear waste disposal program. Nuclear reactors began generating such waste in the 1940's, yet as of 2011, there is still no such program in place.

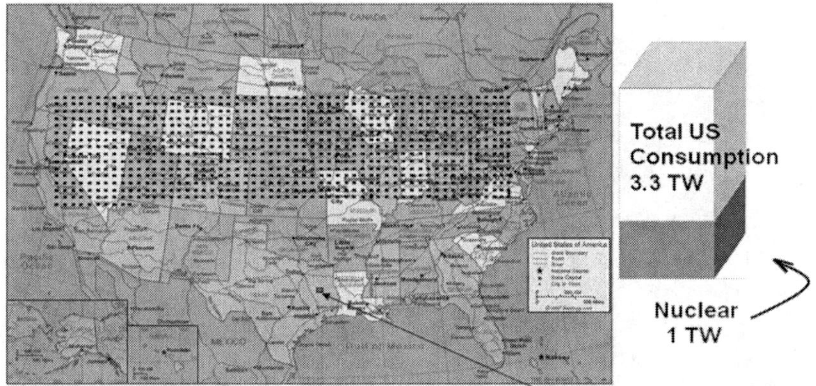

FIGURE 3. US Approximate land area required to produce 1 terawatt of energy from nuclear fission, Black squares represent 1000 reactors deployed across the United States.

Geothermal energy is another interesting option, but an MIT study found that the US Geothermal resource is relatively limited. They estimated that no more than 100 GW of energy could be extracted from areas with suitably high subsurface temperatures at reasonably accessible depths. With current technology, approximately 4000 meters is the maximum feasible depth. The chart shows temperatures at 5500 meters below the surface.

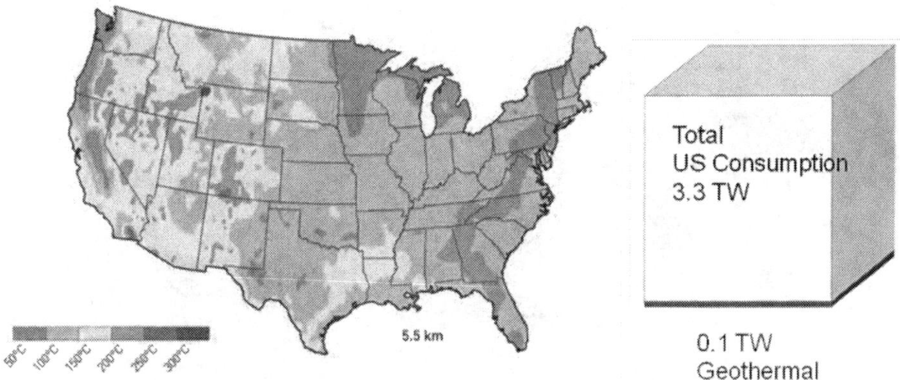

FIGURE 4. US geothermal resource map, subsurface temperatures hot enough to produce steam are desired for today's commercial geothermal technologies.

Wave energy was also evaluated. Wave technology can produce on the order of 14 KW per meter of coastline. The continental US coastline measures 9000 km. Utilizing 100% of the available area would generate approximately 120 GW of carbon-free energy. This seems well worth pursuing, but even with improvements in technology, is unlikely to contribute 1 terawatt to the US energy supply.

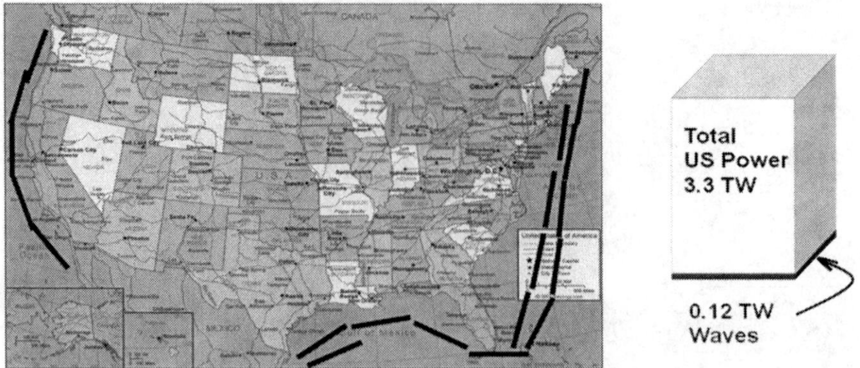

FIGURE 5. Visualizing deployment of wave energy recovery systems across the 9000km US coastline.

Wind Turbines already harvest significant amounts of energy across the United States. Analysis shows that this technology could produce 1 terawatt. There are challenges though. The wind resources are best on the central plains, far from the cities where the most energy is consumed. In addition to a massive investment required to finance the construction of thousands of large scale turbines, a new transmission and distribution system would be needed to carry the power from where it's generated to where it is needed. Offshore locations also provide an excellent wind resource, but there are still significant challenges to achieve acceptable economics for large scale deployment of offshore wind. Finally, the most urgent need for generating capacity is at times of peak demand. The Midwest wind resource actually achieves peak output at night and tends to drop during daylight hours,

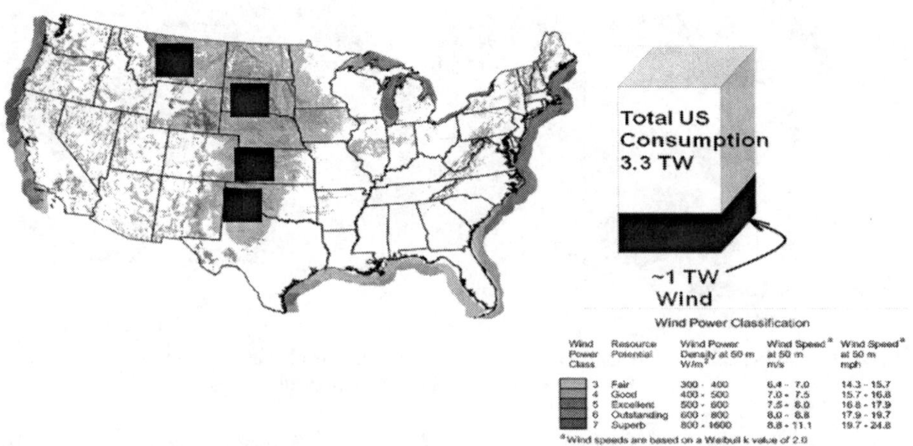

Wind Power Classification

Wind Power Class	Resource Potential	Wind Power Density at 50 m W/m²	Wind Speed[a] at 50 m m/s	Wind Speed[a] at 50 m mph
3	Fair	300 - 400	6.4 - 7.0	14.3 - 15.7
4	Good	400 - 500	7.0 - 7.5	15.7 - 16.8
5	Excellent	500 - 600	7.5 - 8.0	16.8 - 17.9
6	Outstanding	600 - 800	8.0 - 8.8	17.9 - 19.7
7	Superb	800 - 1600	8.8 - 11.1	19.7 - 24.8

[a] Wind speeds are based on a Weibull k value of 2.0

FIGURE 6. US wind resource map and area required to generate 1 Terwatt of electricity using current technology.

This brings us to Photovoltaics (PV). Even after accounting for unproductive areas, commercial systems today will convert a minimum of 5% of the sunlight striking a rooftop into electricity. In much of the US, converting 5% of incident sunlight to electricity in an area about 20,000 km^2 can produce 1 terawatt. This is roughly equivalent to the footprint of all existing buildings and roads in the United States. Producing energy on roofs is attractive, because the energy can be consumed in the building with no additional transmission infrastructure. Equipping roadways with PV requires development of new installation practices, but would place the power near existing transmission and distribution infrastructure, and provide an alternative to developing vast stretches of dessert wilderness for PV installations.

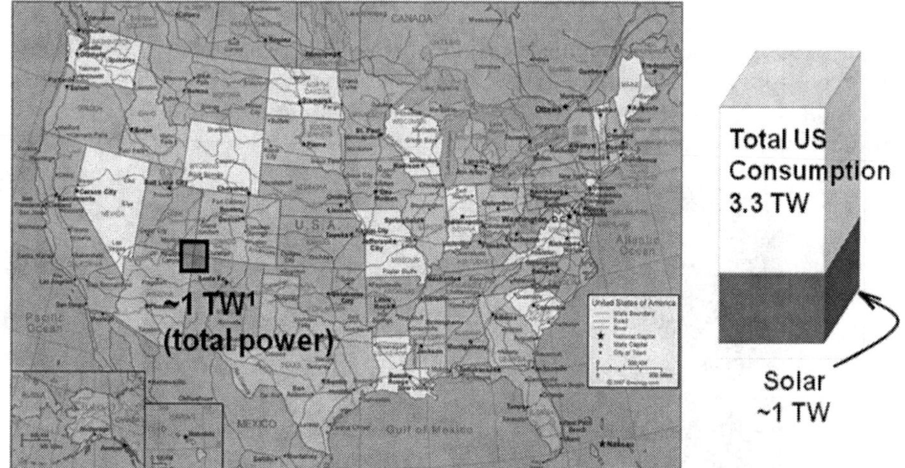

FIGURE 7. Area required to generate 1 Terwatt of electricity using current PV technology.

There is of course a drawback to PV in that the average capacity factor for PV is about 20%. PV does not produce power at night, and production is significantly lower at dawn and dusk, and when there is cloud cover. To provide a constant 1 terawatt of electricity around the clock would require installation of 5 terawatts of capacity, and an investment in storage technology. Fortunately, the power generation from PV typically matches changes in demand, as energy production is highest when the sun is brightest, which usually coincides with peak electrical loads related to air conditioning.

ROOFTOP PHOTOVOLTAICS FOR COMMERCIAL, INDUSTRIAL AND INSTITUTIONAL ROOFTOPS

Opportunity

There are now a large number of commercially available solutions for deploying PV on rooftops. Commercial, industrial and institutional structures typically allow large installations that yield economies of scale that reduce the cost of installations to approximately 65% of residential installations. These rooftops alone total over ten billion square meters worldwide and could easily host nearly 1 terawatt of PV.

In the United States, the forecasted new and reroofing demand for non-residential segment in 2012 is nearly 1 billion square meters. Even with a conservative 5% average installed PV efficiency, the newly applied roofing material could host about 50 GW of new solar installations.

TABLE 1. United States Non-residential Roofing Demand, in millions of m^2. Source: Freedonia 2007 to 2012 global re-roofing report.

	2007	2012	2017
New Construction	200	223	246
Reroof	701	729	771
Total	**901**	**952**	**1017**

In the US, many of these buildings are now being built or re-roofed with reflective white "cool roofs" which reduce HVAC loads by as much as 30% in hot climates. These roofing materials are mandated in the State of California by Title 24 for all new construction and roof replacements.

FIGURE 8. Typical commercial rooftop with white cool roof.

FIGURE 9. Typical California industrial park, showing widespread adoption of white cool roofs and large area available for PV system deployment.

Examining the precedent set by European countries, it becomes apparent that the benefits of rooftop PV have been realized. Early PV efforts were concentrated on the construction of large ground mount systems. Over time, as issues of lost farmland, grid stability and stress on transmission infrastructure became clear, policies changed to discourage, or even ban, ground mount systems and encourage rooftop installations. As a result, the Commercial and Industrial Rooftop market is the fastest growing segment in PV. Analysts predict 100% growth of the overall PV market from 2010 to 2013, forecasting a global demand of 26 GW by 2013. Meanwhile, the commercial and industrial rooftop market is forecast to grow by 150% to over 10 GW by 2013.

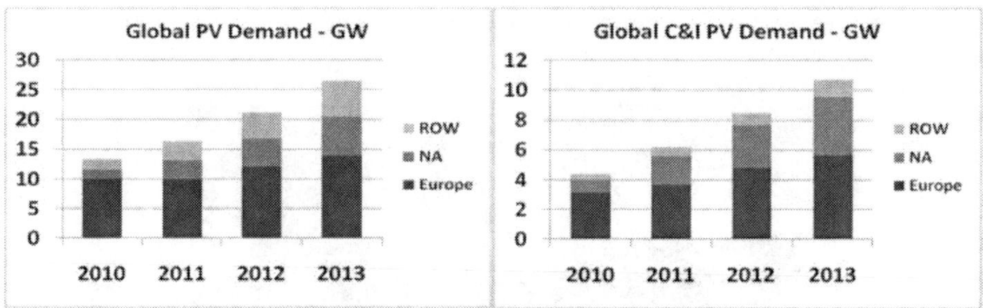

FIGURE 10. Barclays and Think Equity PV demand forecasts for Europe, North America and Rest of World for total market and the Commercial and Industrial segment.

Much effort in PV technology has been focused on the conversion efficiency of solar cells. In developing a commercially attractive solution for deploying PV on large rooftops, higher conversion efficiency is always desirable, but there are many other challenges. High conversion efficiency is necessary, but not sufficient.

Structural Challenges

Most buildings were not designed for solar energy. Typically, architects and engineers work to provide the lowest cost structure that will house the intended activities and meet local building codes. Solar installations bring significant additional loads. The panels themselves may weigh from 2-4 pounds per square foot, adding a deadload of several hundred thousand pounds to typical large structures. Wind can create additional loads. Conventional flat glass panels can generate large lift forces in the typical maximum winds specified by building codes. To keep them in place under high wind loads, flat glass panels are typically ballasted with concrete at up to 5 lbs/sq ft. An alternative is to penetrate the roof membrane and tie the panel mount structure to the building structure, but such installations are undesirable because the numerous penetrations required raise concerns about roof leakage.

There are other load factors as well. In addition to the weight of the panels, the building must sustain dynamic loads created by high winds. Placing large numbers of solar panels on a roof can amplify these loads. Buildings must also sustain snow loads, and are typically designed for a uniform distribution of snow. If a solar installation promotes drifting of snow, point loads can exceed the design margins of the structure, causing the underlying section to collapse. Finally, in seismically active areas, structures must be designed to withstand earthquakes. Often, the largest stresses in these cases come from large masses on the roof resisting lateral shaking of the foundation. Most structures have limited margin for added weight on the roof.

Performance Challenges

An optimized flat panel PV installation is frequently seen in large ground mount systems. Panels are placed in densely packed, continuous rows. The rows are oriented to face due South, and the panels are tilted to place them perpendicular to the sun at the times of maximum insolation. Care is taken to avoid any shading of the panels, even when the sun is low on the horizon. Provisions are made for routine cleaning of the panels, to remove dust and other soiling that can block sunlight and reduce energy yield.

FIGURE 11. Typical flat panel in groundmount installation, showing panels tilted to maximize output and aisles provided to prevent shading and provide cleaning access.

It is usually impossible to duplicate this ideal environment on rooftops. First, obstructions are common. Architects working to maximize natural lighting use large numbers of skylights, which should not be shaded by panels. Cost and energy efficiency considerations often force large amounts of Heating, Ventilation and Air Conditioning equipment not only to be placed on the roof, but to be distributed across the entire building. Electrical distribution, elevators and other systems also compete for roof space. Fire safety and service access dictate large aisles, around the perimeter of the roof, and to access every skylight and piece of equipment placed on the roof. Finally, high walls or parapets are common for aesthetics and for safety. Such walls create shaded areas where PV performance could be compromised. Until buildings are designed with solar as a priority, and cost effective alternatives are developed for current common practices, maximizing PV deployments on rooftops will remain challenging.

PV arrays must be placed clear of rooftop obstructions, as well as the shadows of those obstructions. For crystalline silicon panels, the sensitivity to shadowing is particularly high. Each panel is typically comprised of 72 to 96 solar cells connected in series. These panels are then wired in "strings" of up to 16 panels in series. Individual solar cells that are shadowed will block current flow through an entire string of panels. Thus the shadow of simple standpipe or television antenna can substantially reduce the output of the surrounding PV array.

For PV, it would be best if all buildings faced south. Unfortunately, many other factors dictate the orientation of buildings. As buildings align with street grids, shorelines, rivers or other topographic features, it is relatively uncommon to find buildings with an ideal orientation for PV. As a typical rectangular structure is rotated away from the ideal orientation or "off azimuth", flat panel PV system designers face a number of possible compromises. The rows of tilted panels can be aligned with the building to face southeast or southwest, each having a different impact on both total annual energy production and matching of production to loads. For instance, an array that faces Southeast will have a substantial reduction in energy yield during the afternoon, when air conditioning loads, and time of use electrical rates, are the highest. Alternatively, the designer may choose not to tilt the panels at the ideal angle to maximize light collection, but place them flat, or at small tilt angle such as 5°. This reduces the sensitivity to azimuth, but the annual energy yield will be reduced by approximately 10%, and the panels will be more susceptible to soiling and ponding of dirty runoff water. The designer might also choose to place the panels in rows facing due south, running diagonally across the roof. Such layouts are usually more expensive to install, and result in lower numbers of panels being installed due to interference with the rectilinear layout of the skylights, air handlers and other rooftop equipment.

Finally, PV panels on rooftops face the same soiling from airborne contaminants that ground mounted systems do, but access for cleaning is far more difficult and cleaning costs will be substantially higher on a per panel per year basis. Ideally, a rooftop panel will be soil resistant and self cleaning to the greatest degree possible.

Lifecycle Cost Challenges

Total lifecycle cost drivers on rooftops are also different than for ground mount. There are many additional factors beyond the cost of the PV system. Most of today's PV systems carry a 25 year warranty, yet 25 year roofing systems are rare. This means that most roofs will need to be replaced during the life of the panels. This can dramatically change the economics of a PV system. Not only must the owner budget for removal and replacement of the panels to facilitate the re-roof, they must budget for the loss of energy production during the re-roof period, which will often be during the dry, warm-weather months when the panels will be most productive.

Building owners also must ensure that the manufacturer of the roofing system, and the contractor that installs that system, will honor the warranty on the roof. Installers of PV mounting systems must demonstrate that they can be installed without damaging the roof, and that they will not reduce the life of the roof membrane or seams by adding mechanical, chemical or thermal stresses. Some roof manufacturers will require a second, sacrificial, roofing membrane be installed under the panels to protect the original roof membrane below. This can significantly increase roofing costs.

As mentioned previously, white "cool roofs" are increasingly common. The purpose of such roofs is to reflect solar energy and minimize the thermal loads on the building. There are a number of PV installation methodologies that may compromise this functionality. If the panels are placed close to the roof surface, and the space between the panels and the roof is not well ventilated, the roof can be heated far above the temperature it would normally reach without the solar panels. There are also flexible PV panels which are adhered directly to the roof surface. These devices essentially eliminate the cool roof benefit, and in extreme cases, may increase HVAC loads by more kilowatt hours per year than they will be able to produce.

SOLYNDRA AND ITS ROOFTOP PV SOLUTION

Solyndra is a photovoltaic manufacturer founded in 2005 to develop a PV solution specifically for large commercial, industrial and institutional rooftops. The enabling paradigm shift was the development of tubular solar modules instead of flat panels. Initial proof-of-concept work was performed on borrowed equipment at the National Renewable Energy Lab in Golden, Colorado, using a proven thin film process to deposit a CIGS (Copper Indium Gallium Diselenide) PV device on 6mm diameter tubes approximately 75 mm long. In 2006, a larger facility was built in Santa Clara, California and the process was scaled to produce about 50 tubes per day that were 15 mm in diameter by 30 cm long. These panels proved that the tubular concept was viable, and solved many of the rooftop installation challenges. In 2007, construction began on a production facility with a planned annual production capacity of solar panels rated at 110 MW , and commercial shipments of panels composed of 15mm by 100 cm tubes began in 2008. A follow-on product was released in 2010 using similar modules, but with a new frame and mounting system that eliminated tools and hardware from the installation. This new panel also reduced system installation cost by eliminating the need to run additional wiring to ground the panels.

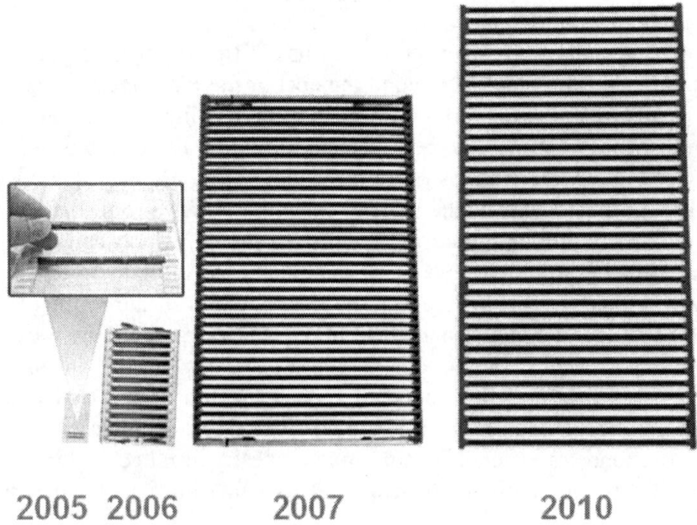

FIGURE 12. Progression of Solyndra Solar Panels from 2005 to 2010.

As of May, 2011, the PV efficiency of the Solyndra CIGS cells average approximately 13%. Efficiency has improved by 37% since 2008. Manufacturers like Solyndra are incorporating sophisticated data acquisition and analysis common in the semiconductor industry to drive the process improvements necessary to elevate production efficiencies toward the CIGS world record efficiency of over 20%.

One significant challenge for the industry when commercializing CIGS PV is encapsulation. The CIGS PV device stack, comprised of thin films of Molybdenum, CIGS, and other trace materials is susceptible to degradation by moisture and oxygen. Solyndra's solution to this problem was to place the 15 mm diameter photovoltaic device inside a second tube with a 22 mm outside diameter. The space between the tubes was filled with an optical coupling agent, a fluid with an index of refraction matched to the glass. The resulting optical effect directs the majority of the incident solar radiation on the outer tube to the surface of the inner tube, increasing the energy output of each tube by 45%. The effect is a consequence of Snell's Law. For two concentric tubes, we see that if $B/A < n$, then all light that hits the outer tube of diameter B will refract to hit the inner tube of diameter A.

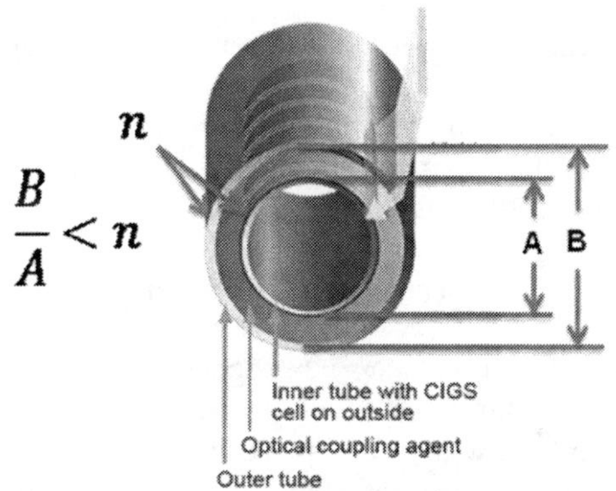

FIGURE 13. Solyndra's Optical Coupling effect.

To exclude water and oxygen, the ends of the tubes are hermetically sealed with a glass-to-metal seal technology similar to that used in the fabrication of light bulbs. Solyndra carefully controls the level of water inside the module to a concentration of less than 30 parts per million. Thousands of tubular modules have been tested at extremes of temperature and humidity to confirm the efficacy of this approach. Packaged in this way, CIGS PV matches or exceeds the reliability of Crystalline Silicon PV technology, which is typically expected to decline in performance by approximately 0.25% per year over a 25 year life.

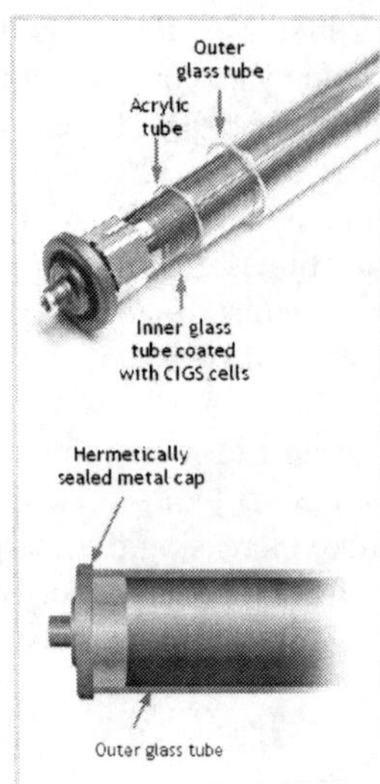

- CIGS inner tube
- Hermetically sealed outer tube
- OCA = 1.5X concentrator effect
- "Passive tracking" of tubes

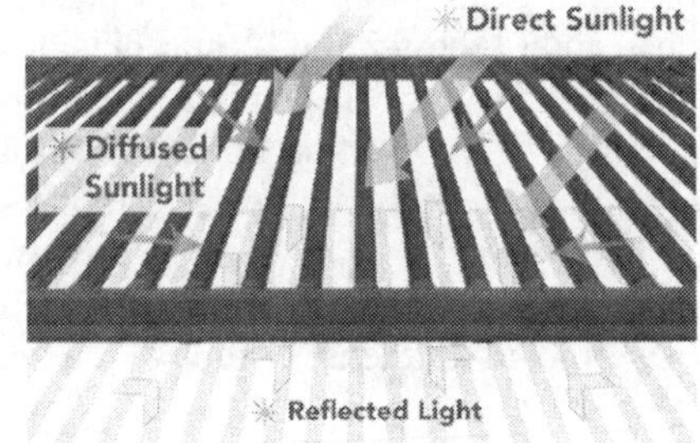

FIGURE 14. Tubular Module architecture and features.

The tubular module collects direct sunlight from any angle. It also collects diffuse sunlight from the rest of the sky more effectively than a flat panel, and captures light reflected off a white cool-roof surface.

Structural Solution

The tubes are spaced such that the open area between tubes comprises over 60% of the total surface area. As a result, wind generated pressure differentials across the panel are very low, and the panels will stay in place in winds as high as 130 mph without additional ballast or roof penetrating anchors.

Solyndra panels are not attached to the roof in any way. Current production panels have an installed load of 2.8 lbs per square foot. Their light weight also minimizes structural loads.

With the addition of ballast or an anchoring system, the panels can be installed in areas with peak gusts up to 180 mph. This is required in many tropical areas subject to cyclonic winds.

The panels are installed flat, parallel to the roof and are not tilted to face the sun. As a result, should snow accumulate on the panels, it typically does not drift or create high point loads.

Taken together, the unique properties of the tubular module, panel and mounting system provide solutions to the problems of installing PV on commercial rooftops.

Performance Solution

As the tubes collect light from all directions, output is not affected significantly as a function of orientation. The series-parallel wiring reduces the impact of shading. Structural loading is seldom a concern. Together, these properties allow the panels to densely "tile" the roof, allowing a much higher percentage of the roof to be utilized for PV, and often allowing annual energy yields higher than could be achieved with flat panels on the same roof, even when these panels have substantially higher individual efficiency.

FIGURE 15. Typical Solyndra Installation. Note panel orientation matches building, and panels are closely packed, covering a large percentage of the roof area.

The tubes tend to shed airborne contaminants more readily than flat panels, especially in the presence of moisture from light rain or condensation. This increases performance between cleanings, and may allow for less frequent cleaning, or elimination of cleaning. When cleaning is needed, the process is simpler than for flat panels.

Lifecycle Cost Solution

As the tubular modules sit above the roof, they not only avoid trapping heat, or directly heating the roof, they may shade the roof and further add to the energy savings of a cool roof. In a warehouse, with minimal energy loads, the combination of the PV system and cool roof may reduce electrical demand by over 50%. The panels perform at their best at the hottest times of the day which is also often the time of the most expensive "peak demand" retail electricity rates.

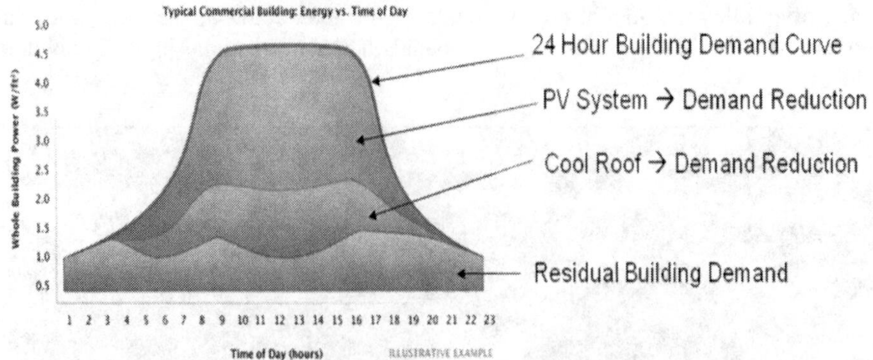

FIGURE 16. Conceptual plot of energy sources and savings for warehouse structure with cool roof and rooftop PV vs. demand profile for same structure with conventional roof.

The elimination of anchors and ballast, as well as tools and hardware, from the installation process means that Solyndra panels can be installed rapidly. This reduces the installation cost, which compensates for the slightly higher individual panel cost, resulting in typical total Solyndra system costs similar to flat panel systems. However, additional benefits accrue over the twenty five year life of the installation as the panels are easily moved to accommodate new equipment installed on the roof, to access the roof for repairs, or to reroof the building.

Finally, the light weight of the system and elimination of tools and hardware allows many roofers to maintain their warranty coverage of the roof membrane, without the need for adding a sacrificial roof membrane or special reinforcements at contact points.

Commercial Status

As of this writing in May of 2011, Solyndra has consolidated operations in a new, 800,000 square foot manufacturing facility with a planned annual capacity to produce solar panels rated at 300MW. Solyndra directly employs over one thousand people, and another thousand are employed providing services to the Fremont operation or in component suppliers across the United States. Over 600,000 panels, containing over twenty four million tubular modules, have been shipped to over 1000 installations in the US, Europe, North Africa, Australia and Asia. Over 100 MW of panels deployed generate over 100 gigawatt-hours of electricity each year. With manufacturing size and scaling, dramatic cost reductions are enabled.

FIGURE 17. Solyndra's thin-film deposition facility in Fremont, California, with 1.2 MW Solyndra PV installation on roof.

While the vast majority of Solyndra installations are on commercial and industrial rooftops, the unique panel architecture has been found useful in additional applications. In sheltered horticulture, shade structures are erected above plants to reduce the amount of sunlight they receive. This has been found to increase crop yields and reduce irrigation requirements, especially in hot, dry climates such as North Africa and Southern Europe. By installing structures with Solyndra panels, a near ideal shade level is achieved, while also allowing production of substantial amounts of energy. This technology application is being demonstrated at prestigious agricultural institutions in California and Italy.

FIGURE 18. 900KW shade structure under construction near Sicily, Italy.

Recalling that one quarter of the United States is under cultivation, it becomes obvious that there is an enormous opportunity to leverage this technology. Many terawatts could be produced, and land productivity would increase while water use fell in many growing regions. The Central and Imperial Valleys of California are two examples of regions where such land use makes sense.

FIGURE 19. Evaluating crop yields under a Solyndra Shade Structure, with added shade cloth, at the CERSAA institute in Albenga, Italy.

ROOFTOP PV SYSTEM ECONOMICS

There are many ways to generate electricity. Some require minimal investment, but have very high operating costs for fuel and maintenance. Diesel generators are an excellent example of this. At the other end of the spectrum, Hydroelectric plants may require massive investment, but have very low operating cost over the many decades that they produce electricity without fuel of any kind. A widely accepted methodology for comparing different technologies is Levelized Cost of Electricity or LCOE. The least complex LCOE models aggregate all acquisition, operating and decommissioning costs over the predicted life of the asset and then divide this cost by the number of kWh produced to calculate an average cost over system life. We'll call this "simple LCOE".

A more sophisticated modeling technique would include the cost of financing, tax benefits and other factors. Such an analysis is mandatory in evaluating alternatives for a specific project in a given location at a certain point in time. However, as financing terms and government policy can change rapidly, the simple LCOE model provides a more consistent means of comparing the actual cost of construction, operation and maintenance.

In Figure 20, we apply the simple LCOE model to rooftop PV systems deployed in California. In 2011, installed system prices will be on the order of \$4.00 per watt, or approximately \$4M for a 1 megawatt system. Average system prices have declined rapidly over the past few years, and are expected to fall below \$3.00 per watt by 2013. The table shows simple LCOE with and without the 30% Federal Investment Tax Credit currently in place through 2016. Current LCOE for rooftop PV is approximately \$0.19/kw-hr, \$0.14 when the 30% ITC is taken into account. By 2013 or 2014, these costs will likely have dropped to \$0.13 and \$0.10 respectively.

Total Projet Cost ($/W)	LCOE (w/ITC)	LCOE (w/o ITC)
4.00	0.14	0.19
3.50	0.12	0.17
3.00	0.11	0.15
2.50	0.10	0.13

Solyndra estimate - 1400 kWh/kWp

FIGURE 20. PV system Levelized Cost of Electricity with and without federal investment tax credit.

The Average Retail Price of Electricity to commercial consumers in California in January 2011 was 12.26 cents/kWh as reported by US Energy Information Administration (EIA). During the same time, the average total electricity rate per kWh for PG&E customers with high electric use and medium to high load factors is 14.52 cents/kWh. Electricity pricing is sensitive to usage, and as consumption of energy increases with HVAC loads during warm, sunny periods, incremental units of electricity under peak rates may cost as much as $0.23/kWh. According to the EIA, residential and commercial electricity rates have historically increased at a CAGR of about 3% between 2005 and 2010. Thus, we can safely assume that the average electricity costs for systems installed in this market will result in a net savings against the prevailing rate, and that the economics will continue to improve as PV system prices decline and commercial retail electric rates increase.

When compared this way, rooftop PV seems an economically attractive alternative, yet adoption in the US has been slow. This is partially explained by the high initial cost of a PV system. Most businesses carefully manage their capital to maximize returns, and tend to focus on investment opportunities to expand their businesses or reduce cost that pay for themselves in a few months or at most, one or two years. PV systems today, in most applications, will take five or more years to pay back the initial investment. Various incentives such as grants in lieu of tax credits have a large effect on these economics. As a result, PV systems are adopted most quickly when regulatory and financial policies put in place improve the economics, and allow favorable financing of the initial investment. Such a structure has been in place for over a decade in the world's largest PV market, Germany. In the next section, we'll compare PV market penetration in the US and Germany.

COMPARING PV MARKETS: THE UNITED STATES VERSUS GERMANY

In Figure 21, the graphics have been scaled to allow the reader to easily compare the land area, electricity demand, typical PV system performance, and the annual energy produced by PV for the US and Germany. Obviously, the US is a far larger country, with roughly eight times the demand for electricity. Note that electrical demand in both countries is far below total energy demand, as much transportation, heating and industrial processing is powered directly by burning fossil fuels.

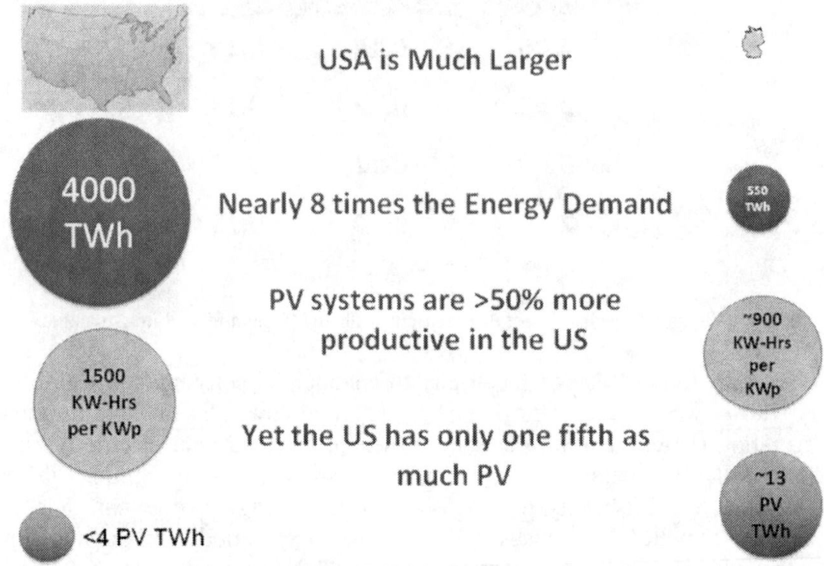

FIGURE 21. Graphical comparison of US and Germany with regard to area, electricity demand, typical PV system performance and annual total PV energy output.

The annual energy yield from a PV system is calculated by multiplying the size of the installation (measured in kilowatts produced under a standard test condition of 25°C and 1000 w/m^2 illumination) by the annual energy yield per KW installed. Due to the lower average latitude, and generally lower cloud cover, PV systems in the US may produce as much as 1800 KW-hrs/KW installed, with systems in large areas of the country exceeding 1500 KW-hrs/KW. In Germany, an average system yield of 900 KW-hrs/KW or less is common. This means that a system of the same size, with the same cost, installed on a roof in the US can be nearly twice as productive, and thus produce electricity at approximately half the cost of the German installation.

These factors clearly favor PV installations in the US, yet the US PV energy supply is less than one third of Germany's. In 2010, approximately 1 GW of PV capacity was installed in the US, while Germany installed nearly seven times as much. This is a direct result of differences in policy and administration that create substantially higher barriers to PV adoption in the US.

BARRIERS TO PV ADOPTION IN THE UNITED STATES

Access to Capital

The financing of PV systems in Germany is straightforward. A national feed-in tariff program mandates that electrical utilities will pay a fixed price per kW-hr produced by a PV system for 20 years from the date of installation. This tariff is reduced periodically to promote cost reduction in the PV industry and to avoid windfall profits for PV installers. Thus, the ultimate end customer for most PV systems is an electrical utility with excellent credit. Guaranteed sales at a known price to a stable, credit-worthy customer gives investors high confidence in revenue forecasts for a PV system. Such security incentivizes investors to make equity investments in PV systems. They leverage this equity with loans from commercial banks. All banks use a standardized documentation set, so applying for such loans is routine. Germany's KFW bank provides commercial banks with long term capital to support these loans at a low cost.

Compare this to the situation in the United States. The primary incentive for investing in PV has been the Federal Investment Tax Credit. This incentive rewards those who invest in PV with a credit against their tax bill equal to 30% of the total PV system cost. Those without a substantial tax exposure can only monetize the incentive by selling it to another party at a significant discount to its value in a complex transaction. After the 2008 financial crisis, few entities had substantial tax exposure, and the market for tax equity shrank such that investment in PV in the US slowed. The American Recovery and Reinvestment Act of 2009 responded to this problem with a "Grant in Lieu" of the tax credit. PV investors were able to receive a check for the full amount of the tax credit from the US Treasury within 60 days of completing a project. This is a very effective incentive, but is temporary, and expires on December 31, 2011, unless Congress acts to extend it.

In addition to the federal incentive, a patchwork of state and local incentives exist. Compensation may be dramatically different from one state to the next, and is often substantially lower than European rates. In the vast majority of cases, for rooftop PV, the electricity is not sold to the utility, but is sold to the occupant of the building through a net-metering process, where the customer only pays for the electricity drawn from the utility. Thus, the value of the PV electricity is calculated as the net savings in electricity cost. This creates two problems. First, the savings are relatively small, and no extra compensation is given to the investor for the societal benefits of producing carbon-free power, reducing stress on the electric grid at time of peak demand and reducing the need for large capital investments to deploy new generating, transmission and distribution assets. Second, the revenue stream for the system will not come from a low risk utility, but from the occupant of the building. This company will likely have worse credit than the utility, and there is added risk as the entity may choose to vacate the building at some point in the future. This makes banks much less likely to finance such projects. When they do finance them, it will usually be at higher interest rates that will increase the total project cost. There are some success stories. As utilities work to reach renewable portfolio standards that dictate a certain percentage of their power that must come from renewable sources, they are developing innovative programs like the New Jersey SREC program. In states with strong programs you find the fastest growth, New Jersey being the best example in the United States today, with strong growth of rooftop power.

There are additional issues. The application process for financing is not standardized in the United States. Documentation requirements vary widely between institutions. The longest commercial loan terms available are typically seven years, requiring very high payments that reduce early returns for investors. By contrast, readily available fifteen and twenty year financing in Germany make PV a much more attractive investment.

Access to capital and lower economic returns are by far the most substantial barriers to PV adoption in the US, but there are other factors that also contribute to the slower adoption of PV.

Utility Policies, Building Codes and Insurance

It is possible to assemble a system of solar panels, batteries and electronics that will allow a building to operate completely independent from the national grid. However, such systems are costly, and few buildings have space for enough solar panels to cover 100% of their energy needs. Virtually all commercial and industrial PV systems are connected to the grid via the local electrical utility.

Utilities have a number of concerns regarding any independent source of energy on their grid. Large sources of power coming on and offline unpredictably complicate the job of delivering power at a stable voltage to the surrounding buildings. This can result from PV systems being turned off for service, or even from shading due to passing clouds. Also, when the utility shuts off power to a section of their transmission infrastructure, they need to be sure that no energy is being fed into their systems from buildings that previously would have been simple, passive loads. Failure to do so could endanger line maintenance personnel.

Without national standards, each utility chooses different strategies and tactics for integrating PV into their electrical systems. The cost and complexity of grid connection can vary widely from one jurisdiction to the next. Particularly onerous regulations in a given region can increase costs to the point where PV becomes less attractive to potential customers.

The utility provides just one of several approvals required to move forward with a PV installation. The installation must also be approved by the local building department, who ensures the structure is not being overloaded, and that panels cannot create a hazard by falling or blowing off the building. The building departments will also check that the wiring meets national and local codes for electrical safety. The local fire marshal will need to approve the configuration of the panels on the roof for safe egress from the building. Safe access for firefighters to the roof and the skylights and equipment on the roof are also required. For larger installations, which may require erection of an electrical enclosure at ground level, local planning departments will need to approve aesthetics, lost parking spaces, and impact to traffic flow.

Again, the rules enforced vary widely. Even where standards, such as the National Electrical Code exist, interpretations of various guidelines can vary. At best, the approval process can increase cost by adding months to the engineering and planning phase of a project. At worst, changes in requirements can force integrators to rework large sections of an installation when it is found deficient during inspection.

The National Electrical Code as typically construed in the US also increases the cost of US systems. European systems are wired to produce a maximum DC voltage of 1000 volts. In the US, systems are typically limited to 600 volts. This reduces the number of panels that can be connected in series, and increases the wiring cost by as much as 50%.

Finally, with all requirements met, projects frequently encounter obstacles with the companies that insure the buildings upon which the panels are to be deployed. When advised of a pending installation that will add to the insured value of a facility, insurers may insist on costly measures to reduce the risk of loss. In the case of newer technologies, insurers may refuse to insure a facility if panels they are uncomfortable with are installed.

All the preceding factors add cost and complexity to the installation of rooftop PV. Furthermore, the conservative nature of the many regulators and approvers tends to dramatically slow the adoption of new technologies. There is another factor that works to slow the rate of change of PV technology and reduce the rate at which the cost of PV can be reduced, this is the certification process.

Certification

As any technology becomes widely available, consumers, regulators, financiers and underwriters will demand some sort of standards and testing to ensure the safety, durability and suitability of that technology for a given application. Such efforts usually begin at a national level. Over a period of many years, international bodies tend to synchronize standards or create reciprocal agreements between regulatory agencies. The PV industry is not yet at that stage.

PV manufacturers today must pursue multiple approvals and listings in order to be able to deploy their products in each market. The design of a product sold internationally is highly constrained by the need to meet so many, often conflicting, sets of regulations.

- UL 1703
- ULC/ORD - C1703-01
- IEC (61646 61730-1, 61730-2)
- EN (61646 61730-1, 61730-2)
- DIN EN (61646 61730-1, 61730-2)
- IECEE CB Scheme
- TUV ISO 9001:2008
- CE Mark
- California DSA – In Progress
- CEC (California)
- FSEC (Florida)
- ACEC (Australia)
- BRE (UK)
- ENV1187-T3 (France)
- GSE (Italy)
- IsEC (Israel)
- JET (Japan)
- FM Global Insurance – In Progress

FIGURE 22. Certifications currently held or being pursued by Solyndra.

Figure 22 shows some of the certifications Solyndra maintains for the solar panel models they currently sell. Any design improvement, change in raw materials or suppliers may require extensive recertification and require a large amount of effort to explain the change to certification authorities and prove that safety and reliability will not be compromised. In the case of a specific panel, this certainly works in the customer's interest. However, across the industry, the process slows the implementation of new technologies that can reduce cost and boost performance.

CONCLUSION

Rooftop PV is a proven technology that could produce a terawatt of peak electrical output with minimal investment in new transmission and distribution infrastructure. This takes advantage of typically wasted space and allows building owners a way to monetize an unused asset. This can be accomplished without taking advantage of new land for solar energy plants.

With the current federal incentive structure and system prices, rooftop PV electricity costs are competitive with peak rates in many markets. PV system costs are expected to decline rapidly over the next few years, while the cost of electricity from fossil fuels continues to rise, and the urgency to reduce our output of CO^2 and other global warming gases continues to increase.

As a result of these trends, industry forecasts predict global commercial and industrial rooftop PV installations in excess of 6 GW in 2011, and 10 GW in 2013. This is the fastest growing segment in the global photovoltaic market.

Solyndra panels are just one example of recent technological advances that have solved problems that limited the adoption of rooftop PV technology.

Deployment of this technology in the US lags behind Europe as a result of differences in energy policy, regulatory structure and access to capital, as well as a patchwork of national, state and local regulations covering all aspects of the PV installation.

The pace of innovation in PV, including improvements in efficiency and reductions in cost, is slowed by the complex regulatory environment. Harmonization of standards will reduce cost and allow improved products to reach the market more quickly.

If the US can implement policies more favorable to PV deployment, we can expect rapid growth in installations, and the US can derive a growing fraction of its energy needs from a source which is renewable, non-polluting, silent and does not emit global warming gases. By further incentivizing the domestic manufacturing of these PV systems, significant gains in employment could result.

ACKNOWLEDGMENTS

The material in the first section regarding US energy demand and evaluating various carbon-free technologies was generously shared by Dr. David Ginger at the University of Washington

Several Solyndra team members helped research and edit this document, including Soumitra Mishra, John Scott, David Miller, Chris Spindt, Tefford Reed, Steve Doll, Neeta Rikhi, and Solyndra's founder and Executive Chairman, Dr. Christian Gronet.

REFERENCES FOR FURTHER READING

1. Antonio Luque and Steven Hegedus, *Handbook of Photovoltaic Science and Engineering* John Wiley & Sons, Ltd, 2003.
2. Many publications can be found at http://www.nrel.gov/publications/popular_publications.html , the NREL website.
3. Website of the David Ginger Lab http://depts.washington.edu/gingerlb/ at the University of Washington

Topics in Nuclear Power*

Robert J. Budnitz

Earth Sciences Division
Lawrence Berkeley National Laboratory
Berkeley, CA 94720

Abstract. The 104 nuclear plants operating in the US today are far safer than they were 20–30 years ago. For example, there's been about a 100-fold reduction in the occurrence of "significant events" since the late 1970s. Although the youngest of currently operating US plants was designed in the 1970s, all have been significantly modified over the years. Key contributors to the safety gains are a vigilant culture, much improved equipment reliability, greatly improved training of operators and maintenance workers, worldwide sharing of experience, and the effective use of probabilistic risk assessment. Several manufacturers have submitted high quality new designs for large reactors to the U.S. Nuclear Regulatory Commission (NRC) for design approval, and some designers are taking a second look at the economies of smaller, modular reactors.

INTRODUCTION

The general public tends to think that today's reactors are about the same as they were 30 years ago, when the Three Mile Island (TMI) nuclear reactor experienced a core melt in 1979. In fact, they are far safer than they were at that time, just as the airline industry is. Today, the risk of a passenger's dying in a plane crash is 30–50 times less than it was 30 years ago.

Figure 1 shows the location of the 104 nuclear plants operating in the US today. All produce more than 500 MW (electric), and some are as large as about 1300 MW. The US has about one quarter of the nuclear power reactors operating in the world today. The youngest of these was designed in the 1970s, so the technology of the operating power reactors is very old. For example, the control rooms generally use analog electronics because almost none of them has switched over in a major way to digital technology.

* **Update:** This paper is an edited and slightly modified version of a spoken talk that was given about a week before the major reactor accident on March 11, 2011 in Japan. On that date, a huge earthquake and tsunami caused serious damage to the six-reactor complex at the Fukushima–Daiichi site in Japan, leading to three melted reactor cores and important radioactive releases. At the time when this paper was being developed from the spoken transcript, the specific root causes of that nuclear accident were still being investigated and were not yet understood in detail. The thrust of this paper, which is that in the U.S. as well as in other advanced countries the overall safety of the operating reactors is substantially better than it was two or three decades ago, is not altered by the fact of the accident at Fukushima—any more than a major aircraft crash that might occur today would alter the conclusion that commercial aircraft safety has generally improved overall during the same period by a substantial factor. RJB.

Physics of Sustainable Energy II: Using Energy Efficiently and Producing it Renewably
AIP Conf. Proc. 1401, 436-444 (2011); doi: 10.1063/1.3653868

Still, the plants are remarkably safe in the sense that the sort of events that might lead to an accident is uncommonly rare. How is such safety achieved in such big, complex systems? One answer is normally "through wonderful designs." The industry did not start with wonderful designs, but over the years, the reactors have been significantly altered for the better. A second answer is "through great attention to operation and maintenance," and that is certainly happening today.

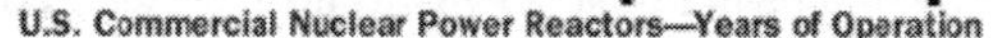

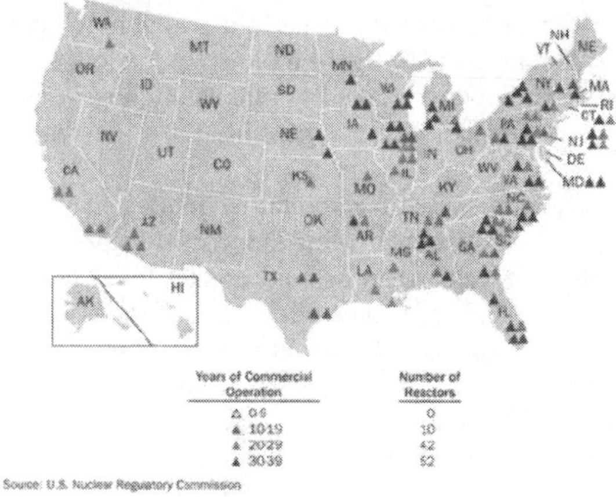

FIGURE 1. Nuclear reactors operating in the US.

HOW IS SAFETY MEASURED?

I will define a safe plant as one whose probability of a major accident in a given year is "acceptably low." But how low is acceptably low? The NRC provided answers over two decades ago: Their safety goals come down in practice to seeking reactors for which the core damage frequency is less than about 10^{-4} per reactor per year. A core damage accident does not lead to a major release if the heat and radioactivity remain contained. Based on analysis, only about one time in 100 would a core damage accident in one of today's operating large reactors lead to a large release of radioactivity. The NRC is generally accepting of a reactor whose likelihood of a large release would occur with a frequency of less than about 10^{-5} per year. If all 104 plants in the US just met these expectations, the US industry could expect about one core-damage accident per century among all of them and one major-release accident per millenium. *[NOTE INSERTED LATER: The Fukushima accident did not result in a "large release" by the common definition of such, because none of the releases came even close to threatening anyone offsite with a prompt radiation-caused fatality.]*

To assess such small accident frequencies requires a safety analysis, which must be intrinsically probabilistic. Hence, analysts have developed a technique known as probabilistic risk analysis (PRA). The method involves the following specific steps:

- Postulate every initiator and determine its frequency
- Work out the contingent probability of "core damage," given the initiator
- Work out the consequences of each sequence.

In the airline industry, there are many accidents to provide data for such an analysis. Since the Three Mile Island accident, however, the US nuclear industry has had no large accidents to use for benchmarking. That's a triumph for engineering, but a "problem" for the analyst charged with figuring out what the low accident frequencies might be. The industry does, however, have lots of data on things that go wrong that—but for engineering—could lead to bigger problems. Such incidents provide a major part of the input to the PRAs.

PRAs are conceptually simple but operationally complex. Analysts have come up with about 100 accident scenarios, each following a particular chain of events: In a typical sequence, something breaks or an operator makes an error, then another piece of equipment has to come in, and then some other human has to take some action, and so forth. If all those steps fail to work, the result can be a loss of core cooling leading to a core meltdown. To analyze such sequences, analysts need to know the probability that, for example, when outside power is lost, a backup pump fails to start or a human fails to turn a dial correctly. Furthermore, the analyst needs to know the probability that these failure events are correlated. The answers come from a large amount of data that has been collected over the years, plus a highly structured analysis. The reliability of many of the major subsystems has been analyzed, and the results have checked out with actual experience, which gives a measure of confidence that the overall PRA analysis is robust.

PRAs calculate that the probability of a core damage accident in the current US fleet of nuclear reactors is typically in the range near or slightly above about 10^{-5} per reactor per year, a factor of three to ten lower than the NRC criterion for acceptably safe. PRA calculations help reactor operators better understand specific systems. They also help identify accident precursors and can uncover trends that are going in the wrong direction, or design flaws or human errors that can affect some subset of operating plants. Plant operators use such safety indictors to know where to intervene.

EMPHASIS ON QUALITY

The nuclear industry has always been intently focused on quality, and in recent years this has been emphasized even more. They have instituted procedures that help operators at all plants learn from their common experiences. For example, there are industry-organized users' groups of operators from different plants concerned with specific common engineered systems, such as the system that injects borated water from a tank when needed. These groups meet about once a year to share experience. They identify the "best practices" and then try to emulate them everywhere. Such learning from experience has led to significant safety record improvements.

In practice, quality begins with the plant design and construction. Most of those in operation today were designed in the 1960s or 1970s. Many had problems in the earlier days, but by now most of those have been corrected. The industry has not discovered very many more problems in recent years. Quality operations are also crucial. The operators in the control rooms today have far better training than at the time of TMI. Then, there were only a few training simulators, the most important being the four operated by each of the four reactor manufacturers at that time. Typically, a given operator went for a week or two of training annually. Today, each plant has its own simulator and operators are trained on them about every five shifts. Every time an odd event happens, the plants simulate that event to train their operators how to respond to it. The significant decrease in operator errors shows up in the data.

The culture of safety has also greatly improved, especially with the implementation of a practice known as "continuous improvement." That practice involves creating an atmosphere in which plant workers can criticize each other with no fault being assigned. They can also report on each other in a constructive, "no fault" way. The practice is rigorously observed because everyone understands how important it is to learn from mistakes. It is generally acknowledged that achieving and maintaining a strong "safety culture" is at the heart of achieving overall safety itself.

Crews also learn from experiences worldwide. Each day, a number of abnormal events occur at one or another power reactor somewhere in the world. All are reported and examined, so that everyone can learn from them. In addition, the industry sponsors peer-to-peer inspection visits between plants.

Also contributing to safety are systems that are redundant and that are designed with extra margins. For example, if a standby pump is required to provide 200 gallons/minute, a plant might be equipped with pumps capable of 400 gal/min. This extra engineering margin, which is of course common in engineered systems of all kinds, is rigorously maintained and analyzed in nuclear plants, in a way that is not typical of most engineered systems—although large aircraft represent another example where this type of attention is given on a continuing and rigorous basis.

The NRC has become better informed about relative risks of different events. They now focus mostly on the big things, rather than on a lot of the little things. As a result, relations between plant operators and regulators have become more cooperative and constructive.

SAFETY INDICATORS

Numerous measures of reactor performance all show impressive safety gains in the last two decades. Figure 2 plots the number of so-called "significant events" per reactor year for US reactors since 1988. At the time of TMI, there were 4 such events per plant per year. In 1988 there was less than one (0.9). By 2006, that number had dropped to 0.01, that is, only one event per year throughout the US. There was only one event in 2007, for example, as the figure shows. These data show a safety improvement in this figure-of-merit by about two orders of magnitude. This US trend is mirrored in other countries as well.

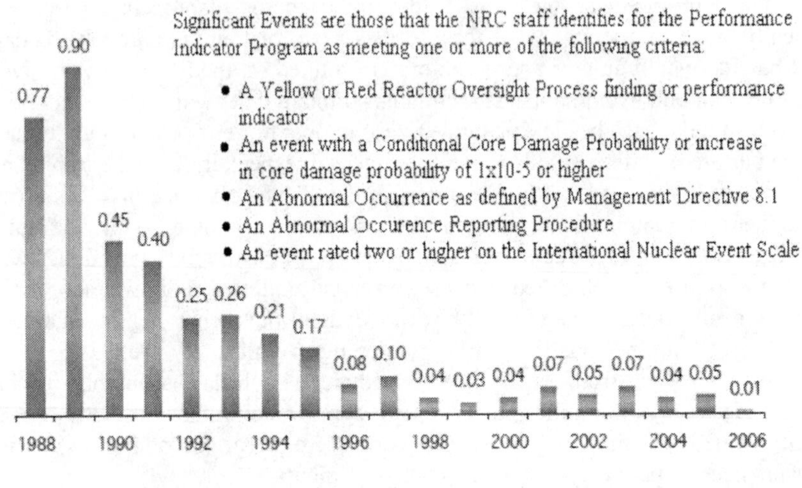

Significant Events at U.S. Nuclear Plants:
Annual Industry Average, Fiscal Year 1988-2006
Significant Events are those that the NRC staff identifies for the Performance Indicator Program as meeting one or more of the following criteria:
• A Yellow or Red Reactor Oversight Process finding or performance indicator
• An event with a Conditional Core Damage Probability or increase in core damage probability of 1x10-5 or higher
• An Abnormal Occurrence as defined by Management Directive 8.1
• An Abnormal Occurence Reporting Procedure
• An event rated two or higher on the International Nuclear Event Scale
0.77
0.90
0.45
0.40
0.25
0.26
0.21
0.17
0.08
0.10
0.04
0.03
0.04
0.07
0.05
0.07
0.04
0.05
0.01
1988
1990
1992
1994
1996
1998
2000
2002
2004
2006
Source: NRC Information Digest. 1988 is the earliest year data is available.
Update: 11/07

FIGURE 2. Significant events at US nuclear plants.

Figure 3 plots another indicator of safe operation: the number of scrams while the reactor is critical: If any of a large number of abnormalities occurs, a signal is generated to shut down the plant, or "scram" it, by inserting the control rods. In 1985, there were about 4 scrams per plant per year. By 2006 that number had dropped by an order of magnitude to 0.32, or only about one scram per year for every three plants. Note that a scram does not necessarily mean that the reactor is in trouble. It is often a minor issue that needs to be fixed to assure overall safety --- you want to stop and fix it.

Yet another safety measure—safety system actuation—is plotted in Figure 4. Most of the time, the backup safety systems are idle, but it's crucial that they work when they are suddenly called into action. The number of times that a safety system fails is plotted in Figure 4. The failure rate has dropped from about 1.4 per plant-year in 1998 to 0.65 in 2007, a factor of two improvement over a decade.

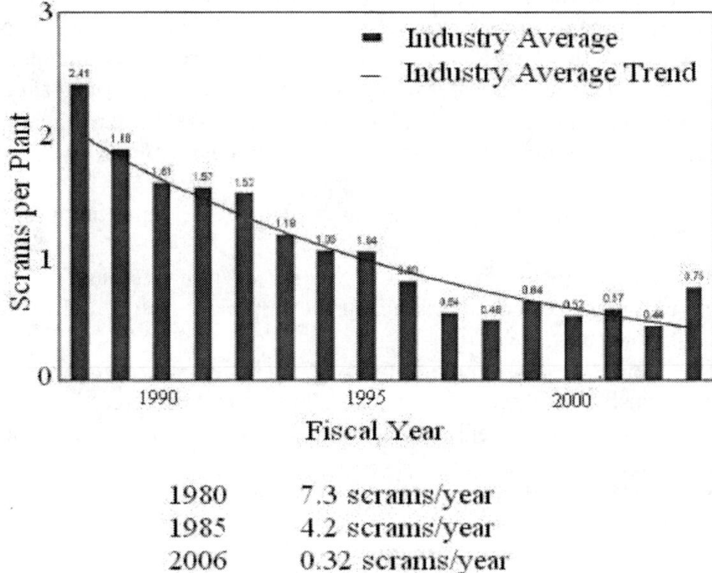

FIGURE 3. Automatic scrams while critical [Nuclear Regulatory Commission].

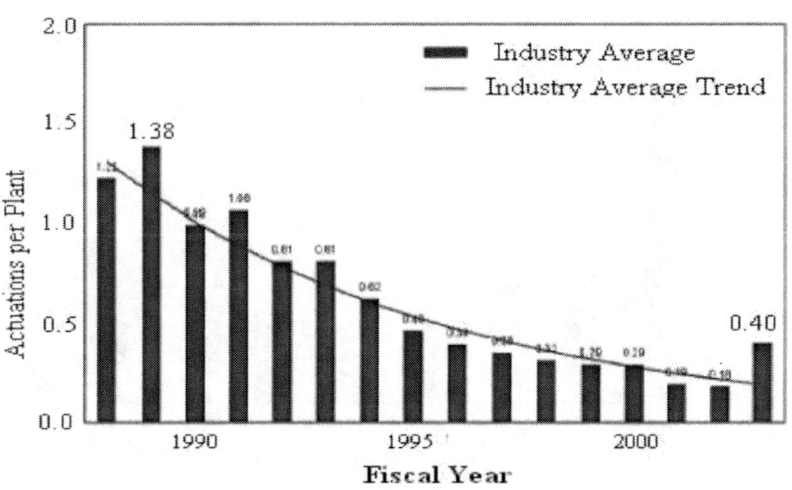

FIGURE 4. Safety systems actuations [Nuclear Regulatory Commission].

Figure 5 shows the forced outage rate, or the percent of time that a plant was shut down as a result of repairs needed in the wake of an event. The outage rate has dropped from about 10% in 1998 to less than 1% in 2010. The smaller outage rate suggests that the repairs are less complicated, enabling the plant to come back online more quickly. There are a dozen or so other safety indicators that mostly tell the same story, that reactors are far safer today on average than 20-30 years ago.

The safety improvements can be attributed to the factors already discussed. All reflect the development and importance of a culture of safety:

- Learning from experience: an industry-wide reporting system and no-fault reporting
- Analysis: major efforts to analyze each event for its causes and implications
- Maintenance: concentrating on the important things and designing for easier maintenance
- Operator errors: simulator training and better procedures
- Industry-wide peer-to-peer inspection visits, task forces
- Design changes: eliminating design flaws and working toward a "forgiving" design
- NRC: Risk-informed enforcement actions (less attention to minor events)

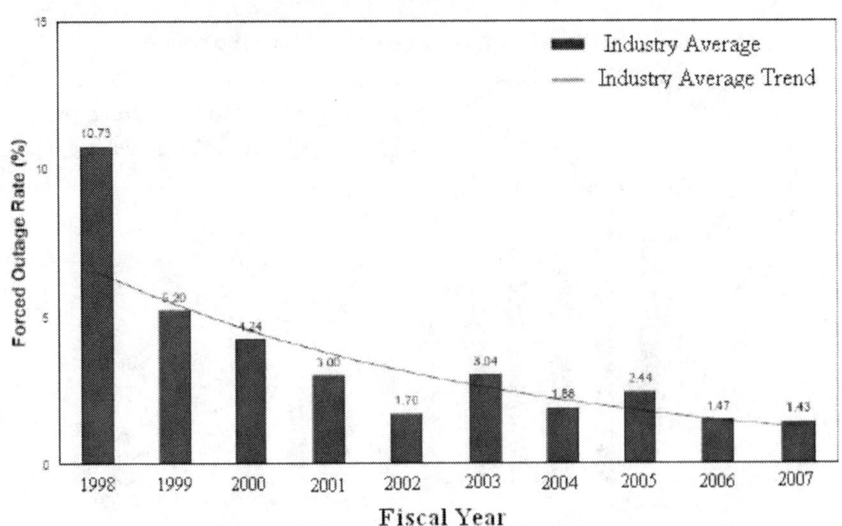

FIGURE 5. Forced outage rate [Nuclear Regulatory Commission].

NEW PLANT DESIGNS

A few years ago, NRC started to receive applications for certification of new plant designs, and also applications from utilities to build new plants using those new designs. The number of such potential new reactors under regulatory review now is about 25. No problems appear to stand in the way of the eventual approval of most of them. Each of the new designs is far better in terms of its safety than the plants now operating, none of which is younger than 20 years old. The new designs build on lessons accumulated from years of experience. Several plants of the newest generation of design have already been built in Japan, and some in China. Of course, the plants at Fukushima–Daiichi in Japan that suffered the recent accident all began operation in the 1970s, so are of a much older design. The most active vendors in the U.S. market are General Electric in collaboration with Hitachi, Westinghouse in collaboration with Toshiba, Areva, and Mitsubishi. The new reactors being proposed are typically about the same size as the largest of the currently operating reactors.

The industry is, however, growing increasingly interested in smaller, modular reactors. A major reason is safety. Today's reactors are so big that a small event can lead to a large thermal release and a large release of radioactivity. It's much easier to keep a smaller core cool, largely thanks to its higher surface-to-volume ratio and certain passive safety features. Some small cores can almost cool themselves. Because the cooling problem is less severe, the plant would not require as much auxiliary equipment, the design could be simpler, there would be fewer things that could go wrong, and the plant would be easier to understand and operate.

There are other reasons to favor small plants. They would not require the utilities to raise as much initial capital for their construction. Additional reactors could be added more easily for increased capacity, and the waste inventory per plant would be reduced.

The main reason not to go with smaller reactors is the economies of scale that are gained in the capital cost with a bigger reactor. Plants of half the size of the new ones we are building are estimated to cost 75% as much to build. Although many manufacturers would love to build plants about one tenth the size of the current generation, the cost per megawatt of doing so using the traditional approach could be two to three times higher. This has led to some innovative design and manufacturing ideas that go well beyond the "traditional approach."

Several factors are coming together now, so much so that even large manufacturers like GE and Westinghouse are putting resources into looking at the small-plant designs. The idea would be to manufacture the reactors at a factory and ship them largely as major modules to the sites where they are needed. This "factory fabrication" scheme is expected to provide major cost savings. Additional savings would come from the reduction in auxiliary safety equipment and the smaller staffing required. The simpler designs should also result in a smaller regulatory burden.

One stumbling block is that smaller reactors represent new territory for the Nuclear Regulatory Commission, whose regulations are currently tailored mainly to the very large light water reactors (LWRs) currently in operation in the US. Even if the smaller designs were also LWRs, the NRC will have to rework many of their regulations to be appropriate to the small reactors. Some of the smaller non-LWR reactor designs, such as gas-cooled and liquid-sodium-cooled reactors, are an even greater departure for the regulators. The NRC is now systematically looking at how to modify their regulations to make them relevant to smaller reactor designs. So far, most other countries are waiting and watching the NRC's lead.

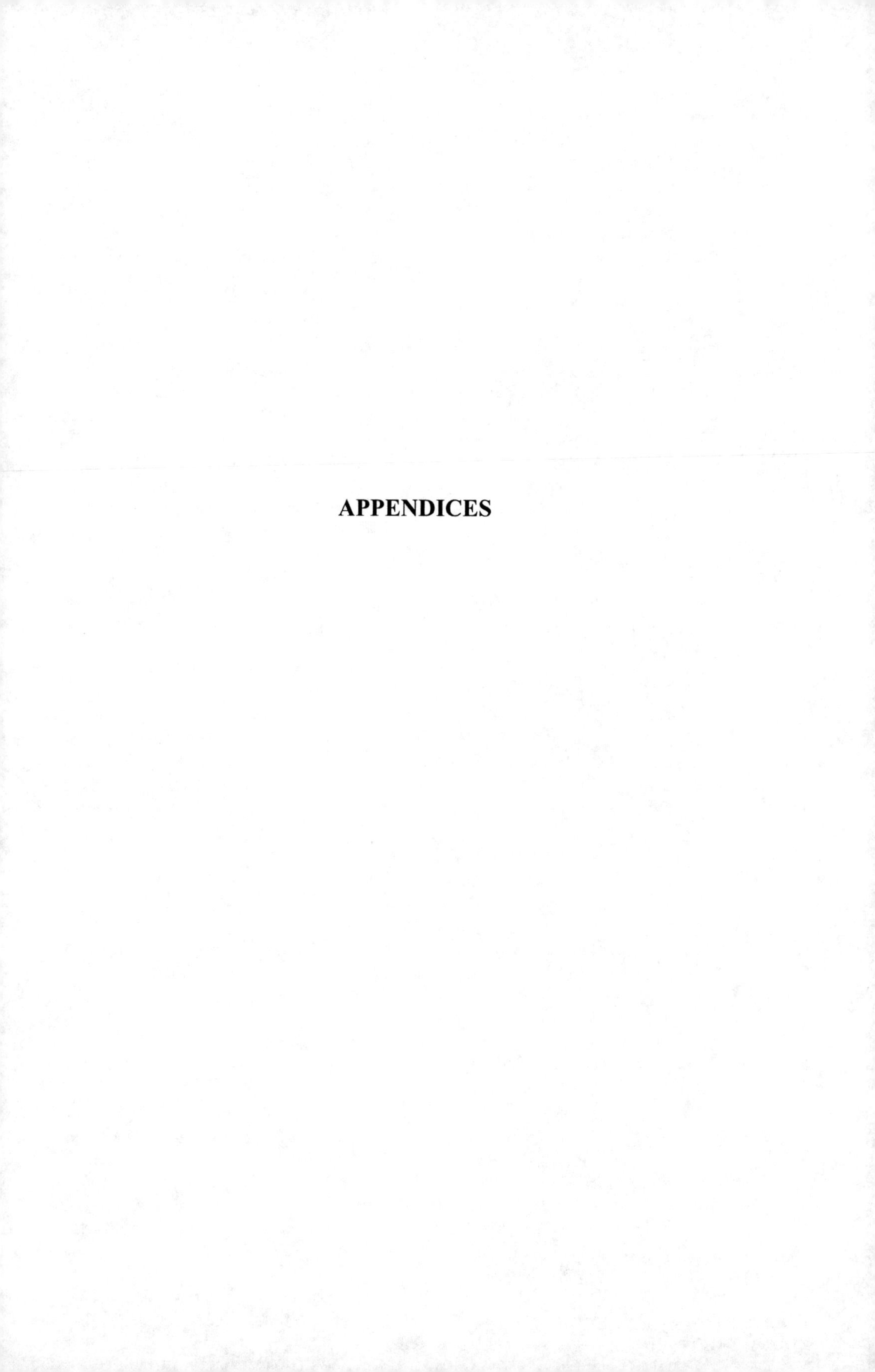

APPENDICES

Energy and Environment Chronology

David Hafemeister

Physics Department
California Polytechnic State University
San Luis Obispo, CA 93407

1859
—Edwin Drake drills 21 meters for 500 barrels of oil at Titusville, Pennsylvania, to begin the *petroleum era.*
—The rechargeable lead–acid battery is developed by Gaston Plante.

1870
—John D. Rockefeller founds Standard Oil.

1879
—Thomas Edison and J. Swan, independently, invent the incandescent lightbulb.

1882
—First US coal-fired power plant lights up Manhattan by Thomas Edison using DC voltage.

1908
—Model–T Ford runs on gas or ethanol, getting 21 miles per gallon.

1911
—Standard Oil Company monopoly is broken by the Supreme Court.

1930s
—General Motors, Firestone and Standard Oil of California buy US electric streetcar systems across and replace them with buses.

1942
—December 2: Enrico Fermi's reactor goes critical at Stagg Field, University of Chicago, Illinois to begin the *nuclear era.*

1945
—August 15: Office of Price Administration (OPA) lifts gasoline rationing.

1946
—May 6: Division of Oil and Gas established in Department of Interior.
—May 21: US President Harry Truman orders US Government to take possession of coal mines during a strike.
—June 18: National Petroleum Council established.

1947
—January 1: Atomic Energy Commission begins operation.

Physics of Sustainable Energy II: Using Energy Efficiently and Producing it Renewably
AIP Conf. Proc. 1401, 447-454 (2011); doi: 10.1063/1.3653869

—March 25: Coal-mine disaster kills 11 In Centralia, Illinois.
—June 16: Federal Power Commission authority extended to all natural gas producers.

1952
—December 5: Severe air pollution (0.7 ppm SO_x and particulates) kills 4000 in London in 4 days.

1953
—August 7: Congress gives US government jurisdiction of ocean floors beyond 3-mile boundary.
—December 8: US President Dwight Eisenhower delivers "Atoms for Peace" speech before the United Nations.

1954
—August 30: Atomic Energy Act of 1954 encourages peaceful use of nuclear energy.
—Bell Labs develops the silicon photovoltaic cell.

1956
—Federal Highway system begins at a cost of $129 billion.

1957
—King Hubbert correctly predicts US petroleum production peak between 1966–71, which happened in 1970 when lower forty-eight produced 9.1 Mbbl/day. Hubbert uses a finite resource in differential equations, but does not use economics.

1959
—March 10: Eisenhower limits oil imports to stimulate domestic production and refining capacity.

1962
—October 11: Congress authorizes the president to impose mandatory oil import quotas.

1963
—December 17: Clean Air Act provides assistance to states for air pollution research.

1965
—October 2: Water Quality Act establishes the Water Control Administration.
—October 20: Solid Waste Disposal Act provides assistance for study, collection, and disposal of solid wastes.
—November 9: First major power blackout covers northeast US.

1967
—November 21: Clean Air Act gives authority to Secretary of Health Education and Welfare to set auto emission standards.

1969
—January 1: National Environmental Policy Act (NEPA) establishes framework for Environmental Impact Statements and the Council of Environmental Quality (CEQ).
—January–February: Major oil spill from offshore drilling near Santa Barbara, CA.
—December 30: Oil depletion allowance reduced from 27.5% to 22%.

1970
—March 5: President Richard Nixon issues executive order requiring federal agencies to evaluate their activities under the National Environmental Policy Act.
—April 22: First "Earth Day" celebration.
—July 9: Nixon requests Congress to create Environmental Protection Agency (EPA).
—October 23: Merchant Marine Act Amendment provides subsidies for oil and liquified natural-gas tankers.

—December 24: Geothermal Steam Act authorizes leases for geothermal steam.
—Clean Air Act sets national air quality and auto emission standards.

1971

—July 23: Supreme Court decision on siting of nuclear power plant at Calvert Cliffs, Maryland, requires the Atomic Energy Commission to comply with the National Environmental Policy Act.

1972

—EPA bans DDT.
—Clean Water Act sets pollution standards for water.
—US and Canada agree to clean up the Great Lakes, source of 95% of US fresh water, used by 25 million persons.

1973

—June 29: White House Energy Policy Office created with former Governor John Love of Colorado as first "energy czar."
—October 17, 1973 to March 17, 1974: Organization of Petroleum Exporting Counties (OPEC) embargoes US and the Netherlands because of their support for Israel.
—November 7: Nixon creates Project Independence to end oil imports by 1980.
—November 27: Emergency Petroleum Allocation Act provides authority for oil allocations.
—December 4: Presidential Federal Energy Office created with William Simon as energy czar.
—December 15: Congress mandates daylight savings to save energy.
—December 28: Nixon signs Endangered Species Act.
—EPA begins phasing out leaded gasoline.
—EPA begins limits on factory pollution discharges.

1974

—June 22: Federal Energy Administration (FEA) authorized to order utilities and industry to convert from oil/gas to coal.
—September 3: Congress authorizes funds for geothermal energy and solar heating demonstrations.
—October 5: Congress repeals daylight savings time in winter to save energy.
—October 11: Energy Reorganization Act abolishes the AEC to create the Energy Research and Development Administration (ERDA, later DOE) and the Nuclear Regulatory Commission (NRC).
—December 31: Congress requires ERDA to submit an annual comprehensive energy plan.
—Arthur Rosenfeld, establishes research on buildings and energy at Lawrence Berkeley National Laboratory, to begin the *enhanced end-use efficiency energy era*. Rosenfeld was the last graduate student of Enrico Fermi, who began the *nuclear reactor era* in 1942. The LBL work was originally based on the 1974–5 study by the American Physical Society.

1975

—January 4: Congress establishes the 55-mph speed limit to save energy.
—March 17: Supreme Court rules that the states do not have jurisdiction over the outer continental shelf.
—October 29: ERDA dedicated its first wind power system at Sandusky, Ohio.
—October: President Gerry Ford halts plans to reprocess US spent fuel at Barnwell, SC.
—December 22: Energy Policy and Conservation Act (EPCA) establishes prices for US crude oil, the Strategic Petroleum Reserve, emergency energy powers for the president, Corporate Average Fuel Economy (CAFE) standards of 27.5 mpg by 1985, and cost-effective, appliance energy standards.
—Introduction of catalytic converter mufflers for cars.

1976

—April 5: Congress authorizes production from naval petroleum reserves.
—August 14: Energy Conservation and Production Act (ECPA) creates incentives for conservation and renewables, funds weatherization for low income homes, establishes a program for energy standards for new buildings and establishes Solar Energy Research Institute (now the National Renewable Energy Laboratory) to begin the *renewable energy era*.

—Toxic Substance Control Act reduces environmental and human health risks. EPA begins phase out of PCBs.

1977

—April 7: President Jimmy Carter indefinitely defers reprocessing of nuclear fuel and stops construction of Clinch River Breeder Reactor.
—July 5: Solar Energy Research Institute (SERI) opens in Golden, Colorado; renamed in 1991 as the National Renewable Energy Laboratory (NREL).
—October 1: Department of Energy created from ERDA and the FEA. Jimmy Carter declares it is "the moral equivalent of war."

1978

—November 9: National Energy Act establishes weatherization grants for low income families, conservation programs for local governments, energy standards for consumer products, programs to convert utilities to coal, and energy tax credits.
—Buried, leaking containers found at Love Canal, NY; cleanup was completed in 1998.
—The Public Utilities Regulatory Policy Act (PURPA) of 1978 allows customers to sell electrical energy to electrical utilities on an avoided cost basis. Residential and commercial electricity producers can sell reliable, timely electrical energy at a rate lower than the utility can produce it.

1979

—March 28: Accident at Three Mile Island nuclear power plant in PA effectively halts purchase of new reactors in the US.
—Spring: Second oil shock from Iran oil curtailments causes gasoline shortages.
—August 17: President Carter begins to lift price controls on domestic crude oil.
—November 3: US embassy in Iran seized by revolutionaries; Carter suspends oil imports from Iran on November 14.

1980

—April 2: Windfall profit tax on crude oil to give assistance for weatherization of homes of low income people.
—June 30: Energy Security Act creates the Synthetic Fuels Corporation and funds renewable energy projects.
—Tax on certain chemicals creates a "superfund" to pay for cleanup when responsible parties fail.

1981

—January 28: President Ronal Reagan completes price decontrol on domestic crude oil.
—National Research Council finds acid rain intensifying in northeast US.
—1981–1985: Reagan and Congress debate funding for conservation programs.

1982

—May 24: Reagan proposes transfer of Department of Energy functions to the Department of Commerce.

1983

—Clean-up of Chesapeake Bay begins.
—EPA encourages homeowners to test for radon.

1984

—December 3: Chemical disaster at Union Carbide plant in Bhopal, India, kills 3000 and partially disables 2700.

1985

—February 4: Department of the Interior reduces estimates for offshore oil (27 to 12 Bbbl) and gas (163 to 91 tcf).

—June 27: EPA modifies mileage test, lowering CAFE's 27.5 mpg by 2 mpg.
—June 28: EPA curbs tall smokestacks to avoid distant pollution.
—July 16: Appellate Court confirms EPCA's appliance standards by voiding DoE's "no-standard" standard. A key issue was the magnitude of the discount rate for future benefits.
—Antarctic Ozone Hole for stratospheric ozone discovered.

1986
—April 26: Accident at Soviet reactor in Chernobyl, Ukraine.
—August 21: Large release of carbon dioxide from the depths of Lake Nyos, Cameron, kill 1800.

1987
—Montreal Protocol for ozone protection bans chlorofluorocarbons.
—December 27: Congress approves Yucca Mountain as the only high-level nuclear waste repository under development.

1988
—Congress bans ocean dumping of sewage sludge and industrial waste.

1989
—March 24: Exxon Valdez spill of 0.3 million barrels of oil in Prince William Sound, Alaska.
—March 23: Stanley Pons and Martin Fleishman announce discovery of "cold fusion." Physicists do not believe them.

1991
—January-February: Iraq sets 700 oil fires in Kuwait in wake of Gulf War.
—Exxon pays $1 billion for Exxon Valdez spill.

1992
—EPA bans ocean dumping of sewage sludge.
—EPA launches Energy Star Program to identify energy efficient products.

1993
—EPA's Common Sense Initiative shifts OSHA regulations from pollution-based to industry-based.
—EPA research finds secondhand indoor cigarette smoke causes 3000 lung-cancer deaths per year to nonsmokers.

1994
—EPA launches Brownfields Program to restore abandoned city sites.

1995
—70% of US metropolitan areas that had unhealthy air in 1990 meet air quality standards in 1995.

1997
—EPA restricts particulate matter air emissions down to 2.5 μ diameter.
—Kyoto Protocol to limit greenhouse gases (GHG) is agreed in principle by most nations, including the US, but the details for trading allotments of GHG are not certain.

1999
—Minivans and sport utility vehicles (SUVs) will have the same emission standards as cars.
—EPA requires cars, SUVs, minivans and trucks to have the same standards by 2004, reducing SUV pollution by 77–95%.
—Honda sells the first commercial hybrid auto.

2000

—Most utilities buy combined-cycle natural gas turbines at 55–60% efficiency, a major supply side break through, establishing the *natural gas era*.
—California hit with rolling blackouts, stimulated by Enron maneuvers.

2001
—January: California's partial deregulation contributes to rolling blackouts caused by many factors.
—July 23: The Kyoto-Bonn Protocol on limiting GHG moved towards implementation, supported by 178 nations and the EU. Without support from the US, a former advocate turned critic, the future of the Kyoto process is in doubt. The US stated it would use voluntary caps on GHG emissions.
—Sales of minivan and SUVs equal sales of cars.
—September 11: The World Trade Center in New York City is destroyed by two hijacked commercial aircraft. Some 3000 are killed at the WTC, the Pentagon and a field in Pennsylvania, and 360 firefighters are placed on mical leave or light-duty work. Mid-east oil will be more difficult to import.

2002
—February: President Bush proposed a *voluntary* 18% carbon reduction by 2012 in terms of *greenhouse gas intensity*, which is the ratio of national total energy use of carbon fuels to GDP dollar.
—February 14: Secretary of Energy Spencer Abraham recommends approval of the Yucca Mountain, NV, geological repository for 77,000 tonnes of nuclear spent fuel, to begin in 2010.
—November: Bush Administration relaxes pollution standards on existing coal-fired power plants and gives managers of national forests more discretion to approve logging and commercial activities.

2003
—April 1: CAFE standards for light-truck/SUVs rise from 20.7 mpg to 21.0 in 2205, 21.6 in 2006 and 22.2 mpg in 2007.
—June 7: The 18-GW_e Three Gorges Dam closes, to be completed by 2009, displacing 1.1 million people.
—August 11: Auto manufacturers no longer contest California's 2005–20 phase-in of low and zero emission vehicles.
—August 11: Gov. Michael Leavitt nominated to replace Gov. Christine Todd Whitman as EPA administrator.
—August 14: Electrical grid failure of 62 GW_e darkens 8 states and Canada (previously on 11-9-65 and 7-13-77).
—August 27: President Bush gives emission exemptions from the Clean Air Act for rehabilitated power plants.

2004
—Natural gas prices double and supplies from Canada become more uncertain. Coal plant orders rise from 2 to 100 and utilities request site approval for 3 nuclear plants.
—July 9: US Court of Appeals in DC rules against the 10,000–year limit on radiation safety at Yucca Mountain. The court concluded that EPA must either issue a revised standard that is "consistent with" the NAS peak-dose standard "or return to Congress and seek legislative authority to deviate from the NAS report."
—September 24: The California Air Resources Board required automobiles to lower carbon emissions by 30%, to be phased in over 2009 and 2016. The Board claimed the extra cost would be $1000 per vehicle, but it would save $2500 in fuel. New York and other states claim they will follow suit.
—October 26: General Motors downsizes the Hummer from 6400 pounds (12 mpg) to 4700 pounds (16 city/20 highway).
—Kyoto climate change treaty goes into effect without US ratification.

2005
—January 19: A billion dollar LNG explosion at Skikda, Algeria clouds plans for 35 LNG projects in the US.
—February 16: The Kyoto Protocol enters into force after Russian ratification. The 120 nations that ratified emitted 61% of greenhouse gases by signatories, over the 55% threshold for ratification. The

US, which emits 37%, did not ratify, while India and China did not sign. The carbon-trading allotments initially were selling for $10/tonne.
—August 29–30: Katrina, a category 3 hurricane, floods 80% of New Orleans, killing 1836 and costing $86 billion (2007$), the US's most expensive natural disaster.
—Energy Policy Act of 2005 modifies the Public Utilities Regulatory Policy Act of 1978 to include rules for net metering, time-based metering and the Smart Grid.

2006
—Al Gore's documentary movie, *An Unfortunate Truth*, heightens concerns on carbon dioxide emissions from burning fossil fuels. Greenland glaciers and the North Polar Cap melt faster than expected.

2007
—January 10: European Union agrees to cut CO_2 emissions by 20% by 2020, compared to 1990 levels. Under the Kyoto protocol, the EU was already committed to an 8% decrease. The EU further agreed to make biofuels at least 10% of vehicle fuel by 2020.
—April 2: US Supreme Court rules that states may regulate green house gasses in *Massachusetts vs. EPA*.
—CAFE auto standards raised from 27.7 mpg to 35 mpg by 2020. California and other states lose in court to more the date forward on the grounds of cleaner air for their cities.
—Intergovernmental Panel on Climate Change (IPCC) 2007 conclusions: "Global mean surface temperatures have risen by 0.74 °C (+/– 0.18 °C) when estimated by a linear trend over the past 100 years (1906-2005)....Average Arctic temperature increased at almost twice the global average rate in the past 100 years....Sea levels rose 0.9 to 1.6 mm/year between 1993 and 2003....Carbon dioxide is increasing at 1.9 ppm/year."

2008
—July 9: The Group of Eight (G-8) of industrialized nations agree to cut greenhouse gases by 50% by 2050.
—December 22: Over a billion gallons of coal fly ash sludge flow from a holding dam near Kingsport, Tennessee. The sludge contains dangerous amounts of arsenic, boron, cadmium, cobalt, mercury, lead, selenium and other elements.
—Gasoline hits $4.50 a gallon in the US, and twice that in many countries. SUV sales plummet.
—General Motors plans to introduce the Chevy Volt by about 2011 with an all–electric range of 40 miles with lithium ion batteries and an extended range from a 3–cycle engine. Toyota states it will follow, perhaps with nickel metal hydride batteries. EPRI points out that the un-used electricity at night can be put to good use on these proposals.
—Utilities propose to build some 30 nuclear power plants.
—China plans to build over 50 coal-fired power plants per year.
—British Petroleum gives Lawrence Berkeley Laboratory $500 million to develop cellulosic fuels from switch grass. Corn food price rises as corn–ethanol for autos increases.
—Polymer photovoltaics hit 5% efficiency, needing further research to enhance robustness.
—Refrigerators use less than 25% of former energy use. Movement towards green buildings expands, with the hope to have carbon–neutral buildings (with some local renewable energy).

2009
—April 17: US EPA rules that the emissions of six greenhouse gases, including CO_2, are a danger to the public and should be regulated under the Clean Air Act. This followed the April 2007 ruling of the Supreme Court.
—July 9: Prior to the Copenhagen negotiations, the G–8 nations (Canada, France, Germany, Italy, Japan, Russia, US and UK) agree to reduce greenhouse gases by 80% by 2050.
—December 18: Climate negotiations collapse in Copenhagen as the Kyoto Protocol emissions treaty is set to expire in 2012. A non-binding agreement between Brazil, China, India, South Africa and the US is announced.

2010

—March 25: The World Meteorological Organization states that the 2000-2009 decade was the warmest decade on record. This finding agrees with the US NASA's finding that the decade was the warmest since the 1850's, when the first systematic records of temperatures were begun.

—An explosion at the West Virginia Upper Big Branch Mine killed 29 miners, the largest loss since the Kentucky Finley Coal Mine explosion in 1970, which killed 78. China, reportedly, losses 5,000 to 20,000 a year in mine explosions.

—April 20: The Deepwater Horizon drill rig in the Gulf of Mexico explodes, killing 11 and releasing 4.9 million barrels of oil over 100 days, devastating fragile coastal environments from Louisiana to Florida. Capping the well at a depth of 4 km proved illusive until partial containment on July 15, and final plugging on September 19. 1.8 million gallons of dispersant were used for the clean up.

—July 7: A panel of scientists at East Anglia University, UK concluded that allegations against the climate scientists were not valid, concluding that they acted with "rigor and honesty," but that they had failed "to display the proper degree of openness." A panel at Penn State University concluded the same result.

2011

—January 13: EPA vetoes the water permit for the massive mountaintop removal at the Spence No. 1 site in West Virginia.

—January 15: EPA approves the use of 15% ethanol blended fuel, produced from corn.

—March 11: A 9.0 earthquake and 15-meter tsunami kill 20,000 and stop the cooling of 3 nuclear reactors, which partially melt. The three-month old low-enriched uranium and MOX spent fuels catch fire in cooling pond #4. Replacement electricity for cooling pumps is not available for nine days. Reportedly, the amount of radiation released was 10% of that released at Chernobyl, which was more widely dispersed. Minor levels of radiation arrive in Tokyo, 200 km away. A massive and lengthy clean-up effort will be needed, with a large political impact on future energy supplies.

—The price of photovoltaic *panels* drops to under $0.80/peak-Watt, perhaps $3-4 *total* cost for large solar farms. About 10 GW of serious proposals are made in the south-west US, with the aid of US loan guarantees.

—Energy Information Administration (EIA) data: Research on hydrodynamic fracturing of Devonian shale began in mid 1970's, and became economically competitive when combined with horizontal drilling in about 2005. US *proved reserves* of 273 trillion cubic feet are now extended with the hydrofracking (or fracking) *technically recoverable reserves* of 862 TCF. The impact on water supplies has yet to be resolved. At least 32 nations have these types of reserves. EIA projects that shale gas will grow from 14% of consumption in 2009 to 45% by 2035, while imported gas will drop from 11% to 1%.

—EIA projects US electricity production will grow from 3.7 trillion kWh/year in 2009 to 4.9 TkWh/yr by 2035. EIA projects only modest fuel shifts over these 26 years, with coal dropping slightly from 45% to 43%, natural gas rising from 23% to 25%, renewables rising from 10% to 14%, and nuclear dropping from 20% to 17%. The 3.7 TkWh/yr in 2009 corresponds to 420 GW of average power, with 1.4 kW/person.

—EIA projects modest shifts in the total US consumption of energy from 2009 to 2035: coal remaining constant at 21%, natural gas dropping from 25% to 24%, petroleum dropping from 37% to 33%, nuclear dropping from 9% to 8%, renewables rising from 7% to 10%, and liquid biofuels rising from 1% to 3%. EIA projects a 5.6% rise in US CO_2 emission from 5.98 billion tones/yr in 2005 to 6.31 in 2035.

—EIA projects that crude oil prices will rise from $60/barrel to $133/barrel within the *wide range* of $51 and $210. In 2009, the total cost of energy per person was $3,400 with $800 for residential and $1,530 for transportation. In 2008, in a time with good employment and higher petroleum costs, transportation cost/person was $2,300.

—In 2010, world-wide clean energy investments rose by 30% to $243 billion, with 90% of this in the G-20 nations, with China investing the most with $54.4 billion, Germany in second place with $41.2 billion, and the US in the third place with $34 billion (Pew Environmental Group).

Energy Outlook, 1990–2035

Energy Information Administration

Department of Energy
Washington, DC 20585

International Energy Outlook 2010

Fig. 2: World marketed energy use by fuel type.

Fig. 3: World liquids production.

Fig. 5: World coal consumption by region.

Fig. 6: World net electricity generation by fuel.

Fig. 9: World delivered energy consumption in the transportation sector.

Fig. 10: World energy-related carbon dioxide emissions.

Fig. 11: Impacts of four Kaya factors on world carbon-dioxide emissions.

Fig. 13: World marketed energy consumption: OECD and Non-OCED.

Fig. 32: World oil prices in three cases.

Fig. 104: World energy-related carbon dioxide emissions by fuel type.

U.S. Annual Energy Outlook 2011

Fig. 1: Shale gas offsets declines in other U.S. supply to meet consumption growth and lower need.

Fig. 2: The projected fuel mix for electricity generation gradually shifts to lower carbon options.

Fig. 3: In the AEO2011 reference case, energy-related carbon dioxide emissions grow to almost 6 percent over 2005 levels by 2035.

Fig. 5: Change in conventional liquids production by top non-OPEC producers.

Fig. 6: Delivered energy consumption by sector.

Fig. 7: Energy consumption by fuel.

Fig. 8: Energy use per capita and per 2005 dollar of gross domestic product.

Fig. 9: Outputs from the industrial and service sectors.

Fig. 10: Total energy production and consumption.

Fig. 11: Energy production by fuel.

Diagram 1a: U.S. energy flow 2009.

Diagram 1b: U.S. energy flow 2006.

International Energy Outlook 2010

Figure 2. World marketed energy use by fuel type

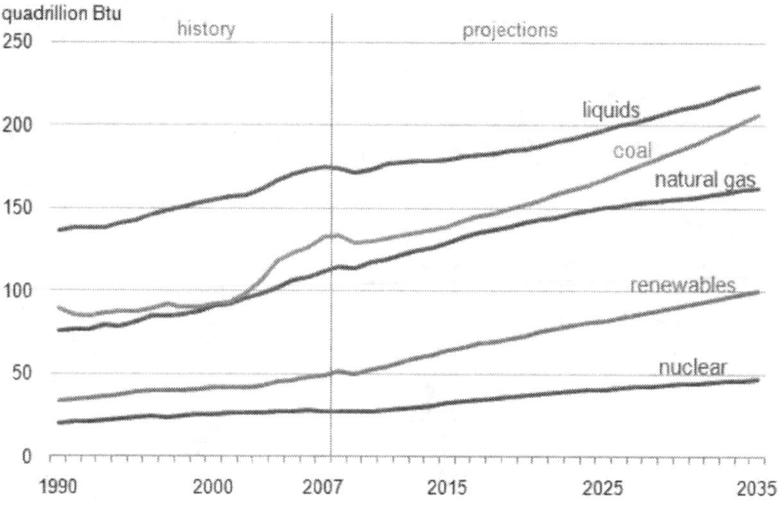

Figure 3. World liquids production

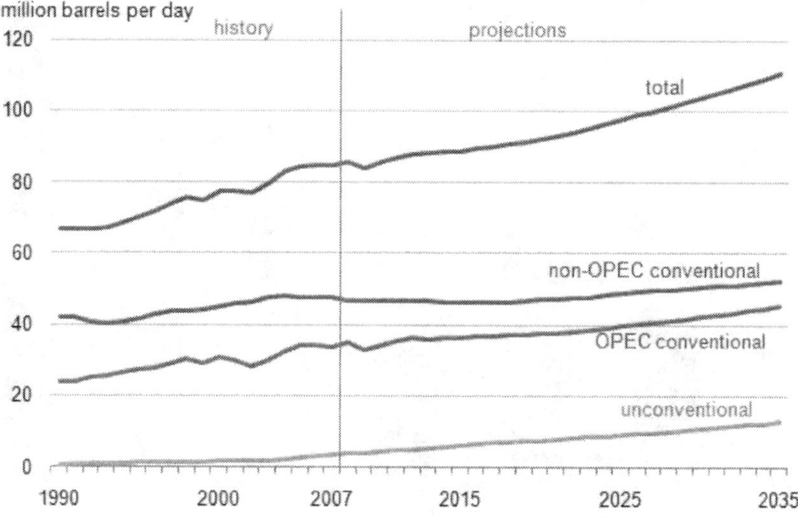

Figure 5. World coal consumption by region

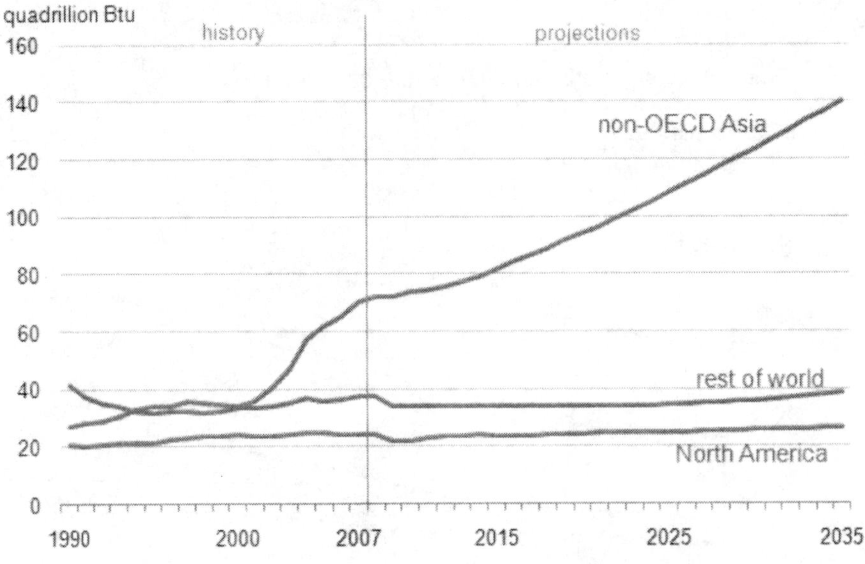

Figure 6. World net electricity generation by fuel

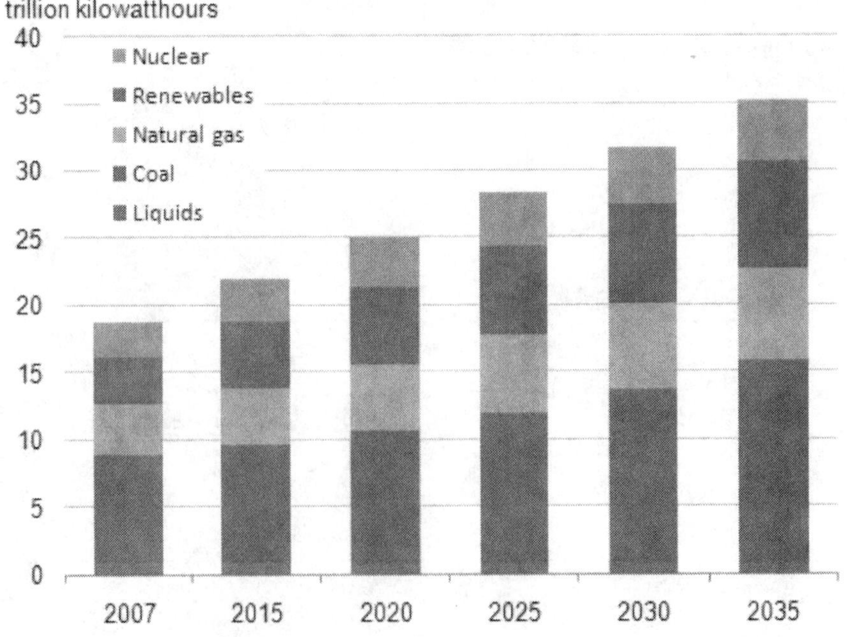

Figure 9. World delivered energy consumption in the transportation sector

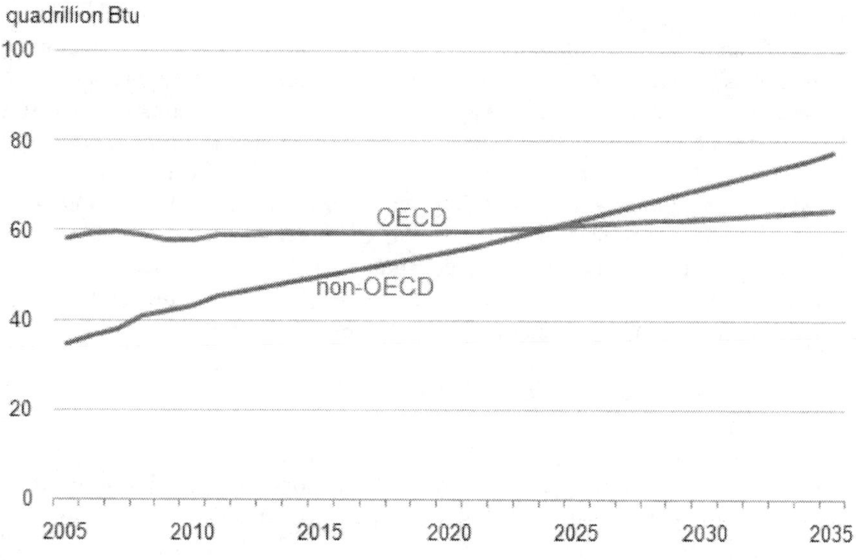

Figure 10. World energy-related carbon dioxide emissions

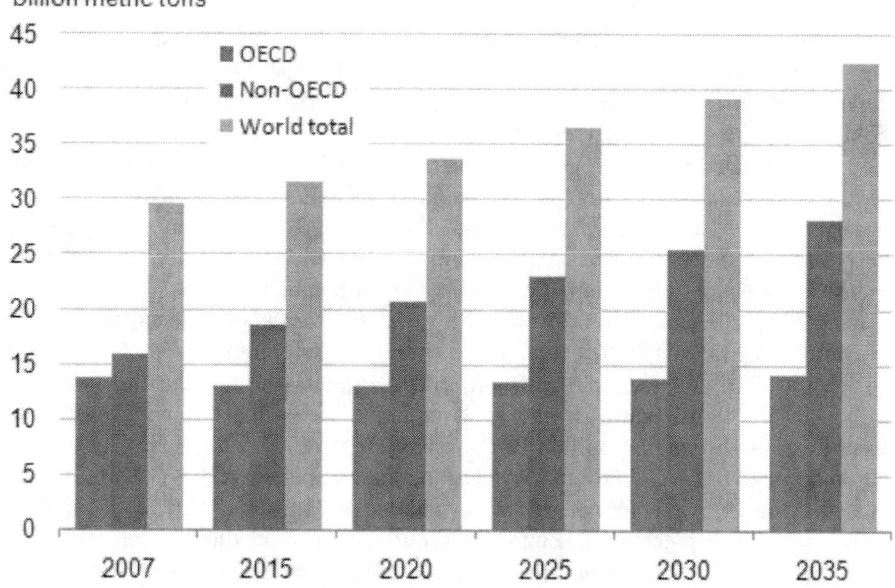

The Kaya decomposition of emissions trends

Yoichi Kaya, a Japanese economist, has provided an intuitive approach to the interpretation of historical trends and future projections of energy-related carbon dioxide emissions. It is used to describe the relationship among the factors that influence trends in energy-related carbon dioxide emissions as seen in Fig. 11.

[D. Hafemeister inserted this paragraph.] The carbon dioxide emission rate (CO_2) can be written as a product of four factors which are obtained from the ratios of three factors: (1) the rate of energy consumption (*E*), (2) the rate of economic activity as measured by gross domestic product (*GDP*), and (3) the population size (*POP*). This gives the following four factors: (1) the ratio of the rate of energy-related carbon dioxide emissions to the rate of energy consumption (*E*), (2) the ratio of the rate of economic activity to the rate of gross domestic product (*GDP*), (3) the ratio of the rate of economic activity to the population size (*POP*) and (4) the population size (*POP*):

$$CO_2 = (CO_2/E) \times (E/GDP) \times (GDP/POP) \times POP. \qquad (1)$$

By taking the differential of Eq. 1 and dividing by Eq. 1, the rate of change of CO_2 is obtained in terms of the rates of change for the 4 factors. The symbol $\tilde{\mathbf{A}}$ in front of each term indicates these are fractional (or percent) change per year in carbon emission, which equals the *sum* of the rates of change $\tilde{\mathbf{A}}$ for the four components.

$$\tilde{\mathbf{A}}(CO_2) = \tilde{\mathbf{A}}(CO_2/E) + \tilde{\mathbf{A}}(E/GDP) + \tilde{\mathbf{A}}(GDP/POP) + \tilde{\mathbf{A}}(POP). \qquad (2)$$

This result can also be derived by taking the differential of the natural logarithm of Eq. 1, and dividing that by a time period dt.

The first two components on the right-hand side of the equation represent the carbon dioxide intensity of energy supply (CO_2/E) and the energy intensity of economic activity (*E/GDP*). When they are multiplied together, the resulting measure is carbon dioxide emissions per dollar of GDP (CO_2/GDP)—i.e., the carbon intensity of the economy, which is another common measure used in analysis. Economic output (*GDP*) is decomposed into output per capita (*GDP/POP*) and population (*POP*). At any point in time, the level of energy-related carbon dioxide emissions can be seen as the product of the four Kaya Identity components—energy intensity, carbon dioxide intensity of energy supply, output per capita, and population.

Using 2007 data as examples, world energy-related carbon dioxide emissions totaled 29.7 billion metric tons in that year, world energy consumption totaled 495 quadrillion Btu, world GDP totaled $63.1 trillion, and the total world population was 6,665 million. Using those figures in the Kaya equation yields the following: 60.1 metric tons of carbon dioxide per billion Btu of energy (CO_2/E), 7.8 thousand Btu of energy per dollar of GDP (*E/GDP*), and $9,552 of income per person (*GDP/POP*).

Of the four Kaya components, policymakers are most actively concerned with energy intensity of economic output (*E/GDP*) and carbon dioxide intensity of the energy supply (CO_2/E), because they correspond to the policy levers most available to them. Reducing growth in per-capita output would also have a mitigating influence on emissions, but governments generally pursue policies to increase rather than reduce output per capita to advance objectives other than greenhouse gas mitigation. Some countries, such as China, have policies related directly to limiting population growth, but most countries pursue policies that only indirectly influence population growth.

Both OECD and non-OECD economies have experienced or are expected to experience declines in energy intensity. These are the only values that are consistently negative across all time periods at the aggregate level. In the historical period, only OECD Asia showed a rise in energy intensity, reflecting an increase in the energy intensity of Japan's economy. However, Japan has the lowest energy intensity among all the fully industrialized OECD economies.

Carbon intensity varies across time and regions, but in no case does it change as much as energy intensity does. Over the 1990-2005 period, the largest annual decline worldwide (0.7 percent) is for non-OECD Europe and Eurasia, where much of the old energy infrastructure was shut down and replaced after the fall of the Soviet Union. The next largest annual decline (0.6 percent) occurred in OECD Europe, where coal consumption fell from 17.7 quadrillion Btu in 1990 to 12.9 quadrillion Btu in 2005 and was replaced by natural gas consumption, which increased from 11.2 quadrillion Btu in 1990 to 19.8 quadrillion Btu in 2005. In many regions, including North America, the carbon intensity of energy supply remained largely unchanged from 1990 to 2005. For the entire world, carbon intensity declined by only 0.1 percent annually from 1990 to 2005, compared with a 1.5-percent average annual decline in energy intensity.

Over the period from 2005 to 2020, carbon intensity declines in the *IEO2010* Reference case in every part of the world. While explicit carbon policies, such as the caps in OECD Europe, are not included in the model, analysts' judgment regarding, for example, nuclear power have taken those policies into account. L In other areas, declining carbon intensity is the result of policies such as renewable portfolio standards and other approaches to promote alternatives to fossil fuels. From 2020 to 2035, there is a slight decrease in carbon intensity of energy supply in OECD economies and a slight increase in non-OECD economies, so that there is virtually no change on a worldwide basis in the absence of additional policies to stem emissions growth, which are not included in the Reference case.

For non-OECD countries, increases in output per capita, coupled with even moderate population growth, overwhelm the improvements in energy and carbon intensity. For example, the combined decrease in carbon intensity and energy intensity in non-OECD economies averages 3.0 percent per year from 2005 to 2020. With output per capita rising by 4.3 percent per year and population growing by 1.2 percent per year, however, the net increase in non-OECD carbon dioxide emissions is 2.3 percent per year. Over the same period, the combined decrease (improvement) in carbon intensity and energy intensity in OECD economies averages 2.1 percent per year—lower than in non-OECD economies—but because OECD output per capita increases by 1.4 percent per year and population growth averages 0.5 percent per year, the net result is that OECD carbon dioxide emissions decline by an average of 0.2 percent per year.

Figure 11. Impacts of four Kaya factors on world carbon dioxide emissions

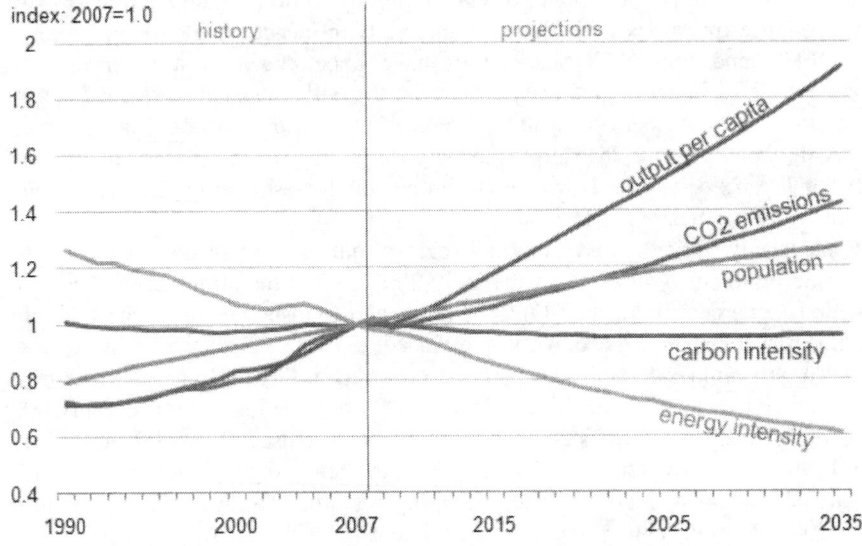

Figure 13. World marketed energy consumption: OECD and Non-OECD,1990-2035

quadrillion Btu

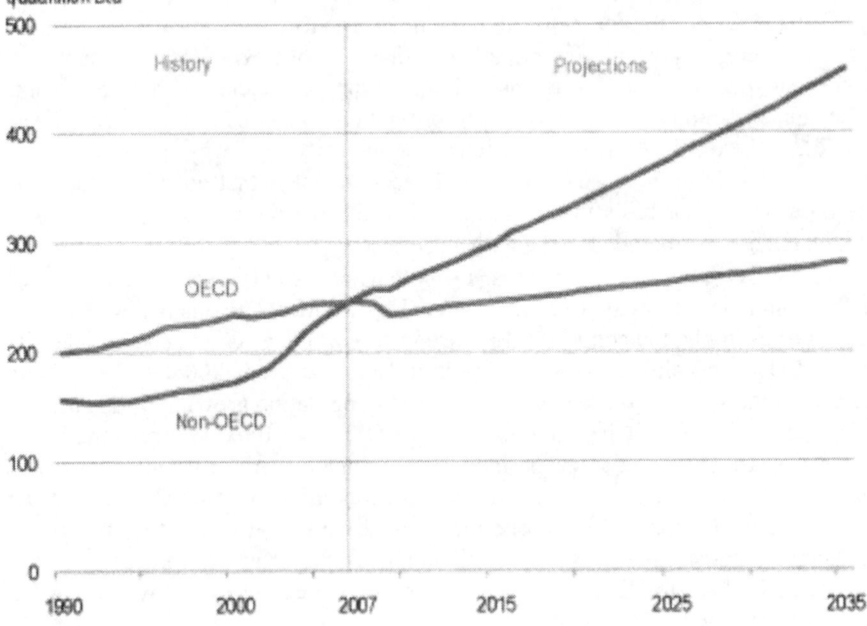

Figure 32. World oil prices in three cases, 1980-2035

2008 dollars per barrel

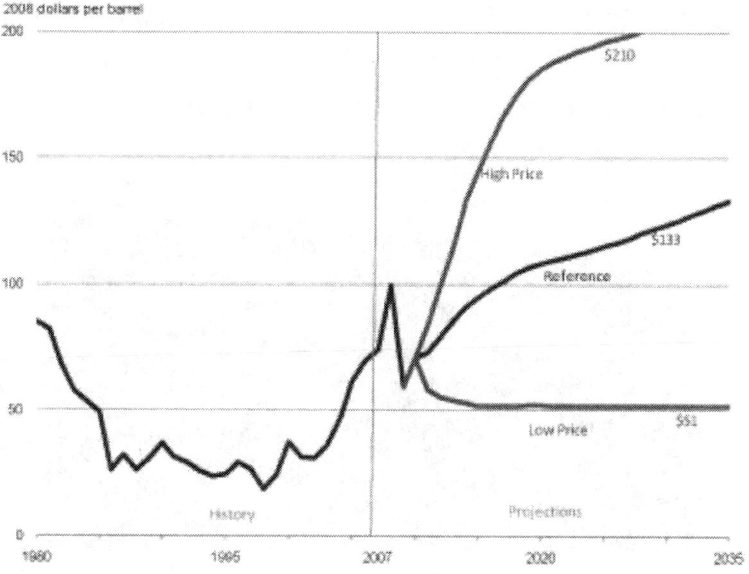

Figure 104. World energy-related carbon dioxide emissions by fuel type, 1990-2035

(billion metric tons)

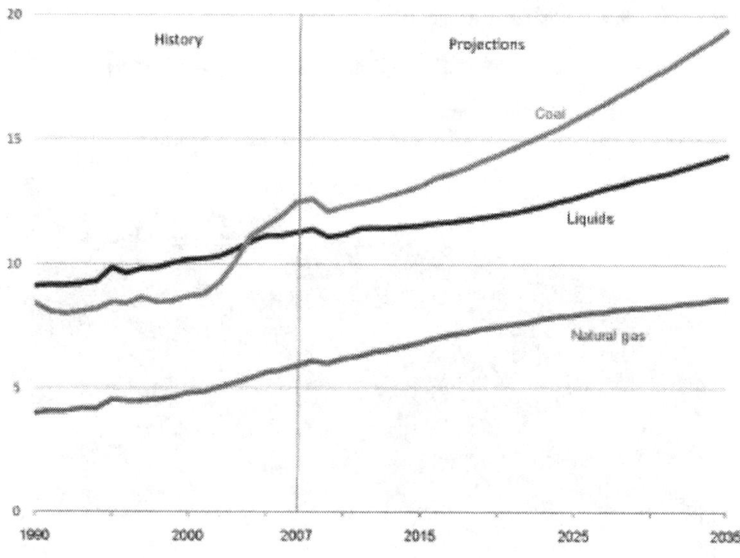

U.S. Annual Energy Outlook 2011

Figure 1. Shale gas offsets declines in other U.S. supply to meet consumption growth and lower need

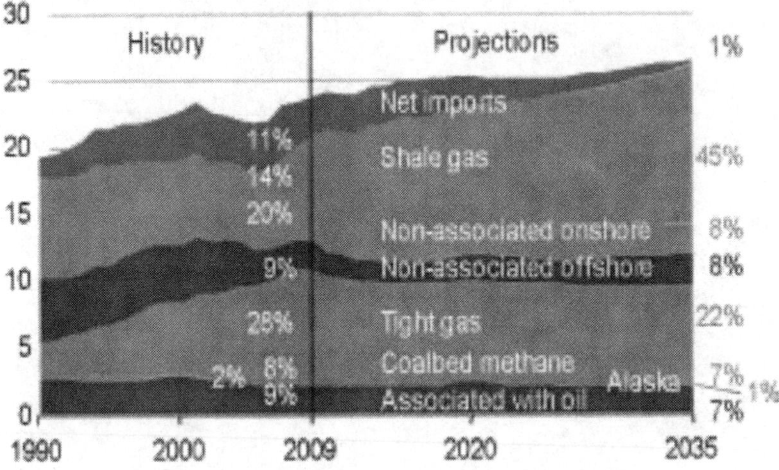

Figure 2. The projected fuel mix for electricity generation gradually shifts to lower carbon options

Net electricity generation (trillion kilowatthours per year)

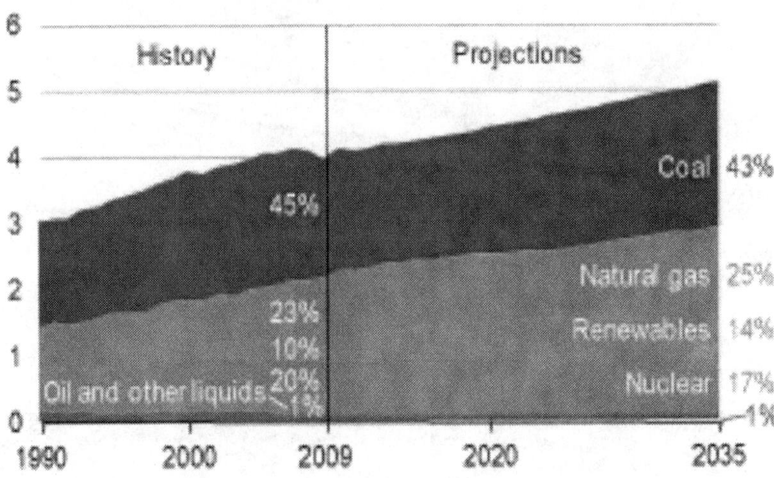

Figure 3. In the AEO2011 reference case, energy-related carbon dioxide emissions grow to almost 6 percent over 2005 levels by 2035

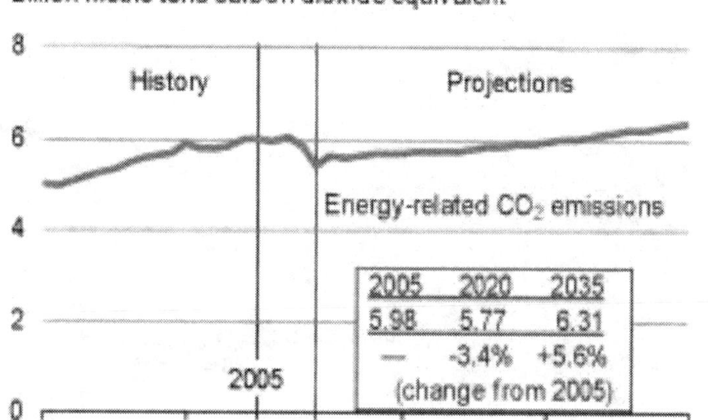

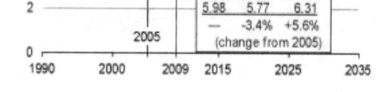

Figure 5. Change in conventional liquids production by top non-OPEC producers, 2009-2035

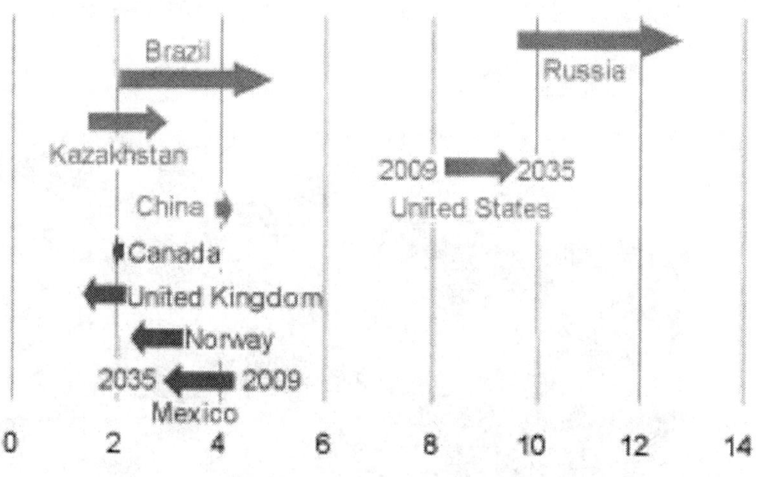

Figure 6. Delivered energy consumption by sector, 1980
Quadrillion Btu

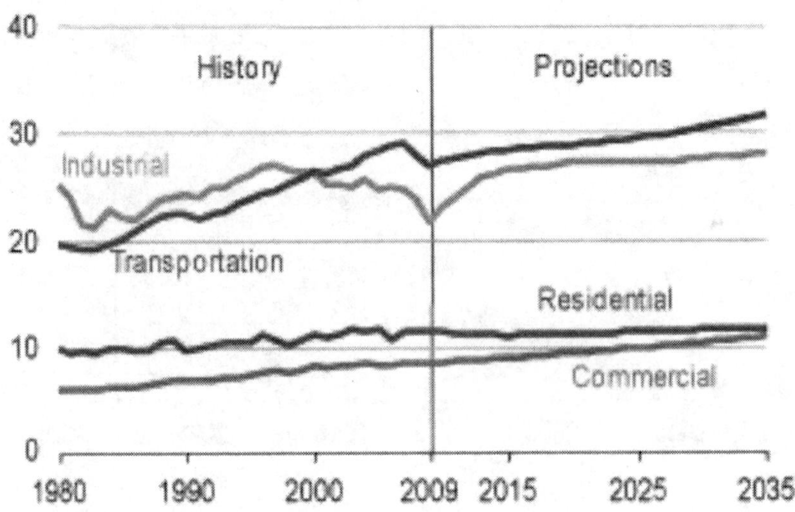

Figure 7. Energy consumption by fuel, 1980-2035

Primary energy consumption (quadrillion Btu per year)

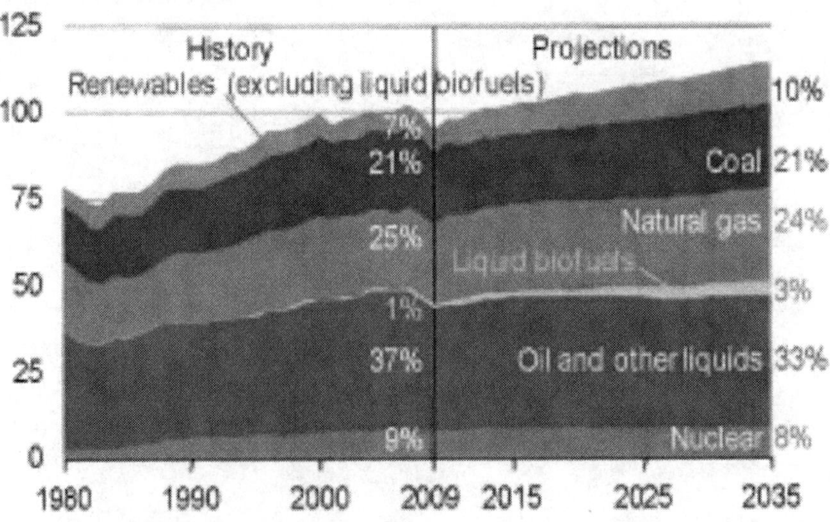

Figure 8. Energy use per capita and per 2005 dollar of gross domestic product, 1980-2035

Index, 2005 = 1

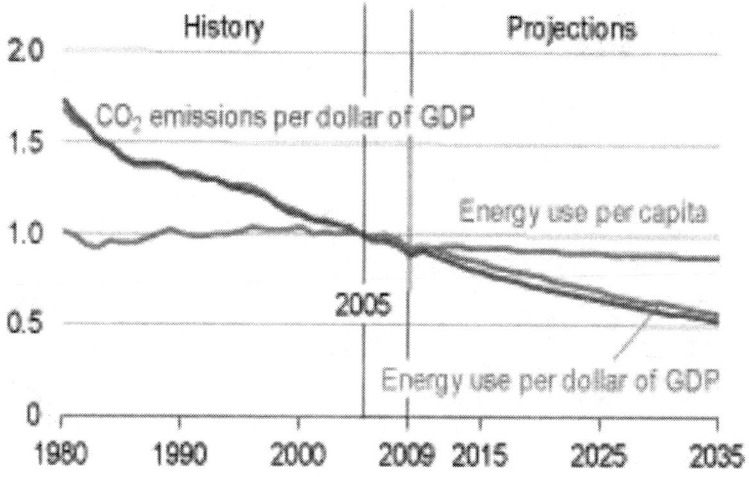

Figure 9. Outputs from the industrial and service sectors, 1990-2035

Trillion 2005 dollars

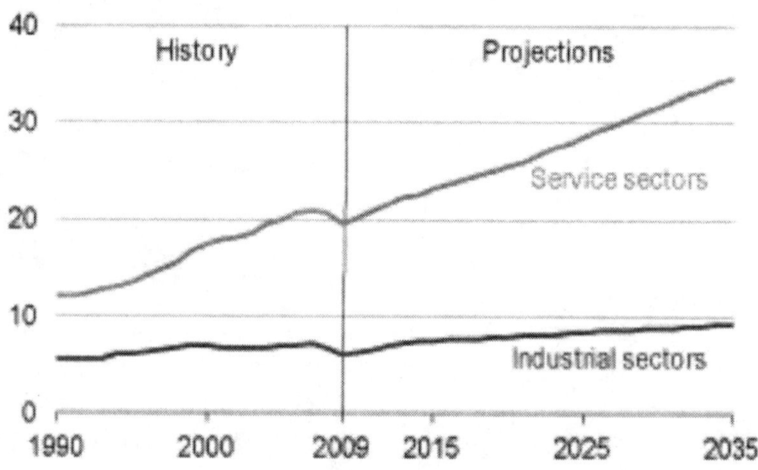

Figure 10. Total energy production and consumption, 1980-2035
Quadrillion Btu

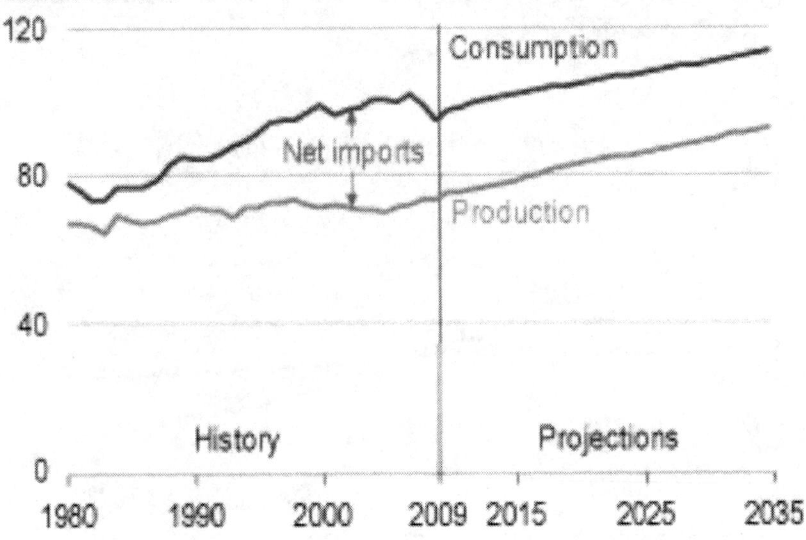

Figure 11. Energy production by fuel, 1980-2035
Quadrillion Btu

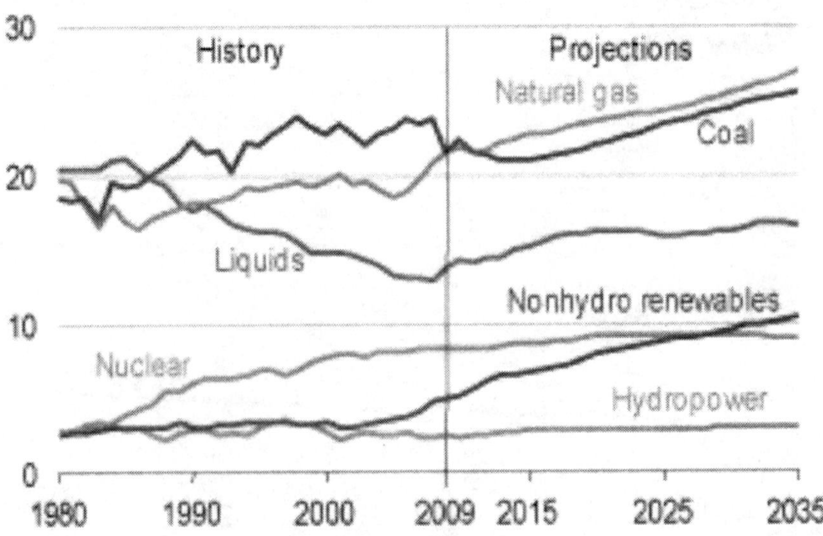

Figure 1.0 Energy Flow, 2009
(Quadrillion Btu)

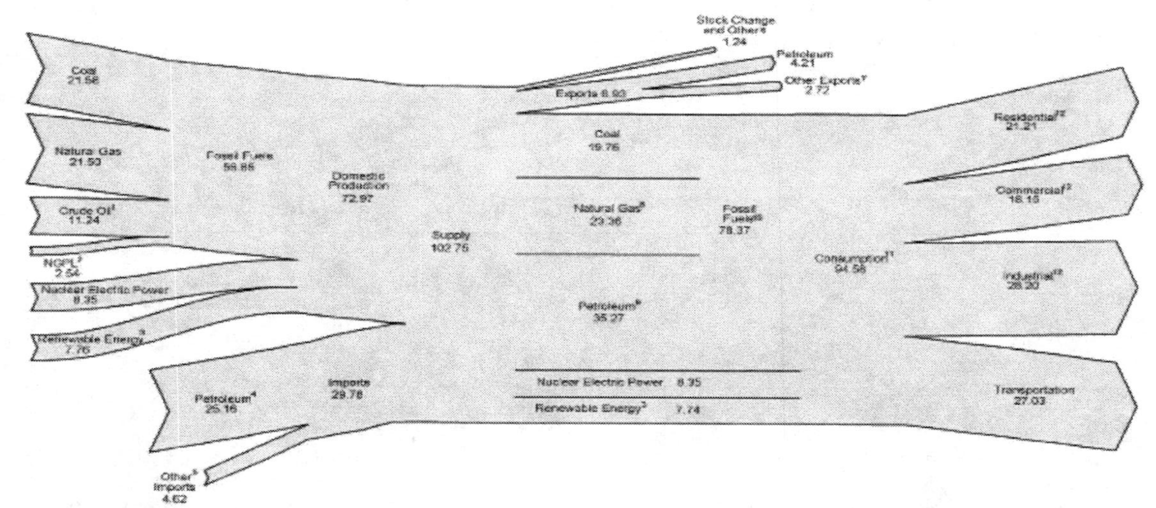

[1] Includes lease condensate.
[2] Natural gas plant liquids.
[3] Conventional hydroelectric power, biomass, geothermal, solar/photovoltaic, and wind.
[4] Crude oil and petroleum products. Includes imports into the Strategic Petroleum Reserve.
[5] Natural gas, coal, coal coke, biofuels, and electricity.
[6] Adjustments, losses, and unaccounted for.
[7] Coal, natural gas, coal coke, electricity, and biofuels.
[8] Natural gas only; excludes supplemental gaseous fuels.
[9] Petroleum products, including natural gas plant liquids, and crude oil burned as fuel.
[10] Includes 0.02 quadrillion Btu of coal coke net exports.
[11] Includes 0.12 quadrillion Btu of electricity net imports.
[12] Total energy consumption, which is the sum of primary energy consumption, electricity retail sales, and electrical system energy losses. Losses are allocated to the end-use sectors in proportion to each sector's share of total electricity retail sales. See Note, "Electrical Systems Energy Losses," at end of Section 2.

Notes: • Data are preliminary. • Values are derived from source data prior to rounding for publication. • Totals may not equal sum of components due to independent rounding.

Sources: Tables 1.1, 1.2, 1.3, 1.4, and 2.1a.

U.S. Energy Information Administration / Annual Energy Review 2009

Diagram 1. Energy Flow, 2006
(Quadrillion Btu)

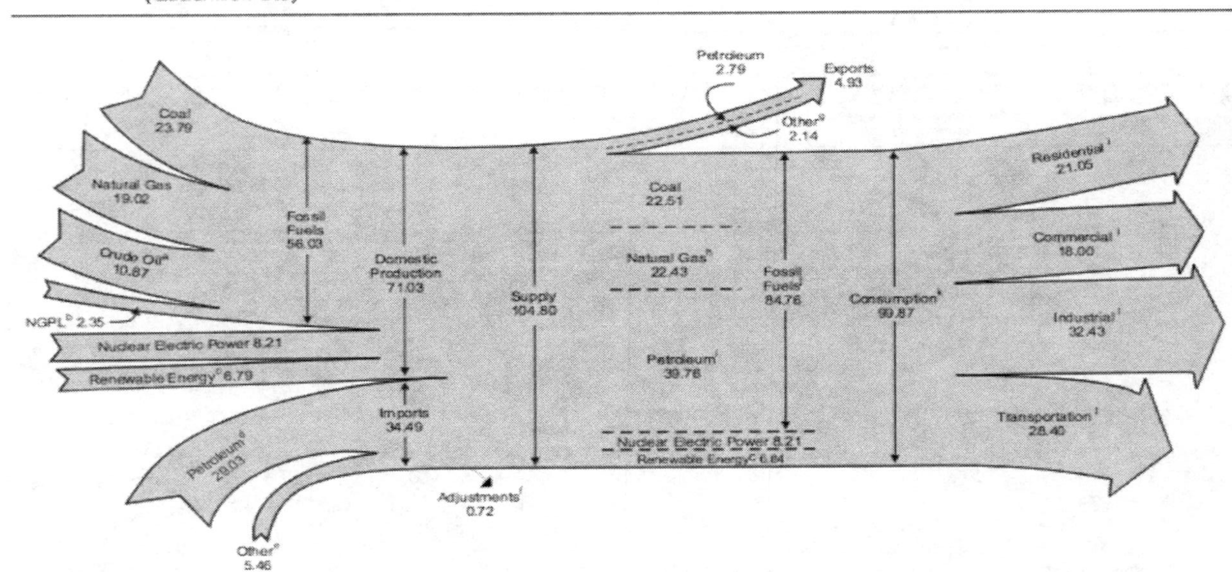

[a] Includes lease condensate.
[b] Natural gas plant liquids.
[c] Conventional hydroelectric power, biomass, geothermal, solar/PV, and wind.
[d] Crude oil and petroleum products. Includes imports into the Strategic Petroleum Reserve.
[e] Natural gas, coal, coal coke, fuel ethanol, and electricity.
[f] Stock changes, losses, gains, miscellaneous blending components, and unaccounted-for supply.
[g] Coal, natural gas, coal coke, and electricity.
[h] Natural gas only; excludes supplemental gaseous fuels.
[i] Petroleum products, including natural gas plant liquids, and crude oil burned as fuel.
[j] Includes 0.06 quadrillion Btu of coal coke net imports.
[k] Includes 0.06 quadrillion Btu of electricity net imports.
[l] Primary consumption, electricity retail sales, and electrical system energy losses, which are allocated to the end-use sectors in proportion to each sector's share of total electricity retail sales. See Note, "Electrical Systems Energy Losses," at end of Section 2.

Notes: • Data are preliminary. • Values are derived from source data prior to rounding for publication. • Totals may not equal sum of components due to independent rounding.

Sources: Tables 1.1, 1.2, 1.3, 1.4, and 2.1a.

Energy Information Administration / Annual Energy Review 2006

Energy Units

Dimensional Prefixes

10	deka (da)	10^{-1}	deci (d)
10^{2}	hecto (h)	10^{-2}	centi (c)
10^{3}	kilo (k	10^{-3}	milli (m)
10^{6}	mega (M)	10^{-6}	micro (m)
10^{9}	giga (G)	10^{-9}	nano (n)
10^{12}	tera (T)	10^{-12}	pico (p)
10^{15}	peta (P)	10^{-15}	femto (f)
10^{18}	exa (E)	10^{-18}	atto (a)

Physics Constants

$a = e^2/\hbar c$ = 1/137.036 (fine structure constant)
$c = 2.998 \times 10^{8}$ m/sec (speed of light in a vacuum)
$e = 1.60 \times 10^{-19}$ coulomb, C (electron/proton charge)
$1/4pe_o = 9.0 \times 10^{-9}$ N-m^2/C^2
$e_o = 8.8 \times 10^{-12}$ C^2/N-m^2 (permittivity of space)
$m_o = 4p \times 10^{-7}$ N/A^2 (permeability of space)
$G = 6.67 \times 10^{-11}$ N-m^2/kg^2 (gravitational)
$g = 9.807$ m/sec^2 (acceleration of gravity at 45° latitude at sea level)
$h = 6.63 \times 10^{-34}$ J-sec = 4.14×10^{-15} eV-sec (Planck)
$\hbar = h/2p = 1.06 \times 10^{-34}$ J-sec = 6.59×10^{-14} eV-sec
$k_B = 1.38 \times 10^{-23}$ J/K = 8.63×10^{-5} eV/K (Boltzman's constant)
k_BT = 0.26 eV = 1/40 eV (at room temperature 300K)
$m_e = 9.110 \times 10^{-31}$ kg (electron mass)
$m_p = 1.673 \times 10^{-27}$ kg = 1.6726 amu (proton mass)
$m_n = 1.675 \times 10^{-27}$ kg = 1.6749 amu, (neutron mass)
m_ec^2 = 511 keV, m_pc^2 = 938.3 MeV, m_nc^2 = 939.6 MeV
$N_A = 6.023 \times 10^{23}$ molecules/gram-mole (Avagodro's number)
volume of mole of gas at STP = 22.4 liter at 2.7×10^{19} molecules/cm^3
R_{gas} = 8.31 J/g-mole-K (ideal gas)
$s = 5.669 \times 10^{-8}$ J/m^2-K^4-sec (Stefan–Boltzman)
p = 3.14159265358279 = the number of letters in the words in *How I need a drink, alcoholic of course, after the heavy lectures on quantum mechanics.*

Miscellaneous Useful Numbers

$a_o = \hbar^2/mc^2$ = 0.053 nm (Bohr radius)
$e^2/4pe_o$ = 1.44 MeV-fermi (1 fm = 10^{-15} m)
$e^2/8pe_oa_o$ = 13.6 eV (hydrogen binding energy)
$E = hc/l$ = 1.24/l (photon energy in eV with wavelength in m)
Wein l = $2.90 \times 10^{6}/T$ (blackbody maximum in nm with T in Kelvin)
$E = hn$ = 2 eV (visible light from 6000 K at 0.6 m)
E = hn = 0.1 eV (IR at 300 K, 12 m)

band gap: Si (1.1 eV), Ge (0.7 eV)
1.3 fm $A^{1/3}$ (nuclear radius, A atomic number)
c_V » $3R_{gas}$ = 24.9 J/g-mole-K (high temperature specific heat with T >> q_D)
$R_{SI} = R_{English}/5.67$ and $U_{SI} = 5.67 \times U_{English}$ (insulation)
0 K = –273.15°C = –459.7°F (absolute zero).

Length
1 mile (mi) = 5280 ft = 1.609 km
1 nautical mile = 1.86 km = 1.16 mi
1 m = 3.281 ft = 39.37 in
1 micron (m) = 10^{-6} m = 10^3 nm = 10^4 angstrom
1 inch (in) = 2.54 cm
1 fermi (fm) = 10^{-15} m
universe expanse 10^{26} m >> human height 1.8 m >> nuclear radius 10^{-15} m.

Area
1 m^2 = 10.8 ft^2
1 km^2 = 0.386 mi^2
1 acre = 43,560 ft^2 = 1 mi^2/640
1 hectare (ha) = 10^4 m^2 = 2.47 acres
1 barn (b) = 10^{-24} cm^2.

Volume
1 m^3 = 1000 liters = 264 US gallons (gal) = 35.3ft^3
1 $mile^3$ = 4.17 km^3
1 acre-foot = 43,560 ft^3 = 326,000 gal = 1234 m^3 = 0.1234 hectare-m
1 liter = 1000 cm^3 = 0.264 gal
1 bbl petroleum = 42 gal = 0.159 m^3.

Time
1 year = 365.25 days = 8766 hr = 3.154×10^7 sec » p x 10^7 sec
1 day = 86,400 sec
1 shake = 10^{-8} sec
1 age of universe = 4×10^{17} sec >> human 2×10^9 sec >> nuclear 10^{-23} sec ($2r/c$).

Mass
1 kg = 2.205 pounds (lb) = 32.3 ounces (oz)
1 lb = 16 oz = 453.6 g, 1 oz = 28.4 g
1 metric tonne (t) = 1000 kg = 1.102 tons
1 English ton = 2000 lb = 907.0 kg = 0.907 t

Force
1 newton (N) = 1 kg m/sec^2 = 10^5 dynes (dyn) = 0.22 lb

Pressure
1 bar (atm) = 76 cm Hg = 760 torr = 14.7 lb/in^2 = 10^5 pascal (1 N/m^2)

Energy/Heat

1 J = 1 W-sec = 1 calorie/4.2 = 1 kilocal/4200 = 6.242 x 10^{18} eV = 1 Btu/1055 = 0.738 ft-lb
1 eV = 1.602 x 10^{-19} eV
1 kWh = 3.6 x 10^6 J = 3412 Btu (electricity at h = 33% uses 10^4 Btu/kWh)
1 bbl crude petroleum = 5.8 MBtu = (42 gal)(138,000 Btu/gal)
1000 ft^3 (STP) natural gas = 10 therms = 1.03 MBtu = 1.09 GJ
1 trillion cubic feet (TCF) natural gas = 1.03 quads, 1 Gt coal = 27.8 quads
1 m^3 (STP) = 39 MJ, 1 TCF = 10^{12} ft^3
1 ton coal = 25.2 MBtu = 0.9 tonne coal
1 quad = 10^{15} Btu = 172 Mbbl = 0.97 TCF = 36 Mt coal = 292 G-kW_th = 1.05 EJ = 1.05 10^{18} J
1 Gbbl = 5.8 quads
1 cubic foot natural gas at STP = 1000 Btu = 1 MJ
1 terawatt-year (TWyr) = 8.76 x 10^{12} kWh = 31.5 EJ
1 kWh/m^2 = 313 Btu/ft^2
1 kiloton TNT (kton) = 4.2 x 10^{12} J = 10^{12} calories
1 kg fission = 17 kton TNT (60 g/kton)
1 kg DT fusion = 85 kt (12 g/kton)
1 $MW_{thermal}$-day = 1 gram ^{235}U = 0.3 $MW_{electric}$-day.

Power

1 W = 1 J/sec = 1 N-M/sec = 1 kg-m^2/sec^3
1 hp = 550 ft-lb/sec = 0.746 kW = 746 J/sec
1 kW = 3412 Btu/hour
1 Mbbl/d = 0.365 Gbbl/year = 2.12 quads/yr = 71 GW_t(thermal)
US 100 quads/yr = 3400 GW_t = 47 Mbbl/d = 17 Gbbl/yr = 100 tcf/yr (equivalent)
US (280 M)/capita = 12 kW_t = 60 bbl/yr = 12.8 tonne/yr coal = 0.35 million ft^3/yr
1 lumen = 1/673 watt visible light
1 lux = 1 lumen/m^2, 1 foot candle = 1 lumen/ft^2 = 0.0929 lux.

Air

molecular weight (28.96)
density (1.293 kg/m^3)
sound speed (331.4 m/sec)
volume fraction (N_2/78%, O_2/21%, A/0.9%, H_2O/0.4%, CO_2/370 ppm)
specific heat (constant pressure, 1004 J/kg-K) and (constant volume, 720 J/kg-K)
viscosity (0.17 millipoise).

Water

density [1000 kg/m^3 (4 °C), 997 (25 °C), 958 (100 °C), 1025 (salt), 900 (ice)]
latent heat of fusion (333 kJ/kg), vaporization (2.26 MJ/kg)
specific heat water (4.2 kJ/kg-K), steam (100 °C, 2.01 kJ/kg-K), ice (2.1 kJ/kg-K)
viscosity (17.5 millipoise at 0°C, 2.8 at 100°C)
flow in Sverdup = 1 M m^3/sec
oceans (1350 x 10^{15} m^3), ice (29 x 10^{15} m^3), ground water (8.3 x 10^{15} m^3), lakes (0.13 x 10^{15} m^3).

Earth
radius R_E = 6357 km polar and 6378 km equatorial
area = 5.10 x 10^{14} km^2 (oceans 71%)
mass = 5.98 x 10^{24} kg
g' = 9.8 m/$sec^2$$(R_E/r)^2$
atmospheric pressure = 10^5 Pa $e^{-h/H}$, atmospheric height H = 8.1 km
atmospheric mass = 5.14 x 10^{18} kg with 1.3 x 10^{16} kg H_2O, oceanic mass = 1.4 x 10^{21} kg.

Sun
solar flux s_o = 1.367 kW/m^2 = 0.13 kW/ft^2 = 2.0 cal/minute-cm^2 = 435 Btu/ft^2-hr
24-hour average horizontal flux (40°N latitude) = 185 W/m^2
mass = 2 x 10^{30} kg
radius = 0.696 Mkm
distance to Earth = 150 Mkm.

Radiation (colloquial and SI units)
Rate of decay
1 curie (radiation of 1 g radium) = 1 Ci = 3.7 x 10^{10} decay/sec
1 bequerel (SI) = 1 Bq = 1 decay/sec.
Absorbed in air
1 Roentgen = 1 R = 87 ergs/g = 0.0087 J/kg.
Physical dose absorbed
1 rad = 100 erg/g = 0.01 J/kg
1 gray (SI) = 1 Gy = 1 J/kg = 100 rad.
Biological Dose Equivalent (absorbed dose times biological effectiveness Q)
x, g and e (Q = 1), n (Q = 5–20), alphas and fission fragments (Q = 20)
1 sievert (SI) = 1 Sv = 1 J/kg = 100 Rem
1 Rem = 0.01 J/kg = 0.1 Sv.
US annual average background dose (1990 BEIR-V) = 360 mRem (3.6 mSv):
radon (200 mRem), body radioactivity (39 mRem), medical x-rays (53 mRem)
cosmic radiation (31 mRem), sea level (28 mRem), Denver (81 mRem).

WWW Energy Sites

GENERAL SITES

American Geophysical Union: www.agu.org/sci_soc/
American Institute of Physics: www.aip.org/history/
American Physics Society: www.aps.org/public_affairs
APS Forum on Physics and Society: www.aps.org/units/fps
Congressional Legislation: //thomas.loc.gov
Congressional Budget Office: www.cbo.gov
Congressional Research Service: www.cnie.org/NLE/CRS
DOE Information Bridge: www.osti.gov/bridge/
DOE Labs: www.XX.gov; XX = anl, bnl, lanl, llnl, ornl, pnl, sandia, y12
Economic Report of the President: www.access.gpo.gov/eop
General Accounting Office: www.gao.gov
Government Printing Office: www.access.gpo.gov/su_docs/
National Academy Press: www.nap.edu
Office of Technology Assessment Legacy: www.wws.princeton.edu/~ota/
Science: www.sciencemag.org
White House: www.whitehouse.gov

ENERGY SITES

American Council for an Energy Efficient Economy: www.aceee.org
American Gas Association: www.aga.org
American Nuclear Society: www.ans.org
American Petroleum Institute: www.api.org
Ballard Fuel Cells: www.ballard.com
Bureau of Transportation Statistics: www.bts.gov/publications/
Clean Energies Future: www.ornl.gov/ORNL/Energy_Eff/CEF.htm
Davis Energy Group:www.davisenergy.com
DOE Efficiency/Renewable Energy (600 links): www.eren.doe.gov.
DOE Alternate Energy Vehicles: www.fleets.doe.gov
Efficient Windows: www.efficientwindows.org
Energy Information Agency: www.eia.doe.gov
Energy Star: www.energystar.gov
EPA fuel economy: www.fueleconomy.gov
First Solar: www.firstsolar.com
Fuel Cells: www.fuelcells.org

Hubbert: www.hubbertpeak.com
Hydrogen: www.hydrogenus.org or www.clean-air.org
Hydrogen Research: www.sc.doe.gov/bes/hydrogen.pdf
International Solar Energy Society: www.ises.org
Lawrence Berkeley National Lab: //enduse.lbl.gov
LBL Buildings: //eetd.lbl.gov/buildings.html
National Renewable Energy Laboratory: www.nrel.gov
National Transportation Statistics: www.bts.gov
Nuclear Power: //web.mit.edu/nuclearpower/
International Energy Agency: www.iea.org
Pacific Gas and Electric: www.pge.com/003_save_energy
Princeton: www.princeton.edu/~cees
Princeton Plasma Physics Lab simulations: //ippex.ppnl.gov
Rocky Mountain Institute: www.rmi.org
US Green Buildings Council: www.usgbc.org
Windpower: www.windpower.org

ENVIRONMENT SITES

Acid Rain: //bqs.usgs.gov/acidrain
Air Quality Management District: www.aqmd.gov
British Medical Journal: www.bmj.com
Chernobyl: www.ic-chernobyl.kiev.ua
Congressional Research Service: www.cnie.org/NLE/CRS
Dr. Everett Koop: www.drkoop.com
DOE Nuclear Waste: //cid.em.doe.gov
DOE Biology/Environment: www.sc.doe.gov/feature/biology_and_environment.htm
DOE Carbon Dioxide Information Analysis Center: //cdiac.esd.ornl.gov
Greening Earth Society: www.greeningearthsociety.org
Indiana Law: www.law.indiana.edu/v-lib/index.html
Intergovernmental Panel on Climate Change: www.ipcc.ch
Earth Data: //personal.cmich.edu/~Franc1m/homepage.htm
Environmental Protection Agency: www.epa.gov
Global Change Research Program: www.usgcrp.gov
NASA: //earthobservatory.nasa.gov
NASA Climate/Radiation: //climate.gsfc.nasa.gov
National Cancer Institute: www.nci.nih.gov
National Center for Atmospheric Research: www.ucar.edu
NOAA Climate: www.ncdc.noaa.gov
NSF: www.geo.nsf.gov/start.htm
Nuclear Waste Technology Review Board: www.nwtrb.gov
Michigan Radiation/Health: www.umich.edu/~radinfo
Pacific Institute: www.pacinst.org
US Geological Survey: www.usgs.gov
World Bank: //publications.worldbank.org
World Resources Institute: www.earthtrends.wri.org

Author Bio Briefs

Katey Walter Anthony is an Aquatic Ecosystem Ecologist and an Assistant Professor in the Water and Environmental Research Center in the Institute of Northern Engineering & International Arctic Research Center at the University of Alaska, Fairbanks She studies methane emissions and biogeochemistry of North Siberian thermokarst lakes. Her work won the 2006 1st place winner of the United States Council of Graduate Schools/ University Microfilms International Distinguished Dissertation Award in the field of Mathematics, Physical Sciences and Engineering. She has worked at the Novosibirsk State University and the Kuban State University in Russia. Her research covers methane in the Arctic, lakes, biogeochemistry, climate change, permafrost/thermokarst, carbon cycling, isotopes. Studies link to greenhouse gas emission estimates from modern lakes, lakes formed throughout the Holocene since the last deglaciation, and modeling of future lake dynamics as permafrost thaws in the Arctic.

Ben Bierman is the Executive Vice President in charge of Operations and Engineering at Solyndra. He is responsible for worldwide engineering and operations at Solyndra, including the design of their innovative tubular Rooftop PV systems and the new, state-of-the-art 800,000 square foot Fab 2 manufacturing facility. Ben joined the company in August 2006 prior to the start of manufacturing operations. By March 2011, Solyndra had produced over 25 million tubular modules and shipped nearly 100 MW of solar panels. Prior to joining Solyndra, Bierman served as the Vice President of Business Management of Coherent, Inc., a laser and laser systems manufacturing company from 2005 until 2006. Mr. Bierman served as Managing Director at Lam Research Corporation, a semiconductor manufacturing equipment company, from 2003 to 2005. Previously, during an eight-year tenure at Applied Materials, Inc. Mr. Bierman served as Director of Engineering and Technology, and Managing Director of two product units that designed and produced chip manufacturing equipment that enabled the 90 and 65 nanometer semiconductor device nodes.

Robert Budnitz is at the UC Lawrence Berkeley National Laboratory, where he was Associate Director and Head of LBNL's Energy & Environment Division (1975-78). He was a senior officer at the U.S. Nuclear Regulatory Commission, serving as Director of the NRC Office of Nuclear Regulatory Research (1978-80). From 1981 to 2002, he had a one-person consulting practice in Berkeley, working on nuclear reactor safety and radioactive waste management safety. His principal work during this period was advancing the methodology of PRA (probabilistic risk assessment), with an emphasis on seismic PRA and PRA for other external hazards. He was awarded the American Nuclear Society "Theos J. Thompson Award for Reactor Safety" in 2005, as well as the Society for Risk Analysis "Outstanding Risk Practitioner Award for 2001." His current research is involved with nuclear-reactor safety, and risk analysis methods and applications.

David E. Claridge is the Director of the Energy Systems Laboratory and also serves as the Leland Jordan Professor of Mechanical Engineering at Texas A&M University. He is internationally known for his work on energy efficiency, particularly for leading out in development and implementation of the Continuous Commissioning® approach to improving energy efficiency in large buildings including higher education facilities, medical facilities, office buildings, and airports. He has taught numerous commissioning workshops, holds five patents and is the author of over 350 journal and conference papers. He had 10 years of experience in the HVAC field prior to coming to Texas A&M in 1986. He is a Fellow of the American Society of Heating, Refrigerating and Air Conditioning Engineers and the American Society of Mechanical Engineers.

George Crabtree is Senior Scientist and Distinguished Fellow in the Materials Science Division at Argonne National Laboratory and Distinguished Professor of Physics, Electrical, and Mechanical Engineering at University of Illinois-Chicago. His research interests include materials science, sustainable energy and nanoscale superconductors and magnets. He has led several workshops for the Department of Energy, most recently on basic science supporting energy technology, and he has co-chaired the Undersecretary of Energy's assessment of DOE's Applied Energy Programs. In 2010 he was the Chair of the American Physical Society Study on Integrating Renewables on the Electricity Grid. He is a member of the National Academy of Sciences and has testified before the U.S. Congress on the hydrogen economy and on meeting sustainable energy challenges.

Elizabeth Deakin is Professor of City and Regional Planning and Urban Design at UC Berkeley, where she also is an affiliated faculty member of the Energy and Resources Group and the Master of Urban Design group. From 1999-2009 she also was Director of the UC Transportation Center, which she helped found in 1989. In addition, from 2004-2008, she served as co-director of UC Berkeley's new Global Metropolitan Studies Initiative, which involves nearly 70 faculty members from 12 departments. Deakin's research focuses on transportation and land use policy, the environmental impacts of transportation, and equity in transportation. She has published over 200 articles, book chapters, and research reports on topics ranging from environmental justice to transportation pricing to development exactions and impact fees. Her recent research projects have addresses these issues in China, the EU, Latin America and the US. She has testified before Congress regarding every transportation bill since ISTEA in 1991, most recently appearing before the House Technology and Infrastructure Committee.

Louis Desroches obtained a Ph.D. in Astrophysics at UC Berkeley, focusing on low-luminosity and low-mass active galactic nuclei, with Luis Ho, Eliot Quataert, and Alex Filippenko. Inspired by physicists like Art Rosenfeld, Steven Chu, Dan Kammen, and John Holdren convinced him to make a career switch. He was an Environmental Energy Policy Postdoctoral Fellow at LBNL in the Energy Efficiency Standards group for a year and a half. Recently he transitioned to Program Manager within the same group.

Christopher Field is the founding director of the Carnegie Institution's Department of Global Ecology, Professor of Biology and Environmental Earth System Science at Stanford University, and Faculty Director of Stanford's Jasper Ridge Biological Preserve. Field's research emphasizes impacts of climate change, from the molecular to the global scale. He has, for nearly two decades, led major experiments on responses of California grassland to multi-factor global change. Field has served on many national and international committees related to global ecology and climate change. He was a coordinating lead author for the fourth assessment report of the Intergovernmental Panel on Climate Change and a member of the IPCC delegation that received the Nobel Peace Prize in 2007. In September, 2008, he was elected co-chair of working group 2 of the IPCC, and will lead the next assessment on climate change impacts, adaptation, and vulnerability. He is a fellow of the American Association for the Advancement of Science and an elected member of the American Academy of Arts and Sciences and the National Academy of Sciences.

Inez Fung is the Richard and Rhoda Goldman Distinguished Professor in the Physical Sciences and the founding director of the Berkeley Atmospheric Sciences Center. She is a Professor of Atmospheric Science in the Department of Earth and Planetary Science and the Department of Environmental Science, Policy and Management. Since 2005, she has also been a Founding Co-Director of the Berkeley Institute of the Environment. Fung is a principal architect of large-scale mathematical modeling approaches and numerical models to represent the geographic and temporal variations of sources and sinks of CO2, dust and other trace substances around the globe. Fung's work in climate modeling predicts the co-evolution of CO2 and climate and concludes that the diminishing capacities of the land and oceans to store carbon act to accelerate global warming. She is leading a new project (the HydroWatch Project) in two of the UC Natural Reserves to track the life-cycle of water using cutting-edge technologies. She is also on the science team of a new satellite – the Orbiting Carbon Observatory – that will measure the abundance of CO2 over the whole globe. She was named one the "Scientific American 50" in 2005 and received the World Technology Network Award for the Environment in 2006.

Ashok Gadgil is Division Director and Faculty Senior Scientist, Environmental Energy Technologies Division, LBNL. He is also Rudd Family Foundation Distinguished Chair of Safe Water and Sanitation, Professor of Civil and Environmental Engineering, University of California, Berkeley. He developed a very energy efficient method to disinfect drinking water with UV light as part of his research on energy in the developing world. At UC–Berkeley, his graduate course, "Technologies for Sustainable Communities," integrates a range of disciplines to understand why most technical attempts to help communities in the developing world fail and what it takes to help them succeed.

Girish Ghatikar is a Program Manager at Lawrence Berkeley National Laboratory, overseeing Demand Response (DR) technologies, Open Auto-DR (OpenADR) standards, international Smart Grid, and energy-related services and markets. Ghatikar's background is in key areas of information technology, standards, technology transfer, business innovation, and policies for Energy Efficiency, DR, Smart Grid, and its applications. Ghatikar serves on the Steering and Technical Committees for OASIS (Organization for Advancement of Structured Information Standards) and user groups to advance open Smart Grid standards and energy technologies. Ghatikar holds Master degrees in Telecommunication Systems/ Computer Technologies, and Infrastructure Planning/ Management.

Dian Grueneich is Former Commissioner of the California Public Utilities Commission (2005-10) with over 30 years experience in energy and environmental issues. While at the Commission, Commissioner Grueneich was the lead assigned Commissioner on energy efficiency, demand response, transmission permitting and planning, Western energy issues, low income energy, several Advanced Metering Infrastructure dockets and served as the

Commissioner representative on the Governor's Climate Action Team. She guided the preparation and adoption of the California Long-Term Energy Efficiency Strategic Plan and the successful permitting of three major renewable transmission lines. She is a leader in the deployment of SmartGrid technologies to optimize the reliability, security and efficiency of the electrical grid and maximize the potential of demand side resources and the electrification of transportation systems. She serves on the U.S. Department of Energy Small Electricity Advisory Committee and its Smart Grid subcommittee. She also chairs the Residential Retrofit Working Group of the National State Energy Energy Efficiency Action Network and serves as an adviser to Stanford University's Precourt Energy Institute and Energy Efficiency Center. She received the 2010 National Association of Regulatory Utility Commissioners Kilmarx Clean Energy Award.

David Hafemeister is Physics Professor (emeritus) at California Polytechnic State University. He spent a dozen years in Washington at the US Senate, State Department, ACDA and National Academy of Sciences. He also did research at MIT, Princeton and Stanford universities and Argonne, Lawrence-Berkeley, and Los Alamos National Labs. He participated in the passage of the 1975 EPCA and 1976 ECPA energy laws. His energy lab at Cal Poly was dedicated to an inspirational physicist as "The Arthur Rosenfeld House Doctor Laboratory." His book, *Physics of Societal Issues: Calculations on National Security, Environment and Energy* quantifies some of the topics of this conference.

Philip Haves is the leader of the Simulation Research group in the Building Technologies Department at Lawrence Berkeley National Laboratory. He has a BA in Physics from Oxford University and a PhD in Radio Astronomy from Manchester University. He is a Fellow of the American Society of Heating, Refrigerating and Air-conditioning Engineers (ASHRAE), the immediate past chair of ASHRAE's Technical Committee 4.7 *Energy Calculations* and a former president of the US affiliate of the International Building Performance Simulation Association.

K. John Holmes is the associate director of the National Research Council's Board on Energy and Environmental Systems. He is responsible for helping to manage the board's activities and directing committee studies on contentious environmental and energy issues, particularly those related to renewable energy, motor vehicles, air quality, and the quantitative analysis of policy impacts. He has published articles on a wide range of energy and environmental topics, including climate change, air quality management, mobile source emissions, stratospheric ozone depletion, carbon emissions trading, water resources management, use of regulatory models, and the history of natural resources management.

Steve Horne is the Chief Technology Officer of SolFocus, a Concentrating Photovoltaics manufacturer. He is an engineering graduate from the University of New South Wales, Australia, where some of the most impressive advances in Silicon cells have been made. Not heeding the call however, he spent several years building and commissioning coal burning power plants in Australia before moving to Silicon Valley and a career in technology. Mr. Horne has worked in development, manufacturing and marketing in several areas including test equipment, robotics and materials science. As co-founder and executive at SolFocus, he is actively atoning for his earlier sins.

Jonathan Jiang is a Research Scientist at NASA's Jet Propulsion Laboratory, California Institute of Technology. As a Principal Investigator, he leads the research of using multiple satellite measurements to study the influence of pollution on clouds, precipitation, and to evaluate the global climate model simulations of clouds and water vapor in the Earth atmosphere. Jiang received the NASA Exceptional Achievement Medal in 2010.

Bruce Johnston is the Program Manager of the University of California's Advanced Solar Technologies Institute, which is located at UC Merced. He established the Solar Institute at Castle Center in Atwater, CA in 2008 with Prof. Roland Winston. In 1996, he joined two colleagues in forming Kontraband Interdiction and Detection Services, Inc., which is currently one of the largest K-9 detection training and service providers in the United States.

Daniel Kammen is the World Bank Group's Chief Technical Specialist for Renewable Energy and Energy Efficiency. He was appointed to this newly-created position in October 2010, in which he provides strategic leadership on policy, technical, and operational fronts. The aim is to enhance the operational impact of the Bank's renewable energy and energy efficiency activities while expanding the institution's role as an enabler of global dialogue on moving energy development to a cleaner and more sustainable pathway. Previously, he was Class of 1935 Distinguished Professor of Energy at the University of California, Berkeley, with parallel appointments in the Energy and Resources Group, the Goldman School of Public Policy, and the department of Nuclear Engineering. He is also the founding director of the Renewable and Appropriate Energy Laboratory (RAEL), Co-Director of the Berkeley Institute of the Environment, and Director of the Transportation Sustainability Research Center. He is a Permanent Fellow of the African Academy of Sciences and a fellow of the American Physical Society. He serves on two US National Academy of Sciences boards and, in April, 2010 was named by US Secretary of State Clinton as the first Energy and Climate Fellow for the Western Hemisphere. He served as a contributing or coordinating lead author on various reports of the Intergovernmental Panel on Climate Change since 1999.

Sila Kiliccote is the Deputy Director of the Demand Response Research Center in the Building Technologies Department at LBL. Sila recently received the 2010 GridWeek Award for Leadership in Smart Grid Acceleration. GridWeek cited Kiliccote for her "leadership, vision, non–traditional approach, ability to create step function vs. incremental change, and willingness to take risk." She has been a part of the automated demand response team developing an automated communication infrastructure, integrating it with building control systems and working with stakeholders to standardize the information model. Her areas of interest include characterization of buildings and demand reduction, demand responsive lighting systems, building systems integration and feedback for demand-side management.

Jonathan Koomey is a Consulting Professor at Stanford University, worked for more than two decades at Lawrence Berkeley National Laboratory, and has been a visiting professor at Yale University (Fall 2009) and Stanford University (2004-5 and Fall 2008). Dr. Koomey holds M.S. and Ph.D. degrees from the Energy and Resources Group at UC Berkeley, and an A.B. in History of Science from Harvard University. He is the author or coauthor of eight books and more than 150 articles and reports, and is one of the leading international experts on the economics of reducing greenhouse gas emissions and the effects of information technology on resource use. His latest solo book is the 2nd edition of *Turning Numbers into Knowledge: Mastering the Art of Problem Solving.* (`http:/www.analyticspress.com`)

Ronald Kwok is a Senior Research Scientist at the Jet Propulsion Laboratory, California Institute of Technology. His research interests include the mass and energy balance of the Arctic and Southern Ocean ice cover and the role of the sea ice in global climate. His current focus is on the analysis of thickness, small-scale sea ice kinematics, time-varying gravity from various spaceborne and airborne remote sensing instruments.

Tristan L'Ecuyer is an assistant professor in the Department of Atmospheric and Oceanic Sciences at the University of Wisconsin-Madison. L'Ecuyer's research centers on using satellite observations to improve our understanding of the global water and energy cycles and determining their role in global climate change.

Barbara Goss Levi has spent most of her career as an editor for Physics Today magazine, writing news stories about current research. In the early 1980s she worked on issues of energy and arms control at Princeton University's Center for Energy and Environmental Studies. She has remained interested in those topics and has been an active member of the Forum on Physics and Society. Under the auspices of FPS, she has helped organize short courses on Energy Sources: *Conservation and Renewables* (1985), *Global Warming: Physics and Facts* (1991) and, this year, on *Physics of Sustainable Energy: Using Energy Efficiently and Producing It Renewably*.

Timothy Lipman is Co-Director for the University of California Berkeley Transportation Sustainability Research Center, based at the Institute of Transportation Studies, Director of the U.S. DOE Pacific Region Clean Energy Application Center (PCEAC), and a lecturer with the UC-Berkeley Department of Civil and Environmental Engineering. He is an energy and environmental technology, economics, and policy research engineer and lecturer, focusing on electric vehicles, fuel cell technology, combined heat and power systems, renewable energy, and electricity and hydrogen infrastructure. He completed a Ph.D. degree in Environmental Policy Analysis with the Graduate Group in Ecology at UC Davis (1999).

Michael Lubell is the Director of Public Affairs at the American Physical Society and Professor of Physics at the City College of the City University of New York, where he was the Physics Department Chair from 1999 to 2006. He has carried out research in atomic, molecular, optical, nuclear and high-energy physics. He is credited as being one of the pioneers of science lobbying in Washington and has served on many scientific advisory committees inside and outside of government.

Tom Murphy is an associate professor of physics at the University of California, San Diego. In the field of astrophysics, he built a cryogenic integral field spectrograph for the Palomar 200-inch telescope and used it to understand the histories of colliding galaxies. More recently, he developed and heads a project to test General Relativity by millimeter-accuracy laser ranging to the reflectors left on the moon by the Apollo astronauts. Murphy's keen interest in energy topics began with his teaching a course in energy and the environment to non-science majors at UCSD. Motivated by the unprecedented challenges we face, he has applied his instrumentation skills to exploring alternative energy and associated measurement schemes.

Lawrence T. Papay is currently a consultant with a variety of clients in electric power and other energy areas. His expertise and knowledge range across a wide variety of electric system technologies, from production, to transmission and distribution, utility management and systems, and end-use technologies. He has held positions including senior vice president for the Integrated Solutions Sector, Science Applications International Corporation, and senior vice president and general manager of Bechtel Technology and Consulting. He also held several positions at Southern California Edison, including senior vice president, vice president, general superintendent, and director of research and development (R&D), with responsibilities for areas including bulk power generation, system planning, nuclear power, environmental operations, and development of the organization and plans for the company's R&D efforts. He is a member of the National Academy of Engineering. He received a B.S. degree in physics from Fordham University and S.M. and Sc.D. degrees in nuclear engineering from the Massachusetts Institute of Technology.

Raymond Pierrehumbert is the Louis Block Professor in Geophysical Sciences at the University of Chicago. His research concerns the physics of climate, especially the long-term evolution of the climates of the solar system and extrasolar planets. He was a lead author on the Third Assessment Report of the Intergovernmental Panel on Climate Change and he is the author of Principles of Planetary Climate (Cambridge University Press, 2010).

Mary Ann Piette is Research Director of the Demand Response Research Center and the Deputy of the Building Technologies Department at LBNL. Ms. Piette has extensive experience developing and evaluating low-energy and demand response technologies for buildings. She specialized in commissioning, energy information systems, benchmarking, and diagnostics. She has authored over 100 papers on efficiency and demand response and received the Benner Award at the National Conference on Building Commissioning for contributions to making commissioning "business as usual". Ms. Piette has a Master's of Science Degree in Mechanical Engineering from UC Berkeley and a Licentiate in Building Services Engineering from the Chalmers University of Technology in Sweden.

Burton Richter is a Nobel Laureate (Physics 1976), member of the National Academy of Science, and a past president of both the American Physical Society and the International Union of Pure and Applied Physics. He is the Paul Pigott Professor Emeritus at Stanford University and the former Director of the Stanford Linear Accelerator Center. For the past decade he has spent most of his time on energy issues; his recent book is Beyond Smoke and Mirrors: Climate Change and Energy in the 21st Century (Cambridge University Press, 2010).

Arthur Rosenfeld was appointed in 2000 California Governor Gray Davis to be a Commissioner at the California Energy Commission (CEC), and he was re-appointed by Governor Arnold Schwarzenegger for 2005 to 2010. He formerly was a Professor of Physics at UC-Berkeley, formed the Center for Building Science at Lawrence Berkeley National Laboratory, which he led until 1994. From 1994 -1999 Dr. Rosenfeld served as Senior Advisor to DOE Assistant Secretary for Energy Efficiency and Renewable Energy. At the CEC he was responsible for the Public Interest Energy Research program for energy efficiency, and as chair of the Energy Efficiency Committee he was responsible for the California energy efficiency standards for buildings and for appliances. He was the Assigned Commissioner to collaborate with the Public Utilities Commission Proceeding on demand response, critical peak pricing, advanced metering, and energy efficiency programs, with a budget of $1 billion/year. Rosenfeld received the APS Szilard Award for Physics in the Public Interest in 1986, the DOE Carnot Award for Energy Efficiency in 1993, and the DOE Enrico Fermi Award in 2006. The University of California at Davis has endowed the Arthur Rosenfeld Chair for Enhanced Energy Use Efficiency. He was Enrico Fermi's last graduate student.

Benjamin Santer is at Lawrence Livermore National Laboratory Program for Climate Model Diagnosis and Intercomparison, where as an atmospheric scientist he focuses on climate model evaluation, the use of statistical methods in climate science, and identification of natural and anthropogenic "fingerprints" in observed climate records. Santer's early research on the climatic effects of combined changes in greenhouse gases (GHGs) and sulfate aerosols contributed to the historic "discernible human influence" conclusion of the 1995 Report by the Intergovernmental Panel on Climate Change (IPCC). He spent much of the last decade addressing the contentious issue of whether model-simulated changes in tropospheric temperature are in accord with satellite-based temperature measurements. His recent work has attempted to identify anthropogenic fingerprints in a number of different climate variables, such as tropopause height, atmospheric water vapor, the temperature of the stratosphere and troposphere, and ocean surface temperatures in hurricane formation regions. From 1987-1992 he was at the Max-Planck Institute for Meteorology in Germany, working on the development and application of climate fingerprinting methods. He was awarded the Norbert Gerbier–MUMM International Award (1998), a MacArthur Fellowship (1998), the U.S. Department of Energy E.O. Lawrence Award (2002), and a Distinguished Scientist Fellowship from the U.S. Dept. of Energy, Office of Biological and Environmental Research (2005). He is a member of the National Academy of Sciences (2011).

Peter Schwartz is Associate Professor of Physics at Cal Poly University, San Luis Obispo. After 10 years of nanotechnology research he changed his research to *energy sustainability* in 2006 and spent a year at UC Berkeley's Energy and Resources Group. His research interests include Appropriate Technology, Concentrated Solar Power, Electric Transportation, and Economic Analysis of Increased Efficiency and Renewable Energy

Substitution. At Cal Poly, Schwartz is teaching courses on "Energy, Society and Environment" and "Appropriate Technology for The World's People." The later consists of three courses and a collaborative, appropriate technology summer school in Guatemala, described in the April-issue of Physics News ("Developing Hands-On-Sustainable Energy Solutions"). (http://appropriatetechnology.wikispaces.com/)

Geoff Sharples advises energy technology companies on fund raising and commercial operations. He is currently raising capital for advanced wind power companies that are particular suited to offshore deployment. Previously, Sharples held roles in corporate venture in the renewable sector with Électricité de France (EDF) and most recently at Google's renewable energy practice. He has also held key commercial operations roles in solar and at Clipper Windpower a wind turbine manufacturer and developer of wind projects. At Clipper he helped launch Clipper's 2.5 MW turbine into the market by negotiating and closing a $300 million transaction with energy company British Petroleum. His energy career started in 1997 with the development and project-financing of independent power plants in Europe and North Africa for PSEG Global.

Abigail Swann is a post-doctoral fellow in the Sustainability Science Program at the Kennedy School of Government and the Organismic and Evolutionary Biology Depart at Harvard University. She is presently analyzing the potential thresholds in the ecoclimate of the Amazon forest and agricultural regions in Brazil with respect to deforestation and the spread of agriculture within the forest. Her PhD thesis work at UC Berkeley, *Ecoclimate: Variations, Interactions, and Teleconnections*, studied the interactions between the ecosystems and climate on global scales, particularly with respect to changes in forest cover.

Oleksandr Tanskyi is a Senior Research Associate at the Energy Systems Laboratory and a Mechanical Engineering PhD student at Texas A&M University. He has three years experience in the design of building power supply systems and one year in implementation of the Continuous Commissioning® approach to improve energy efficiency in buildings. He has an M.S. in Electrical Engineering from National Technical University of Ukraine and an M.S. in Mechanical Engineering from Texas A&M, with a Fulbright Scholarship.

Norbert Untersteiner is an emeritus professor in the Department of Atmospheric Sciences and Geophysics at the University of Washington in Seattle. He chaired the department from 1988 to his retirement in 1997. His career has been focused on the physics of sea ice, especially its thermodynamics. From 1971-1979 he directed the Arctic Ice Dynamics Joint Experiment (AIDJEX) and, in 1981, a NATO Advanced Study Institute that resulted in a comprehensive textbook "The Geophysics of Sea Ice".

Michael Webber is Associate Director of the Center for International Energy and Environmental Policy, Co-Director of the Clean Energy Incubator, and Assistant Professor of Mechanical Engineering at the University of Texas. He is on the Board of Advisors of *Scientific American* and has authored more than 125 articles, columns, and book chapters. He has given more than 150 lectures, speeches, and invited talks in the last few years, including testimony for a U.S. Senate hearing. Webber holds four patents and is an originator of the Pecan Street Project, a multi-institutional public-private partnership in Austin to create smart electricity and water utilities of the future.

Christopher Yang is a researcher and the co-leader of Infrastructure System Analysis research group within the Sustainable Transportation Energy Pathways (STEPS) research program at the Institute of Transportation Studies at UC Davis. His research interests lie in understanding the role of advanced vehicles and fuels in helping to reduce transportation greenhouse gas emissions. Yang's main areas of research are understanding alternative fuel infrastructure through system modeling, including hydrogen infrastructure systems and electric grid interactions with plug-in vehicles, and scenarios and options for long-term reductions in greenhouse gases from the transportation sector. He completed his PhD in Mechanical Engineering from Princeton University.

Sonia Yeh is an associate research scientist at the Institute of Transportation Studies and a faculty member of Graduate Group in Transportation Technology and Policy, UC-Davis. Yeh's research interest is to advance the understanding of future energy systems and their environmental and social impacts, and to seek policy solutions to improve their sustainability performance. Her fields of expertise include energy system modeling, policy analysis, lifecycle assessment and learning-by-doing. She serves as a policy adviser for the design and implementation of California's Low Carbon Fuel Standard (LCFS) and served on advisory panels for climate policies considered in other states. She is a committee member of the Transportation Research Board of the National Research Council.

Sustainable Energy II Participants

Jamie Alessio
Elizabeth Anderson
Katey Walter Anthony
Negin Aryaee
Deb Banerjee
Polly Baranco
Keith Bardin
Sage Bauers
Kathryn Bay
Joseph Becker
Ernie Behringer
Peter Berg
John Berger
Benjamin Bierman
Jeremy Black
David Blackman
William Blackmon
Julio Blanco
Chrissa Blattner
Nate Blumenkrantz
Gina Bochicchio
Ulrike Boesenberg
William Boggs
Matthew Bomberg
Dan Bowman
George Breznay
George Brown
Robert Budnitz
Adam Burgasser
Julie Burlage
Jennifer A Burney
Paola Cadau
Pamela Campos
John Caraher
Kevin Carley
Christine Carmichael
David Claridge
Thomas Carter
Richard Cohen
Jonathan Cole
Pierce Corden
George Crabtree
Paul Craig
Vikram Dalal
Elizabeth Deakin
Paul Debevec
Jessie E Denver
Louis Desroches
Steven Diesburg
Sean Duncan
Douglas Epperson
Tyler Espinoza
Bernard Feldman
Christopher Field
Jane Flood
Kyle Forinash
Mike Foster
Marc Fountain
Inez Fung
Ashok Gadgil
Chris Gaffney
Uri Ganiel
Frank Garcia
James Gates
Adnan Ghribi
David Gibbs
Judith Gibbs
Peter Gollon
Esteban Gonzalez
Anamika Gopal
Nathan Norman Haese
David Hafemeister
Gina Hafemeister
Wilson Hago
Linda Halabi
Jonathan E Hardis
Gabriel Harmon
Hugh Haskell
Richard Haskell
Philip Haves
Erik Helgren
Shu Chin Ho
Art Hobson
Bettina Hodel
K. John Holmes
Daniel Holz
John Holzrichter
Steven Horne
Robert Hosbach
Eric Isaacs
Mary James
Bruce Johnston
Daniel Kammen
Miron Kaufman
Sila Kiliccote
Sang Kim
Na Kim
Tim Kirkpatrick
Lynda Klein
Malcolm Knapp
Rob Knapp
Alex Koberle
Carl Kocher
Joseph Lach
Estella Lai
Emily Leslie
Barbara Levi
Ilan Levi
Todd Levin
Wei Li
Lawrence Lim
Lungyee Lin
Michael Lin
Timothy Lipman
Michael Lucibella
Peter Lydon
Yolanda Ma
Anupam Madhukar
Todd Maki
Jeffrey Marque

Ronald McKnight
Thomas Meyer
Chris Miller
Nader Mirabolfathi
Blayne Morgan
William Morrow, III
Alexandra Moskaleva
Rich Muller
Elizabeth Muller
Thomas Murphy Jr
Mary Murphy-Waldorf
Parveen Mustansir
James Neff
Danielle Nicholas
Rebecca Nie
Charles Niederriter
Brian Hoyt Nordstrom
Michitaka Ohtaki
J. David Osorio
Tyler Otto
Joseph Pacold
Demetri Papamichalis
M. Hossein Partovi
Jason Pawelczyk
Alex Perl
Stephen Portis
Devesh Ranjan
Brian Rasnow
Keith Ray
Chris Ringer
Robert Rohde
Arthur Rosenfeld
Robert Michael Ryan
Peter Saeta
Ansu Sahoo
Benjamin Santer
Alfred Schlachter
Peter Schwartz
Bob Shanbrom
Daniel Sheehan
Yezhou Shi
John Smedley
Paul Spencer
Julian Sproul
Craig Stephan
Alan Sweedler
Julian Sweet
Jorge Rivera
Marc Thomas
Valerie Thomas
Marshall Thomsen
Fatima Toor
Reza Toossi
Greg Trautman
Sam Vigil
Shelby Vorndran
James Waite
Michael Webber
Christopher Weber
Max Wei
Maya Wheelock
Brian P. Wilfley
Thomas W Williams
Gerald Witt
Gordon Wozniak
Hai-Sheng Wu
Kuan-Chuen Wu
Jonathan S Wurtele
Seong Yap
Sonia Yeh
Anna Zaniewski
Li Zhang
Dustin Zubke

AUTHOR INDEX